International Association of Fire Chiefs

Fire Inspector
Principles and Practice

First Edition Revised

JONES & BARTLETT
LEARNING

Jones & Bartlett Learning
World Headquarters
5 Wall Street
Burlington, MA 01803
978-443-5000
info@jblearning.com
www.jblearning.com

International Association of Fire Chiefs
4025 Fair Ridge Drive
Fairfax, VA 22033
www.IAFC.org

National Fire Protection Association
1 Batterymarch Park
Quincy, MA 02169-7471
www.NFPA.org

Jones & Bartlett Learning books and products are available through most bookstores and online booksellers. To contact Jones & Bartlett Learning directly, call 800-832-0034, fax 978-443-8000, or visit our website, www.jblearning.com.

Substantial discounts on bulk quantities of Jones & Bartlett Learning publications are available to corporations, professional associations, and other qualified organizations. For details and specific discount information, contact the special sales department at Jones & Bartlett Learning via the above contact information or send an email to specialsales@jblearning.com.

08727-7

Production Credits

Chief Executive Officer: Ty Field
President: James Homer
Chief Product Officer: Eduardo Moura
Vice President, Publisher: Kimberly Brophy
VP of Sales, Public Safety Group: Matthew Maniscalco
Executive Editor—Fire: William Larkin
Senior Content Developer: Jennifer Deforge-Kling
Senior Acquisitions Editor: Janet Maker
Production Editor: Sandy Paparisto
Senior Marketing Manager: Brian Rooney
V.P., Manufacturing and Inventory Control: Therese Connell
Cover Design: Kristin E. Parker
Rights and Media Manager: Joanna Lundeen
Rights and Media Research Coordinator: Ashley Dos Santos
Composition: diacriTech, Chennai, India
Cover Image: © Dennis Wetherhold, Jr.
Printing and Binding: LSC Communications
Cover Printing: LSC Communications

Copyright © 2016 by Jones & Bartlett Learning, LLC, an Ascend Learning Company

All rights reserved. No part of the material protected by this copyright may be reproduced or utilized in any form, electronic or mechanical, including photocopying, recording, or by any information storage and retrieval system, without written permission from the copyright owner.

The procedures and protocols in this book are based on the most current recommendations of responsible sources. The National Fire Protection Association (NFPA), International Association of Fire Chiefs (IAFC), and the publisher, however, make no guarantee as to, and assume no responsibility for, the correctness, sufficiency, or completeness of such information or recommendations. Other or additional safety measures may be required under particular circumstances.

Notice: The individuals described in "You Are the Fire Inspector" and "Fire Inspector in Action" throughout the text are fictitious.

Additional illustration and photographic credits appear on page 350, which constitutes a continuation of the copyright page.

To order this product, use ISBN: 978-1-284-13774-3

The Library of Congress has cataloged the first printing as follows:
Fire inspector : principles and practice / International Association of Fire Chiefs ; National Fire Protection Association ; Jones & Bartlett Learning, LLC.
  p. cm.
 Includes index.
 ISBN-13: 978-0-7637-4939-2
 ISBN-10: 0-7637-4939-7
 1. Fire prevention—Handbooks, manuals, etc. 2. Building inspection—Handbooks, manuals, etc.
I. International Association of Fire Chiefs. II. National Fire Protection Association. III. Jones & Bartlett Learning.
 TH9176.F574 2012
  628.9'22—dc23
                    2011025587
6048

Printed in the United States of America
21 20 19   10 9 8

Brief Contents

1	Introduction to Fire Inspector	2
2	Building Construction	12
3	Types of Occupancies	40
4	Fire Growth	56
5	Performing an Inspection	76
6	Reading Plans	100
7	Occupancy Safety and Evacuation Plans	116
8	Fire Alarm and Detection Systems	148
9	Fire Flow and Fire Suppression Systems	168
10	Portable Fire Extinguishers	202
11	Electrical and HVAC Hazards	220
12	Ensuring Proper Storage and Handling Practices	242
13	Safe Housekeeping Practices	272
14	Writing Reports and Keeping Records	288
15	Life Safety and the Fire Inspector	306
	Appendix A: Changing Codes and Standards	316
	Appendix B: NFPA 1031 Correlation Guide	319
	Appendix C: NFPA 1031, Standard for Professional Qualifications for Fire Inspector and Plan Examiner, 2014 Edition, ProBoard Matrix	320
	Appendix D: Fire and Emergency Service Higher Education (FESHE) Correlation Guide	328
	Glossary	330
	Index	344
	Credits	350

Contents

1 Introduction to Fire Inspector 2
   Introduction 4
   What Is a Fire Inspection?............ 4
   Fire Inspection and the Fire Service 4
      Additional Fire Prevention Roles
         within a Fire Department 5
   Roles and Responsibilities for Fire Inspector I . 5
   Roles and Responsibilities for Fire Inspector II . 6
   Codes and Standards 6
      Model Code Organizations 7
      Prescriptive Codes and
         Performance-Based Codes 7
      Code and Standard Adoption 7
   Understanding the Legal Processes 8
      Authority Having Jurisdiction 8
      State and Local Law 8
      Range of Authority 8
      Legal Proceedings 8
   Permits 8
   Ethics and the Fire Inspector 9
   Career Development 9

2 Building Construction 12
   Introduction 14
   Types of Construction Materials 14
      Masonry 14
      Concrete 15
      Steel 15
      Other Metals 16
      Glass 16
      Gypsum Board 16
      Wood 16
      Plastics 17
   Types of Construction 18
      Type I Construction: Fire Resistive 18
      Type II Construction: Noncombustible 19
      Type III Construction: Ordinary 19
      Type IV Construction: Heavy Timber 20
      Type V Construction: Wood Frame 20
   Building Components 22
      Foundations 22
      Floors and Ceilings 22
      Roofs 23
      Walls 26
      Columns 28
      Beams, Girders, Joists, and Rafters 29
      Doors and Windows 29
      Windows 31
      Fire Doors and Fire Windows 34
      Interior Finishes and Floor
         Coverings 35

3 Types of Occupancies 40
   Introduction 42
   Occupancy Classification 42
   One- and Two-Family Dwelling Units 43
   Lodging or Rooming Houses 43
   Hotels 44
      Dormitories 44
   Apartment Buildings 45
   Residential Board and Care Occupancy ... 45
   Health Care Occupancy 46
   Ambulatory Health Care Occupancies 46
   Day-Care Occupancy 47
   Educational Occupancy 47
   Business Occupancies 48
   Industrial Occupancies 48
   Mercantile Occupancy 49
   Storage Occupancies 49
   Assembly Occupancies 50
      Small Assembly Uses 51
   Detention and Correctional Occupancies ... 51
   Multiple Occupancies 52
      Mixed Occupancy 52
      Separated Occupancy 52
   Special Structures and High-Rise
     Buildings 52

4 Fire Growth 56
   Introduction 58
   The Chemistry of Fire 58
      What Is Fire? 58
      States of Matter 58
      Fuels 58
      Types of Energy 59
      Conservation of Energy 59
      Conditions Needed for Fire 60
      Products of Combustion 60

Fire Spread.................... 61
   Methods of Extinguishment.............. 62
   Classes of Fire 63
 Characteristics of Solid-Fuel Fires........ 64
   Solid-Fuel Fire Development 64
   Characteristics of a Room-
      and-Contents Fire 66
   Special Considerations 68
 Characteristics of Liquid-Fuel Fires 68
 Characteristics of Gas-Fuel Fires........ 69
   Vapor Density................... 69
   Flammability Limits............... 69
   Energy Required for Ignition of Vapors...... 69
   Boiling Liquid, Expanding-Vapor
      Explosions................... 70
 Fire Growth...................... 70
   Building or Contents Hazard 70
   Hidden Building Elements 70
   Interior Finish 71
   Furnishings 72
   Building Construction Elements.......... 72
 Flame Spread and Smoke Development
   Ratings 72

5 Performing an Inspection.......... 76

 Introduction 79
 Types of Inspections................. 79
 When is the Best Time to Inspect?....... 80
 Pre-Inspection Process 80
   Classification of Construction
      and Occupancy 81
   Codes......................... 81
   Tools and Equipment 82
   Standard Forms 83
   Scheduling and Introductions 86
 The Fire Inspection Process........... 86
   Presentation 86
   Conducting the Exterior Inspection 87
   Conducting the Interior Inspection......... 88
   Preplan Sketch 92
   New Construction Considerations 92
   Remodeling Considerations 92
 Post-Inspection Meeting 93
 Documentation 93
 Code Violations 94

 Investigating Complaints 95
 Improving the Inspection Process 95

6 Reading Plans 100

 Introduction 103
 Authority to Review 103
 Types of Plans 103
   Code Analysis................... 104
 Type of Drawings in Plan Sets 105
   Site Plans 105
   Structural Plans 105
   Architectural Plans................ 105
   Electrical Plans.................. 106
   Specifications Book 106
 Type of Views 107
 Plan Review Process............... 108
   Application Phase................ 108
   Review Phase 108
   Approval 112
   Site Visit..................... 112
   Commissioning 113
 Performance Based Design........... 113
 Technical Assistance 113
 Equivalencies and Alternatives......... 113

7 Occupancy Safety and Evacuation Plans................. 116

 Introduction 119
 Occupant Load................... 119
   Occupant Load Factors 120
   Calculate the Occupant Load for
      a Single-Use Occupancy............ 122
   Calculate the Occupant Load for a
      Multiple-Use Occupancy 122
   Calculate Occupant Load Increases 123
   Sufficient Capacity for Occupant Load 123
 Means of Egress.................. 124
   Exit Access.................... 124
   Exit 125
   Exit Discharge 125
 Evaluation of the Means of Egress 126
   Exits Serving More than One Story........ 127
   Egress Capacity from a Point
      of Convergence 128

Measurement of Means of Egress 128
Corridor Capacity 128
Minimum Width 128
Number of Means of Egress 129
Remoteness of Exits 129
Measurement of Travel Distance to Exits.... 129
Common Path of Travel 133
Measuring Travel Distance 133
Computing Required Egress Capacity...... 134
Maximum Egress Capacity 134
Means of Egress Elements and Arrangements 134
Doors................................ 134
Panic Hardware 135
Horizontal Exits....................... 135
Stairs 136
Smokeproof Enclosure 136
Ramps 136
Exit Passageways 137
Fire Escape Stairs 137
Escalators, Moving Walkways, and Elevators 138
Areas of Refuge....................... 138
Ropes and Ladders 138
Windows 138
Exit Lighting 139
Emergency Lighting 139
Exit Signs 139
Maintenance of the Means of Egress 140
Evacuation and Emergency Plans 140
Plan Requirements 140
Alert................................. 140
Get Out and Stay Out.................. 142
Meeting Place........................ 142
Evacuation Leaders.................... 142
Evacuation Plan Success............... 143
Evacuation Drills 143

8 Fire Alarm and Detection Systems . 148

Introduction 150
Fire Alarm and Detection Systems 150
Plan Review 150
Fire Alarm and Detection System Components 151
Residential Smoke Alarms 153
Alarm Initiation Devices................ 155
Alarm Notification Appliances 159
Other Fire Alarm Functions 159

Fire Alarm Systems.................... 159
Fire Department Notification 160
Wiring Concerns 162
Fire Alarm and Detection System Maintenance 162
Fire Alarm and Detection System Testing .. 162
Documentation and Reporting of Issues 163

9 Fire Flow and Fire Suppression Systems...................... 168

Introduction 171
Plan Review of Fire Suppression Systems........................ 171
Water Supply..................... 172
Municipal Water Systems 172
Types of Fire Hydrants............. 174
Wet-Barrel Hydrants................... 174
Dry-Barrel Hydrants................... 174
Fire Hydrant Locations 176
Measuring Flow Rates............. 176
Flow and Pressure 176
Obtaining a Static and Residual Pressure Readings 177
Determining Hydrant Flow Test Results 178
Automatic Sprinkler Systems......... 178
Occupancy Hazards 179
Water Supply 180
Types of Sprinkler Systems......... 184
Wet-Pipe Sprinkler Systems 184
Dry-Pipe Sprinkler Systems 184
Preaction Sprinkler Systems 185
Combined Dry-Pipe and Preaction Systems . 185
Deluge Sprinkler Systems.............. 185
Special Types 186
Automatic Sprinkler Heads......... 188
Types of Sprinkler Heads 189
Release Temperature.................. 191
Standpipe Systems 192
Class I Standpipe 193
Class II Standpipes.................... 193
Class III Standpipes................... 194
Water Flow in Standpipe Systems........ 194
Water Supplies....................... 194
Fire Suppression System Testing....... 194
Underground Flush Test............... 194
Hydrostatic Test...................... 195

Air Test. 195
Main Drain Test 195
Hood and Duct System Testing. 195
Documentation and Reporting of Issues 195

10 Portable Fire Extinguishers 202

Introduction 204
Portable Fire Extinguishers. 204
Purposes of Fire Extinguishers 204
Classes of Fires 204
Class A Fires. 205
Class B Fires. 205
Class C Fires. 205
Class D Fires 206
Class K Fires. 206
Classification of Fire Extinguishers 207
Labeling of Fire Extinguishers 207
Traditional Lettering System. 207
Pictograph Labeling System 208
Fire Extinguisher Placement 208
Classifying Area Hazards 209
Determining the Most Appropriate
  Placement of Fire Extinguishers 209
Types of Extinguishing Agents 210
Water 210
Dry Chemicals 210
Carbon Dioxide 212
Foam 212
Wet Chemicals 213
Dry-Powder 214
Fire Extinguisher System Readiness. 214
Documentation and Reporting of Issues 215

11 Electrical and HVAC Hazards 220

Introduction 222
Electrical Systems Overview 223
The Basics of Electricity 223
Potential Electrical Hazards 224
Arcing and Overheating 224
Protective Practices for Electrical
  Systems. 224
Overcurrent Protection. 224
Grounding 225
Generators 226
Transformers. 227

What to Look for During a
  Fire Inspection 228
Common Problems with Wiring. 228
Common Problems with Electrical
  Equipment. 228
Cables, Conduits, and Raceways. 228
Extension Cords. 229
Outlets and Switches 229
Electrical System Boxes 229
Lamps and Light Fixtures 230
Motors 230
Hazardous Areas. 230
Static Electricity 230
Heating, Ventilation, and Air Conditioning
  Systems Overview 231
Energy Conversion 231
Distribution Systems 233
Safety Systems 233
Exhaust Systems. 233
Smoke Management Systems 234
Stairwell Smoke Management Systems . . . 235
Testing 235
Potential Hazards 235
Smoke Distribution through the Ducts 235
What to Look for During Inspection 236

12 Ensuring Proper Storage and Handling Practices 242

Introduction 245
Classification of Flammable and
  Combustible Liquids. 245
Class I Liquids 245
Class II and Class III Liquids 246
Gases 246
Classification by Physical Hazard 246
Classification by Health Hazard 247
Physical States of Gases 247
Classification by Usage 248
Fuel Gases 248
Industrial Gases 248
Medical Gases 248
Hazardous Materials 249
Labeling 249
Department of Transportation System. 249
Labels and Placards 249
NFPA 704 Marking System. 251

Hazardous Materials Information System
Marking System 251
Military Hazardous Materials/WMD
Markings 251
Safety Data Sheet (SDS) 252
**Storage and Handling of Flammable
Liquids** 253
Flammable Liquid Tank Storage 253
Other Storage of Flammable Liquids. 254
Container Storage in Buildings 254
Handling Methods 255
Fire Prevention Methods 255
Storage of Gases. 256
Gas Containers. 256
Storage Safety Considerations. 257
Care in Handling 257
Safeguards for Escaping Gas 257
**Hazardous Materials Storage
and Handling** 258
Containers 258
Container Type. 258
Container Volume 259
Nonbulk Storage Vessels. 260
Hazardous Material Storage
Lockers 261
Transporting Hazardous
Materials 262
Railroad Transportation 262
Pipelines 264
Fire Protection Systems. 265
Fire Protection Systems and Equipment
for Flammable Liquids. 265
Fire Protection Systems and Equipment
for Flammable Gases 265
Fire Protection Systems for Hazardous
Materials 266
Ensuring Code Requirements. 266

13 Safe Housekeeping Practices. 272
Introduction 274
Housekeeping Overview 274
Exterior Issues 275
Blocked, Obstructed, or Impaired
Access 275
Obstructions to Fire Protection
Equipment. 275
Protection of Flammable Liquid and
Gas Equipment 276
Fire Exposure Threats. 276
Wildland Interface 277

Interior Issues 277
Oily Waste, Towels, or Rags 278
Dust and Lint Accumulation. 278
Mechanical Equipment. 278
Combustible Materials and Storage. 279
Trash or Recycling Issues 279
Packing and Shipping Materials 280
Flammable/Combustible Liquids 280
Obstructions to Fire Protection Equipment . 281
Kitchen Cooking Hoods, Exhaust Ducts, and
Equipment. 281
Compressed Gas Cylinders. 281
Control of Smoking 282
Correction of Housekeeping Issues 283

14 Writing Reports and Keeping Records. 288
Introduction 290
Written Documentation. 290
Barriers that Affect Communication 291
Field Notes. 291
Sketches, Diagrams, and Photographs 293
Documenting the Fire Inspection. 293
Fire Inspection Report 293
Reinspection Reports 294
Final Notice 294
**Choosing the Right Type of
Documentation.** 294
Letters 294
Email 296
Checklists 296
Detailed Reports. 296
Recording Complaints. 296
Code References. 299
Record-Keeping Practices. 299
Records Retention Requirements 299
Freedom of Information Requests. 300
Organizational Practices 301
Reporting Systems 302
Legal Proceedings. 302
Presenting Evidence 302

15 Life Safety and the Fire Inspector. 306
Introduction 308
**Role of the Fire Inspector in Fire and Life
Safety Education.** 308
Public Education, Public Information,
and Public Relations. 308

Creating a Fire and Life Safety Education Program 309
   Indentifying the Problem 309
Developing the Fire and Life Safety Education Program. 311
   Support 311
   Identifying Stakeholders. 311
   Developing Relationships 312
   Developing Goals and Objectives 312
   Content Development 312
   Budget 312
   Presenting the Material. 313
Effectiveness Metrics 313

Appendix A: Changing Codes and Standards. 316
Appendix B: NFPA 1031 Correlation Guide 319
Appendix C: NFPA 1031, Standard for Professional Qualifications for Fire Inspector and Plan Examiner, 2014 Edition, ProBoard Matrix 320
Appendix D: Fire and Emergency Service Higher Education (FESHE) Correlation Guide 328
Glossary 330
Index 344
Credits 350

Resource Preview

Fire Inspector: Principles and Practice

Jones & Bartlett Learning, the National Fire Protection Association®, and the International Association of Fire Chiefs have joined forces to raise the bar for fire service once again with the release of *Fire Inspector: Principles and Practice, First Edition Revised*.

Chapter Resources

Fire Inspector: Principles and Practice, First Edition Revised thoroughly supports instructors and prepares future fire inspectors for the job. This text meets and exceeds the Fire Inspector I and II requirements as outlined in Chapters 4 and 5 of NFPA 1031, *Standard for Professional Qualifications for Fire Inspector and Plan Examiner*, 2014 Edition. It also addresses course outcomes from the National Fire Academy's Fire and Emergency Services Higher Education (FESHE) Discipline-Specific (Fire Preventions) Principles of Code Enforcement course.

Fire Inspector: Principles and Practice, First Edition Revised serves as the core of a highly effective teaching and learning system. Its features reinforce and expand on essential information and make information retrieval a snap. These features include:

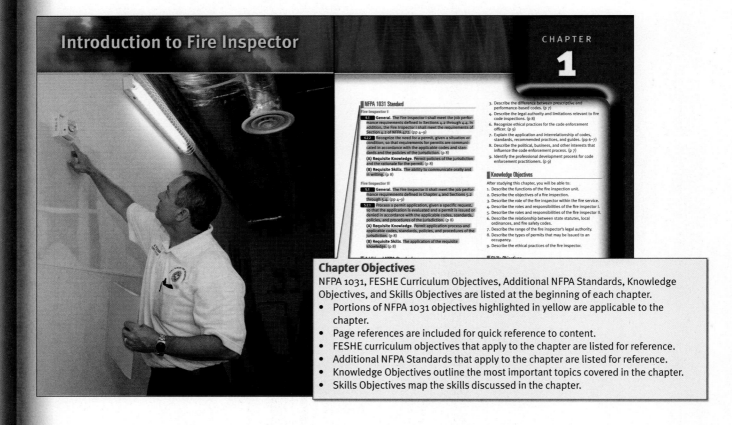

Chapter Objectives
NFPA 1031, FESHE Curriculum Objectives, Additional NFPA Standards, Knowledge Objectives, and Skills Objectives are listed at the beginning of each chapter.
- Portions of NFPA 1031 objectives highlighted in yellow are applicable to the chapter.
- Page references are included for quick reference to content.
- FESHE curriculum objectives that apply to the chapter are listed for reference.
- Additional NFPA Standards that apply to the chapter are listed for reference.
- Knowledge Objectives outline the most important topics covered in the chapter.
- Skills Objectives map the skills discussed in the chapter.

Resource Preview

You Are the Fire Inspector
Each chapter opens with a case study intended to stimulate classroom discussion, capture students' attention, and provide an overview for the chapter. An additional case study is provided in the end-of-chapter Wrap-Up material.

Hot Terms
Hot Terms are easily identifiable within the chapter and define key terms that the student must know. A comprehensive glossary of Hot Terms also appears in the Wrap-Up.

Fire Inspector Tips
This series of tips provides advice from masters of the trade.

Comprehensive Measurements
Both U.S. Imperial units and Metric units are used throughout the text.

Wrap-Up
End-of-chapter activities reinforce important concepts and improve students' comprehension. Additional instructor support and answers for all questions are available on the Instructor's Toolkit CD-ROM.

Chief Concepts
Chief Concepts highlight critical information from the chapter in a bulleted format to help students prepare for exams.

Fire Officer II Bar
A bar is used to flag content that applies only to Fire Inspector II students. Students preparing for the Fire Inspector I level should skip over content marked by the Fire Inspector II icon, or incorporate it at the instructor's discretion for further study.

Resource Preview

Wrap-Up

Hot Terms

Ambulatory health care occupancy A building or portion thereof used to provide services or treatment simultaneously to four or more patients that, on an outpatient basis. (NFPA 101, *Life Safety Code*)

Apartment building is a building or portion thereof containing three or more dwelling units with independent cooking and bathroom facilities. (NFPA 5000)

Assembly occupancies Buildings (1) used for a gathering of 50 or more persons for deliberation, worship, entertainment, eating, drinking, amusement, awaiting transportation, or similar uses; or (2) used as a special amusement building regardless of occupant load. (NFPA 101, *Life Safety Code*)

Business occupancy An occupancy used for the transaction of business other than mercantile. (NFPA 101, *Life Safety Code*)

Day-care occupancy An occupancy in which four or more clients receive care, maintenance, and supervision, by other than their relatives or legal guardians, for less than 24 hours per day. (NFPA 101, *Life Safety Code*)

Detention and correctional occupancy An occupancy used to one or more persons under varied degrees of restraint or security where such occupants are mostly incapable of self-preservation because of security measures not under the occupant's control. (NFPA 101, *Life Safety Code*)

Dormitory A building or space in a building in which group sleeping accommodations are provided for more than 16 persons who are not members of the same family in one room, or a series of closely associated rooms, under joint occupancy and single management, with or without meals, but without individual cooking facilities. (NFPA 101, *Life Safety Code*)

Educational occupancies Buildings used for educational purposes through the twelfth grade by six or more persons for 4 or more hours per day or more than 12 hours a week. (NFPA 101, *Life Safety Code*)

Health care occupancy An occupancy used for purposes of medical or other treatment or care of four or more persons where such occupants are mostly incapable of self-preservation due to age, physical or mental disability, or because of security measures not under the occupant's control. (NFPA 101, *Life Safety Code*)

Hotel A building or group of buildings under the same management in which there are sleeping accommodations for more than 16 people and is primarily used by transients for lodging with or without meals. (NFPA 101, *Life Safety Code*)

Industrial occupancy An occupancy in which products are manufactured or in which processing, assembling, mixing, packaging, finishing, decorating, or repair operations are conducted. (NFPA 101, *Life Safety Code*)

Mercantile occupancy An occupancy used for the display and sale of merchandise. (NFPA 101, *Life Safety Code*)

Mixed occupancy A multiple occupancy where the occupancies are intermingled. (NFPA 101, *Life Safety Code*)

Multiple occupancy A building or structure in which two or more classes of occupancy exist. (NFPA 101, *Life Safety Code*)

Lodging or rooming house Building or portion thereof that does not qualify as a one- or two-family dwelling, that provides sleeping accommodations for a total of 16 or fewer people on a transient or permanent basis, without personal care services, with or without meals, but without separate cooking facilities for individual occupants. (NFPA 101, *Life Safety Code*)

Occupancy The intended use of a building

One- or two-family dwelling A building that contains no more than two dwelling units with independent cooking and bathroom facilities. (NFPA 5000)

Residential board and care occupancy A building or portion thereof that is used for lodging and boarding of four or more residents, not related by blood or marriage to the owners or operators, for the purpose of providing personal care services. (NFPA 101, *Life Safety Code*)

Storage occupancy An occupancy used primarily for the storage or sheltering of goods, merchandise, products, vehicles, or animals. (NFPA 101, *Life Safety Code*)

Separated occupancy A multiple occupancy where the occupancies are separated by fire resistance-rated assemblies. (NFPA 101, *Life Safety Code*)

Fire Inspector *in Action*

You prepare to make an inspection at an existing surf board shop in a strip mall. You look in the file and locate the folder which classifies the occupancy as mercantile. When you enter the shop, you smell a strong chemical odor. You find that the business now has begun repairing surf boards in the back room in addition to sales. The operation involves using different flammable resins, paints, and solvents. Also, different power tools that produce large amounts combustible dusts are in use. You make notes in preparation of correcting the occupancy classification.

1. What single factor would change the classification of this occupancy?
   A. Use of flammable liquids
   B. Combustible dusts
   C. Repair operations
   D. All of the above

2. What is the correct classification of this occupancy?
   A. Assembly
   B. Business
   C. Industrial
   D. Storage

3. What can the business owner do to keep the mercantile occupancy classification?
   A. The business location must be changed.
   B. Eliminate the repair shop.
   C. Move all operations except sales to another location.
   D. Either B or C.

4. If you select the wrong occupancy class, what is the consequence?
   A. The incorrect licensing fees will be charged.
   B. An inadequate level of fire and life safety may be required.
   C. Insurance coverage may be denied.
   D. Pre-fire planning will be incorrect.

Instructor Resources

A complete teaching and learning system developed by educators with an intimate knowledge of the obstacles that instructors face each day supports *Fire Inspector: Principles and Practice*. These resources provide practical, hands-on, time-saving tools such as PowerPoint presentations, customizable lesson plans, test banks, and image/table banks to better support instructors and students.

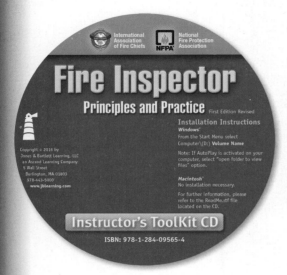

Instructor's ToolKit CD
ISBN: 978-1-284-09565-4

Preparing for class is easy with the resources on this CD. The CD includes the following resources:

- Adaptable PowerPoint Presentations: Provides instructors with a powerful way to create presentations that are educational and engaging to their students. These slides can be modified and edited to meet instructors' specific needs.
- Detailed Lesson Plans: The lesson plans are keyed to the PowerPoint presentations with sample lectures, lesson quizzes, and teaching strategies. Complete, ready-to-use lecture outlines include all of the topics covered in the text. The lecture outlines can be modified and customized to fit any course.
- Image and Table Bank: Offers a selection of the most important images and tables found in the text. Instructors can use these graphics to incorporate more images into the PowerPoint presentations, make handouts, or enlarge a specific image for further discussion.
- The Test Bank contains over 300 multiple-choice questions and allows instructors to create tailor-made classroom tests and quizzes quickly and easily by selecting, editing, organizing, and printing a test along with an answer key, including page references to the text. Each question includes NFPA 1031 references and the level being tested.

Student Resources

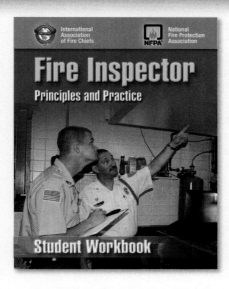

Student Workbook
ISBN: 978-0-7637-9857-4

This resource is designed to encourage critical thinking and aid comprehension of the course material through use of the following materials:
- Case studies and corresponding questions
- Figure-labeling exercises
- Crossword puzzles
- Matching, fill-in-the-blank, short-answer, and multiple-choice questions
- Answer key with page references

Technology Resources

Navigate TestPrep: Fire Inspector I & II
ISBN: 978-0-7637-9860-4

Navigate TestPrep: Fire Inspector I & II is a dynamic program designed to prepare students to sit for Fire Inspector certification examinations by including the same type of questions they will likely see on the actual examination.

It provides a series of self-study modules, organized by chapter and level, offering practice examinations and simulated certification examinations using multiple-choice questions. All questions are page referenced to Fire Inspector: Principles and Practice, Enhanced Revised First Edition for remediation to help students hone their knowledge of the subject matter.

Students can begin the task of studying for Fire Inspector certification examinations by concentrating on those subject areas where they need the most help. Upon completion, students will feel confident and prepared to complete the final step in the certification process—passing the examination.

Acknowledgements

Editorial Board

Shawn Kelley (IAFC)
Director of Strategic Services/GPSS
International Association of Fire
  Chiefs
Fairfax, Virginia

Ken Holland
National Fire Protection Association
Quincy, Massachusetts

William F. Jenaway
Executive Vice President of Volunteer Firemen's Insurance Services, Inc. (VFIS)
King of Prussia, Pennsylvania

Robert Morris, Director
Fire Prevention Bureau
Darien-Woodridge Fire District
Darien, Illinois

Lead Author

Dr. William F. Jenaway, CFO, CFPS, CSP, CHCM, CPP, MIFE has over 40 years of fire service and fire inspection experience and holds AS, BS, MA, and PhD degrees. Dr. Jenaway began his career as a volunteer fire fighter in East Bethlehem Township in Washington County, Pennsylvania where he ultimately served as the Fire Chief and Fire Marshal. He subsequently served as the Fire Chief and President of the King of Prussia Volunteer Fire Company in Montgomery County, Pennsylvania. He also served for 15 years as the Chairman of the municipality's Fire and Rescue Services Board.

In 2001, Dr. Jenaway was named "Volunteer Fire Chief of the Year" by *Fire Chief Magazine* at the annual conference of the International Association of Fire Chiefs (IAFC). In 2009, he was elected as Township Supervisor in Upper Merion Township, Pennsylvania where he served as the liaison to the Planning Commission and the Fire and Rescue Services. In 2011, he served as the Vice Chairman of Upper Merion Township, Pennsylvania.

Dr. Jenaway's over 30 year professional career focused on safety and fire protection engineering. He first served as an insurance field engineer and branch manager, where he was responsible for the inspections of properties, training clients in various loss prevention issues, and testing fire and safety equipment and systems. Dr. Jenaway later served as the Director of Training for the loss control department of CIGNA Property Casualty insurance companies, followed by service as their Vice President of Technical Services. He also served as the Senior Vice President of Risk Control for the Reliance Insurance Companies. In all of these roles, fire and life safety inspection practices were a daily part of his activities.

In addition, Dr. Jenaway served as the Chairman of the Pennsylvania Senate Resolution 60 Commission which studied the fire and emergency medical service (EMS) delivery system in Pennsylvania and developed legislative and operational recommendations to improve the fire and EMS delivery system. He also served as the Chairman for two National Fire Protection Association (NFPA) Standards: NFPA 1250: *Recommended Practice in Emergency Service Organization Risk Management* and NFPA 1201: *Standard for Providing Emergency Services to the Public*. He has also testified numerous times before local, state, and federal legislative bodies.

Currently, Dr. Jenaway sits on the Underwriters Laboratories Casualty Council. He is a member of the distinguished "Gilmore Commission," appointed by President Clinton and President Bush, which is studying the readiness of the United States to deal with domestic terrorism involving weapons of mass destruction. He is also the Vice Chairman on the International Commission on Fire Service Accreditation.

Dr. Jenaway is the author of more than 200 articles on fire, safety, and management-related topics. He has also written six texts on fire and safety discipline and four children's books. For many years, Dr. Jenaway authored the monthly "Inspect-O-Gram" series for the International Society of Fire Service Instructors (ISFSI) and taught classes for the ISFSI on fire inspection practices.

Dr. Jenaway is currently employed as the Executive Vice President of VFIS. Dr. Jenaway is an adjunct professor in the Public Safety Graduate School at St. Joseph's University in Philadelphia, Pennsylvania, in the Legal Studies Graduate School at California University of Pennsylvania, and in the Fire Science Program at Columbia Southern University, an online university.

Acknowledgments

Authors

Tim Capehart
Fire Engineer, Retired
Bakersville Fire Department
Bakersville, California

Riley Caton
Gresham Fire and Emergency Services
Gresham, Oregon

Robert M. Coleman, EFO
Chief of Department, Retired
North Attleborough Fire Department
North Attleborough, Massachusetts
Bristol Community College/Adjunct Instructor
Fall River, Massachusetts

Arthur E. Cote PE, FSPE
Chief Engineer, Retired
National Fire Protection Association
Quincy, Massachusetts

Bradford T. Cronin, CFPS
Harbor Fire Protection
Providence, Rhode Island

Robert Drennen
Director of Emergency Services
Upper Moreland Township
Willow Grove, Pennsylvania
Battalion Chief, Retired
Philadelphia Fire Department
Philadelphia, Pennsylvania

Dr. Robert S. Fleming, EdD, CFPS, CFO, EFO
Professor of Management
Rowan University
Glassboro, New Jersey

William Galloway
Assistant State Fire Marshal
Office of State Fire Marshal
Columbia, South Carolina

Daniel B C Gardiner, CFPS
Fire Chief, Retired
Fairfield Fire Department
Fairfield, Connecticut.

Jim Goodloe, BA, EFO
Florida Division of State Fire Marshal
Tallahassee, Florida

David A. Hall
Fire Chief
Springfield Fire Department
Springfield, Missouri

Mike Montgomery
Fire Marshal
Humble, Texas

Wm. E. "Bill" Moran
Central Florida Fire Academy
Orlando, Florida

Robert P. Morris
Fire Prevention Bureau
Darien-Woodridge Fire District
Darien, Illinois

Peter J. Mulvihill, PE
North Lake Tahoe Fire Protection District
Incline Village/Crystal Bay, Nevada

Jon Nisja
Fire Safety Supervisor
Minnesota State Fire Marshal Division
St. Paul, Minnesota

Reviewers

Chris Chadwick
Shreveport Fire Academy
Shreveport, Louisiana

Michael D. Chiaramonte
National Fire Academy
Clayton, North Carolina

Richard Connelly
Boston Fire Department
Boston, Massachusetts

Patrick G. Collier
Orland Fire Protection District
Orlando, Florida

Jason D'Eliso
Scottsdale Fire Department
Scottsdale Community College
Scottsdale, Arizona

Rick Fultz
Fresno Fire Department
Fresno, California

Wayne Hamilton
City of Asheville
Asheville, North Carolina

Chris Hofmann
Northland Community
and Technical College
East Grand Forks, Minnesota

Jim Iammatteo
Mesabi Range Community
and Technical College
Virginia, Minnesota

Mike Julazadeh
Charleston Fire Department
Charleston, South Carolina

Frederick J. Knipper
Duke University/Duke Health
 System Fire & Life Safety Division
Sanford, North Carolina

Paul A. Paquette
Somerset Fire Department
Somerset, Massachusetts

Gerald H. Phipps II
Wyoming Fire Academy – WY Dept.
 of Fire Prevention and Electrical
 Safety
Riverton, Wyoming

Rudy Ruiz
Sandusky Career Center
Sandusky, Ohio

Edsel Smith, Junior
West Virginia University Fire
 Academy and Regional Education
 Services Agency
Buckhannon, West Virginia

James Smith
City of Kinston
Kinston, North Carolina

Tim Swaim
Kernersville Fire Rescue
Kernersville, North Carolina

Robert Swiger
Raleigh Fire Department
Raleigh, North Carolina

David Telban
Cleveland Fire Department
Cuyahoga Community College
Cleveland, Ohio

Stephen Thomas
Guilford Technical Community College
Jamestown, North Carolina

William Trisler
Commission on Fire Prevention and
 Control, Connecticut Fire Academy
Windsor Locks, Connecticut

Acknowledgments

Photographic Contributors

We would like to extend a huge "thank you" to Glen E. Ellman, the photographer for this project. Glen is a commercial photographer and fire fighter based in Fort Worth, Texas. His expertise and professionalism are unmatched!

We would also like to thank Fire Marshal Landon Stallings who opened up his Fire Prevention Bureau, Fort Worth Fire Department, for many of the photographs in this text book. Specifically we would like to acknowledge the following Fire Inspectors:

Marshal Allen
Fire Captain – Commercial Inspection

Gwen Barnes
Admin Assist to Fire Marshal

Tony Blythe
Fire Captain – Addressing

Robert Broadwater
Fire Engineer – Technical Inspections

Debra Clarke
Admin Tech – Revenue

Robert Creed
Fire Lieutenant – Addressing

Brian Hannah
Fire Captain – Technical Inspections

James Horton
Fire Lieutenant – Hazmat

Don Isaacs
Building Permits – Customer Services

Bob Jenkins
Fire Engineer – Technical Inspections

Bob Morgan
Sr. Protection Engineer

Greg Nelson
Fire Protection Engineer

Chip Paiboon
Fire Protection Specialist

Liezel Surel
Admin Tech - Permits

Introduction to Fire Inspector

CHAPTER 1

NFPA 1031 Standard

Fire Inspector I

4.1 General. The Fire Inspector I shall meet the job performance requirements defined in Sections 4.2 through 4.4. In addition, the Fire Inspector I shall meet the requirements of Section 4.2 of NFPA 472. (pp 4–9)

4.2.2 Recognize the need for a permit, given a situation or condition, so that requirements for permits are communicated in accordance with the applicable codes and standards and the policies of the jurisdiction. (p 8)

(A) Requisite Knowledge. Permit policies of the jurisdiction and the rationale for the permit. (p 8)

(B) Requisite Skills. The ability to communicate orally and in writing. (p 8)

Fire Inspector II

5.1 General. The Fire Inspector II shall meet the job performance requirements defined in Chapter 4 and Sections 5.2 through 5.4. (pp 4–9)

5.2.1 Process a permit application, given a specific request, so that the application is evaluated and a permit is issued or denied in accordance with the applicable codes, standards, policies, and procedures of the jurisdiction. (p 8)

(A) Requisite Knowledge. Permit application process and applicable codes, standards, policies, and procedures of the jurisdiction. (p 8)

(B) Requisite Skills. The application of the requisite knowledge. (p 8)

Additional NFPA Standards

NFPA 1 *Fire Code*
NFPA 101 *Life Safety Code*
NFPA 1033 *Standard for Professional Qualifications for Fire Investigator*

FESHE Objectives

Principles of Code Enforcement

1. Explain the code enforcement system and the fire inspector's role in that system. (pp 4–8)
2. Describe the development and adoption processes for codes and standards. (pp 6–7)
3. Describe the difference between prescriptive and performance-based codes. (p 7)
4. Describe the legal authority and limitations relevant to fire code inspections. (p 8)
5. Recognize ethical practices for the code enforcement officer. (p 9)
6. Explain the application and interrelationship of codes, standards, recommended practices, and guides. (pp 6–7)
7. Describe the political, business, and other interests that influence the code enforcement process. (p 7)
8. Identify the professional development process for code enforcement practitioners. (p 9)

Knowledge Objectives

After studying this chapter, you will be able to:

1. Describe the functions of the fire inspection unit.
2. Describe the objectives of a fire inspection.
3. Describe the role of the fire inspector within the fire service.
4. Describe the roles and responsibilities of the fire inspector I.
5. Describe the roles and responsibilities of the fire inspector II.
6. Describe the relationship between state statutes, local ordinances, and fire safety codes.
7. Describe the range of the fire inspector's legal authority.
8. Describe the types of permits that may be issued to an occupancy.
9. Describe the ethical practices of the fire inspector.

Skills Objectives

There are no skills objectives for this chapter.

You Are the Fire Inspector

You have been making joint inspections with a seasoned fire inspector for six months. Most of your inspections have been of restaurants, bars, taverns, and night clubs, with an occasional auto repair shop. You now have been given your first assignment. You are assigned to inspect the local community college, which consists of numerous buildings including the administration building, library, dormitories, classrooms, dining hall, student center, and gymnasium.

1. What is the first priority of fire inspection?
2. Which codes and standards should you review prior to the fire inspection?

Introduction

Effective fire inspections, plan reviews, and public education efforts conducted by fire inspectors can prevent fires and the loss of life. Just as crime prevention has assisted police forces to perform more effectively, fire prevention aids all aspects of the fire service. If fires are prevented from starting, then the true duty of the fire service to the public, protecting lives and property, is accomplished. The fire inspector provides a valuable service to both the municipality and the property owner by assisting in carrying out the fire department's job of protecting life and property Figure 1.

The fire inspection unit may be part of the fire department or an independent unit. The fire inspection unit typically focuses on the "Three E's of Fire Prevention," which involve:
- Engineering to eliminate hazards through design, correcting hazards, or reducing hazards
- Educating the public to develop a positive attitude toward fire safety and educate people to be fire safe
- Enforcing the codes, standards, and regulations that are designed to help prevent fires

What Is a Fire Inspection?

A fire inspection is a visual inspection of a building and its property that is conducted to determine if the building complies with all pertinent statutes and regulations of the jurisdiction Figure 2. Fire inspections are conducted to reduce the risk of fire and maintain a reasonable level of protection of life and property from the hazards created by fire, explosion, and hazardous materials. Fire inspections are conducted at businesses, apartment buildings, schools, hospitals, places of public assembly, and other occupancies with the exception of one- or two-family dwellings. Safety surveys may be conducted in homes at the request of the owner, but these are typically done simply as a courtesy.

A fire inspection includes testing fire suppression, fire alarm, and detection systems, conducting occupant evacuation drills, evaluating the fire retardant quality of materials used within the building, and providing guidelines to the building owners on fire safe materials. In addition to standard fire inspections, there are also general fire safety or process inspections that are performed to check for overcrowding, blocked exits during business hours, or other types of unsafe practices that may be occurring while the building or space is occupied, placing the lives of the occupants at risk. Examples include a new movie release at the local movie theatre resulting in larger than normal crowds, or overflowing storage in mercantile properties during the holiday season. General fire safety or process inspections may be the result of a complaint, such as overcrowding in a bar, or are initiated by the fire inspector.

Fire Inspection and the Fire Service

Depending upon the community's size, complexity, state and local legislative requirements, and multiple other factors, fire inspectors may work within the community's fire department, the community's independent code enforcement agency, or in a business that performs fire inspections as a contracted service by the community. For many years, the responsibilities of

Fire Inspector Tips

Often the term fire safety inspection is used to describe the activities being performed during the fire inspection. The term is also regularly applied to inspections that involve the enforcement of the NFPA 101, *Life Safety Code,* which expands the scope of the inspection to include general safety issues. The terms fire inspection and fire safety inspection may be used separately or interchangeably.

Figure 1 The fire inspector provides a valuable service to both the municipality and the property owner by assisting in carrying out the fire department's job of protecting life and property.

fire inspection lay strictly within the fire department; however, the complexities of the task, coupled with legislative, training, and mandated services resulted in some communities creating an independent agency or contracting the service to a private enterprise. In some communities, local fire department engine or ladder companies continue to conduct fire inspections in addition to pre-planning inspections. In the end, the local government determines where and who performs fire inspections in the community.

Additional Fire Prevention Roles within a Fire Department

If a fire inspector works within a fire department, the fire inspector may hold multiple roles during a profession, particularly in smaller fire departments. Some of the more common positions that fire inspectors may also assume include:

- **Fire marshal**: Fire marshals inspect businesses and enforce public safety laws and fire codes. They may respond to fire scenes to help investigate the cause of a fire.
- **Fire investigator**: Fire investigators may have full police powers to investigate and arrest suspected arsonists and people causing false alarms.
- **Fire and life safety education specialist**: These individuals educate the public about fire safety and injury prevention and present juvenile fire safety programs.
- **Fire protection engineer**: The fire protection engineer usually has an engineering degree, reviews plans, and works with building owners to ensure that their fire suppression and detection systems will meet code and function as needed. Some fire protection engineers actually design these systems.

Roles and Responsibilities for Fire Inspector I

According to NFPA 1031, *Standard for Professional Qualifications for Fire Inspector and Plan Examiner*, a **Fire Inspector I** conducts basic fire inspections and applies codes and standards **Figure 3**. The Fire Inspector I must also meet the job performance requirements specified in NFPA 1031, including:

- Inspect structures in the field and write reports based on observations and findings.
- Identify the need for a permit and how to obtain the permit.
- Recognize the need for a plan review and when to send it out to an expert for further review.
- Investigate common complaints and act accordingly to resolve any compliance or safety issues.
- When presented with a fire protection, fire prevention, or life safety issue, identify the applicable code or standard that is being violated.
- Participate in legal proceedings and provide testimony or written comments as required.
- Identify the occupancy classification of a single-use occupancy.
- Compute the allowable occupant load of a single-use occupancy, post that number publically, and take corrective action if over-crowding occurs.

Figure 2 A fire inspection is a visual inspection of a building and its property that is conducted to determine if the building complies with all pertinent statutes and regulations of the jurisdiction.

Figure 3 A Fire Inspector I conducts basic fire inspections and applies codes and standards.

- Evaluate the exits of an existing building to ensure that a safe building evacuation may take place during an emergency.
- Verify the type of construction for an addition or remodeling project and direct compliance as needed.
- Determine if existing fixed fire suppression systems are working properly through testing and observation.
- Determine if existing fire detection and alarm systems are working properly through testing and observation.
- Determine if existing fixed portable fire extinguishers are working properly through testing and observation.
- Recognize any hazardous conditions involving equipment, processes, and operations in an occupancy.
- Compare an approved plan to an existing fire protection system and identify and act on any deficiencies found.
- Verify that emergency planning and preparedness measures are in place and have been practiced.
- Inspect the emergency access for the fire department to an existing site.
- Verify that hazardous materials, flammable and combustible liquids and gases are stored, handled, and used in accordance with local codes and laws.
- Recognize a hazardous fire growth potential in a building or space.
- Determine code compliance, given the codes, standards, and policies of the jurisdiction and a fire protection issue.
- Verify that a building has a sufficient water supply in case of a fire.

Roles and Responsibilities for Fire Inspector II

According to NFPA 1031, a **Fire Inspector II** conducts most types of inspections and interprets applicable codes and standards Figure 4. The Fire Inspector II must also meet the job performance requirements specified in NFPA 1031, including:
- Process a permit application.
- Process a plan review application.
- Investigate complex complaints, given a reported situation or condition, and bring the issue to a resolution.
- Recommend modifications to codes and standards of the jurisdiction based on a fire safety issue.
- Recommend policies and procedures for the delivery of inspection services, given management objectives.
- Compute the maximum allowable occupant load of a multi-use building based on in-the-field observations or a description of its uses.
- Identify the occupancy classifications of a mixed-use building, based on in-the-field observations or a description of its uses.
- Determine the building's area, height, occupancy classification, and construction type by examining an approved plan, a description of a building, or the construction features.
- Evaluate the fire protection systems and equipment to ensure that they can protect the life safety of occupants from any hazards present in the structure.
- Analyze the elements of the exits in a building or portion of a building to ensure that they meet with applicable codes and standards.
- Evaluate hazardous conditions involving equipment, processes, and operations in a building.
- Evaluate emergency planning and preparedness procedures using copies of existing or proposed plans and procedures.
- Verify code compliance for storage, handling, and use of hazardous materials and flammable and combustible liquids and gases.
- Determine the fire growth potential in a building or space based on the contents, interior finish, and construction elements.
- Inspect emergency access for the fire department to a site.
- Verify compliance with construction documents and ensure that life safety systems and building services equipment are installed, inspected, and tested to perform as described in the engineering documents and the operation and maintenance manuals.
- Classify the occupancy type of a building based on a set of plans, specifications, and a description of a building.
- Compute the maximum allowable occupant load in accordance with applicable codes and standards.
- Review the proposed installation of fire protection systems based on shop drawings and system specifications and ensure that the system is code compliant and installed in accordance with approved drawings.
- Verify that the means of exit elements are provided and meet all applicable codes and standards.
- Verify the construction type of a building based on a set of approved plans and specifications.

Figure 4 A Fire Inspector II conducts most types of inspections and interprets applicable codes and standards.

Codes and Standards

Codes and standards are regulatory or industry-adopted documents that provide guidance for the safe construction and use of buildings. A <u>code</u> is a rule or law established for enforcement

> **Fire Inspector Tips**
>
> Local and state officials will determine the code or codes they believe are the most appropriate and should be enforced in their community. The fire inspector's role is to enforce the codes adopted by these officials. The code enforcement system is intended to keep the people of the community safe—at work, at home, and when out and about in the community.
>
> The codes to be enforced are typically construction and fire-based; however, some communities have integrated fire inspection into the entire code enforcement unit for the community. This may require fire inspectors to conduct building inspections, fire inspections, life safety inspections, plumbing inspections, property maintenance inspections, and health inspections. Each of these inspection requirements typically calls for specific educational and possibly certification requirements for the inspector. These tasks are locally defined and the fire inspector's job description will generally outline the educational, certification, and performance requirements of the position.

of a fire protection, life safety, building construction, or property maintenance issue. Codes relate to many types of equipment used in construction, such as heating, ventilating, and air-conditioning equipment; construction materials; and similar products. A **standard** is a document in which are specifications or requirements defining the minimum levels of performance, protection, or construction. The standard may or may not be a legally mandated document, but considered an industry best practice to follow.

Model Code Organizations

Model code organizations write or develop codes and standards for adoption by governmental and other regulatory agencies such as insurance organizations. Model code organizations include the National Fire Protection Association (NFPA), International Code Council, and the American Society of Mechanical Engineers. To ensure a high level of acceptance of codes, model code organizations adopt stringent guidelines for developing codes. Technical advisory committees are made up of the different interest groups that will be directly affected by the codes. For example, a sprinkler manufacturer may sit on the committee revising the code on fire suppression systems. A fire code technical advisory committee may include representatives from the fire service, the insurance industry, equipment manufacturers, and contractors. The various members of the technical advisory committees provide a balance of perspectives and views in the development of the code, assuring that all users are involved in the decision making process.

Prescriptive Codes and Performance-Based Codes

Prescriptive codes and performance-based codes are used during the fire inspection process. A **prescriptive code** lists the specific details the installation or construction must meet, such as the type of electrical wiring to be used in an occupancy. A **performance-based code** outlines a requirement that a design has to meet, but does not state that a particular method or material must be used to meet the requirement. An example of a performance-based code is the requirement that a fire pump serving a high-rise building must be protected by 2-hour fire rated construction or be physically separated by 50 feet from the building. The fire inspector uses both types of codes to ensure that a building is safe to be occupied.

Code and Standard Adoption

Local, county, district, or state officials may select for adoption the codes or standards that meet the specific needs of the jurisdiction. The code or standard adoption process typically follows this path:

- The need to adopt a new code arises when a new code is created by a model code agency, a fire inspector expresses the need for a code to enforce, or a code needs modification.
- The proposed modified language is prepared by technically competent staff or by consultants or the new code is collected.
- The proposed modified code or new code is provided to elected officials for review.
- Upon review and any corrections, the modified or new code are posted for review by the general public.
- A hearing is held to discuss the new or modified code by the elected officials with the general public **Figure 5**.
- Input is provided, considerations for changes are taken, modifications may be made, and the elected officials then vote the code into statute.
- There may be an appeals process depending on the level of government enacting the regulations, state law governing adoption of safety regulations, and related oversight agencies such as State Fire Marshal's office, Public Utilities Commission, etc.

As a rule, codes and standards are adopted under most state laws or statutes or by local ordinances. State statutes will generally provide guidance on how to apply codes and standards in your local jurisdiction. However, there are some federal laws governing fire safety issues. For example, the 1990 *Hotel and Motel Safety Act* requires that federal government employees traveling on official business must stay in hospitality properties equipped with a fire sprinkler system.

Figure 5 A hearing is held to discuss the new or modified code by the elected officials with the general public.

> **Fire Inspector Tips**
>
> An effective fire inspector is trained to recognize fire hazards and to correct problem situations. You also must be capable of communicating information to building occupants and owners both orally and in written form. As a fire inspector, you must be able to translate codes and standards into language that the layperson can understand in order to explain to property owners why certain conditions present a fire hazard and how to correct the situation. For example, you may be asked to explain the difference between a smoke detector and carbon monoxide detector to a homeowner, and why both may be needed.

> **Fire Inspector Tips**
>
> Recommended practices or guides are manufacturer or industry developed techniques to perform an activity in a safe manner. They are typically not adopted into a code, ordinance, or law, but are simply the best practice to follow.

Understanding the Legal Processes

The legal authority for fire inspection varies by state to state and, in many cases, varies within each state. The job description of the fire inspector should list the limits of legal authority that the position holds in the local jurisdiction, and how to gain the next level of expertise or authority when needed. As part of the introductory training process for the fire inspector, a specific session must be included on the authority of the fire inspector and the process that must be used when citations are required or when legal action must be taken upon the occupant or owner as the process will vary by community.

Authority Having Jurisdiction

Who is the Authority Having Jurisdiction (AHJ) for fire inspections? According to the NFPA, the AHJ is the organization, office, or individual responsible for enforcing the requirements of a code or standard, or for approving equipment, materials, an installation, or a procedure. The AHJ receives its authority through local or state government. For example, in Florida, the AHJ for fire inspections is the local fire chief of the fire department. If there is not a local fire department, then a government official is designated by the local government. The Office of State Fire Marshal has direct responsibility for all state-owned facilities. However, authority is granted to the State Fire Marshal to inspect other occupancies when there is reason to believe that a fire safety violation exists, if a local fire inspector is unavailable, or if a higher level of expertise or authority is necessary.

The authority to perform fire inspections differs from state to state. Some states require state oversight of specific occupancies such as nursing homes, day care centers, and certain personal care facilities. You must be familiar with the laws and regulations of your state and local government prior to engaging in fire inspection and code enforcement activities.

State and Local Law

State and local law affects every aspect of your job. For example, your fire inspection unit may have specific requirements that every written notice must follow. In order to avoid legal disputes, a standardized notice of violation that has been reviewed for legal sufficiency by the attorney for the jurisdiction should be used by all personnel.

Range of Authority

The range of authority of the fire inspector is defined by state or local law. The range of authority typically allows for greater authority for a Fire Inspector II. The range of authority may deal with types of properties that the fire inspector can inspect, the size of properties that the fire inspector may inspect, or types of inspection that the fire inspector may perform, for example, hazard-based or fire-protection-equipment-based.

Legal Proceedings

As a fire inspector, you may be involved in legal proceedings. The results of a fire inspection, failure by a building owner to comply with corrective actions, and resulting fires may require you to appear in court. These legal proceedings may be criminal or civil in nature. Legal proceedings may require you to present documentation of your training and expertise, the local procedures used in conducting fire inspections in your jurisdiction, and certification to support any information you provide as testimony.

Permits

Many fire inspection activities result in the issuance of a **permit**. A fire inspection may be a step in the permit application process, depending on the type of permit and local procedures. A permit is a document stating the compliance with or restrictions for use based upon applicable codes. The permit may be a:

- Use and occupancy permit—A permit to occupy a structure. For example, a nail salon in a mall.
- Special use permit—A permit to perform specified activities for a specified time period. For example, a Halloween costume store in a strip mall.
- Fire protection equipment permit—A permit to install or use a particular fire protection equipment device. For example, a permit to install a sprinkler system in a storeroom of a shoe store or to use a fire hydrant.
- Hazardous material use permit—A permit to use a specific product for a specific purpose for a specific time period. For example, a permit to use a chemical during a manufacturing process.

The objective of the permit application process is to make certain that buildings are safe for the public to occupy and fire fighters to operate. The permit application process mirrors the plan review process, which is covered in detail in Chapter 6, *Reading Plans*. The owner of the occupancy submits a standard local permit application, it is reviewed to ensure that the occupancy meets all local codes and standards, and the permit is either approved or denied. Many agencies combine the permit application process with a plan review for a new occupancy for greater efficiency.

Ethics and the Fire Inspector

Ethics are critical in fire inspection. A fire inspector must not be influenced by business or political interests. Inappropriate behavior by individuals in power or those holding the public trust gets frequent attention by the media. The fire service is not exempt. Most of the time, fire inspectors make ethical decisions; however, some make unethical choices that often appear in the newspaper and have very negative consequences for the individual and the organization.

Ethical choices are based on a value system. The fire inspector has to consider each situation, often subconsciously, and make a decision based on his or her values. If the organization has clear values that are part of a strong organizational culture, the fire inspector uses the organization's value system. If the values are not clear, the individual substitutes his or her own value system.

The key to improving ethical choices is to have clear organizational values. This can be accomplished by:

- Having a code of ethics that is well known throughout the organization
- Selecting employees who share the values of the organization
- Ensuring that top management exhibit ethical behavior
- Having clear job goals
- Having performance appraisals that reward ethical behavior
- Implementing an ethics training program

Even at the company level, these values can be implemented to help prevent undesirable ethical choices. One way to help judge a decision is to ask yourself three questions:

- What would my parents and friends say if they knew?
- Would I mind if the paper ran it as a headline story?
- How does it make me feel about myself?

Asking these questions can help prevent an event that could devastate the department's and the fire inspector's reputation for years to come.

Career Development

The fire inspector in today's world is expected to continually develop his or her professional skills and knowledge. The required skills and knowledge are defined not only in NFPA standards but also various state laws. Professional organizations or state agencies offer continuing education and professional credentialing. Depending upon the state, fire inspectors may be required to complete certifications or tests to document performance capabilities.

Wrap-Up

■ Chief Concepts

- Effective fire inspections, plan reviews, and public education efforts conducted by fire inspectors can prevent fires and the loss of life.
- A fire inspection is a visual inspection of a building and its property that is conducted to determine if the building complies with all pertinent fire and safety statutes and regulations of the jurisdiction.
- Fire inspections are conducted to reduce the risk of fire and maintain a reasonable level of protection of life and property from the hazards created by fire, explosion, and hazardous materials.
- Depending upon the size, complexity, state and local legislative requirements, and multiple other factors, fire inspectors may work within the community's fire department, the community's independent code enforcement agency, or in a business that performs fire inspections as a contracted service by the community.
- Codes and standards are regulatory or industry adopted documents that provide guidance for the safe construction and use of buildings.
- A code is a rule or law established for enforcement of a fire protection, life safety, building construction, or property maintenance issue. Codes relate to many types of equipment used in construction, such as heating ventilating and air-conditioning equipment, construction materials and similar products.
- A standard is a document containing specifications or requirements defining the minimum levels of performance, protection, or construction. The standard may or may not be a legally mandated document, but is considered an industry best practice to follow.
- The job description of the fire inspector should list the limits of legal authority that the position holds in the local jurisdiction, and how to gain the next level of expertise or authority when needed.
- Many fire inspection activities result in the issuance of a permit. A permit is a document stating the compliance with or restrictions for use based upon applicable codes.

■ Hot Terms

<u>Code</u> A standard that is an extensive compilation of provisions covering broad subject matter or that is suitable for adoption into law independently of other codes and standards.

<u>Fire and life safety education specialist</u> A member of the fire department who deals with the public on education, fire safety, and juvenile fire safety programs.

<u>Fire inspection</u> A visual inspection of a building and its property to determine if the building complies with all pertinent statutes and regulations of the jurisdiction.

<u>Fire inspector I</u> An individual at the first level of progression who has met the job performance requirements specified in this standard for Level I. The fire inspector I conducts basic fire inspections and applies codes and standards. (NFPA 1031)

<u>Fire inspector II</u> An individual at the first level of progression who has met the job performance requirements specified in this standard for Level II. The Fire Inspector II conducts most types of inspections and interprets applicable codes and standards. (NFPA 1031)

<u>Fire investigator</u> An individual who has demonstrated the skills and knowledge necessary to conduct, coordinate, and complete an investigation. (NFPA 1033)

<u>Fire marshal</u> A member of the fire department who inspects businesses and enforces laws that deal with public safety and fire codes.

<u>Fire protection engineer</u> A member of the fire department who works with building owners to ensure that their fire suppression and detection systems will meet code and function as needed.

<u>Performance-based code</u> Outlines the requirement that a design has to meet, but does not state that a particular method or material must be used to meet the requirement.

<u>Permit</u> A document issued by the authority having jurisdiction for the purpose of authorizing performance of a specified activity. (NFPA 1)

<u>Prescriptive code</u> Defines the specifics of a material of construction or action to be taken; such as the type of electrical wiring to use, based on the anticipated usage or requirement to conduct evacuation drills in a structure.

<u>Standard</u> A document, the main text of which contains only mandatory provisions using the word "shall" to indicate requirements and that is in a form generally suitable for mandatory reference by another standard or code or for adoption into law. Nonmandatory provisions shall be located in an appendix or annex, footnote, or fineprint note and are not to be considered a part of the requirements of a standard.

Fire Inspector *in Action*

A developer wants to build a three-story commercial office building with residential condominiums on the second and third floors. Their first stop is at your office to find out "What do you require?" Their second question is, "What do I have to do as far as plans are concerned?" What the developer is really asking is, "What are the fire protection requirements for the building that is going to be built?" and "How often do we have to check with you during the building phase?" Additionally, the developer will be asking about the coordination between the fire protection requirements and any other code requirements such as the building department or zoning, and the timelines of such compliance efforts.

To formulate an answer, you must know the occupancy type, construction type, and materials being used. Once these questions are answered you will be able to give the developer an idea of when you need to see the plans, and at what point you will be visiting the construction site.

1. A code is:
    A. determined by the fire inspector based on what he/she feels is best for the specific location or occupancy.
    B. a legal requirement, a building feature, or an operating feature at a certain location.
    C. the manner that life safety is accomplished in a building.
    D. a legal mandate allowing the inspector access to buildings.

2. A standard is:
    A. how a code requirement is implemented.
    B. a set of specifications or requirements that is considered industry best practice, and meets a minimum level of performance guidelines.
    C. a customary requirement that is required by all occupancies of a specific class.
    D. an everyday way of accomplishing something regardless of the occupancy.

3. The most important tool the fire inspector possesses is:
    A. the ability to document the inspection.
    B. maps and previous inspection reports of the occupancy.
    C. support from the building's owner.
    D. knowledge.

4. Any non-compliance (code violation) that has been noticed during the fire inspection:
    A. must be documented in the written report of the inspection.
    B. must be told to the person in authority of the occupancy during the closing conference.
    C. may be overlooked by the fire inspector if there is good reason at the time.
    D. shall be shared with the personnel at the nearest fire station for their safety.

Building Construction

CHAPTER 2

NFPA 1031 Standard

Fire Inspector I

4.3.4 Verify the type of construction for an addition or remodeling project, given field observations or a description of the project and the materials being used, so that the construction type is identified and recorded in accordance with the applicable codes and standards and the policies of the jurisdiction. (pp 14–18)

(A) Requisite Knowledge. Applicable codes and standards adopted by the jurisdiction, types of construction, rated construction components, and accepted building construction methods and materials. (pp 18–36)

(B) Requisite Skills. The ability to read construction plans, make decisions, and apply codes and standards.

Additional NFPA Standards

NFPA 80 *Standard for Fire Doors and Other Opening Protectives*

NFPA 220 *Standard on Types of Building Construction*

FESHE Objectives

There are no FESHE objectives for this chapter.

Knowledge Objectives

After studying this chapter, you will be able to:
1. Describe the characteristics of the following building materials: masonry, concrete, steel, glass, gypsum board, and wood.
2. Identify the types of building construction.
3. List the characteristics of each of the following types of building construction: fire-resistive construction, noncombustible construction, ordinary construction, heavy timber construction, and wood-frame construction.
4. Describe the function of each of the following building components: foundations, floors, ceilings, roofs, trusses, walls, doors, windows, interior finishes, and floor coverings.

Skills Objectives

There are no skills objectives for this chapter.

You Are the Fire Inspector

The fire chief asks you to accompany a fire officer during a pre-planning walkthrough of an old mill building that now holds offices and shops. He asks you to determine the construction features of the mill building to help determine what potential problems may occur during a fire. With the mix of original and renovated construction, this walkthrough may pose a few challenges.

1. What type of construction do you expect to find during the walkthrough?
2. What are the potential problems that may occur with this type of construction during a fire?

Introduction

Knowing the varieties or kinds of building construction and the types of materials used to construct a building is vital for a fire inspector. Building construction affects how fires grow and spread and determines the special safety features required to control fire growth and protect the building's occupants. Building construction will also determine the codes that the building needs to comply with.

An understanding of building construction begins with the materials used. Building components are usually made of different materials. The properties of these materials and the details of their construction determines the basic fire characteristics of the building itself.

Types of Construction Materials

Function, appearance, price, and compliance with building and fire codes are all considerations when selecting building materials and construction methods. Architects often place a priority on functionality and aesthetics when selecting materials, whereas builders are often more concerned about price and ease of construction. For their part, building owners are usually interested in durability and maintenance expenses, as well as the materials' initial costs and aesthetics.

Fire inspectors use a completely different set of factors to evaluate construction methods and materials. As a fire inspector, your chief concerns are fire prevention and building behavior under fire conditions. A building material or construction method that is attractive to architects, builders, and building owners can create serious problems or deadly hazards for occupants and for fire fighters.

The most common building materials are wood, masonry, concrete, steel, aluminum, glass, gypsum board, and plastics. Within these basic categories, hundreds of variations are possible. The key factors that affect the behavior of each of these materials under fire conditions are outlined here:

- **Combustibility**: Whether or not a material will burn determines its combustibility. Materials such as wood will burn when they are ignited, releasing heat, light, and smoke, until they are completely consumed by the fire. Concrete, brick, and steel are noncombustible materials that cannot be ignited and are not consumed by a fire.
- **Thermal conductivity**: This characteristic describes how readily a material will conduct heat. Heat flows very readily through metals such as steel and aluminum. By contrast, brick, concrete, and gypsum board are poor conductors of heat.
- Decrease in strength at elevated temperatures: Many materials lose strength at elevated temperatures. For example, steel loses strength and will bend or buckle when exposed to fire temperatures, and aluminum melts in a fire. By contrast, bricks and concrete can generally withstand high temperatures for extended periods of time.
- Thermal expansion when heated: Some materials—steel, in particular—expand significantly when they are heated. A steel beam exposed to a fire will stretch (elongate); if it is restrained so that it cannot elongate, it will sag, warp, or twist. As a general rule, a steel beam will elongate at a rate of 1″ (25.4 mm) per 10′ (304 cm) of length at a temperature of 1000°F (538°C).

Masonry

Masonry includes stone, concrete blocks, and brick. The individual components are usually bonded together into a solid mass with mortar, which is produced by mixing sand, lime, water, and

Fire Inspector Tips

When steel beams are heated during a violent fire, they can expand and push walls out of alignment. This can lead to a total or partial collapse of the wall.

Figure 1 Masonry materials are inherently fire resistive.

Portland cement. Masonry materials are inherently fire resistive Figure 1. They do not burn or deteriorate at the temperatures normally encountered in building fires. Masonry is also a poor conductor of heat, so it will limit the passage of heat from one side through to the other side. For these reasons, masonry is often used to construct **fire walls**. A fire wall helps prevent the spread of a fire from one side of the wall to the other.

All masonry walls are not necessarily fire walls, however. A single layer of masonry may be used over a wood-framed building to make it appear more substantial. If there are unprotected openings in a masonry wall, a fire can often spread through them. If the mortar has deteriorated or the wall has been exposed to fire for a prolonged time, a masonry wall can collapse during a fire.

A masonry structure also can collapse under fire conditions if the roof or floor assembly collapses. In such a case, the masonry falls because of the mechanical action of the collapse. Aged, weakened mortar can also contribute to a collapse. Regardless of the cause, a collapsing masonry wall can be a deadly hazard to occupants and to fire fighters.

Concrete

Like masonry, concrete is a naturally fire-resistive material. It does not burn or conduct heat well, so it is often used to insulate other building materials from fire. Concrete does not have a high degree of thermal expansion—that is, it does not expand greatly when exposed to heat and fire—nor will it lose strength when exposed to high temperatures.

Concrete is made from a mixture of Portland cement and aggregates such as sand and gravel. Different formulations of concrete can be produced for specific building purposes. Concrete is not necessarily inexpensive; however, when fully cured, it is one of the strongest construction materials that is easily formed and shaped at the point of installation. It is often used for foundations, columns, floors, walls, roofs, and exterior pavement.

Under compression, concrete is strong and can support a great deal of weight; under tension, however, it is weak. When this material is used in building construction, steel reinforcing rods are often embedded in the concrete to strengthen it under tension. In turn, the concrete insulates the steel reinforcing rods from heat.

Although it is inherently fire resistive, concrete can be damaged by exposure to a fire. For example, a fire may convert trapped moisture in the concrete to steam. As the steam expands, it creates internal pressure that can cause sections of the concrete surface to break off in a process called **spalling**. Severe spalling in reinforced concrete can expose the steel reinforcing rods to the heat of the fire. If the fire is hot enough, the steel might weaken, resulting in a structural collapse, though this is a rare event.

Steel

Steel is the strongest building material in common use. It is very strong in terms of both tension and compression. It can also be produced in a wide variety of shapes and sizes, ranging from heavy beams and columns to thin sheets. Steel is often used in the structural framework of a building to support floor and roof assemblies. It is resistant to aging and does not rot, although most types of steel will rust unless they are protected from exposure to air and moisture.

Steel is an alloy of iron and carbon. Additional metals may be added to the mix to produce steel with special properties, such as stainless steel or galvanized steel.

When considered by itself, steel is not fire resistive. It will melt at extremely high temperatures, although these extraordinary temperatures are rarely encountered at structure fires Figure 2. Steel conducts heat well, so it tends to expand and lose strength as the temperature increases. For this reason, other materials—such as masonry, concrete, or layers of gypsum board—are often used to protect steel from the heat of a fire. Sprayed-on coatings of mineral or cement-like materials are also used to insulate steel members. The amount of heat absorbed by steel depends on the mass of the object and the amount of protection surrounding it. Smaller, lighter pieces of steel absorb heat more easily than larger and heavier pieces.

Steel both expands and loses strength as it is heated. Therefore, an unprotected steel roof beam directly exposed to a fire may elongate sufficiently to cause a supporting wall to collapse. Heated steel beams will often sag and twist, whereas steel columns tend to buckle as they lose strength. The bending and distortion are caused by the uneven heating that occurs in actual fire situations. Some residential occupancies are now using metal

Figure 2 Steel will melt at extremely high temperatures.

studs instead of wood. These metal studs will fail early when exposed to fire conditions.

■ Other Metals

A variety of other metals, including aluminum, copper, and zinc, are also used in building construction. Aluminum is often used for siding, window frames, door frames, and roof panels. Copper is used primarily for electrical wiring and piping; it is sometimes used for decorative roofs, gutters, and down spouts. Zinc is used primarily as a coating to protect metal parts from rust and corrosion.

Aluminum is occasionally used as a structural material in building construction. It is more expensive and not as strong as steel, so its use is generally limited to light-duty applications such as awnings and sunshades. Aluminum expands more than steel when heated and loses strength quickly when exposed to a fire. This material has a lower melting point than steel, so it will often melt and drip in a fire.

■ Glass

Almost all buildings contain glass—in windows, doors, skylights, and sometimes walls. Ordinary glass breaks very easily, but glass can be manufactured to resist breakage and to withstand impact or high temperature.

Glass is noncombustible, but it is not fire resistive. Although ordinary glass will usually break when exposed to fire, specially formulated glass can be used as a fire barrier in certain situations. The thermal conductivity of glass is rarely a significant factor in the spread of fire.

Many different types of glass are available:

- Ordinary window glass will usually break with a loud pop when heat exposure to one side causes it to expand and creates internal stresses that fracture the glass. The broken glass forms large shards, which usually have sharp edges.
- <u>Tempered glass</u> is much stronger than ordinary glass and more difficult to break. Some tempered glass can be broken with a spring-loaded center punch; it will shatter into small pieces that do not have the sharp edges of ordinary glass.
- <u>Laminated glass</u> is manufactured so that a thin sheet of plastic is placed between two sheets of glass. The resulting product is much stronger than ordinary glass. When exposed to a fire, laminated glass windows are likely to crack and remain in place. Laminated glass is sometimes used in buildings to help soundproof areas.
- <u>Glass blocks</u> are thick pieces of glass similar to bricks or tiles. They are designed to be built into a wall with mortar so that light can be transmitted through the wall. Glass blocks have limited strength and are not intended to be used as part of a load-bearing wall, but they can usually withstand a fire. Some glass blocks are approved for use with fire-rated masonry walls.
- <u>Wired glass</u> is made by molding tempered glass with a reinforcing wire mesh **Figure 3**. When wired glass is subjected to heat, the wire holds the glass together and prevents it from breaking. This material is often used in fire doors and windows designed to prevent fire spread.

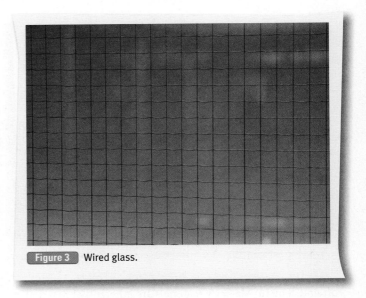

Figure 3 Wired glass.

■ Gypsum Board

<u>Gypsum</u> is a naturally occurring mineral composed of calcium sulfate and water molecules that is used to make plaster of Paris. Gypsum is a good insulator and noncombustible; it will not burn even in atmospheres of pure oxygen.

<u>Gypsum board</u> (also called drywall, Sheetrock, or plasterboard) is commonly used to cover the interior walls and ceilings of residential living areas and commercial spaces. Gypsum board is manufactured in large sheets typically 4′ × 8′ (1.2 m × 2.4 m) or 4′ × 12′ (2.4 m × 7.3 m) panels consisting of a layer of compacted gypsum sandwiched between two layers of specially produced paper. The sheets are nailed or screwed in place on a framework of wood or metal studs, and the edges are secured with a special tape. The nail or screw heads and tape are then covered with a thin layer of plaster.

Gypsum board has limited combustibility, because the paper covering will burn slowly when exposed to a fire. It does not conduct or release heat to an extent that would contribute to fire spread, so this material is often used to create a firestop or to protect building components from fire. Gypsum blocks are sometimes used as protective insulation around steel members or to create fire-resistive enclosures or interior fire walls.

When gypsum board is heated, some of the water present in the calcium sulfate will evaporate, causing the board to deteriorate. If it is exposed to a fire for a long time, the gypsum board will fail. Water will also weaken and permanently damage gypsum board.

Although gypsum board is a good finishing material and an effective fire barrier, it is not a strong structural material. It must be properly mounted on and supported by wood or steel studs. Gypsum board mounted on wood framing will protect the wood from fire for a limited time.

■ Wood

Wood is probably the most commonly used building material in our environment. It is inexpensive to produce, is easy to use, and can be shaped into many different forms, from heavy structural supports to thin strips of exterior siding. Both soft woods (such as pine) and hard woods (such as oak) are used for building construction.

A wide variety of wood products are used in building applications:

- Solid lumber is squared and cut into uniform lengths. Examples of solid lumber include the heavy timbers used in churches, mills, and barns and the lightweight boards used for siding and decorative trim.
- <u>Laminated wood</u> consists of individual pieces of wood glued together. Lamination is used to produce beams that are longer and stronger than solid lumber and to manufacture curved beams.
- <u>Wood panels</u> are produced by gluing together thin sheets of wood. Plywood is the most common type of wood panel used in building construction. Small chips (chip board) or particles of wood (particle board) can also be used to make wood panels, although these panels are usually much weaker than those constructed from plywood or solid lumber.
- <u>Wood trusses</u> are assemblies of pieces of wood or wood-and-metal combinations; they are often used to support floors and roofs. The structure of a truss enables a limited amount of material to support a heavy load.
- <u>Wooden beams</u> are efficient load-bearing members assembled from individual wood components. The shape of a wooden I-beam or box beam enables it to support the same load that a solid wood beam of its size could support.

For fire inspectors, the most important characteristic of wood is its high combustibility. Wood acts as fuel in a structural fire and can provide a path for the fire to spread. It ignites at fairly low temperatures and is gradually consumed by the fire, weakening and eventually leading to the collapse of the structure. Great quantities of heat and hot gases are created by the wood burning process, until eventually all that remains is a small quantity of residual ash.

The rate at which wood ignites, burns, and decomposes depends on several factors:

- Ignition: A small ignition source contains less energy and will take a longer time to ignite a fire. Using an accelerant will greatly speed this process.
- Moisture: Damp or moist wood takes longer to ignite and burn. New lumber usually contains more moisture than lumber that has been in a building for many years. A higher relative humidity in the atmosphere also makes wood more difficult to ignite.
- Density: Heavy, dense wood is harder to ignite than lighter or less dense wood.
- Preheating: The more the wood is preheated, the faster it ignites.
- Size and form: The rate of combustion is directly related to the surface area of the wood. During the combustion process, high temperatures release flammable gases from the surface of the wood. A large solid beam is difficult to ignite and burns relatively slowly; by contrast, a lightweight truss ignites more easily and is consumed rapidly.

Exposure to the high temperatures generated by a fire can also decrease the strength of wood through the process of <u>pyrolysis</u>. This chemical change occurs when wood or other materials are heated to a temperature high enough to release some of the volatile compounds in the wood, albeit without igniting these gases. The wood then begins to decompose without combustion.

Fire-Retardant–Treated Wood

Wood cannot be treated to make it completely noncombustible. Nevertheless, impregnating wood with mineral salts makes it more difficult to ignite and slows the rate of burning. Fire-retardant treatment can significantly reduce the fire hazards of wood construction. On the downside, the treatment process can also reduce the strength of the wood.

■ Plastics

Plastics are synthetic materials that are used in a wide variety of products today **Figure 4** . Plastics may be transparent or opaque, stiff or flexible, tough or brittle. They are often combined with other materials to produce building products.

Although plastics are rarely used for structural support, they may be found throughout a building. Building exteriors may include vinyl siding, plastic window frames, and plastic panel skylights. Foam plastic materials may be used as exterior or interior insulation. Plastic pipe and fittings, plastic tub and shower enclosures, and plastic lighting fixtures are commonly used. Even carpeting and floor coverings often contain plastics.

The combustibility of plastics varies greatly. Some plastics ignite easily and burn quickly, whereas others will burn only while an external source of heat is present. Some plastics will withstand high temperatures and fire exposure without igniting.

Many plastics produce quantities of heavy, dense, dark smoke and release high concentrations of toxic gases as they

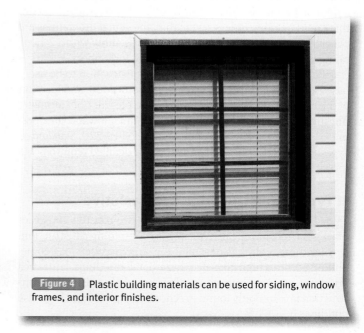

Figure 4 Plastic building materials can be used for siding, window frames, and interior finishes.

Fire Inspector Tips

The glue that is used to create laminated wood is *flammable*.

burn. This smoke resembles smoke from a petroleum fire—not surprising given that most plastics are made from petroleum products.

Thermoplastic materials melt and drip when exposed to high temperatures, even those as low as 500°F (260°C). Although heat can be used to shape these materials into a variety of desired forms, the dripping, burning plastic can rapidly spread a fire. By contrast, **thermoset materials** are fused by heat and will not melt as the plastic burns, although their strength will decrease dramatically.

Types of Construction

Fire inspectors need to understand and recognize five types of building construction—designated as Types I through V. These classifications are directly related to fire protection and fire behavior in the building.

Buildings are classified based on the combustibility of the structure and the fire resistance of its components. Buildings using Type I or Type II construction are assembled with primarily noncombustible materials and limited amounts of wood and other materials that will burn. This does not mean that Type I and Type II buildings cannot be damaged by a fire. If the contents of the building burn, the fire could seriously damage or destroy the structure.

In buildings using Type III, Type IV, or Type V construction, both the structural components and the building contents will burn. Wood and wood products are used to varying degrees in these buildings. If the wood ignites, the fire will weaken and consume both the structure and its contents. Structural elements of these buildings can be damaged or destroyed in a very short time.

Fire resistance refers to the length of time that a building or building component can withstand a fire before igniting. Fire resistance ratings are stated in hours based on the time that a test assembly withstands or contains a standard test fire. For example, walls are rated based on whether they stop the progress of the test fire for periods ranging from 20 minutes to 4 hours. A floor assembly could be rated based on whether it supports a load for 1 hour or 2 hours. Because ratings are based on a standard test fire that could be more or less severe than the actual fire, ratings are considered merely guidelines; that is, an assembly with a 2-hour rating will not necessarily withstand a fire for 2 hours.

Building codes specify the type of construction to be used, based on the height, area, occupancy classification, and location of the building. Other factors, such as the presence of automatic fire sprinklers, will also affect the maximum allowable area permitted in each construction classification. NFPA 220, *Standard on Types of Building Construction,* provides the detailed requirements for each type of building construction.

Type I Construction: Fire Resistive

Type I construction, (also referred to as fire-resistive construction) is the most fire-resistive category of building construction. It is used for buildings designed for large numbers of people, buildings with a high life-safety hazard, tall buildings, large-area buildings, and buildings containing special hazards. Type I construction is commonly found in schools, hospitals, and high-rise buildings **Figure 5**.

Figure 5 Type I construction is commonly found in schools, hospitals, and high-rise buildings.

A building with Type I construction can withstand and contain a fire for a longer period of time than buildings with Type II, III, IV, or V construction. The fire resistance and combustibility of all construction materials are carefully evaluated, and each building component must be engineered to contribute to fire resistance of the entire building to warrant this classification.

All of the structural members and components used in Type I construction must be made of noncombustible materials, such as steel, concrete, or limited combustible materials, such as gypsum board. In addition, the structure must be constructed or protected so that there are at least 2 hours of fire resistance. Building codes specify the fire-resistance requirements for different components. For example, columns and load-bearing walls in multistory buildings could have a fire-resistance requirement of 3 or 4 hours, whereas a floor might be required to have a fire-resistance rating of only 2 hours. Some codes allow Type I buildings to use a limited amount of combustible materials as part of the interior finish.

If a Type I building exceeds specific height and area limitations, codes generally require the use of fire-resistive walls and/or floors to subdivide it into compartments. A compartment might consist of a single floor in a high-rise building or a part of a floor in a large-area building. In any event, a fire in one compartment should not spread to any other parts of the building. To ensure that fire is contained, stairways, elevators shafts, and utility shafts should be enclosed in construction that prevents fire from spreading from floor to floor or from compartment to compartment.

Type I buildings typically use reinforced concrete and protected steel-frame construction. As mentioned earlier, concrete is noncombustible and provides thermal protection around the steel reinforcing rods. Reinforced concrete can fail, however, if it is subjected to a fire for a long period of time or if the building contents create an extreme fire load. Given its tendency to lose strength in the face of high temperatures, structural steel framing must be protected from the heat of a fire. In Type I construction, the structural steel members are generally encased in concrete, shielded by a fire-resistive ceiling, covered with multiple layers of gypsum board, or protected by a sprayed-on insulating material **Figure 6**. An unprotected steel beam exposed to a fire could fail and cause the entire building to collapse.

Figure 6 Sprayed-on fireproofing materials are often used to protect structural steel.

Type I building materials should not provide enough fuel by themselves to create a serious fire. It is the contents of the building that determine the severity of a Type I building fire. In theory, a fire could consume all of the combustible contents of a Type I building yet leave the structure basically intact.

Although Type I construction may provide the highest degree of safety, it does not eliminate all of the risks to occupants and fire fighters. Serious fires can still occur in fire-resistive buildings, because the burning contents may produce copious quantities of heat and smoke. For this reason, fire-resistive construction is typically combined with automatic sprinklers in modern high-rise buildings.

Type II Construction: Noncombustible

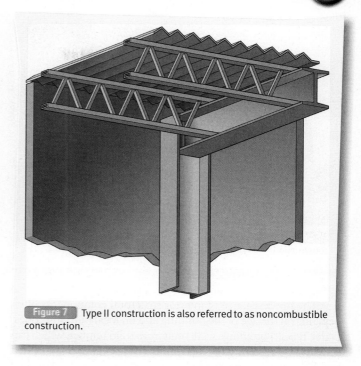

Figure 7 Type II construction is also referred to as noncombustible construction.

Type II construction is also referred to as noncombustible construction Figure 7 . All of the structural components in a Type II building must be made of noncombustible materials. The fire-resistive requirements, however, are less stringent for Type II construction than for Type I construction. In some cases, there are no fire-resistance requirements. In other cases, known as protected noncombustible construction, a fire-resistance rating of 1 or 2 hours may be required for certain elements.

Noncombustible construction is most common in single-story warehouse or factory buildings, where vertical fire spread is not an issue. Unprotected noncombustible construction is generally limited to a maximum of two stories. Some multistory buildings are constructed using protected noncombustible construction.

Steel is the most common structural material in Type II buildings. Insulating materials can be applied to the steel when fire resistance is required. A typical example of Type II construction is a large-area, single-story building with a steel frame, metal or concrete block walls, and a metal roof deck. Fire walls are sometimes used to subdivide these large-area buildings and prevent catastrophic losses. Undivided floor areas must be protected with automatic sprinklers to limit the fire risk.

Fire severity in a Type II building is determined by the contents of the building because the structural components contribute little or no fuel and the use of combustible interior finish materials is limited. If the building contents provide a high fuel load, a fire could collapse and destroy the structure. Automatic sprinklers should be used to protect combustible and valuable contents.

Type III Construction: Ordinary

Type III construction is also referred to as ordinary construction because it is used in a wide variety of buildings, ranging from commercial strip malls to small apartment buildings Figure 8 . Ordinary construction is usually limited to buildings of no more than four stories, but it can sometimes be found in buildings as tall as six or seven stories.

Type III construction buildings have masonry exterior walls, which support the floors and the roof structure. The interior structural and nonstructural members—including the walls, floor, and roof—are all constructed of wood. In most ordinary construction buildings, gypsum board or plaster is used as an interior finish material, covering the wood framework and providing minimal fire protection.

Fire-resistance requirements for interior construction of Type III buildings are limited. The gypsum or plaster coverings over the interior wood components provide some fire resistance. Key interior structural components may be required to have fire resistance ratings of 1 or 2 hours, or there may be no requirements at all. Some Type III buildings use interior masonry load-bearing walls to meet the requirements for fire-resistant structural support.

A building of Type III construction has two separate fire loads: the contents and the combustible building materials used. Even a vacant building will contain a sufficient quantity of wood and other combustible components to produce a large fire. A fire involving both the contents and the structural components can quickly destroy the building.

Figure 8 Type III construction is also referred to as ordinary construction.

Figure 9 Type IV construction is also known as heavy timber construction.

The fire resistance of interior structural components often depends on the age of the building and on local building codes. Older Type III buildings were built from solid lumber, which can contain or withstand a fire for a limited time. Newer or renovated buildings may have lightweight assemblies that can be damaged much more quickly and are prone to early failure.

Type IV Construction: Heavy Timber

Type IV construction is also known as heavy timber construction Figure 9. A heavy timber building has exterior walls that consist of masonry construction, and interior walls, columns, beams, floor assemblies, and roof structure that are made of wood. The exterior walls are usually brick and are extra thick to support the weight of the building and its contents.

The wood used in Type IV construction is much heavier than that used in Type III construction. It is more difficult to ignite and will withstand a fire for a much longer time before the building collapses. A typical heavy timber building might have 8-inch-square wood posts supporting 14″ (355 mm) deep wood beams and 8″ (203 mm) floor joists. The floors could be constructed of solid wood planks, 2 or 3″ (50 mm or 76 mm) thick, with a top layer of wood serving as the finished floor surface.

Heavy timber construction, if constructed to meet code, has no concealed spaces (voids). This structure helps reduce the horizontal and vertical fire spread that often occurs in ordinary construction buildings. Unfortunately, many heavy timber buildings do have vertical openings for elevators, stairs, or machinery, which can provide a path for a fire to travel from one floor to another.

In the past, heavy timber construction has been used to construct buildings as tall as six to eight stories with open spaces suitable for manufacturing and storage occupancies. Similar modern buildings are usually built with either fire-resistive (Type I) or noncombustible (Type II) construction. New buildings of Type IV construction are rare, except for special structures that feature the construction components as architectural elements.

The solid wood columns, support beams, floor assemblies, and roof assemblies used in heavy timber construction will withstand a fire much longer than the smaller wood members used in ordinary and lightweight combustible construction. Nevertheless, once involved in a fire, the structure of a heavy timber building can burn for many hours. A fire that ignites the combustible portions of heavy timber construction is likely to burn until it runs out of fuel and the building is reduced to a pile of rubble. As the fire consumes the heavy timber support members, the masonry walls will become unstable and collapse.

Mill Construction

Mill construction was common during the 1800s in the United States, especially in northeastern states. In particular, large mill buildings were used as factories and warehouses. In many communities, dozens of these large buildings were clustered together in industrial areas, creating the potential for huge fires involving multiple multistory buildings. The radiant heat from a major fire in one of these buildings could spread the fire to nearby buildings, resulting in the loss of several surrounding structures.

Mill construction was state-of-the-art for its time. Automatic sprinkler systems were developed to protect buildings of mill construction; as long as the sprinklers are properly maintained, these buildings have a good safety record. Without sprinklers, a heavy timber mill building is a major fire hazard.

Only a few of these original mill buildings are still used for manufacturing. Many more have been converted to small shops, galleries, office buildings, and residential occupancies. These conversions tend to divide the open spaces into smaller compartments and create void spaces within the structure. The revamped buildings may contain either noncombustible contents or highly flammable contents, depending on their use. Occupancies may range from unoccupied storage facilities to occupied loft apartments and office space. Appropriate fire protection and life-safety features, such as modern sprinkler systems, must be built into the conversions to ensure the safety of occupants.

Type V Construction: Wood Frame

In Type V construction, all of the major components are constructed of wood or other combustible materials Figure 10. Type V construction is often called wood-frame construction and is the most common type of construction used today. Wood-frame construction is used not only in one- and two-family dwellings and small commercial buildings, but also in larger structures

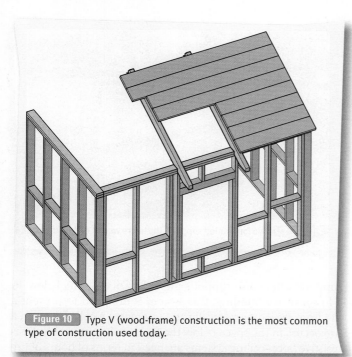

Figure 10 Type V (wood-frame) construction is the most common type of construction used today.

> **Fire Inspector Tips**
>
> A self-supporting brick wall erected along side the bearing wood wall is considered a veneer wall. If it carries part of the load of the building, then it is a composite wall.

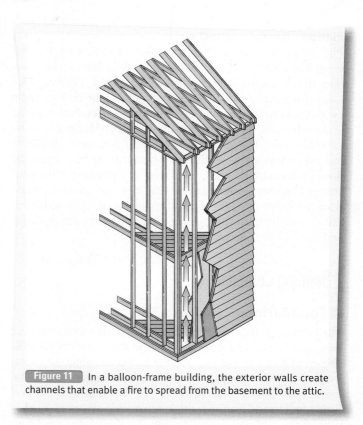

Figure 11 In a balloon-frame building, the exterior walls create channels that enable a fire to spread from the basement to the attic.

such as apartment and condominium complexes and office buildings up to four stories in height.

Many wood-frame buildings do not have any fire-resistive components. Some codes require a 1-hour fire resistance rating for limited parts of wood-frame buildings, particularly those of more than two stories. Plaster or gypsum board barriers are often used to achieve this rating.

Because all of their structural components are combustible, wood-frame buildings can rapidly become totally involved in a fire. In addition, Type V construction usually creates voids and channels that allow a fire within the structure to spread quickly. A fire that originates in a Type V building can easily extend to other nearby buildings.

Wood-frame buildings often collapse and suffer major destruction from fires. Smoke detectors are essential to warn building residents early if a fire occurs. Although compartmentalization can help limit the spread of a fire, automatic sprinklers are the most effective way of protecting lives and property in Type V buildings.

Wood-frame buildings are constructed in a variety of ways. Older wood-frame construction was assembled from solid lumber, which relied on its mass to provide strength. To reduce costs and create the largest building with the least amount of material, lightweight construction makes extensive use of wooden I-beams and wooden trusses. Structural assemblies are engineered to be just strong enough to carry the required load. As a result, there is little built-in safety margin, and these buildings can collapse early, suddenly, and completely during a fire.

The presence of flammable building contents may greatly increase the severity of a fire and accelerate the speed with which the fire destroys a wood-frame building. Indeed, the structural elements of such a building can be damaged or destroyed in a very short time. Fires that start to consume structural parts of a Type V building present a high risk of building collapse.

Wood-frame buildings can be covered with wood siding, vinyl siding, aluminum siding, brick veneer, or stucco. The covering on a building does not reflect the type of construction of that building. Just because you see a brick covering, it does not guarantee that the building is constructed of solid brick walls.

Two systems are used to assemble wood-frame buildings: balloon-frame construction and platform construction. Because these systems developed in different eras, you can anticipate the type of construction based on the age of a building.

Balloon-Frame Construction

Balloon-frame construction was popular between the late 1800s and the mid-1900s. This type construction is not permitted by today's newer codes without fire blocking. In a balloon-frame building, the exterior walls are assembled with wood studs that run continuously from the basement to the roof Figure 11. In a two-story building, the floor joists that support the first and second floors are nailed to these continuous studs.

As a result, an open channel between each pair of studs extends from the foundation to the attic. Each of these channels provides a path that enables a fire to spread from the basement to the attic without being visible on the first- or second-floor levels.

> **Fire Inspector Tips**
>
> A thorough review of NFPA 220, *Standard on Types of Building Construction*, and your building code should be undertaken to grasp a full understanding of the relationship between construction types and the permitted building's height and area.

Platform-Frame Construction

Platform-frame construction is used for almost all modern wood-frame construction. In a building with platform construction, the exterior wall studs are not continuous. The first floor is constructed as a platform, and the studs for the exterior walls are erected on top of it. The first set of studs extends only to the underside of the second-floor platform, which blocks any vertical void spaces. The studs for the second-floor exterior walls are erected on top of the second-floor platform. At each level, the floor platform blocks the path of any fire rising within the void spaces. Platform framing prevents fire from spreading from one floor to another through continuous stud spaces. Although a fire can eventually burn through the wood platform, the platform will slow the fire spread.

Building Components

Foundations

Figure 12   The foundation supports the entire weight of the building.

The primary purpose of a building foundation is to transfer the weight of the building and its contents to the ground Figure 12. The weight of the building itself is called the dead load; the weight of the building's contents is called the live load. The foundation ensures that the base of the building is planted firmly in a fixed location, which helps keep all other components connected. The design and construction of the foundation is essential to the long-term stability of the building. Likewise, any damage to the foundation can affect the structural stability of the building.

Modern foundations are usually constructed of concrete or masonry, although wood may be used in some areas. Foundations can be shallow or deep, depending on the type of building and the soil composition. Some buildings are built on concrete footings or piers; others are supported by steel piles or wooden posts driven into the ground.

Most foundation problems are caused by circumstances other than a fire, such as improper construction, shifting soil conditions, or earthquakes. Although fires can damage timber post and other wood foundations, most foundations remain intact even after a severe fire has damaged the rest of the building.

When inspecting a building, take a close look at the foundation. Look for cracks that indicate movement of the foundation. If the building has been modified or remodeled, look for areas where the support could be compromised.

Floors and Ceilings

Fire-Resistive Floors

In multistory buildings, floors and ceilings are generally considered a combined structural system—that is, the structure that supports a floor also supports the ceiling of the story below. In a fire-resistive building, this system is designed to prevent a fire from spreading vertically and to prevent a collapse when a fire occurs in the space below the floor–ceiling assembly. Fire-resistive floor–ceiling systems are rated in terms of hours of fire resistance based on a standard test fire.

Concrete floors are common in fire-resistive construction. The concrete can be cast in place or assembled from panels or beams of precast concrete, which are made at a factory and then transported to the construction site for placement. The thickness of the concrete floor depends on the load that the floor needs to support.

Concrete floors can be either self-supporting or supported by a system of steel beams or trusses. The steel can be protected by sprayed-on insulating materials or covered with concrete or gypsum board. If the ceiling is part of the fire-resistive rating, it provides a thermal barrier to protect the steel members from a fire in the area below the ceiling.

The ceiling below a fire-resistant floor can be constructed from plaster or gypsum board, or it can be a system of tiles suspended from the floor structure. In many cases, a void space between the ceiling and the floor above contains building systems and equipment such as electrical or telephone wiring, heating and air-conditioning ducts, and plumbing and fire sprinkler system pipes Figure 13. If the space above the ceiling is not subdivided by fire-resistant partitions or protected by automatic sprinklers, a fire can quickly extend horizontally across a large area.

Wood-Supported Floors

Wood floor structures are common in non-fire-resistive construction. Wooden floor systems range from heavy timber construction, which is often found in old mill buildings, to modern, lightweight construction.

Heavy timber construction can provide a huge fuel load for a fire, but it can also contain and withstand a fire for a considerable length of time without collapsing. Heavy timber construction uses posts and beams that are at least 8″ (203 mm) on the smallest side and often as large as 14″ (355 mm) in depth. The floor decking is often assembled from solid wood boards, 2 or 3″ (50 mm or 76 mm) thick, which are covered by an additional 1″ of finished wood flooring. The depth of the wood in this

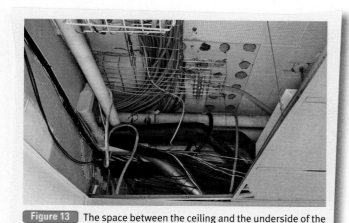

Figure 13 The space between the ceiling and the underside of the floor above often holds electrical and communications wiring and other building systems.

system will often contain a fire for an hour or more before the floor fails or burns through.

Conventional wood flooring, which was widely used for many decades, is much lighter than heavy timber but uses solid lumber as beams, floor joists, decking, and finished flooring. It burns readily when exposed to a fire, but generally takes about 20 minutes to burn through or reach structural failure. This time estimate is only a general, unscientific guideline; the actual burning rate depends on many other factors.

Modern lightweight construction uses structural elements that are much less substantial than conventional lumber. Lightweight wooden trusses or wooden I-beams are often used as supporting structures. Thin sheets of plywood are used as decking, and the top layer often consists of a thin layer of concrete or wood covered by carpet. This floor construction provides little resistance to fire. The lightweight structural elements can fail or the fire can burn through the decking quickly.

Ceiling Assemblies

From a structural standpoint, the ceiling is considered to be part of the floor assembly for the story above. Ceilings are mainly included in buildings for appearance reasons. Their primary function is to hold light fixtures and to diffuse light. Ceilings also conceal heating, ventilation, and air-conditioning (HVAC) distribution systems. In addition, they conceal wiring and fire sprinkler systems.

Ceilings can be part of the fire-resistive package. They can be covered with plaster or gypsum board, or they can consist of dropped ceilings covered with mineral tiles. Some ceilings are fire rated as part of the floor assembly.

■ Roofs

The primary purpose of a roof is to protect the inside of a building from the weather. Often, roofs are not designed to be as strong as floors, especially in warmer climates. If the space under the roof is used for storage, or if extra HVAC equipment is mounted on the roof, the load could exceed the designer's expectations. Also, this space may contain different types of insulation to help protect the structure from climate changes. In older homes, for example, this insulation often contains ground-up paper that could lead to increased smoldering and fire spread.

Several methods and materials are used for roof construction. Roofs are constructed in three primary designs: pitched, curved, and flat. The major components of a roof assembly are the supporting structure, the roof deck, and the roof covering.

Pitched Roofs

A <u>pitched roof</u> has sloping or inclined surfaces. Pitched roofs are used on many houses and some commercial buildings. The pitch or angle of the roof can vary depending on local architectural styles and climate. Variations of pitched roofs include gable, hip, mansard, gambrel, and lean-to roofs **Figure 14**.

Pitched roofs are usually supported by either rafters or trusses. <u>Rafters</u> are solid wood joists mounted in an inclined position. Pitched roofs supported by rafters usually have solid wood boards as the roof decking.

Modern lightweight construction uses manufactured wood trusses to construct most pitched roofs. Many lightweight trusses are manufactured using gusset plates that penetrate the wood no more than 3/8" (9.5 mm). The decking usually consists of thin plywood or a sheeting material such as wood particleboard. When these trusses and decking are exposed to heat and fire, they often fail after a short period of time with no warning beforehand.

Steel trusses also are used to support pitched roofs. The fire resistance of steel trusses is directly related to how well the steel is protected from the heat of the fire. Several roof-covering materials are used on pitched roofs, usually in the form of shingles or tiles **Figure 15**. Shingles are usually made from felt or mineral fibers impregnated with asphalt, although metal and fiberglass shingles are also used. Wood shingles, often made from cedar, are popular in some areas. In older construction, wood shingles were often mounted on individual wood slats instead of continuous decking; these types of shingles would burn rapidly in dry weather conditions.

Shingles are generally durable, economical, and easily repaired. A shingle roof that has aged and deteriorated should be removed completely and replaced. Some older buildings may have newer layers of shingles on top of older layers, which can make it difficult to cut an opening for ventilation.

Roofing tiles are usually made from clay or concrete products. Clay tiles, which have been used for roofing since ancient times, can be either flat or rounded. Rounded clay tiles are sometimes called mission tiles. Clay tiles are both durable and fire resistant.

Slate tiles are produced from thin sheets of rock. Slate is an expensive, long-lasting roofing material, but it is very brittle and becomes slippery when wet.

Metal panels of galvanized steel, copper, and aluminum also are used on pitched roofs. Expensive metals such as copper are often used for their decorative appearance, whereas lower-cost galvanized steel is used on barns and industrial buildings. Corrugated, galvanized metal panels are strong enough to be mounted on a roof without roof decking. Metal roof coverings will not burn, but they can conduct heat to the roof decking.

Curved Roofs

<u>Curved roofs</u> are often used for supermarkets, warehouses, industrial buildings, arenas, auditoriums, bowling alleys, churches, airplane hangars, and other buildings that require

Figure 14 Examples of pitched roofs: **A.** A gable roof. **B.** A hip roof. **C.** A mansard roof. **D.** A gambrel roof.

large, open interiors. Curved roofs are usually supported by steel or wood bowstring trusses or arches.

The decking on curved roof buildings can range from solid wooden boards or plywood to corrugated steel sheets. The decking material must be identified before ventilation openings can be made. Often, the roof covering consists of layers that include felt, mineral fibers, and asphalt, although some curved roofs are covered with foam plastics or plastic panels.

Flat Roofs

<u>Flat roofs</u> are found on houses, apartment buildings, shopping centers, warehouses, factories, schools, and hospitals Figure 16. Most flat roofs have a slight slope so that water can drain off the structure. If the roof does not have the proper slope or if the drains are not maintained, water may pool on the roof, overloading the structure and causing a collapse.

The support systems for most flat roofs are constructed of either wood or steel. A wood support structure uses solid wood beams and joists; laminated wood beams may be included to provide extra strength or to span long distances. Lightweight construction uses wood trusses or wooden I-beams as the supporting members. Such lightweight assemblies are much less fire resistant and collapse more quickly than do solid-wood systems.

Steel also is used to support flat roofs. Open-web steel trusses (sometimes called bar joists) can span long distances and remain stable during a fire if they are not subjected to excessive heat. Automatic sprinklers can protect the steel from the heat, but without such protection, the steel over a fire will soon weaken and sag. Heated steel support members can even elongate enough to push out the exterior walls of the building.

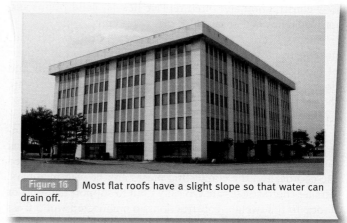

Figure 16 Most flat roofs have a slight slope so that water can drain off.

Figure 15 Several roof-covering materials are used on pitched roofs. **A.** Tile roof. **B.** Wood shingle roof. **C.** Slate roof.

The roof covering is applied on top of the deck. Most flat roofs are covered with multiple-layer, built-up roofing systems that help prevent leakage. A typical built-up roof covering contains five layers: a vapor barrier, thermal insulation, a waterproof membrane, a drainage layer, and a wear course. The outside of the flat roof reveals only the top layer, which is often gravel; it serves as the wear course and protects the underlying layers from wear and tear.

Most flat roof coverings contain highly combustible materials, including asphalt, roofing felt, tarpaper, rubber or plastic membranes, and plastic insulation layers. These materials can be difficult to ignite. Once lit, however, they burn readily, releasing great quantities of heat and black smoke.

Flat roofs can present unique problems during vertical ventilation operations. A roof with a history of leaks could have several patches where additional layers of roofing material make for an extra-thick covering. Sometimes an entirely new roof is constructed on top of the old one. A similar problem may be encountered in dealing with remodeled buildings or additions. An old flat roof may be found under a new pitched section, resulting in two separate void spaces.

Trusses

A truss is a structural component that is composed of smaller components in a triangular configuration or a system of triangles. Trusses are used extensively in support systems for both floors and roofs. They are common in modern construction for several reasons. The triangular geometry creates a strong, rigid structure that can support a load much greater than its own mass. For example, both a solid beam and a simple truss with the same overall dimensions can support the same load. The truss requires much less material than the beam, is much lighter, and can span a long distance without supports. Trusses are often prefabricated and transported to the building location for installation there.

Trusses are used in residential construction, apartment buildings, small office buildings, commercial buildings, warehouses, fast-food restaurants, airplane hangars, churches, and even firehouses. They may be clearly visible, or they may be concealed within the construction. Trusses are widely used in

Flat roof decks can be constructed of wood planking, plywood, corrugated steel, gypsum, or concrete. The material used depends on the building's age and size, the climate, the cost of materials, and the type of roof covering.

new construction and often replace heavier solid beams and joists in renovated or modified older buildings.

Trusses can be used for many different purposes. In building construction, those made of wood, steel, or combinations of wood and steel are used primarily to support roofs and floors. Three of the most commonly used types of trusses are the parallel chord truss, the pitched chord truss, and the bowstring truss.

A **parallel chord truss** has two parallel horizontal members connected by a system of diagonal and sometimes vertical members **Figure 17**. The top and bottom members are called the chords, and the connecting pieces are the web members. Parallel chord trusses are typically used to support flat roofs or floors. In lightweight construction, an engineered wood truss is often assembled with wood chords and either wood or light steel web members. A steel bar joist is another example of a parallel chord truss.

A **pitched chord truss** is typically used to support a sloping roof **Figure 18**. Most modern residential construction uses a series of prefabricated wood pitched chord trusses to support the roof. The roof deck is supported by the top chords, and the ceilings of the occupied rooms are attached to the bottom chords. In this way, the trusses define the shape of the attic.

A **bowstring truss** has the same shape as an archery bow, where the top chord represents the curved bow and the bottom chord represents the straight bowstring **Figure 19**. The top chord resists compressive forces and the bottom chord resists tensile forces. Bowstring trusses are usually quite large and widely spaced in the structure. They were popular in warehouses, supermarkets, and similar buildings with large, open floor areas. The roof of a building with bowstring trusses has a distinctive curved shape.

A building with a bowstring truss roof will have a distinctive curved roof **Figure 20**. In some buildings, the trusses are exposed and readily visible from the inside of the building. In other buildings, the floor or roof trusses are not visible because they are enclosed in attic or floor spaces **Figure 21**.

Walls

Walls are the most readily visible parts of a building because they shape the exterior and define the interior. Walls may be constructed of masonry, wood, steel, aluminum, glass, and many other materials.

Walls are either load bearing or nonbearing. **Load-bearing walls** provide structural support **Figure 22**. A load-bearing wall supports a portion of both the building's weight (dead load) and its contents (live load), transmitting that load down to the building's foundation. Damaging or removing a load-bearing wall can result in a partial or total collapse of the building **Figure 23**. Load-bearing walls can be either exterior or interior walls.

Nonbearing walls, by contrast, support only their own weight **Figure 24**. Many nonbearing walls are interior partitions that divide the building into rooms and spaces. The exterior walls of a building can also be nonbearing, particularly when a system of columns supports the building. Frequently, contractors will remove and replace nonbearing walls due to deterioration or a

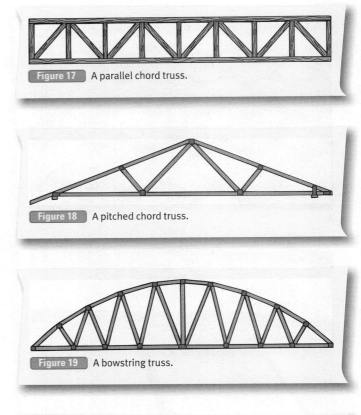

Figure 17 A parallel chord truss.

Figure 18 A pitched chord truss.

Figure 19 A bowstring truss.

Figure 20 A building with a bowstring truss.

Figure 21 A fast-food restaurant with hidden trusses for roof support.

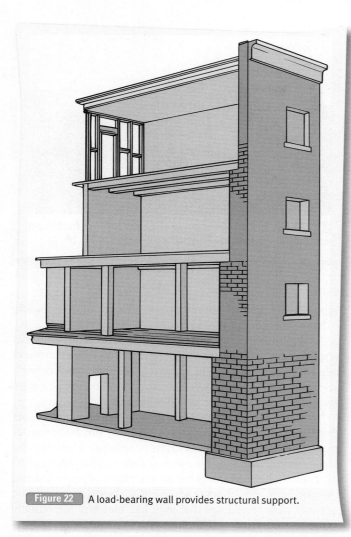

Figure 22 A load-bearing wall provides structural support.

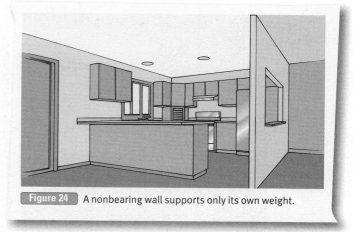

Figure 24 A nonbearing wall supports only its own weight.

In addition to load-bearing and nonbearing walls, several specialized walls may be used in construction:

- <u>Party walls</u> are constructed on the line between two properties and are shared by a building on each side of the line. They are almost always load-bearing walls. A party wall is often—but not always—constructed as a fire wall between the two properties. In ordinary construction, a party wall supports the floor joist of both structures. Frequently, the pocket in the brick wall that supports the floor joists is common to both buildings and can be a horizontal channel for fire spread. Failure of a party wall can result in structural collapse of both properties.
- Fire walls are designed to limit the spread of fire from one side of the wall to the other side. A fire wall might divide a large building into sections or separate two attached buildings. Fire walls usually extend from the foundation up to and through the roof of a building. They are constructed of fire-resistant materials and may be fire rated. A fire wall is a free-standing and nonbearing wall. Frequently found in Type V construction, the fire wall is designed to prevent the extension of fire from one section of the building to another. To be effective, it should be a solid wall with no openings. Since this is not practical, fire doors are installed at points where passage is required and the doors should remain in a closed position or close automatically if fire is detected. Not only is the wall nonbearing, it is not tied into either section of the building. In the event of a major fire involvement and collapse of a section of the building, the fire wall will not only prevent radiant heat from impacting on the section of the building uninvolved in fire, but it will not be affected by the collapse because it is not connected to either section of the building.
- <u>Fire partitions</u> are interior walls that extend from a floor to the underside of the floor above. They often enclose fire-rated interior corridors or divide a floor area into separate fire compartments.
- <u>Fire enclosures</u> (often called exit enclosures) are fire-rated assemblies that enclose interior vertical openings, such as stairwells, elevator shafts, and chases for building utilities. A fire enclosure prevents fire and smoke from spreading from floor to floor via the vertical opening. In multistory buildings, fire enclosures also protect the occupants using the exit stairways.

Figure 23 Damaging or removing a load-bearing wall can result in a partial or total collapse of the building.

desire to change the appearance of the building. This should in no way affect the structural integrity of the rest of the building. Nonbearing walls are less stable than load-bearing walls. The load applied to a load-bearing wall makes it more stable than an identical nonbearing wall. As such, the load placed on a load-bearing wall actually stabilizes the wall if the building were to be affected by an earthquake or an explosion.

Figure 25 Parapet walls extend above the roof line of a building.

Figure 26 A coping stone or cap stone sits on top of the parapet wall and its purpose is to keep out the rain and snow.

- **Curtain walls** are nonbearing exterior walls attached to the outside of the building. These walls often serve as the exterior skin on a steel-framed high-rise building.
- **Partition walls** are interior nonbearing walls that can easily be removed and replaced. This is common in office buildings where the floor configuration and shape of the offices can change as tenants move or as tenant needs change.
- **Parapet walls** extend above the roof line of a building Figure 25 . In newer structures, the parapet wall may sit on all four walls of a structure. In older structures, it frequently is free-standing and is only located above the front wall of a structure. In a deteriorated state, it can be prone to collapse. It is exposed on three sides to elements of weather such as rain and snow, and is regularly subjected to the lateral pressure exerted by wind. It can rest on either a load-bearing wall or a nonbearing wall. A coping stone or cap stone sits on top of the parapet wall and its purpose is to keep out the rain and snow Figure 26 . Although it may have originally been installed with a thin coat of mortar to set it in place, the mortar frequently deteriorates over time and the stone is essentially held in place by gravity. The stones can weigh between 5 and 50 pounds and any type of lateral load can dislodge them.

Solid, load-bearing masonry walls, at least 6 to 8″ (152–203 mm) thick, can be used for buildings up to six stories high. Nonbearing masonry walls can be almost any height. Masonry walls provide a durable, fire-resistant outer covering for a building, so they are often used as fire walls. A well-designed masonry fire wall is often completely independent of the structures on either side. Even if the building on one side burns completely and collapses, the fire wall should prevent the fire from spreading to the building on the other side.

Older buildings often had masonry load-bearing walls that were several feet thick at the bottom and then decreased in thickness as the building's height increased. Modern masonry walls are often reinforced with steel rods or concrete to provide a more efficient and more durable structural system.

When properly constructed and maintained, masonry walls are strong and can withstand a vigorous assault by fire. If the interior structure begins to collapse and exerts unanticipated forces on the exterior walls, however, solid masonry walls may fail during a fire.

Even though a building has an outer layer of brick or stone, fire inspectors should never assume that the exterior walls are masonry. Many buildings that look like masonry are actually constructed using wood-frame techniques and materials. A single layer of brick or stone, called a veneer layer, is applied to the exterior walls to give the appearance of a durable and architecturally desirable outer covering.

Wood framing is used to construct the walls in many small commercial buildings. The wood framing used for exterior walls can be covered with a variety of materials, including wooden siding, vinyl siding, aluminum siding, stucco, and masonry veneer. Moisture barriers, wind barriers, and other types of insulation are usually applied to the outside of the vertical studs before the outer covering is applied.

Vertical wooden studs support the walls and partitions inside the building. If fire resistance is critical, steel studs are used to frame walls. The wood framing is usually covered with gypsum board and any of a variety of interior finish materials. The space between the two wall surface coverings may be empty, it may contain thermal and sound insulating materials, or it may contain electrical wiring, telephone wires, and plumbing. These spaces often provide pathways through which fire can spread.

Columns

Columns are load-bearing structural members that perform the same function as a load-bearing wall. Columns, however, are used to support internal walls and elements. What then is the purpose of using a column as opposed to a load-bearing wall? The advantage of using a column and an I-beam within a building is that it provides an open space within the structure. Lines of sight and better natural lighting result from a column configuration as compared to load-bearing walls.

Contractors frequently are hired to replace load-bearing masonry walls with a column and I-beam configuration. This can be a challenge and if not done properly, can result in collapse of the structure.

■ Beams, Girders, Joists, and Rafters

Beams, girders, joists, and rafters are structural members that sit on walls or columns and provide horizontal or diagonal support for the other loads within the structure. The differences between these elements are their size and function within the building although these terms are frequently used interchangeably. The critical factor is to understand the relationship between the support system that is holding up these members and the loads that these members are supporting.

A beam is the most common. It is usually supported at both ends (__simple beam__) or at more than two points if the loads require (__continuous beam__). It is relatively large (4″ × 6″ to 10″ × 10″ [101 mm × 152 mm to 254 mm × 254 mm]) and is common in heavy mill construction. A __girder__ is a beam that supports other beams. Frequently we think of girders being constructed of steel. But with the development of engineered wood, laminated wood beams are prevalent as a support system in newer construction. A __joist__ is a wooden beam, usually recognized as part of a floor or roof assembly. A rafter is a beam that is specifically used to support a roof system.

The strength of a beam decreases as its length increases. The dimensions (height and depth) must increase as the length increases, so the practical limit of a wood beam is approximately 24′ (732 cm). Spans longer than this would require greater dimensions as well as stronger supporting elements and additional supporting elements to carry the weight of the larger sized beams.

The beam's limitation to span distances longer than 24′ (732 cm) has been overcome by the development of the truss. The development of the truss has given architects and builders the flexibility to span wide distances without the need for support systems. Lightweight metal truss has been utilized in a variety of occupancies from commercial strip shopping centers to the World Trade Center. Loads must still be calculated to determine the proper type and size of truss to be used.

Although the failure of a few beams should not result in a catastrophic failure of the building, a synergy resulting from other factors combined with a localized collapse can present a challenge to the overall structural stability.

■ Doors and Windows

Doors and windows are important components of any building. Although they generally have different functions—doors provide entry and exit, whereas windows provide light and ventilation—in an emergency, doors and windows are almost interchangeable. A window can serve as an entry or an exit, for example, whereas a door can provide light and ventilation.

There are hundreds of door and window designs, with many different applications. Of particular concern to fire inspectors are fire doors and fire windows.

Basic Door Construction

Doors can be categorized both by their construction material and by the way they open. Both interior and exterior doors have the same basic components **Figure 27**:

- Door—the entryway itself
- __Jamb__—the frame that secures the door to the studs of a wall
- __Hardware__—the handles, hinges, and other components
- Locking device—either in the handle or encased in the door itself

> **Fire Inspector Tips**
>
> Fire inspectors must constantly determine the criticality of a structural element relative to the overall stability of the building. If a load-bearing wall is compromised, the overall structural stability of the building can be in jeopardy. Conversely, interior nonbearing partition walls are regularly moved or completely removed to meet tenant needs in a high-rise office building.
>
> The integrity of the foundation system, load-bearing walls and vertical columns is critical to the structural stability of all buildings. Determining when these elements have been compromised or damaged is a major component of the inspection process. Damage to any of these load-bearing elements can cause localized collapse or potentially catastrophic failure of the entire structure. How can damage occur? It can be caused over a period of time by neglect and deterioration or it can be caused instantaneously by an explosion or a hurricane. Fire inspectors are called upon to determine if a building is still sound after it has been affected by any of these forces.

Figure 27 The parts of a door.

Doors are generally constructed of wood, metal, or glass. Wood doors are commonly used in residences and may be found in some commercial buildings. Three types of wood swinging doors are distinguished: slab, ledge, and panel. Slab doors come in solid-core or hollow-core designs and are attached to wood-frame construction with normal hardware Figure 28.

<u>Solid-core</u> doors are constructed of solid wood core blocks covered by a face panel. Some have an Underwriters Laboratories (UL) or Factory Mutual fire rating. Such doors are typically used for entrance doors. Solid-core doors are heavy and their construction enables them to contain fire better than hollow-core doors do.

<u>Hollow-core</u> doors have a lightweight, honeycomb interior, which is covered by a face panel. They are often used as interior doors, such as for bedrooms. Hollow-core doors are easy to force and will burn through quickly.

Ledge doors are simply wood doors with horizontal bracing. These doors, which are often constructed of tongue-and-groove boards, may be found on warehouses, sheds, and barns.

Panel doors are solid wood doors that are made from solid planks to form a rigid frame with solid wood panels set into the frame. Panel doors are used as both exterior and interior doors and may be made from a variety of types of wood. These doors resist fire longer than hollow-core slab doors do.

Metal doors may be either decorative (for residential use) or utilitarian (for warehouses and factories). They also may have either a hollow-core or a solid-core construction. Hollow-core metal doors have a metal framework interior so they are as lightweight as possible. By contrast, solid-core metal doors have a foam or wood interior that is intended to reduce the door's weight without affecting its strength. Metal doors may be set in either a wood frame or a metal frame. Residential metal doors may appear to be panel doors and are often used as entry doors.

Glass doors generally have a steel frame with tempered glass; alternatively, they may be simply tempered glass and not require a frame, but have metal supports to attach hardware. Glass doors are easy to force, but can be dangerous owing to the large number of small broken pieces that are produced when glass is broken.

Types of Doors

Doors also can be classified by how they open. The five most common ways that doors open are inward, outward, sliding, revolving, and overhead. Overhead, sliding, and revolving doors are the most readily identified. Inward-opening and outward-opening doors can be differentiated based on whether the hinges are visible. If you can see the hardware, the door will swing toward you (outward; Figure 29A); if the hinges are not visible, the door will swing away from you (inward; Figure 29B).

Door frames may be constructed of either wood or metal. Wood-framed doors come in two styles: stopped and rabbet. Stopped door frames have a piece of wood attached to the frame that stops the door from swinging past the latch. Rabbeted door frames are constructed with the stop cut built into the frame so it cannot be removed. Metal frames look like rabbeted door frames.

Inward-Opening Doors

Inward-opening doors of wood, steel, or glass can be found in most structures. They have an exterior frame with a stop or rabbet that keeps the door from opening past the latch. The locking mechanisms may range from standard door knob locks to deadbolt locks or sliding latches.

Outward-Opening Doors

Outward-opening doors are used in commercial occupancies and for most exits Figure 30. They are designed so that people can leave a building quickly during an emergency. Outward-opening doors may be constructed of wood, metal, or glass. They usually have exposed hinges, which may present an entry opportunity. More frequently, however, these hinges will be sealed so that the pins cannot be removed. Several types of locks, including handle-style locks and deadbolts, may be used with these doors.

Sliding Doors

Most sliding doors are constructed of tempered glass in a wooden or metal frame. Such doors are commonly found in residences and hotel rooms that open onto balconies or patios Figure 31. Sliding doors generally have two sections and a double track; one side is fixed in place, while the other side slides. A weak latch on the frame of the door secures the movable side. Many people prop a wood or metal rod in the track to provide additional security. Sliding doors are also used as main entrance doors on retail buildings; these doors are often equipped with an automatic opening feature.

Revolving Doors

Revolving doors are most commonly found in upscale buildings and buildings in large cities Figure 32. They are usually made of four glass panels with metal frames. The panels are designed to collapse outward with a certain amount of pressure to allow for rapid escape during an emergency. Revolving doors are generally secured by a standard cylinder lock or slide latch lock.

Solid Core **Hollow Core**

Figure 28 A slab door may have either solid-core or hollow-core construction.

Figure 29 **A.** An outward-opening door will have hinges showing. **B.** A door with the hinges not showing will open inward.

Because of *Life Safety Code®*, fire prevention, and building code requirements, standard outward-opening doors are often found adjacent to the revolving doors.

Overhead Doors

Overhead doors come in many different designs—from standard residential garage doors to high-security commercial roll-up doors Figure 33 . Most residential garage doors have three or four panels, which may or may not include windows. Some residential overhead garage doors come in a single section and tilt rather than roll up. These doors, which may be made of wood or metal, usually have a hollow core that is filled with insulation or foam. By contrast, commercial security overhead doors are made of metal panels or hardened steel rods. They may use solid-core or hollow-core construction, depending on the amount of security needed. Overhead doors can be secured with cylinder-style locks, padlocks, or automatic garage door openers.

■ Windows

Windows come in many shapes, sizes, and designs for different buildings and occupancies. You must become familiar with the specific types of windows found in local occupancies and learn to recognize them.

Fire Inspector Tips

You must be vigilant in locating and controlling security bars on windows used as an exit or means of escape. Residential, schools, day care and similar occupancies frequently have security bars installed. This is a life-safety hazard because it can delay or prevent occupants from escaping a burning building.

Glass Construction

The **glazed** (transparent) part of the window is most commonly made of glass. Window glass comes in several configurations: regular glass, double-pane glass, plate glass (for large windows), laminated glass, and tempered glass. Plastics such as Plexiglas and Lexan may also be used in windows. The window may contain one or more panes of glass. Insulated glass, for example, usually has two or more pieces of glass in the window and will be discussed in more detail later.

Figure 30 An outward-opening door enables rapid exit from a building.

Figure 31 Most sliding doors are constructed of tempered glass in a wooden or metal frame.

Figure 32 Building codes may require outward-opening doors next to a revolving door.

Regular or Annealed Glass

Single-pane glass, also called regular or **annealed** glass, is normally used in construction because it is relatively inexpensive; larger pieces are called plate glass. This type of glass is easily broken with a firefighting tools. When broken, plate glass creates long, sharp pieces called shards, which can penetrate helmets, boots, and other protective gear, causing severe lacerations and other injuries.

Double-Pane Glass (Insulated Windows)

Double-pane glass is used in many homes because it improves home insulation by using two panes of glass with an air pocket between them. Some double-pane windows may include an inert gas such as argon between the panes for additional insulation value. These windows are sealed units, which makes them more expensive to replace. However, replacing the glass alone is less expensive than replacing the entire window assembly. Forcing entry through insulated windows is basically the same as forcing entry through single-pane windows, except that the two panes may need to be broken separately. These windows also will produce dangerous glass shards.

Plate Glass

Commercial **plate glass** is a stronger, thicker glass used in large window openings. Although it is being replaced by tempered glass in modern construction for safety reasons, commercial plate glass can still be found in older large buildings, storefronts, and residential sliding doors. It can easily be broken with a sharp object. When broken, commercial plate glass will create many large, sharp pieces.

Figure 33 **A.** Commercial overhead doors are made of metal panels or steel roads. **B.** Residential overhead garage doors may tilt or roll up.

Figure 34 Double-hung windows allow the inner and outer sashes to move freely up and down.

Figure 35 A single-hung window has only one movable sash.

Laminated Glass

Laminated glass, also known as safety glass, is used to prevent windows from shattering and causing injury. Laminated glass is molded with a sheet of plastic between two sheets of glass. This type of glass is most commonly used in vehicle windshields, but it may also be found in other applications such as doors or building windows.

Tempered Glass

Tempered glass is specially heat treated, making it four times stronger than regular glass. This type of glass is commonly found in side and rear windows in vehicles, in commercial doors, in newer sliding glass doors, and in other locations where a person might accidentally walk into the glass and break it. Tempered glass breaks into small pellets without sharp edges to help prevent injury during accidents.

Wired glass is tempered glass that has been reinforced with wire. This kind of glass may be clear or frosted, and it is often used in fire-rated doors that require a window or sight line from one side of the door to the other. Wired glass is difficult to break and force.

Window Frame Designs

Double-Hung Windows

<u>Double-hung windows</u> have two movable sashes, usually made of wood or vinyl, that move up and down **Figure 34** . They are common in residences and have wood, plastic, or metal runners. Newer double-hung window sashes may be removed or swung in for cleaning. They may have either one locking mechanism that is found in the center of the window or two locks on each side of the lower sash that prevent the sashes from moving up or down.

Single-Hung Windows

Single-hung windows are similar to double-hung windows, except that the upper sash is fixed and only the lower sash moves **Figure 35** . The locking mechanism is the same as on double-hung windows as well. It may be difficult to distinguish between single-hung and double-hung windows from the exterior of a building.

Jalousie Windows

A <u>jalousie window</u> is made of adjustable sections of tempered glass encased in a metal frame that overlap each other when closed **Figure 36** . This type of window is often found in mobile homes and is operated by a small hand-wheel or crank located in the corner of the window.

Figure 36 Jalousie windows are opened and closed with a small hand-crank.

Figure 37 An awning window has larger panels than a jalousie window does.

Awning Windows

Awning windows are similar in operation to jalousie windows, except that they usually have one large or two medium-sized glass panels instead of many small panes **Figure 37**. Awning windows are operated by a hand crank located in the corner or in the center of the window. These kinds of windows can be found in residential, commercial, or industrial structures. Residential awning windows may be framed in wood, vinyl, or metal, whereas commercial and industrial windows will usually have metal frames.

Commercial and industrial awning windows often use a lock and a notched bar to hold the window open, rather than a crank. These so-called projected windows are discussed later in this chapter.

Horizontal-Sliding Windows

Horizontal-sliding windows are similar to sliding doors **Figure 38**. The latch is similar as well and attaches to the window frame. People often place a rod or pole in the track to prevent break-ins. Newer sliding windows have latches between the windows, similar to those found on double-hung windows.

Casement Windows

Casement windows have a steel or wood frame and open away from the building with a crank mechanism **Figure 39**. Although they are similar to jalousie or awning windows, casement windows have a side hinge, rather than a top hinge. Several types of locking mechanisms can be used with these windows.

Projected Windows

Projected windows, also called factory windows, are usually found in older warehouse or commercial buildings **Figure 40**. They can project inward or outward on an upper hinge. Screens are rarely used with these windows, but forcing entry may not be easy, depending on the integrity of the frame, the type of locking mechanism used, and the window's distance off the ground. These windows may have fixed, metal-framed wire glass panes above them.

Figure 38 Horizontal-sliding windows work like sliding doors.

Fire Doors and Fire Windows

Fire doors and fire windows are constructed to prevent the passage of flames, heat, and smoke through an opening during a fire. They must be tested and meet the standards set in NFPA 80, *Standard for Fire Doors and Other Opening Protectives*. Fire doors and fire windows come in many different shapes and sizes, and they provide different levels of fire resistance. For example, fire doors can swing on hinges, slide down or across an opening, or roll down to cover an opening.

The fire rating on a door or window covers the actual door or window, the frame, the hinges or closing mechanism, the latching hardware, and any other equipment that is required to operate the door or window. All of these items must be tested and approved as a combined system. All fire doors must have a mechanism that keeps the door closed or automatically closes the door when a fire occurs. Doors that are normally open can be closed by the release of a fusible link, by a smoke or heat detector, or by activation of the fire alarm system.

Figure 39 Casement windows open to the side with a crank mechanism. Used with permission of Andersen Corporation.

Figure 40 Projected windows may open inward or outward.

Fire doors and fire windows are rated for a particular duration of fire resistance to a standard test fire. This is similar to the fire-resistance rating system used for building construction assemblies. A 1-hour rating, however, does not guarantee that the door will resist any fire for 60 minutes; rather, it simply establishes that the door will resist the standard test fire for 60 minutes. In any given fire situation, a door rated at 1 hour will probably last twice as long as a door rated 30 minutes.

Fire doors and windows are labeled and assigned the letters A, B, C, D, or E, based on their approved-use locations Figure 41 . For example, doors designated as "A" are approved for use as part of a fire wall; doors designated as "B" are approved for use in stair shafts and elevator shafts. Table 1 shows the approved uses for each of the letter designations.

Fire windows are used when a window is needed in a required fire-resistant wall. These windows are often made of wired glass, which is designed to withstand exposures to high temperatures without breaking. Fire-resistant glass without wires is available for some applications; in such cases, special steel window frames are required to keep the glass firmly in place.

Wired glass is also used to provide vision panels in or next to fire doors. Vision panels allow a person to view conditions on the opposite side of a door without opening the door, or to make sure no one is standing in front of the door before opening it. In these configurations, the entire assembly—including the door, the window, and the frame—must all be tested and approved together.

When a window is required only for light passage, glass blocks can sometimes be used instead of wired glass. Glass blocks will resist high temperatures and remain in place during a fire (assuming they are properly installed). The size of the opening is limited and depends on the fire-resistance rating of the wall.

■ Interior Finishes and Floor Coverings

The term interior finish commonly refers to the exposed interior surfaces of a building. These materials affect how a particular building or occupancy reacts when a fire occurs. Interior finish considerations include whether a material will ignite easily or resist ignition, how quickly a flame will spread across the

Figure 41 An approved fire door has a label indicating its classification and rating.

Table 1 NFPA 80 Designations for Fire Doors and Fire Windows

Class A	Openings in fire walls and in walls that divide a single building into fire areas
Class B	Openings in enclosures of vertical communications through buildings and in 2-hour-rated partitions providing horizontal fire separations
Class C	Openings in walls or partitions between rooms and corridors having a fire-resistance rating of 1 hour or less
Class D	Openings in exterior walls subject to severe fire exposure from outside the building
Class E	Openings in exterior walls subject to moderate or light fire exposure from outside the building

material surface, how much energy the material will release when it burns, how much smoke it will produce, and what the smoke will contain.

A room with a bare concrete floor, concrete block walls, and a concrete ceiling has no interior finishes that will increase the fire load. But if the same room had an acrylic carpet with rubber padding on the floor, wooden baseboards, varnished wood paneling on the walls, and foam plastic acoustic insulation panels on the ceiling, the situation would be vastly different. These interior finishes would ignite quickly, flames would spread quickly across the surfaces, and significant quantities of heat and toxic smoke would be released.

Different interior finish materials contribute in various ways to a building fire. Each individual material has certain characteristics, and fire fighters must evaluate the particular combination of materials in a room or space. Typical wall coverings include painted plaster or gypsum board, wallpaper or vinyl wall coverings, wood paneling, and many other surface finishes. Floor coverings might include different types of carpet, vinyl floor tiles, or finished wood flooring. All of these products will burn to some extent, and each one involves a different set of fire risk factors. Interior finish has been listed as a contributing factor in many large life-loss fires. Fire inspectors should ask for documentation that the material has the proper flame spread rating. If not, the material should be removed, covered with a non-combustible material, or treated with a product to improve its flame spread rating. Foam plastics or carpeting on ceiling or walls should be a "red flag" to fire inspection that further Inspection is warranted.

Fire Inspector Tips

Buildings that are under construction and buildings that are being renovated or demolished are at high risk for major fires, for several reasons. For example, workers may be using torches for certain tasks. These buildings are often unoccupied for many hours during the construction, so alarms may be delayed. Fire detection and suppression systems may be inoperable. Removal of doors and windows allows the entry of sufficient oxygen for a small fire to grow into a major blaze in a short period of time, and for the fire to spread.

Clearly, buildings under construction or demolition are at high risk for a major fire. Such was the case at the Deutsche Bank building in New York City, where two fire fighters lost their lives. This building was damaged in the terrorist attacks of September 11, 2001, and was set for demolition. The Fire Department of New York (FDNY) responded to a working fire in the building and attempted to use the standpipe system to fight the fire. However, a part of the standpipe system had been removed and was inoperable.

Wrap-Up

Chief Concepts

- Knowing the types of building construction and the types of materials used to construct a building is vital for a fire inspector.
- Building construction affects how fires grow and spread and determines the special safety features required to control fire growth and protect the building's occupants.
- Building construction will also determine the codes that the building needs to comply with.
- The most common building materials are wood, masonry, concrete, steel, aluminum, glass, gypsum board, and plastics.
- Buildings are classified based on the combustibility of the structure and fire resistance of its components.
- The major components of a building are the foundation, floors, ceilings, roofs, trusses, walls, doors, windows, interior finishes, and flooring coverings.
- Buildings that are under construction and buildings that are being renovated or demolished are at high risk for major fires.

Hot Terms

Annealed The process of forming standard glass.

Awning windows Windows that have one large or two medium-size panels, which are operated by a hand crank from the corner of the window.

Balloon-frame construction An older type of wood frame construction in which the wall studs extend vertically from the basement of a structure to the roof without any fire stops.

Bowstring trusses Trusses that are curved on the top and straight on the bottom.

Casement windows Windows in a steel or wood frame that open away from the building via a crank mechanism.

Combustibility The property describing whether a material will burn and how quickly it will burn.

Continuous beam A beam supported at three or more points. Structurally advantageous because if the span between two supports is overloaded, the rest of the beam assists in carrying the load.

Curtain walls Nonbearing walls that are used to separate the inside and outside of the building, but that are not part of the support structure for the building.

Curved roofs Roofs that have a curved shape.

Dead load The weight of a building. It consists of the weight of all materials of construction incorporated into a building, including but not limited to walls, floors, roofs, ceilings, stairways, built-in partitions, finishes, cladding, and other similarly incorporated architectural and structural items, as well as fixed service equipment, including the weight of cranes.

Double-hung windows Windows that have two movable sashes that can go up and down.

Double-pane glass A window design that traps air or inert gas between two pieces of glass to help insulate a house.

Fire enclosure A fire-rated assembly used to enclose a vertical opening such as a stairwell, elevator shaft, and chase for building utilities.

Fire partition An interior wall extending from the floor to the underside of the floor above.

Fire wall A wall with a fire-resistive rating and structural stability that separates buildings or subdivides a building to prevent the spread of fire.

Fire window A window or glass block assembly with a fire-resistive rating.

Flat roofs Horizontal roofs often found on commercial or industrial occupancies.

Glass blocks Thick pieces of glass that are similar to bricks or tiles.

Girder A beam that supports other beams.

Gypsum A naturally occurring material composed of calcium sulfate and water molecules

Gypsum board The generic name for a family of sheet products consisting of a noncombustible core primarily of gypsum with paper surfacing.

Glazed Transparent glass.

Hardware The parts of a door or window that enable it to be locked or opened.

Hollow-core A door made of panels that are honeycombed inside, creating an inexpensive and lightweight design.

Horizontal-sliding windows Windows that slide open horizontally.

Interior finish Any coating or veneer applied as a finish to a bulkhead, structural insulation, or overhead, including the visible finish, all intermediate materials, and all application materials and adhesives.

Jalousie windows Windows made of small slats of tempered glass, which overlap each other when the window is closed. Often found in trailers and mobile homes, jalousie windows are held together by a metal frame and operated by a small hand wheel or crank found in the corner of the window.

Wrap-Up

Jamb The part of a doorway that secures the door to the studs in a building.

Joist A beam.

Laminated glass Glass manufactured with a thin vinyl core covered by glass on each side of the core.

Laminated wood Pieces of wood that are glued together.

Live load The weight of the building contents.

Load-bearing wall A wall that is designed to provide structural support for a building.

Masonry A built-up unit of construction or combination of materials such as brick, clay tiles, or stone set in mortar.

Nonbearing wall A wall that is designed to support only the weight of the wall itself.

Parallel chord trusses Trusses in which the top and bottom chords are parallel.

Parapet wall Walls on a flat roof that extend above the roofline.

Partition wall A non-load-bearing wall that subdivides spaces within any story of a building or room.

Party wall A wall constructed on the line between two properties.

Pitched chord truss Type of truss typically used to support a sloping roof.

Pitched roof A roof with sloping or inclined surfaces.

Platform-frame construction Construction technique for building the frame of the structure one floor at a time. Each floor has a top and bottom plate that acts as a firestop.

Pyrolysis The destructive distillation of organic compounds in an oxygen-free environment that converts the organic matter into gases, liquids, and char.

Plate glass A type of glass that has additional strength so it can be formed in larger sheets, but will still shatter upon impact.

Projected windows Windows that project inward or outward on an upper hinge; also called factory windows. They are usually found in older warehouses or commercial buildings.

Rafters Joists that are mounted in an inclined position to support a roof.

Simple beam Supported at two points neat its ends. In simple beam construction, the load is delivered to the two reaction points and the rest of the structure renders no assistance in an overload.

Spalling Chipping or pitting of concrete or masonry surfaces.

Solid-core A door design that consists of wood filler pieces inside the door. This construction creates a stronger door that may be fire rated.

Tempered glass Glass that is much stronger and harder to break than ordinary glass.

Thermal conductivity A property that describes how quickly a material will conduct heat.

Thermoplastic material A plastic material capable of being repeatedly softened by heating and hardened by cooling and, that in the softened state, can be repeatedly shaped by molding or forming.

Thermoset material A plastic material that, after having been cured by heat or other means, is substantially infusible and cannot be softened and formed.

Truss A collection of lightweight structural components joined in a triangular configuration that can be used to support either floors or roofs.

Type I construction (fire resistive) Buildings with structural members made of noncombustible materials that have a specified fire resistance.

Type II construction (noncombustible) Buildings with structural members made of noncombustible materials without fire resistance.

Type III construction (ordinary) Buildings with the exterior walls made of noncombustible or limited-combustible materials, but interior floors and walls made of combustible materials.

Type IV construction (heavy timber) Buildings constructed with noncombustible or limited-combustible exterior walls, and interior walls and floors made of large-dimension combustible materials.

Type V construction (wood frame) Buildings with exterior walls, interior walls, floors, and roof structures made of wood.

Wired glass Glass made by molding tempered glass around a special wire mesh.

Wooden beams Load-bearing members assembled from individual wood components.

Wood panels Thin sheets of wood glued together.

Wood trusses Assemblies of small pieces of wood or wood and metal.

Fire Inspector *in Action*

You have recently been assigned to a fire inspection unit. While becoming familiar with the buildings in the area, you find that there are three buildings that need to be inspected that have no records of any kind in the files. All of the other buildings have records containing information about the initial construction of the building and all fire inspections. It appears that no one knows the whereabouts of the missing files. You have no alternative but to assemble a new file from your personal observations and research.

You start creating new files by contacting the city hall about the three buildings. Unfortunately, these government agencies do not list the type of building construction of these buildings in familiar terminology. When these buildings were constructed at the turn of the century, the city hall used a different system of classifying them that is no longer in use. You must visit each property and collect information in order to correctly classify the buildings for your new files.

1. A Type II building is:
   A. masonry walls and wood roof.
   B. heavy timber.
   C. same as a Type I but with less fire resistance.
   D. ordinary construction.

2. Dry wall is considered:
   A. combustible.
   B. non-combustible.
   C. limited-combustible.
   D. none of the above.

3. Masonry exterior bearing walls are usually found in:
   A. type I and V construction.
   B. type II and V construction.
   C. type III and IV construction.
   D. type IV and V construction.

4. What can be said about structural steel members?
   A. Being non-combustible, they need not be protected.
   B. They can melt in a severe fire.
   C. When considered by themselves, they are very fire-resistive.
   D. Heavy members absorb heat faster than lightweight members.

Types of Occupancies

CHAPTER 3

NFPA 1031 Standard

Fire Inspector I

4.3.1 Identify the occupancy classification of a single-use occupancy, given a description of the occupancy and its use, so that the classification is made according to the applicable codes and standards. (pp 42–52)

(A) Requisite Knowledge. Occupancy classification types; applicable codes, regulations, and standards adopted by the jurisdiction; operational features; and fire hazards presented by various occupancies. (pp 42–52)

(B) Requisite Skills. The ability to make observations and correct decisions.

4.3.2 Compute the allowable occupant load of a single-use occupancy or portion thereof, given a detailed description of the occupancy, so that the calculated allowable occupant load is established in accordance with applicable codes and standards.

(A) Requisite Knowledge. Occupancy classification; applicable codes, regulations, and standards adopted by the jurisdiction; operational features; fire hazards presented by various occupancies; and occupant load factors. (pp 42–52)

(B) Requisite Skills. The ability to calculate occupant loads, identify occupancy factors related to various occupancy classifications, use measuring tools, and make field sketches. (pp 43–52)

Fire Inspector II

5.3.1 Compute the maximum allowable occupant load of a multi-use building, given field observations or a description of its uses, so that the maximum allowable occupant load calculation is in accordance with applicable codes and standards. (p 52)

(A) Requisite Knowledge. How to calculate occupant loads for an occupancy and for building use; and code requirements, regulations, operational features, and fire hazards presented by various occupancies. (p 52)

(B) Requisite Skills. The ability to calculate occupant loads, identify occupancy factors related to various occupancy classifications, use measuring tools, read plans, and use a calculator. (p 52)

5.3.2 Identify the occupancy classifications of a mixed-use building, given a description of the uses, so that each area is classified in accordance with applicable codes and standards. (p 52)

(A) Requisite Knowledge. Occupancy classification, applicable codes and standards, operational features, and fire hazards presented by various occupancies. (p 52)

(B) Requisite Skills. The ability to interpret code requirements and recognize building uses that fall into each occupancy classification.

Additional NFPA Standards

NFPA 101 *Life Safety Code*
NFPA 5000 *Building Construction and Safety Code*, or the *International Building Code*

FESHE Objectives

There are no FESHE objectives for this chapter.

Knowledge Objectives

After studying this chapter, you will be able to:

1. Describe how the occupancy classification of a building is determined.
2. List the fifteen specific occupancy groupings.
3. Describe the identifying characteristics of one- and two-family dwellings.
4. Describe the identifying characteristics of lodging or rooming houses.
5. Describe the identifying characteristics of hotels and dormitories.
6. Describe the identifying characteristics of apartment buildings.
7. Describe the identifying characteristics of residential board and care occupancies.
8. Describe the identifying characteristics of health care occupancies.
9. Describe the identifying characteristics of ambulatory health care occupancies.
10. Describe the identifying characteristics of day-care occupancies.
11. Describe the identifying characteristics of educational occupancies.
12. Describe the identifying characteristics of business occupancies.
13. Describe the identifying characteristics of industrial occupancies.
14. Describe the identifying characteristics of mercantile occupancies.
15. Describe the identifying characteristics of storage occupancies.
16. Describe the identifying characteristics of assembly occupancies.
17. Describe the identifying characteristics of detention and correctional occupancies.
18. Describe the identifying characteristics of multiple occupancies.
19. Describe the identifying characteristics of mixed occupancies.

Skills Objectives

There are no skills objectives for this chapter.

You Are the Fire Inspector

You are preparing to inspect a recently renovated historic structure in your town's center. In the past, this two-story Victorian was used as a private residence and fell into disrepair. You read in the newspaper that a local history museum purchased the property and renovated it to include small exhibit rooms, a function room, and a gift shop. The first thing that you must determine is the current use of the building because the codes separate building requirements by occupancy. In some cases, such as a strip mall, the occupancy is very clear while in others, such as this historic structure, it is not so obvious. Before knowing what section of the code you should familiarize yourself with in order to make a proper inspection, you visit the building and observe its daily operations.

1. Based on what you know about the building, what type of occupancy do you think it is?
2. What specific features would be necessary to change an occupancy from a one- or two-family dwelling to a business occupancy?

Introduction

In the model building codes, the term **occupancy** refers to the intended use of a building. The occupancy classification of a building is determined by the current use of that building. A building that was constructed and used for one purpose, such as an elementary school, may be later used for a different purpose, such as a community center. Prior to conducting a fire inspection, you will determine the appropriate occupancy class based on the current use of the building. Because a building can change occupancy classifications over time, it is critical that you correctly identify the current use of the building and the appropriate occupancy classification under the applicable building codes and regulations.

Codes and regulations will dictate the requirements necessary to conduct the fire inspection accurately and provide standards that you will use to evaluate the structure. Depending upon your jurisdiction, you may enforce NFPA 101, *Life Safety Code,* NFPA 5000, *Building Construction and Safety Code* or the *International Building Code*. Both NFPA 101 and NFPA 5000 are published by the National Fire Protection Association and address construction, protection, and occupancy features necessary to minimize the danger to life and property from the effects of fire. The *International Building Code* is a model building code developed by the International Code Council. While the various model codes in use today vary in terms of their specific requirements, they are fairly similar in terms of how they classify occupancies. However, due to specific differences in model codes, it is critical to become familiar with the specifics of the model code used within your jurisdiction.

Occupancy Classification

It is extremely important that the correct occupancy classification be determined because the model code requirements differ for each type of occupancy. Improperly classifying a building prior to inspecting it can result in an inadequate level of fire and life safety, such as a failure to require necessary fire protection systems or exits. Similarly, an incorrect occupancy classification could result in requiring fire protection systems that are not required for the occupancy and increasing the costs for the owner.

According to the NFPA 101, *Life Safety Code,* there are fifteen specific occupancy groupings:

1. One- and two-family dwellings
2. Lodging or rooming houses
3. Hotels and dormitories
4. Apartment buildings
5. Residential board and care
6. Health care
7. Ambulatory health care
8. Day-care
9. Educational
10. Business
11. Industrial
12. Mercantile
13. Storage
14. Assembly
15. Detention and correctional

> **Fire Inspector Tips**
>
> A dwelling unit is expected to be occupied by a single family and up to three outsiders, not related to the occupants by blood or marriage.

> **Fire Inspector Tips**
>
> In a condominium building containing multiple dwelling units, each dwelling unit is considered a private dwelling for entry purposes. An individual unit is usually protected by the Fourth Amendment relative to inspecting the unit without first obtaining the occupants' permission; however, common areas of the condominium building may be inspected at all reasonable times as permitted by code.

One- and Two-Family Dwelling Units

According to NFPA 5000, *Building Construction and Safety Code*, a <u>one- or two-family dwelling</u> is a building that contains no more than two dwelling units with independent cooking and bathroom facilities **Figure 1**.

The way in which the living units are separated from each other also helps to determine the occupancy classification. For example, a row of townhouses incorporating complete vertical wall separation between the individual units and an independent means of egress from each unit are considered to be separate dwelling units.

The regulations with respect to inspection of these residential units vary by municipality. A jurisdiction may require an occupancy inspection at the time a home is built, sold, or refinanced. Residents may also request these inspections in the interest of fire and life safety.

In conducting an inspection of a one- or two-family home, you will typically focus on the means of escape, the interior finishes, and fire protection. The placement and functionality of fire protection devices and/or systems should be examined, including residential sprinklers if provided and/or required. Hazards such as utilities, alternative heating devices, and storage should also be addressed during the inspection.

As a fire inspector, you have an essential role as a fire safety educator. You will always want to provide fire safety education when your inspection reveals that it is needed. The focus of these educational activities should be on maintaining working smoke detection systems, creating a predetermined escape plan, and ensuring proper housekeeping. Flammable materials, for example, should not be stored in close proximity to a gas stove or electric range.

Lodging or Rooming Houses

According to NFPA 101, a <u>lodging or rooming house</u> is a building or portion thereof, not categorized as a one- or two-family dwelling, that provides sleeping accommodations for a total of 16 or fewer people on a transient or permanent basis **Figure 2**. A lodging house may provide meals but should not include separate cooking facilities for individual occupants. Personal care services are not provided in lodging houses. Guest houses, bed and breakfasts, small inns or motels, and foster homes fall under this category. Small sleeping accommodations in other occupancies, such as a fire station, can also fall under this category.

Figure 1 A one- or two-family dwelling unit only contains no more than two independent cooking and bathroom facilities.

Figure 2 A lodging house will have room for only 16 guests.

The nature of these occupancies may result in occupants being unaware of their surroundings and thus dependent on the building's fire and life safety features. Of particular concern as you inspect these buildings will be the building compartmentation and the provision of adequate escape routes. The adequacy of the escape routes and its proper signage should be carefully considered during your inspection.

Additional inspection considerations include compliance with use and code requirements, interior finishes, protection of exits, building services, and fire protection systems. Inspections should address housekeeping, electrical installations, heating appliances, cooking operations, fire detection and alarm systems, and sprinkler systems. Your goal is to help prevent the occurrence of a fire, as well as ensuring that necessary building construction features and fire protection systems are in place to quickly alert the building occupants in the event of a fire and provide the means to safely exit the building.

Hotels

According to NFPA 101, a **hotel** is a building or group of buildings under the same management that has sleeping accommodations for more than 16 people and is primarily used by transients for lodging, with or without meals Figure 3. This category includes modern fire-resistant high-rise hotels, motels, inns, and clubs. Hotels present a diverse range of fire and life safety issues that you must be aware of and thoroughly address. In addition to sleeping rooms, a hotel may have restaurants, meeting rooms, ballrooms, stages, theatres, retail stores, garages, offices, storage rooms, and maintenance shops, each of which may fall under a different occupancy class. While the guest rooms would be residential occupancies, the ballrooms, restaurants, and theatres would be assembly occupancies. Offices would be business occupancies, while parking garages and storage areas would be storage occupancies. Retail stores would be mercantile occupancies, and maintenance shops would be industrial occupancies.

When these various occupancies are separated with their own means of egress, they can be treated as separate occupancies. If they share a means of egress, the fire protection requirements are based on the most restrictive requirements of either type of occupancy. If the hotel had a freestanding maintenance shop it would be treated as a separate facility and subject to the requirements of that occupancy class, whereas a maintenance shop housed within the hotel building itself would likely fall under the more restrictive provisions of the other occupancies housed within the building.

All interior areas of the building including guest rooms, top floors, front desk areas, lobbies, assembly rooms, and service areas should be inspected. In this type of occupancy, the building occupants are likely to be unfamiliar with their surroundings and will depend on the building design and fire and life safety protection to ensure their life safety. Fire alarm systems should be clearly heard in all rooms of the building, and clearly marked and lighted exit routes should be available throughout the building.

You should verify that the facility has an appropriate fire and emergency plan and that building personnel are capable of administering its procedures. Personnel should have received the necessary training and practice to know the plan and enact it successfully. In reviewing documentation of staff training and participation in drills, you will want to take into account the impact of staff turnover, which can be fairly significant within the hospitality industry.

Fire Inspector Tips

Laundry rooms with large industrial washers and dryers are common in hotels. These areas present fire prevention issues, including a large fire load, lint buildup inside and outside of the dryers, linen carts blocking the means of egress, storage, and appliances that have some type of heating element. Ensure that these areas have a daily cleanup and maintenance plan to decrease the likelihood of a fire.

Dormitories

According to NFPA 101, a **dormitory** is a building or space in a building in which group sleeping accommodations are provided for more than 16 persons who are not members of the same family. This lodging may be provided in one room or a series of closely associated rooms, under joint occupancy and single management, with or without meals, but without individual cooking facilities Figure 4. Inspection practices of dormitories focus on fire detection systems and occupancy evacuation and must also address issues that are unique to dormitory living.

The inspection should include means of egress, interior finishes, building contents, housekeeping, and fire protection systems. The use of alternative heating devices and overloading electrical circuits can be a concern. Many jurisdictions, whether at the local or state level, may have specific fire and life safety provisions with respect to college and university dormitories. It is important that you ensure that there is a fire and emergency

Figure 3 A hotel has room to house more than 16 occupants.

Chapter 3 Types of Occupancies

Figure 4 Dormitories provide unique challenges for the fire inspector.

Fire Inspector Tips

"Apartment hotels" should be classified as hotels, because they are potentially subject to the same transient occupancy as hotels. Transients are those who occupy accommodations for less than 30 days.

you should consider the occupant load, dwelling units, means of egress, interior finishes, the protection of openings, waste chutes, and hazardous areas.

Inspection of fire and life safety equipment should include fire alarm systems, portable fire extinguishers, sprinkler and/or standpipe systems, and lighting. The importance of the proper placement and maintenance of smoke alarms should be emphasized during any inspection of an apartment building to both building staff and residents.

plan including appropriate escape routes in place and that building occupants and staff have received the necessary fire safety training and have drilled in accordance with the emergency plan.

Apartment Buildings

According to NFPA 5000, an **apartment building** is a building or portion thereof containing three or more dwelling units with independent cooking and bathroom facilities Figure 5 . The nature and character of the occupancy may be based on the location and design of the building as well as the age and social status of the occupants. You must be aware of the code requirements that were in place at the time the building was constructed. Based on the adoption of new codes and requirements over time, the code requirements are often different for existing apartment buildings and new apartment buildings, and you must be familiar with these differences when inspecting buildings and enforcing codes and regulations. An example of this could involve the permitted travel distances to exits in an apartment building built in 1940 versus 2010.

When inspecting an apartment building, it is recommended that you begin by conducting a general observation of the exterior of the building. Upon moving to the interior of the building,

Residential Board and Care Occupancy

NFPA 101 defines a **residential board and care occupancy** as a building or portion thereof that is used for lodging and boarding of four or more residents, not related by blood or marriage to the owners or operators, for the purpose of providing personal care services Figure 6 . These occupancies provide lodging, boarding and personal care services and include residential care homes, personal care homes, assisted living facilities, and group homes.

Examples of residential board and care occupancies include:
- Group housing arrangement for physically or mentally handicapped persons who normally attend school in the community, attend worship in the community, or otherwise use community facilities
- Group housing arrangement for physically or mentally handicapped persons who are undergoing training in preparation for independent living, for paid employment, or for other normal community activities
- Group housing arrangement for the elderly that provides personal care services but that does not provide nursing care
- Facilities for social rehabilitation, alcoholism, drug abuse, or mental health problems that contain a group housing arrangement and that provide personal care services but do not provide acute care

Figure 5 Apartment buildings are often the most difficult occupancies to inspect.

Figure 6 An example of a residential board and care occupancy.

- Assisted living facilities
- Other group housing arrangements that provide personal care services but not nursing care

You must recognize the difference between personal care and health care occupancies. Personal care occupancies include assisting occupants with many of the activities of daily living, such as bathing and dressing. Personal care does not include medical care. If nursing care is provided, then the facility is considered to be a health care occupancy; however, if nursing care is not provided, it is considered to be a residential board and care occupancy. It is important that you accurately determine the classification of the facility and corresponding requirements for compliance with relevant codes and/or regulations.

A primary concern in inspecting these occupancies is the capability to evacuate residents in the event of a fire or other emergency. The characteristics of the occupants, means of egress, and egress capacity should be reviewed. Additional considerations when conducting the inspection include building compartmentation, protection of vertical openings, hazardous areas, fire detection and alarm systems, and fire suppression equipment. The occupants of these facilities rely in varying degrees on staff assistance and building fire and life safety provisions. It is imperative that you verify that all fire protection and life safety equipment is in operational condition and has been properly maintained and tested.

Given the role that facility staff personnel play in ensuring the life safety of residents, it is crucial that you validate that an appropriate fire and emergency plan is in place. You must ensure that facility staff and residents are aware of its procedures and that all staff members have received the necessary fire and life safety training. Staff members must also prove their knowledge and skills through fire drills.

Health Care Occupancy

NFPA 101 defines a **health care occupancy** as an occupancy used for purposes of medical or other treatment or care of four or more persons on an inpatient basis where such occupants are mostly incapable of self-preservation due to age, physical or mental disability, or because of security measures not under the occupant's control. These occupants also include infants, convalescents, or infirmed aged persons Figure 7 . Note that the definition of health care occupancy specifies inpatients. A health care facility used only for outpatients is addressed as an ambulatory health care occupancy.

Included within this occupancy classification are hospitals or other medical institutions, nurseries, nursing homes, and limited care facilities. Because patients within these occupancies are presumed to be incapable of self-preservation in terms of seeking safe refuge should a fire occur, their life safety is based on the incorporation of fire and life safety features into the building and the staff response to a fire or other emergency situation. Life safety features include building construction features such as compartmentalization, the provision of automatic sprinklers, and the training of facility staff in the procedures that should be followed in the event of a fire. These occupancies are often designed to provide for "defending in place" in the event of a fire or other emergency. Under this approach, building features such as compartmentalization and protection of openings are designed to provide protection of patients during a fire.

Figure 7 An example of a health care occupancy.

Essential elements of the inspection of a health care occupancy involve ensuring appropriate means of egress in compliance with code and that all fire detection and suppression systems are properly maintained and fully functional. Inspection and maintenance documentation for these systems should be thoroughly reviewed. It is also imperative that you ascertain that the facility has an appropriate facility safety plan that incorporates fire safety training. Documentation regarding staff training and participation in drills and exercises should be examined.

Ambulatory Health Care Occupancies

An **ambulatory health care occupancy** is defined in NFPA 101 as an occupancy used to provide outpatient services or treatment to four or more patients simultaneously. These services include one or more of the following:

- Treatment for patients that renders the patients incapable of taking action for self-preservation under emergency conditions without the assistance of others.
- Anesthesia that renders the patients incapable of taking action for self-preservation under emergency conditions without the assistance of others.
- Emergency or urgent care for patients who, due to the nature of their injury or illness, are incapable of taking action for self-preservation under emergency conditions without the assistance of others Figure 8 .

Facilities falling under this occupancy classification do not provide overnight sleeping accommodations, while those under the health care occupancy do. This occupancy class includes freestanding emergency medical units, hemodialysis centers, and outpatient surgical units, often referred to as day surgery centers, where general anesthesia is administered. Oral surgery centers also fall under this occupancy classification. Practices, such as the typical physician's or dentist's office, where four or more patients are not rendered incapable of self-preservation are not classified as ambulatory health care occupancies. Physician and dental offices are classified as business occupancies.

Considerations when inspecting an ambulatory health care occupancy include means of egress and protection features such as building construction, smoke barriers, fire detection and alarm systems, and fire protection systems including

Chapter 3 Types of Occupancies

Figure 8 An example of an ambulatory health care occupancy.

portable fire extinguishers. As with health care occupancies, the compartmentation of the building and necessary provisions to ensure patient movement and evacuation are essential elements of ensuring occupant life safety. A thorough inspection of a facility falling under this occupancy class should also include determination of any hazardous areas, such as laboratories, as well as the life safety provisions that have been implemented to address corresponding life safety issues. Your inspection should also include verification of an appropriate written fire safety plan and staff fire safety training and drills.

Day-Care Occupancy

NFPA 101 defines a **day-care occupancy** as an occupancy in which four or more clients receive care, maintenance, and supervision, by other than their relatives or legal guardians, for less than 24 hours per day Figure 9 . Examples of day-care occupancies include adult day-care occupancies, child day-care occupancies, day-care homes, kindergarten classes that are conducted within a day-care occupancy, and nursery schools. Day-care occupancies are not subject to the requirements governing educational occupancies. Day-care facilities within houses of worship are exempt during services.

The three subclasses of day-care occupancies are:
- Day-care occupancies
- Group day-care homes
- Family day-care homes

The three subclasses within the day-care occupancy category are distinguished based on the number of clients serviced. Day-care occupancies service the most clients, followed by group day-care homes, and family day-care homes. As a fire inspector, it is important to recognize that the requirements for each of these occupancy subclasses may differ within the applicable code(s) within your jurisdiction.

Considerations when inspecting a day-care occupancy include the occupant load, means of egress components: capacity of means of egress, the number of exits, the arrangement of means of egress, the travel distance to exits, discharge from exits, proper illumination of the means of egress, emergency lighting, the clear marking of the means of egress, and special means of egress features. Additional considerations when inspecting a day-care occupancy include the protection of vertical openings, protection from hazards, interior finishes, fire detection systems, alarm and communication systems, extinguishment requirements, special provisions, building services, and operating features.

Given the nature of these occupancies, it is important to ensure that the facility has appropriate operational and emergency plans in place that provide for the life safety of building occupants and that staff are properly trained to execute these plans. The chapter *Occupancy Safety and Evacuation Plans* covers this topic in detail.

Educational Occupancy

Educational occupancies are defined by NFPA 101 as buildings used for educational purposes through the twelfth grade by six or more persons that are occupied at least four hours per day or 12 hours per week Figure 10 . Examples of educational occupancies include academies, kindergartens, and schools. Schools for levels beyond the twelfth grade are not classified as educational occupancies. These facilities must comply with the requirements for assembly, business, or other appropriate occupancy. Day-care facilities are not classified as educational occupancies.

When the occupancy threshold of fifty or more persons is met, an educational occupancy may also be classified as an assembly

Figure 9 An example of a day-care occupancy.

Figure 10 An example of an educational occupancy.

occupancy. Educational facilities may contain assembly areas, such as auditoriums, cafeterias, and gymnasiums that typically present a low fire hazard but present a high concentration of occupants.

It is important to develop an understanding of the activities conducted within the building, the number of occupants, and the age of occupants as these factors may determine the occupancy classification. Educational activities can vary from the low fire hazards posed by the typical lecture classroom to the moderate or high fire hazards that a science laboratory presents. It should be noted that under some codes, the requirements for educational occupancies will be different for new buildings than for existing buildings. An example of this might be the requirements with respect to interior floor finishes.

Business Occupancies

According to NFPA 101, a <u>business occupancy</u> is as an occupancy used for the transaction of business other than mercantile. Businesses included under this occupancy classification are general offices, doctor's offices, government offices, city halls, municipal office buildings, courthouses, outpatient medical clinics where patients are ambulatory, college and university classroom buildings with less than 50 occupants, air traffic control towers, and instructional laboratories. Business occupancies typically have a large number of occupants during the business day and limited occupants during non-business hours Figure 11.

Business occupancies generally have a lower occupant density than mercantile occupancies, and the occupants are usually more familiar with their surroundings. However, confusing and indirect egress paths are often developed due to office layouts and the arrangement of tenant spaces. When conducting an inspection of a business occupancy, you should consider means of egress, protection of openings, and hazardous areas. Computer rooms and the protection of the organization's records are business concerns that you will frequently encounter when inspecting a business occupancy. Building services such as waste disposal and the utilization of shafts and chases can contribute to fire protection problems. As always, you should inspect all fire protection systems for proper maintenance, testing, and operation. This would include alarm systems, sprinkler systems, and portable fire extinguishers.

Industrial Occupancies

An <u>industrial occupancy</u> is defined in NFPA 101 as an occupancy in which products are manufactured or in which processing, assembling, mixing, packaging, finishing, decorating, or repair operations are conducted Figure 12. Industrial occupancies include, but are not limited to: chemical plants, factories, food processing plants, furniture manufacturers, hangars (for service and maintenance), laboratories involving hazardous chemicals, laundry and dry cleaning plants, metal-working plants, plastics manufacture and molding plants, power plants, refineries, semiconductor manufacturing plants, sawmills, pumping stations, telephone exchanges, and wood-working plants.

Industrial occupancies expose occupants to a wide range of processes and materials of varying hazard. Special-purpose industrial occupancies, which are characterized by large installations of equipment that dominate the space, such as a power plant, are addressed separately from general-purpose industrial facilities, which have higher densities of human occupancy. Industrial occupancy buildings, along with storage occupancy buildings, are more likely than any other occupancy to have contents with a wide range of hazards, often including an array of hazardous materials that can present fire and life safety hazards for building occupants and emergency response personnel.

Each building or separated portion should be inspected in accordance with the requirements of its primary use. The complexity of inspections of industrial occupancies can prove time consuming. Prior to conducting an inspection of an industrial occupancy, you should first determine and become familiar with all pertinent requirements under your jurisdiction's codes or regulations. Considerations during the inspection should include occupant load, means of egress, protection of openings, hazardous materials use and storage, inside storage, outside storage, housekeeping, and maintenance.

When inspecting these facilities, you should verify the maintenance and operational status of all fire protection systems including fire alarm systems, portable fire extinguishers, water supplies, fire pumps, standpipe systems, sprinkler systems, and special extinguishing systems.

Figure 11 An example of a business occupancy.

Figure 12 An example of an industrial occupancy.

Safety Tips

For a key industrial location in your community, identify the hazardous materials that are resident on-site. Using material safety data sheets for the various hazardous materials identify their classification, properties, fire behavior, labeling, hazards, safe handling practices and use, applicable codes and standards, appropriate fire protection systems and equipment approved for the material, safety procedures, and storage compatibility.

Mercantile Occupancy

A mercantile occupancy is defined under NFPA 101 as an occupancy used for the display and sale of merchandise. Included under this occupancy classification are shopping centers, supermarkets, drug stores, department stores, auction rooms, restaurants with fewer than 50 persons, and any occupancy or portion thereof that is used for the display and sale of merchandise Figure 13.

Mercantile occupancies, as in the case of assembly occupancies, are characterized by large numbers of people who gather in a space that is relatively unfamiliar to them. In addition, mercantile occupancies often contain sizable quantities of combustible contents and use circuitous egress paths that are deliberately arranged to force occupants to travel around displays of merchandise that is available for sale. Bulk merchandising retail buildings, which characteristically consist of a warehouse-type building occupied for sales purposes, are a subclass of mercantile occupancy with a greater potential for hazards than more traditional mercantile operations.

Mercantile occupancies vary in size and in terms of fire load. They encompass many different types of materials and operations. The separation between mercantile occupancies and other occupancies can be an issue that you must address. An example of this would be the shopping mall which contains stores, restaurants, and assembly areas. Covered malls and bulk merchandising retail centers present additional issues, such as fire spread, smoke travel, and challenges of evacuating building occupants, which must be considered during a fire inspection.

Fire Inspector Tips

Although storage occupancies have low occupant loads, they may have very high fuel loads.

When inspecting a mercantile occupancy, you should consider occupant load, means of egress, interior finish, protection of openings, and protection from hazards. Fire protection systems are essential fire and life safety measures in mercantile occupancies. You should verify the operational status, maintenance, and testing of these systems.

Storage Occupancies

NFPA 101 defines a storage occupancy as an occupancy used primarily for the storage or sheltering of goods, merchandise, products, or vehicles. Storage occupancies are characterized by relatively low human occupancy in comparison to building size and by varied hazards associated with the materials stored. Storage occupancies include warehouses, freight terminals, parking garages, aircraft storage hangers, truck and marine terminals, bulk oil storage, cold storage, grain elevators, and barns. These facilities may be separate and distinct facilities or part of a multi-use occupancy Figure 14.

When storage is incidental to the main use of the structure, it should be classified as part of the primary occupancy when determining requirements. This determination may require considerable judgment on your part. One consideration that you can use in determining the appropriate occupancy classification is

Figure 13 An example of a mercantile occupancy.

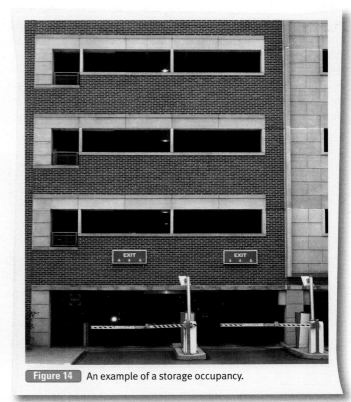

Figure 14 An example of a storage occupancy.

Fire Inspector Tips

Bulk merchandising retail buildings, which characteristically consist of a warehouse-type building occupied for sales purposes, are a subclass of mercantile occupancy rather than a storage occupancy.

the hazard classification of the contents being stored. This would include considering such properties of a hazardous material as its flammability, combustibility, reactivity, and explosive potential.

Additionally, if occupant load is greater than normal for a storage occupancy, then the facility should be classified as an industrial occupancy. Examples of this are operations involving packaging, labeling, sorting, or special handling. Parking garages are classified as industrial occupancies if they contain an area in which repairs are performed.

When conducting an inspection of a storage occupancy, you should consider building contents, protection of openings, indoor and outdoor storage practices, hazardous materials and processes, and housekeeping. The occupant load and means of egress must be reviewed including: exit access and travel distances, exits and locations, identification of exits, and exit discharge.

The hazards associated with the operation of material-handling equipment, such as industrial fork lifts, and the storage and changing of associated fuel should be addressed in the inspection. In inspecting the fire protection systems, you should examine the fire alarm system, portable fire extinguishers, fire pumps, standpipe system, and sprinkler system.

Assembly Occupancies

NFPA 101 defines <u>assembly occupancies</u> as buildings that used for a gathering of 50 or more persons for deliberation, worship, entertainment, eating, drinking, amusement, awaiting transportation, or similar uses; or buildings that are used as a special amusement building regardless of occupant load Figure 15 . Assembly occupancies are characterized by the presence or potential presence of crowds with attendant panic hazard in the case of a fire or other emergency. They are generally or occasionally open to the public, and the occupants, who are present voluntarily, are not ordinarily subject to discipline or control. Such buildings are ordinarily occupied by able-bodied persons and are not used for sleeping purposes.

Assembly occupancies span a wide range of uses, each of which necessitates different considerations Table 1 . There are numerous legal ways an assembly occupancy can be utilized. An example of this could be a church, normally utilized as a place of worship, at times being used for dinning or entertainment. Just as the world is constantly changing, so too are the specific uses of the various buildings within your jurisdiction. Therefore it is important to recognize that both the character of an assembly occupancy and its occupant load may change over time and may have done so since the last time the building was inspected.

The primary considerations when inspecting assembly occupancy include the occupant loads, means of egress, interior finish requirements, building services, smoking provisions, and fire protection systems. As a fire inspector, examine the flame spread ratings and flammability of the building's interior finish, the adequacy of the building's exits, and the building's electrical, heating, and fire protection systems.

In situations involving unique occupancies, such as stages, projection rooms, exhibits and trade shows, or special amusement

Figure 15 An example of an assembly occupancy.

Table 1 Examples of Assembly Occupancies

- Armories
- Assembly halls
- Auditoriums
- Bowling lanes
- Club rooms
- College and university classrooms (50 persons and over)
- Conference rooms
- Courtrooms
- Dance halls
- Drinking establishments
- Exhibition halls
- Gymnasiums
- Libraries
- Mortuary chapels
- Motion picture theatres
- Museums
- Passenger stations and terminals of air, surface, underground, and marine public transportation facilities
- Places of religious worship
- Pool rooms
- Recreation piers
- Restaurants
- Skating rinks
- Special amusement buildings (regardless of occupant load)
- Theatres

Fire Inspector Tips

Some places of assembly are multi-use to the extent that the occupant density varies greatly from event to event. In hotels, the meeting rooms may be set up for classroom use during the day then change to a reception room with a dinner banquet at night. You must assure that the safeguards are adequate for the maximum number of people that may be expected in any such space under any conditions.

buildings, it is important during the inspection to review any special safeguards that have been implemented, such as automatic sprinklers and exhaust systems. This is especially true in assembly occupancies where open flames or pyrotechnics may be present.

It must be recognized that while the probability of fire is usually low with assembly occupancies, the potential for loss of life can be high. Adequate exit provisions are thus a critical requirement for any assembly occupancy. The fire that occurred on February 20, 2003 at The Station, a nightclub in West Warwick, Rhode Island, illustrates the importance of the adequacy of fire exits in assembly occupancies. One hundred individuals perished in this tragic fire.

Small Assembly Uses

NFPA 101 defines small assembly uses as those situations where a room or space contained within another type of occupancy is used for assembly purposes by fewer than 50 persons and is incidental to that occupancy. Small assembly uses are subject to the same provisions that apply to the occupancy.

Detention and Correctional Occupancies

A <u>detention and correctional occupancy</u> is defined in NFPA 101 as an occupancy used to house one or more persons under varied degrees of restraint or security where such occupants are mostly incapable of self-preservation because of security measures not under the occupant's control **Figure 16**. Facilities that fall under the detention and correctional occupancy classification include:

- Jails
- Detention centers
- Correctional institutions
- Houses of correction
- Prerelease centers

Figure 16 An example of a detention and correctional occupancy.

- Reformatories
- Work camps
- Training schools
- Substance abuse centers

There are five categories under this occupancy that correspond to the degree of restraint of occupants within the facility. These categories are:

- Use Condition I—Free Egress—Occupants are permitted to move freely to the exterior.
- Use Condition II—Zoned Egress—Occupants are permitted to move from sleeping areas to other smoke compartments.
- Use Condition III—Zoned Impeded Egress—Occupants are permitted to move within any smoke compartment.
- Use Condition IV—Impeded Egress—Occupants are locked in their rooms or cells, but they can be remotely unlocked.
- Use Condition V—Contained—Occupants are locked in their rooms or cells with manually operated locks that can only be opened by facility staff members.

Detention and correctional facilities present unique and challenging fire and life safety issues. The maintenance of a predefined level of security is essential within these facilities, thus facility personnel may be reluctant to initiate any actions, including evacuation procedures that may compromise security. Facility design, operation, and maintenance are crucial to ensuring fire and life safety, as well as prisoner security.

Within detention and correctional facilities, uses other than residential housing are classified according to the appropriate occupancy classification. The facility may consist of a number of buildings, some of which should be appropriately classified under different occupancies. Examples of this would be an infirmary which would be classified as a health care occupancy, an auditorium or gymnasium which would fall under the assembly occupancy classification, and a business office, storage area, or industrial shop which would be categorized under the appropriate occupancy classification.

Detention and correctional occupancies present the unique challenges associated with sleeping occupants and the occupant's lack of or limited control with respect to egress. Your role as a fire inspector in ensuring the life safety of facility occupants cannot be overstated. The ability of the facility staff to implement a "defend in place" strategy successfully is based upon the design features of the building including its compartmentation, the maintenance and operational readiness of fire protection and other building systems, the existence of a comprehensive emergency plan, and the preparedness of staff members to execute the emergency plan based on their training, drills, and exercises. Under a defend in place strategy, which is typically used in those situations where conducting a total building evaluation would be problematic, building occupants are moved to and "sheltered" in protected and safe areas of refuge within the building.

Elements that must be considered when inspecting a facility falling under this occupancy classification include: construction type, capacity, means of egress, interior finish, protection of openings, contents, and hazardous areas. In reviewing means of egress the inspector should consider horizontal exits, sliding doors, remote controlled release mechanisms, and exit discharge.

Your inspection must include a thorough review of fire protection systems, including sprinkler and standpipe systems, building subdivision, and building services. In addition to inspecting the building fire protection systems, you should review accompanying documentation of system maintenance and testing.

It is imperative that detention and correctional facilities have appropriate emergency plans that address fire and other emergency situations that could endanger facility occupants. Facility personnel must be thoroughly familiar with all aspects of these emergency plans and their roles and responsibilities during an emergency situation.

Multiple Occupancies

A multiple occupancy is defined in NFPA 101 as a building or structure in which two or more classes of occupancy exist. Multiple occupancies are required to comply with the general requirements for multiple occupancies and the requirements for either mixed occupancies or separated occupancies. In those cases where areas of a building used for mercantile, business, industrial, storage, or nonresidential use with an occupant load less than the occupancy threshold specified under the relevant code are incidental to the primary use of the building, they may be considered part of the predominant occupancy and subject to its requirements.

Classifying a building simply as a multiple occupancy is an incomplete classification, as the options for occupancy classification are limited to assembly, educational, day-care, health care, ambulatory health care, residential, residential board and care, mercantile, business, industrial, and storage. Rather, a classification of multiple occupancy needs to include mention of the occupancy types involved. For example, a building with spaces used for storage and spaces used for sales should be classified as a multiple occupancy that is part storage occupancy and part mercantile occupancy.

Mixed Occupancy

A mixed occupancy is a multiple occupancy where the occupancies are intermingled Figure 17 . Each portion of the building is classified according to its use.

Often multiple occupancies exist within an educational complex. Another example of a mixed occupancy would be a building that is used for both manufacturing and warehousing. With mixed occupancies, the means of egress, facilities, types of construction, protection, and other safeguards in the building shall comply with the most restrictive fire and life safety requirements of the occupancies involved.

Separated Occupancy

A separated occupancy is a multiple occupancy where the occupancies are separated by fire resistance-rated assemblies. Where separated occupancies are provided, each part of the building comprising a distinct occupancy, is completely separated from the other occupancies by fire-resistive assemblies. If this level of protection was provided between a manufacturing and warehousing areas of the building, then it could be considered a separated occupancy.

Figure 17 An example of a mixed occupancy building.

Special Structures and High-Rise Buildings

Special structures and high-rise buildings present interesting challenges. High-rise buildings are typically considered buildings having occupied stories 75 feet (2286 cm) or more above the lowest level of fire department access to the building. Special structures include open structures, piers, towers, windowless buildings, underground structures, vehicles and vessels, water-surrounded structures, membrane structures, and tents Figure 18 .

A typical occupancy in a special structure will fall under one of the occupancy classifications previously covered. The means of egress and fire protection requirements for that occupancy apply, as do the additional requirements for the special structure. In addition to the requirements mandated by the applicable occupancy class, you must apply requirements unique to each special type of structure. Guidance on the requirements for special structures can be found in NFPA 101 and other relevant standards. When inspecting a special structure, you should first determine the general occupancy classification and then determine the special requirements that apply. Occupancies in special structures, while often not easy to inspect, must be inspected in accordance with all pertinent requirements.

Figure 18 An example of a special structure.

Wrap-Up

Chief Concepts

- In the model building codes, the term occupancy refers to the intended use of a building. The occupancy classification of a building is determined by the current use of that building.
- A one- or two-family dwelling unit is a building that contains no more than two dwelling units with independent cooking and bathroom facilities.
- A lodging or rooming house is a building or portion thereof that does not qualify as a one- or two-family dwelling, that provides sleeping accommodations for a total of 16 or fewer people on a transient or permanent basis, without personal care services, with or without meals, but without separate cooking facilities for individual occupants.
- A hotel is a building or group of buildings under the same management in which there are sleeping accommodations for more than 16 people and is primarily used by transients for lodging with or without meals.
- An apartment building is a building or portion thereof containing three or more dwelling units with independent cooking and bathroom facilities.
- A residential board and care occupancy as a building or portion thereof that is used for lodging and boarding of four or more residents, not related by blood or marriage to the owners or operators, for the purpose of providing personal care services.
- A health care occupancy as an occupancy used for purposes of medical or other treatment or care of four or more persons on an inpatient basis where such occupants are mostly incapable of self-preservation due to age, physical or mental disability, or because of security measures not under the occupant's control. These occupants also include infants, convalescents, or infirmed aged persons.
- Assembly occupancies are buildings that used for a gathering of 50 or more persons for deliberation, worship, entertainment, eating, drinking, amusement, awaiting transportation, or similar uses; or buildings that are used as a special amusement building regardless of occupant load.
- A day-care occupancy is an occupancy in which four or more clients receive care, maintenance, and supervision, by other than their relatives or legal guardians, for less than 24 hours per day.
- Educational occupancies are buildings used for educational purposes through the twelfth grade by six or more persons for four or more hours per day or more than 12 hours a week.
- A business occupancy is as an occupancy used for the transaction of business other than mercantile.
- An industrial occupancy is an occupancy in which products are manufactured or in which processing, assembling, mixing, packaging, finishing, decorating, or repair operations are conducted.
- A mercantile occupancy is an occupancy used for the display and sale of merchandise.
- A storage occupancy as an occupancy used primarily for the storage or sheltering of goods, merchandise, products, vehicles, or animals.
- Assembly occupancies are buildings that used for a gathering of 50 or more persons for deliberation, worship, entertainment, eating, drinking, amusement, awaiting transportation, or similar uses; or buildings that are used as a special amusement building regardless of occupant load.
- A detention and correctional occupancy is an occupancy used to house one or more persons under varied degrees of restraint or security where such occupants are mostly incapable of self-preservation because of security measures not under the occupant's control.
- A multiple occupancy is a building or structure in which two or more classes of occupancy exist. Multiple occupancies are required to comply with the general requirements for multiple occupancies and the requirements for either mixed occupancies or separated occupancies.
- A mixed occupancy is a multiple occupancy where the occupancies are intermingled. Each portion of the building is classified according to its use.

Wrap-Up

Hot Terms

Ambulatory health care occupancy A building or portion thereof used to provide services or treatment simultaneously to four or more patients that, on an outpatient basis. (NFPA 101, *Life Safety Code*)

Apartment building is a building or portion thereof containing three or more dwelling units with independent cooking and bathroom facilities. {NFPA 5000}

Assembly occupancies Buildings (1) used for a gathering of 50 or more persons for deliberation, worship, entertainment, eating, drinking, amusement, awaiting transportation, or similar uses; or (2) used as a special amusement building regardless of occupant load. (NFPA 101, *Life Safety Code*)

Business occupancy An occupancy used for the transaction of business other than mercantile. (NFPA 101, *Life Safety Code*)

Day-care occupancy An occupancy in which four or more clients receive care, maintenance, and supervision, by other than their relatives or legal guardians, for less than 24 hours per day. (NFPA 101, *Life Safety Code*)

Detention and correctional occupancy An occupancy used to one or more persons under varied degrees of restraint or security where such occupants are mostly incapable of self-preservation because of security measures not under the occupant's control. (NFPA 101, *Life Safety Code*)

Dormitory A building or space in a building in which group sleeping accommodations are provided for more than 16 persons who are not members of the same family in one room, or a series of closely associated rooms, under joint occupancy and single management, with or without meals, but without individual cooking facilities. (NFPA 101, *Life Safety Code*)

Educational occupancies Buildings used for educational purposes through the twelfth grade by six or more persons for 4 or more hours per day or more than 12 hours a week. (NFPA 101, *Life Safety Code*)

Health care occupancy An occupancy used for purposes of medical or other treatment or care of four or more persons where such occupants are mostly incapable of self-preservation due to age, physical or mental disability, or because of security measures not under the occupant's control. (NFPA 101, *Life Safety Code*)

Hotel A building or group of buildings under the same management in which there are sleeping accommodations for more than 16 people and is primarily used by transients for lodging with or without meals. (NFPA 101, *Life Safety Code*)

Industrial occupancy An occupancy in which products are manufactured or in which processing, assembling, mixing, packaging, finishing, decorating, or repair operations are conducted. (NFPA 101, *Life Safety Code*)

Mercantile occupancy An occupancy used for the display and sale of merchandise. (NFPA 101, *Life Safety Code*)

Mixed occupancy A multiple occupancy where the occupancies are intermingled. (NFPA 101, *Life Safety Code*)

Multiple occupancy A building or structure in which two or more classes of occupancy exist. (NFPA 101, *Life Safety Code*)

Lodging or rooming house Building or portion thereof that does not qualify as a one- or two-family dwelling, that provides sleeping accommodations for a total of 16 or fewer people on a transient or permanent basis, without personal care services, with or without meals, but without separate cooking facilities for individual occupants. (NFPA 101, *Life Safety Code*)

Occupancy The intended use of a building

One- or two-family dwelling A building that contains no more than two dwelling units with independent cooking and bathroom facilities. {NFPA 5000}

Residential board and care occupancy A building or portion thereof that is used for lodging and boarding of four or more residents, not related by blood or marriage to the owners or operators, for the purpose of providing personal care services. (NFPA 101, *Life Safety Code*)

Storage occupancy An occupancy used primarily for the storage or sheltering of goods, merchandise, products, vehicles, or animals. (NFPA 101, *Life Safety Code*)

Separated occupancy A multiple occupancy where the occupancies are separated by fire resistance-rated assemblies. (NFPA 101, *Life Safety Code*)

Fire Inspector *in Action*

You prepare to make an inspection at an existing surf board shop in a strip mall. You look in the file and locate the folder which classifies the occupancy as mercantile. When you enter the shop, you smell a strong chemical odor. You find that the business now has begun repairing surf boards in the back room in addition to sales. The operation involves using different flammable resins, paints, and solvents. Also, different power tools that produce large amounts combustible dusts are in use. You make notes in preparation of correcting the occupancy classification.

1. What single factor would change the classification of this occupancy?
   A. Use of flammable liquids
   B. Combustible dusts
   C. Repair operations
   D. All of the above

2. What is the correct classification of this occupancy?
   A. Assembly
   B. Business
   C. Industrial
   D. Storage

3. What can the business owner do to keep the mercantile occupancy classification?
   A. The business location must be changed.
   B. Eliminate the repair shop.
   C. Move all operations except sales to another location.
   D. Either B or C.

4. If you select the wrong occupancy class, what is the consequence?
   A. The incorrect licensing fees will be charged.
   B. An inadequate level of fire and life safety may be required.
   C. Insurance coverage may be denied.
   D. Pre-fire planning will be incorrect.

Fire Growth

CHAPTER 4

NFPA Objectives

Fire Inspector I

4.3.14 Recognize a hazardous fire growth potential in a building or space, given field observations, so that the hazardous conditions are identified, documented, and reported in accordance with the applicable codes and standards and the policies of the jurisdiction.

(A) Requisite Knowledge. Basic fire behavior; flame spread and smoke development ratings of contents, interior finishes, building construction elements, decorations, decorative materials, and furnishings; and safe housekeeping practices. (pp 58–72)

(B) Requisite Skills. The ability to observe, communicate, apply codes and standards, recognize hazardous conditions, and make decisions.

4.3.15 Determine code compliance, given the codes, standards, and policies of the jurisdiction and a fire protection issue, so that the applicable codes, standards, and policies are identified and compliance is determined.

(A) Requisite Knowledge. Basic fire behavior; flame spread and smoke development ratings of contents, interior finishes, building construction elements, life safety systems, decorations, decorative materials, and furnishings; and safe housekeeping practices. (pp 58–72)

(B) Requisite Skills. The ability to observe, communicate, apply codes and standards, recognize hazardous conditions, and make decisions.

FESHE Objectives

There are no FESHE objectives for this chapter.

Fire Inspector II

5.3.10 Determine fire growth potential in a building or space, given field observations or plans, so that the contents, interior finish, and construction elements are evaluated for compliance, and deficiencies are identified, documented, and corrected in accordance with the applicable codes and standards and the policies of the jurisdiction.

(A) Requisite Knowledge. Basic fire behavior; flame spread and smoke development ratings of contents, interior finishes, building construction elements, decorations, decorative materials, and furnishings; and safe housekeeping practices. (pp 58–72)

(B) Requisite Skills. The ability to observe, communicate, interpret codes and standards, recognize hazardous conditions, and make decisions.

Knowledge Objectives

After studying this chapter, you will be able to:

1. Describe the ignition phase, growth phase, fully developed phase, and decay phase of a fire.
2. Describe the characteristics of a room-and-contents fire through each of the four phases of a fire.
3. Describe how interior finishes affect fire growth.
4. Describe how decorative materials may affect fire growth.
5. Describe how furnishings may affect fire growth.
6. Describe how building construction elements may affect fire growth.
7. Describe how to obtain the flame spread and smoke development ratings for the contents of an occupancy.

Skills Objectives

There are no skills objectives for this chapter.

You Are the Fire Inspector

You are inspecting a nightclub that has been redecorated to match the theme implied by the nightclub's name, Disco Lives. In the VIP lounge, you are surprised to see wall-to-wall and floor-to-ceiling carpeting in a windowless room. In your experience, the owners will be more receptive to your suggestions if you explain why a change needs to be made. You clear your throat and prepare to describe in simple and understandable terms how fire growth is affected by interior finishes.

1. In your own words, describe the concept of the fire tetrahedron.
2. Explain how heat travels and why.
3. Describe the how the interior finishes of this room will contribute to fire growth.

Introduction

The fire inspector is expected to ensure code compliance and identify situations where the fire triangle or the fire tetrahedron can become operational and start a fire. Once that fire occurs, the materials of construction and occupancy become potential enhancement to fire growth. As the fire inspector, you are expected to be able to analyze situations, identify potential adverse situations, and act accordingly.

The Chemistry of Fire

To be effective as a fire inspector, you need to understand the conditions needed for a fire to ignite and grow. Understanding these conditions will enable you to identify potential fire hazards in the structures you inspect.

What Is Fire?

Fire is a rapid chemical process that produces heat and usually light. We all know it when we see it, but we really do not know much about how fire works. A unique phenomenon, fire comes in different colors, ranging from red to yellow to blue. Flames can be steady, they can flicker, or they can just glow. We can see flames, but we can also see through them. Fire is neither solid nor liquid. Wood is a solid, gasoline is a liquid, and propane is a gas—but they all burn.

States of Matter

Matter is a term used to describe any physical substance that has mass and volume. Matter is made up of atoms and molecules. Matter exists in three states: solid, liquid, and gas **Figure 1**. An understanding of these states of matter is helpful in understanding fire behavior.

We all know what makes an object solid: it has a definite shape. Most uncontrolled fires are stoked by solid fuels. In structure fires, the building and most of the contents are solids. A solid does not flow under stress. Instead, it has a definite capacity for resisting forces and under ordinary conditions retains a definite size and shape. One characteristic of a solid is its ability to expand when heated and to contract when cooled. In addition, a solid object may change to a liquid state or a gaseous state when heated. Cold makes most solids more brittle, whereas heat makes them more flexible. Because a solid is rigid, only a limited number of molecules are on its surface; the majority of molecules are cushioned or insulated by the outer surface of the solid.

A liquid will assume the shape of the container in which it is placed. Most liquids contract when cooled and expand when heated. In addition, most will turn into gases when sufficiently heated. Liquids, for all practical purposes, do not compress. This characteristic allows fire fighters to pump water for long distances through pipelines or hoses. A liquid has no independent shape, but it does have a definite volume. The liquid with which fire fighters are most concerned is water.

A gas is a type of liquid that has neither independent shape nor independent volume, but rather tends to expand indefinitely. The gas we most commonly encounter is air, the mix of invisible odorless, tasteless gases that surrounds the earth. The mixture of gases in air maintain a constant composition—21 percent oxygen, 78 percent nitrogen, and 1 percent other gases such as carbon dioxide. Oxygen is required for us to live, but it is also required for fires to burn. We will explore the reaction of fuels with oxygen as we look at the chemistry of burning. Some fuels exist in the form of a gas.

Fuels

A fuel is any material that stores potential energy. Think of the vast amount of heat that is released during a large fire. The energy released in the form of heat and light has been stored

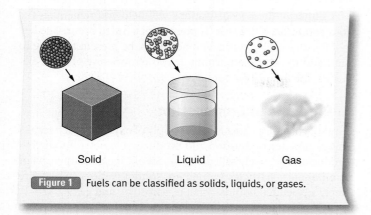

Figure 1 Fuels can be classified as solids, liquids, or gases.

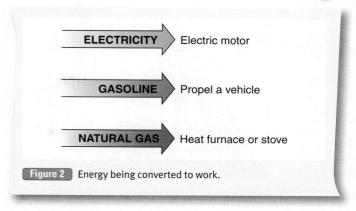

Figure 2 Energy being converted to work.

in the fuel before it is burned. The release of the energy in a gallon of gasoline, for example, can move a car miles down the road Figure 2 .

Types of Energy

Energy exists in the following forms: chemical, mechanical, electrical, light, or nuclear. Regardless of the form in which the energy is stored, it can be changed from one form to another. For example, electrical energy can be converted to heat or to light. Mechanical energy can be converted to electrical energy through a generator.

Chemical Energy

Chemical energy is the energy created by a chemical reaction. Some chemical reactions produce heat (exothermic); others absorb heat (endothermic). The combustion process is an exothermic reaction, because it releases heat energy. Most chemical reactions occur because bonds are established between two substances, or bonds are broken as two substances are chemically separated. Heat is also produced whenever oxygen combines with a combustible material. If the reaction occurs slowly in a well-ventilated area, the heat is released harmlessly into the air. If the reaction occurs very rapidly or within an enclosed space, the mixture can be heated to its ignition temperature and can begin to burn. A bundle of rags soaked with linseed oil will begin to burn spontaneously, for example, because of the heat produced by oxidation that occurs within the mass of rags. The combustion potential of gasoline is one example of chemical energy.

Mechanical Energy

Mechanical energy is converted to heat when materials rub against each other and create friction. For example, a fan belt rubbing against a seized pulley produces heat. Water falling across a dam is one example of mechanical energy based on position: the water releases energy as it moves to a lower area.

Electrical Energy

Electrical energy is converted to heat energy in several different ways. For example, electricity produces heat when it flows through a wire or any other conductive material. The greater the flow of electricity and the greater the resistance of the material, the greater the amount of heat produced. Examples of electrical energy that can produce enough heat to start a fire include heating elements, overloaded wires, electrical arcs, and lightning. Electrical energy is carried through the electrical wires inside homes and is stored in batteries that convert chemical energy to electrical energy.

Light Energy

Light energy is caused by electromagnetic waves packaged in discrete bundles called photons. This energy travels as thermal radiation, a form of heat. When light energy is hot enough, it can sometimes be seen in the form of visible light. If it is of a frequency that we cannot see, the energy may be felt as heat but not seen as visible light. Candles, fires, light bulbs, and lasers are all forms of light energy. Another example of light energy is the radiant energy we receive from the sun.

Nuclear Energy

Nuclear energy is created by splitting the nucleus of an atom into two smaller nuclei (nuclear fission) or by combining two small nuclei into one large nucleus (fusion). Nuclear reactions release large amounts of energy in the form of heat. These reactions can be controlled, as in a nuclear power plant, or uncontrolled, as in an atomic bomb explosion. In a nuclear power plant, the nuclear reaction releases carefully controlled amounts of heat, which then convert water to steam which powers a steam turbine generator. Both explosions and controlled reactions release radioactive material, which can cause injury or death. Nuclear energy is stored in radioactive materials and converted to electricity by nuclear power-generating stations.

Conservation of Energy

The law of the conservation of energy states that energy cannot be created or destroyed; however, it can be converted from one

Fire Inspector Tips

A liquid expands when heated and contracts when cooled. Water expands when heated and also when it freezes. This is why frozen pipes and fittings burst when they freeze.

Fire Inspector Tips

A photo cell in a standard solar backyard light collects energy from sunlight and converts it into electricity to power the light.

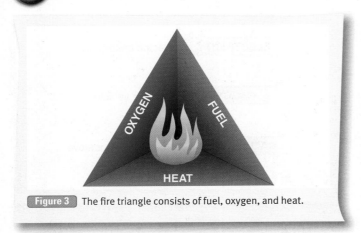

Figure 3 — The fire triangle consists of fuel, oxygen, and heat.

form to another. Think of an automobile: chemical energy in the gasoline is converted to mechanical energy when the car moves down the road. If you apply the brakes to stop the car, the mechanical energy is converted to heat energy by the friction of the brakes, and the car slows. Similarly, in a house fire, the stored chemical energy in the wood of the house is converted into heat and light energy during the fire.

Conditions Needed for Fire

To understand the behavior of fire, you need to consider the three basic elements needed for a fire to occur: fuel, oxygen, and heat. First, a combustible fuel must be present. Second, oxygen must be available in sufficient quantities. Third, a source of ignition, often heat, must be present. If we graphically place these three together, the result is a <u>fire triangle</u> Figure 3 .

Once a fire has ignited, four factors are required to maintain a self-sustaining fire. These are fuel, oxygen, heat, and chain reaction. That is, the fuel, oxygen, and heat continue to interact so that a chain reaction keeps the fire going.

Another way of visualizing all four conditions needed for a fire is the <u>fire tetrahedron</u> Figure 4 . A tetrahedron is a four-sided, three-dimensional figure. Each side of the fire tetrahedron represents one of the four conditions needed for a fire to occur.

Either of these visualizations will help you remember the four conditions that must be present for a fire to occur. The key point is that fuel, oxygen, and heat must be present for a fire to start and to continue burning. If you remove any of these elements, the fire will go out.

Products of Combustion

Fires burn without an adequate supply of oxygen, which results in incomplete combustion and produces a variety of toxic by-products, collectively called <u>smoke</u>. Smoke includes three major components: particles (which are solids), vapors (which are finely suspended liquids—that is, aerosols), and gases.

Smoke particles include unburned, partially burned, and completely burned substances. The unburned particles are lifted in the <u>thermal column</u> produced by the fire. The thermal column is cylindrical area above a fire in which heated air and gases rise and travel upward Some partially burned particles become part of the smoke because inadequate amount of oxygen is available to allow for their complete combustion. Completely burned particles are primarily ash. Some particles in smoke are small enough that they can get past the protective mechanisms of the respiratory system and enter the lungs, and most of these are very toxic to the body.

Smoke may also contain small droplets of liquids Figure 5 . The fog that is formed on a cool night consists of small water droplets suspended in the air. When water is applied to a fire, small droplets may also be suspended in the smoke or haze that forms. Similarly, when oil-based compounds burn, they produce small oil-based droplets that become part of the smoke. Oil-based or lipid compounds can cause great harm to a person when they are inhaled. In addition, some toxic droplets cause poisoning if absorbed through the skin.

Smoke contains a wide variety of gases. The composition of gases in smoke will vary greatly, depending on the amount of oxygen available to the fire at that instant. The composition of the substance being burned also influences the composition of the smoke. In other words, a fire fueled by wood will produce a different composition of gases than a fire fueled by petroleum-based fuels, including plastics.

Almost all of the gases produced by a fire are toxic to the body, including carbon monoxide, hydrogen cyanide, and phosgene.

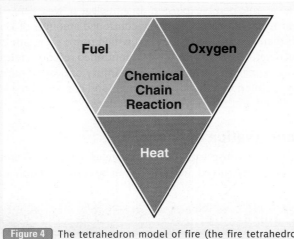

Figure 4 — The tetrahedron model of fire (the fire tetrahedron) includes chemical chain reactions as an essential part of the combustion process.

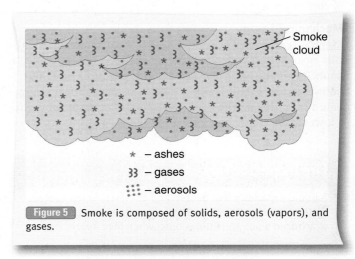

Figure 5 — Smoke is composed of solids, aerosols (vapors), and gases.

Carbon monoxide is deadly in small quantities and was used to kill people in gas chambers. Hydrogen cyanide was also used to kill convicted criminals in gas chambers. Phosgene gas was used in World War I as a poisonous gas to disable soldiers. Even carbon dioxide, which is an inert gas, can displace oxygen and cause <u>hypoxia</u>. Hypoxia is a state of inadequate oxygenation of the blood and tissue.

A discussion of the by-products of combustion would be incomplete without considering heat. Because smoke is the result of fire, it is hot. The temperature of smoke will vary depending on the conditions of the fire and the distance the smoke travels from the fire. Injuries from smoke may occur because of the inhalation of the particles, droplets, and gases that make up smoke. The inhalation of superheated gases in smoke may also cause injury and burns in the respiratory tract. Intense heat from smoke can also severely burn the skin.

■ Fire Spread

Fires grow and spread by three primary mechanisms: conduction, convection, and radiation.

Conduction

<u>Conduction</u> is the process of transferring heat through matter by movement of the kinetic energy from one particle to another `Figure 6`. Conduction transfers energy directly from one molecule to another, much as a billiard ball transfers energy from one ball to the next. Objects vary in their ability to conduct energy. Metals generally have a greater ability to conduct heat than wood does, whereas a substance such as fiberglass (which is used for insulation) has almost no ability to conduct heat or fire.

Objects that are good conductors tend to absorb heat and conduct it throughout the object. For example, heat applied to a steel beam will be readily conducted along the beam. Because the heat spreads out over the beam, the area of the steel beam to which the heat is being applied will not get as hot as it would if the beam were a poor conductor.

Applying heat to a poor conductor such as wood will result in the heat energy staying in the area of the wood to which the heat is applied. Because the wood has a fairly low ignition temperature, the wood will ignite. Heat applied to the wood acts on a small part of the wood; the heat is not conducted to other parts of the wood. By comparison, the same amount of heat energy applied to the steel beam will result in less intense heating at the point where the heat is applied because much of the heat dissipates to other parts of the steel beam.

This behavior has serious consequences for fire inspectors: If two substances have the same ignition temperature, it will be easier to ignite the substance that is a poor conductor. The most important fact to remember about conductivity and fire spread is that poor conductors may ignite more easily but, once ignited, do not spread fire through conduction. Materials that are good conductors are rarely the primary means of spreading a fire.

Convection

<u>Convection</u> is the circulatory movement that occurs in a gas or fluid with areas of differing temperatures owing to the variation of the density and the action of gravity `Figure 7`. To understand the movement of gases in a fire, consider a container of water being heated on a stove. If you put a drop of food coloring in one part of the water and start to heat the water, you can readily see the circulation of the water within the container. As the water heats up, it expands and becomes lighter than the surrounding water. This causes a column of water to rise to the top; the cooler water then falls to the bottom. Thus a cycle of warmer water rises to the top, pushing the cooler water to the bottom. The convection currents have the same effect of a small pump pushing the water around in a set pathway.

The convection currents in a fire involve primarily gases generated by the fire. The heat of the fire warms the gases and particles in the smoke. A large fire burning in the open can generate a <u>plume</u> of heated gases and smoke that rises high in the air. This convection stream can carry smoke and large bands of burning fuel for several blocks before the gases cool and fall back to the earth. If winds are present during a large fire, they may influence the direction in which the convection currents travel.

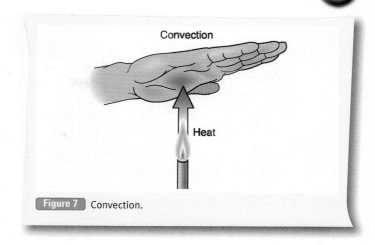

Figure 7 Convection.

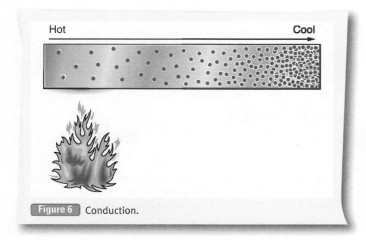

Figure 6 Conduction.

Fire Inspector Tips

The vertical movement of the heated gases is commonly referred to as mushrooming due to the shape of the smoke. When the hot, burning gases travel horizontally along the ceiling, it is commonly referred to as rollover or flameover.

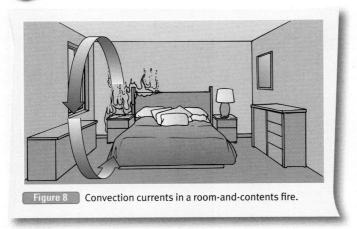

Figure 8 Convection currents in a room-and-contents fire.

large logs. The growth of a fire in a wastebasket to a room-and-contents fire is due in part to the effect of thermal radiation. A building that is fully involved in fire radiates a tremendous amount of energy in all directions. Indeed, the radiant heat from a large building fire can travel several hundred feet to ignite an unattached building.

When a fire occurs in a building, the convection currents generated by the fire rise in the room and travel along the ceiling. These currents carry superheated gases, which may ultimately heat flammable materials enough to ignite them Figure 8 . If the fire room has openings, convection may then carry the fire outside the room of origin and to other parts of the building.

Radiation

Radiation is the transfer of heat through the emission of energy in the form of invisible waves Figure 9 . The sun radiates energy to the earth over the vast miles of outer space; the electromagnetic radiation readily travels through the vacuum of space. When this energy is absorbed, it become converted to heat—that is, the heat you feel as the sun touches your body on a warm day. Of course, the direction in which the radiation travels can be changed or redirected, as when a sheet of shiny aluminum foil reflects the sun's rays and bounces the energy in another direction.

Thermal radiation from a fire travels in all directions. The effect of thermal radiation, however, is not seen or felt until the radiation strikes an object and heats the surface of the object. Thermal radiation is a significant factor in the growth of a campfire from a small flicker of flame to a fire hot enough to ignite

■ Methods of Extinguishment

Although there are many variations on the methods used to extinguish fires, they boil down to four main methods: cooling the burning material, excluding oxygen from the fire, removing fuel from the fire, and interrupting the chemical reaction with a flame inhibitor Figure 10 . There are many variations on the way that these methods can be implemented, and sometimes a combination of these methods is used to achieve suppression of fires.

The method most commonly used to extinguish fires is to cool the burning material. Pouring water onto a fire is one of the most common ways that fire fighters decrease fire temperature.

A second method of extinguishing a fire is to exclude oxygen from the fire. One example is placing the lid on an unvented charcoal grill. Likewise, applying foam to a petroleum fire excludes oxygen from the burning fuel.

Removing fuel from a fire will extinguish the fire. For example, shutting off the supply of natural gas to a fire being fueled by this gas will extinguish the fire. In wildland fires, a firebreak cut around a fire puts further fuel out of the fire's reach.

The fourth method of extinguishing a fire is to interrupt the chemical reaction with a flame inhibitor. Halon-type fire extinguishers, often used to protect computer systems, work in this way. These agents can be applied with portable extinguishers or through a fixed system designed to flood an enclosed space.

Figure 9 Radiation.

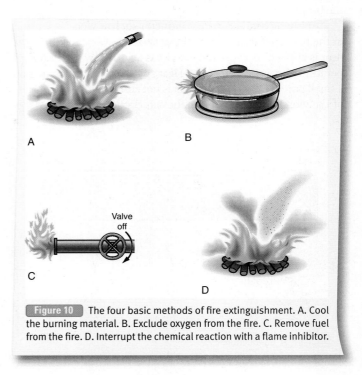

Figure 10 The four basic methods of fire extinguishment. A. Cool the burning material. B. Exclude oxygen from the fire. C. Remove fuel from the fire. D. Interrupt the chemical reaction with a flame inhibitor.

Figure 11 A Class A fire involves wood, paper, or other ordinary combustibles.

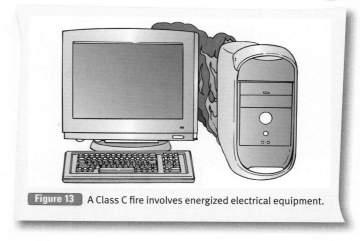

Figure 13 A Class C fire involves energized electrical equipment.

Classes of Fire

Fires are generally categorized into one of five classes: Class A, Class B, Class C, Class D, and Class K. Any structure may be at risk for any of the classes of fire depending upon the type of construction, occupancy, and materials or process in use in the structure.

Class A Fires

Class A fires involve solid combustible materials such as wood, paper, and cloth Figure 11 . Natural vegetation such as the grass that burns in brushfires is also considered to be part of this group of materials. The method most commonly used to extinguish Class A fires is to cool the fuel with water to a temperature that is below the ignition temperature.

Class B Fires

Class B fires involve flammable or combustible liquids such as gasoline, kerosene, diesel fuel, and motor oil Figure 12 . Fires involving gases such as propane or natural gas are also classified as Class B fires. These fires can be extinguished by shutting off the supply of fuel or by using foam to exclude oxygen from the fuel.

Class C Fires

Class C fires involve energized electrical equipment Figure 13 . They are listed in a separate class because of the electrical hazard they present. Attacking a Class C fire with an extinguishing agent that conducts electricity can result in injury or death to the fire fighter. Once the power is cut to a Class C fire, the fire can be treated as a Class A or Class B fire depending on the type of material that is burning.

Class D Fires

Class D fires involve combustible metals such as sodium, magnesium, and titanium Figure 14 . These fires are assigned to a special class because the application of water to fires involving these metals will result in violent explosions. These fires must be attacked with special agents to prevent explosions and to achieve fire suppression.

Class K Fires

Class K fires involve combustible cooking oils and fats in kitchens Figure 15 . Heating vegetable or animal fats or oils in appliances such as deep-fat fryers can result in serious fires that are difficult to fight with ordinary fire extinguishers. Special Class K extinguishers are available to handle this type of fire.

A fire may fit into more than one class. For example, a fire involving a wood building could also involve petroleum products. Likewise, a fire involving energized electrical circuits—a Class C fire—might also involve Class A or Class B materials. If a fire involves live electrical sources, it should be treated as a Class C fire until the source of electricity has been disconnected.

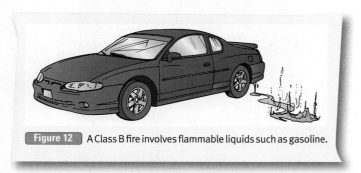

Figure 12 A Class B fire involves flammable liquids such as gasoline.

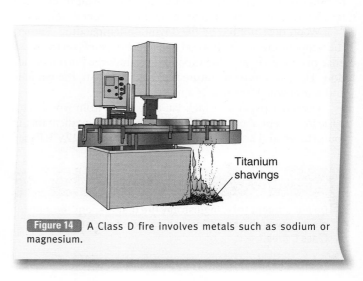

Figure 14 A Class D fire involves metals such as sodium or magnesium.

Figure 15 A Class K fire involves a deep-fat fryer in a kitchen.

Characteristics of Solid-Fuel Fires

This section uses wood as an example of a solid fuel because it is the most common solid fuel and because it shares many characteristics with other solid fuels.

Solid fuels have a definite form and a defined shape. Wood conducts little heat, so the heat acts only on its surface. As a result, the fuel reaches its ignition temperature quickly and ignites. A thin piece of wood burns quickly because of the large surface area exposed to the heat and air. The thinner the wood is cut, the faster it ignites and burns. Thin wood is exposed to heat from both sides, which explains why it is easier to ignite a matchstick than a large log.

Solid fuels do not actually burn in the solid state. Instead, a solid fuel must be instead heated or pyrolyzed to decompose it into a vapor before it will burn. Many solid fuels change directly from a solid to a gas; others change from the solid state to a liquid before they are vaporized.

Because solid fuels do not generally conduct heat, a heat source applied to the surface of the fuel will cause decomposition of only the surface of the material. The ignition of any solid fuel is related to the temperature of the surface of the material as well as the vapors released through pyrolysis. The temperature of the interior of the material, however, does not affect the speed with which the material can be ignited. If the fuel conducted heat to other parts of the fuel, it would take more heat to raise the temperature of the material to the ignition temperature.

Wood does not have a fixed ignition temperature. Rather, the temperature at which wood will ignite depends on the presence of enough flammable vapors to produce the burning reaction. The ignition temperature varies according to the manner and rate at which the heat is applied.

Consider paper as a fuel. Paper is made from wood. The basic ingredient of paper is cellulose, the same ingredient that is found in wood. The ignition temperature of paper is 425°F. Paper

Safety Tips

Do not attempt to extinguish any fire that is energized with electricity with any extinguishing agent not suitable for Class C fires.

Fire Inspector Tips

The thermal conductivity of wood varies with the species and the temperature. The thermal conductivity values for engineered wood is higher than for conventional wood.

Fire Inspector Tips

The configuration of the fuel determines how rapidly the fuel will yield its stored energy. An one pound log contains approximately 8000 BTUs. One pound of pine needles also contains approximately 8000 BTUs. Which fuel will release its heat fastest? This is called the rate of heat release. The material with the greatest surface area is heated more easily reaches its ignition temperature faster and is converted to gas faster. At the same time, oxygen in the air can readily combine with the flammable gas to create a combustible mixture and undergo combustion. The fire will be brief but severe.

ignites because its surface temperature increases rapidly, which allows for rapid burning and a high but brief heat release rate.

Solid-Fuel Fire Development

Solid-fuel fires progress through four phases: the ignition phase, the growth phase, the fully developed phase, and the decay phase. To illustrate the behavior of a fire through these phases, we will use a wooden dollhouse as an example. We will make three assumptions:

- The dollhouse is constructed on a sandy surface that will not burn.
- There are no exposures to which the fire can spread.
- There is no wind to influence the behavior of the fire.

Ignition Phase

The <u>ignition phase</u>, in which the fire is limited to its point of origin, begins as a lighted match is placed next to a crumpled piece of paper. The heat from the match ignites the paper, which sends a small plume of fire upward Figure 16 . The heat generated from the paper sets up a small convection current, and the flame produces a small amount of radiated energy. The combination of convection and radiation serve to heat the fuel around the paper. Because the convection acts more in an upward direction, the

Figure 16 The ignition phase of a fire.

> **Fire Inspector Tips**
>
> In a house fire, the growth phase is the point in the fire development that a smoke detector would operate to alert the occupants to evacuate. If sprinklers are installed, the sprinkler would operate, discharging water on the fire, either extinguishing the fire or controlling it until the fire department arrives.

kindling above the paper will be heated to the ignition temperature first. It is small, so its surface can be easily heated enough to release flammable vapors from the wood.

Growth Phase

The <u>growth phase</u>, when the fire spreads to nearby fuel, occurs as the kindling starts to burn, increasing the convection of hot gases upward **Figure 17**. The hot gases and the flame both act to raise the temperature of the wood located above the kindling. Energy generated by the growing fire starts to radiate in all directions. The convection of hot gases and the direct contact with the flame cause major growth to occur in an upward direction. The energy radiated by the flame will also cause some growth of the fire in a lateral direction. The major growth will be in an upward direction, however, because the radiation and convection both act in an upward direction, whereas growth beside the flame is primarily a function of radiation alone. The net result of the growth phase is that the fire grows from a tiny flame to a fire that involves the heavier wood.

Fully Developed Phase

The <u>fully developed phase</u> produces the maximum rate of burning **Figure 18**. All available fuel has ignited and heat is being produced at the maximum rate. At the fully developed phase, thermal radiation extends in all directions around the fire. The fire can spread downward by radiation and by flaming material that falls on unburned fuel. The fire can spread upward through the thermal column of hot gases and flame above the fire. In fact, a large free-burning fire can carry a thermal column several hundred feet into the air.

A fully developed fire burning outside is limited only by the amount of fuel available. By comparison, a fully developed fire in a building may be limited by the amount of oxygen that is available to it. The fully developed phase lasts as long as a large supply of fuel is available. The temperatures of the fire will vary from one part of the flame to another.

Decay Phase

The final phase of a fire is the <u>decay phase</u>, the period when the fire is running out of fuel **Figure 19**. During the decay phase, the rate of burning slows down because less fuel is available. The rate of thermal radiation decreases. The amount of hot gases rising above the fire also decreases because less heat is available to push the gases aloft and smaller amounts of gases are being produced by the decreasing volume of fire. Eventually, the flames become smoldering embers and the fire goes out because of a lack of fuel.

Key Principles of Solid-Fuel Fire Development

Several important principles of fire behavior are illustrated by the dollhouse fire:

- Hot gases and flame are lighter and tend to rise.
- Convection is the primary factor in spreading the fire upward.
- Downward spread of the fire occurs primarily from radiation and falling chunks of flaming material.
- If there is not more fuel above or beside the initial flame that can be ignited by convection or radiated heat, the fire will go out.
- Variations in the direction of upward fire spread will occur if air currents deflect the flame.
- The total material burned reflects the intensity of the heat and the duration of the exposure to the heat.
- An adequate supply of oxygen must be available to fuel a free-burning fire, although some parts of the flame may have a limited supply of oxygen.

The dollhouse example focused on a relatively simple form of fire. Fire behavior becomes more complex, of course, when fires occur within an actual building.

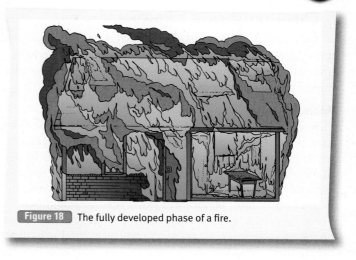
Figure 18 The fully developed phase of a fire.

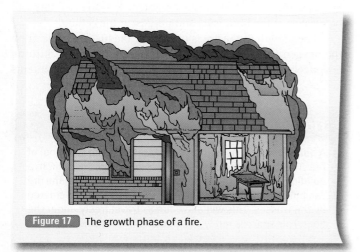
Figure 17 The growth phase of a fire.

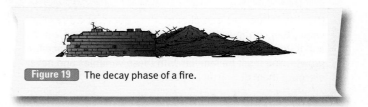
Figure 19 The decay phase of a fire.

Characteristics of a Room-and-Contents Fire

A fire in a building is not just a fire brought inside. The construction of the building and the contents of the room have a major impact on the behavior of the fire.

Room Contents

Fifty years ago, rooms contained many products made from natural fibers and wood. Today, synthetic products are all around us. This change in room contents has changed the behavior of fires. The synthetic products that are so prevalent today—especially plastics—are usually made from petroleum products. When heated, most plastics will pyrolyze to volatile products. These by-products of the combustion process are usually not only flammable, but also toxic. Some will melt and drip. If these droplets are on fire, they can contribute to the spread of the fire. Burning plastics generate dense smoke that is rich in flammable vapors, which makes fire suppression both more difficult and more dangerous.

Walls and ceilings are often painted. Newer paints are generally emulsions of latex, acrylics, or polyvinyl. When these paints are applied, they form a plastic-like coating. This coating has the characteristics of the plastic from which it was formed, and burns readily. Similarly, varnishes and lacquers are very combustible. Most paints and coatings will add to the fire present in a building, and their presence may aid in the spread of the fire from one room to another.

Carpets made today have low melting points and ignite readily. The backing of carpets is often polypropylene; polyurethane foam is commonly used as well. Carpet is readily ignitable by radiant heat, even when it is some distance away from a fire.

Furniture manufactured today has improved resistance to heat from glowing sources such as cigarettes, but it has almost no resistance to ignition from flaming sources. Such furniture can be completely involved in fire in 3 to 5 minutes, and it may be reduced to a burning frame in just 10 minutes.

In summary, the increased use of plastics in furniture, bedding, paints, and carpets sets the stage for complex fires. To see how such a fire might evolve, we will consider a fire in an ordinary bedroom that has painted walls and wall-to-wall carpeting. The room is furnished with a bed, two nightstands, a lamp, a vanity, and a wastebasket.

Ignition Phase

The ignition phase begins when a fallen candle ignites the contents crumpled in the plastic wastebasket sitting next to the vanity . The flame begins small, with a localized flame. This phase typically produces an open flame. The percentage

Figure 20 The ignition phase of a typical room-and-contents fire.

of oxygen present at this point is 21 percent—the same as the percentage in the room air.

As more of the contents in the wastebasket are ignited, a plume of hot gases rises from the wastebasket. Combustible materials in the path of the flame begin to ignite, which increases the extent and intensity of the flame. The convection of hot gases is the primary means of fire growth. Some energy is radiated to the area close to the flame, so that the plastic wastebasket starts to melt and may ignite. Oxygen is drawn in at the bottom of the flame above the wastebasket. At this point, the fire could probably be extinguished with a portable fire extinguisher.

Growth Phase

During the growth phase, additional fuel is drawn into the fire **Figure 21**. As more fuel is ignited, the size of the fire increases. The plume of hot gases and flames creates a convection current that carries hot gases to the ceiling of the room. Next, flammable materials in the path of this plume ignite. Flames begin to spread upward and outward. Hot gases and smoke rise because they are lighter; they hit the ceiling and spread out to form a layer at the ceiling. These products of combustion are trapped in the room and continue to affect the growth of the fire, unlike in a free-burning fire that occurs outdoors. Radiation starts to play a greater role in the growth of the room-and-contents fire.

The temperature of the room continues to increase as the fire grows. The room temperature is highest at the ceiling and lowest at the floor level. As the fire intensifies, the visibility will be greatest at the floor level and poorest at the ceiling.

The growth of the fire is limited by the fuel available or by the oxygen available. If the room in which the fire is burning is noncombustible, the only fuel available will be the contents

---

Fire Inspector Tips

Outside influences can change the course of a fire. Fuel value, fuel configuration, oxygen levels, drafts and winds, and humidity are just a few variables that can influence fire growth. You must be familiar with these variables and never discount them during an inspection.

Fire Inspector Tips

A burning wastebasket that melts creates a burning pool of fire. This type of fire deposits fuel directly on the floor. In a building with lightweight parallel chord floor trusses, the fire can burn through the floor rapidly and enter the void created by the trusses. This will allow fire to travel freely under the floor. This space is rarely equipped with sprinklers.

Figure 21 The growth phase of a typical room-and-contents fire.

Figure 22 The fully developed phase of a typical room-and-contents fire.

of the room. Likewise, if the doors and windows of the room are closed, a limited amount of oxygen will be present. Either of these conditions may limit the growth of the fire. If flames reach the ceiling, however, they are likely to trigger involvement of the whole room.

Fully Developed Phase

As the fire develops, temperatures increase to the point where the flammable materials in the room are undergoing pyrolysis. Large amounts of volatile gases are being released. If the temperature becomes high enough to ignite the materials in the room, a condition called <u>flashover</u> can occur. In flashover, the temperature in the room reaches a point where the combustible contents of the room ignite all at once. This temperature varies depending on the ignition temperature of the room contents. Flashover is the final stage in the process of fire growth.

Flashover is not a specific moment, but rather the transition from a fire that is growing by igniting one type of fuel to another, to a fire where all of the exposed fuel in the room is on fire. If a fire does not have enough ventilation to supply sufficient oxygen, it cannot flash over. If the ventilation openings are too large, the fire may not reach a temperature high enough to bring the whole room to the flashover point. The critical temperature for a flashover to occur is approximately 1000°F (538°C). Once this temperature is reached, all fuels in the room are involved in the fire, including the floor coverings on the floor. Indeed, the temperature at the floor level may be as high as the temperature at the ceiling was before flashover. Fire fighters, even with full personal protective equipment (PPE), cannot survive for more than a few seconds in a flashover.

Once the room flashes over, the fire is fully developed. All of the combustible materials are involved in the fire, and the burning fuels are releasing the maximum amount of heat. This condition is sometimes referred to as steady-state burning. The burning gases at the ceiling level radiate heat to combustible materials in the room. Carpets and other objects can ignite from this radiant heat; melting plastics can ignite and drop flaming particles onto materials below them. An average-sized furnished room with an open door will be completely involved in a flashover within 5 to 10 minutes of the ignition of the fire.

The amount of fire generated depends on the amount of oxygen available. For the fire to generate the maximum amount of heat, ventilation must be adequate to supply the fire with sufficient oxygen. If openings in the room let oxygen in, those same openings will serve as an escape route for hot gases. In this way, the hot gases can spread the fire outside the original fire room. The total fire damage to the room is the result of the intensity of the heat applied and the amount of time the room is exposed to the heat Figure 22 .

Decay Phase

The last phase of a fire is the decay phase Figure 23 . During the decay phase, open-flame burning decreases to the point where there is just smoldering fuel. A large amount of heat build-up means that the room will remain very hot, even though the amount of heat being produced decreases. Although a large amount of heat is available, the amount of fuel available to be pyrolyzed decreases, so a smaller amount of combustible vapors is available. The amount of fuel available dwindles as the supply of fuel becomes exhausted.

A fire in the decay phase will become a smoldering fire and eventually go out when the supply of fuel is exhausted. During this phase, the compartment temperature will start to decrease, though it may remain high for a long period of time. A smoldering fire may continue to produce a large volume of toxic gases and the compartment may remain dangerous even though the fire appears to be under control.

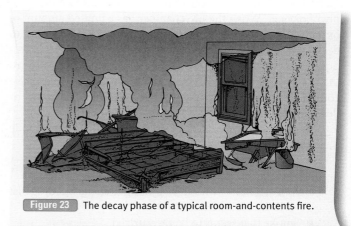

Figure 23 The decay phase of a typical room-and-contents fire.

Special Considerations

Flameover

<u>Flameover</u> (also known as <u>rollover</u>) is the flaming ignition of hot gases that are layered in a developing room or compartment fire. During the growth phase, the hot gases rise to the top of the room. As these gases cool, they fall and are pulled back into the fire by convection currents. If one of the gases in this mixture reaches its ignition temperature, it will ignite from the radiation from the flaming fire or from direct contact with open flames. In other words, the layer of flammable vapor catches fire. These flames can extend throughout the room at ceiling level. Flames can move at speeds of 10 to 15′ (305 to 457 cm) per second. The result of flameover is to increase the temperature in the room. Note that flameover is not the same as flashover, which was described earlier in this chapter.

Thermal Layering

<u>Thermal layering</u> is a property of gases such that the gases rise as they are heated and form layers within a room. The hotter gases are, the lighter they become owing to the decreased density and increased speed of their molecules. As a consequence, the hottest gases will travel by convection currents to the top of the room.

If this normal thermal balance is upset, severe injury can occur. When water is sprayed into the upper part of a room, for example, some of the water will be converted to steam. Steam takes up many times more space than the same amount of liquid water. The steam generated by the application of the water will flood the room and can result in severe burns to fire fighters even when they are in full PPE, because the ensuing thermal imbalance replaces the normal layering of gases with superheated steam. The energy in the superheated gases at the top of the room converts the water to steam, and the superheated steam expands and can fill the whole room with steam.

Backdraft

<u>Backdraft</u> is the sudden explosion of heated, oxygen-deprived fire gases when oxygen is reintroduced. The development of a backdraft requires a unique set of conditions. When a fire generates quantities of combustible gases, these gases can become heated above their ignition temperatures. Backdrafts require a "closed box"—that is, a room or building that has a limited supply of oxygen. When the fire chamber has a limited amount of oxygen, the oxygen concentration will be reduced, leading to decreased combustion. If a supply of new oxygen is then introduced into the room, explosive combustion can occur owing to the presence of the superheated gases. This kind of explosive combustion may exert enough force to cause great injury or death to fire fighters.

Signs and symptoms of an impending backdraft include the following:

- Any confined fire with a large heat build-up
- Little visible flame from the exterior of the building
- A "living fire," with smoke puffing from the building that looks like it is breathing
- Smoke that seems to be pressurized

Fire Inspector Tips

Since the mid 1980s, residential home design, construction materials and methods, and changes in the materials used to manufacture home furnishings have greatly impacted fire development, grow, ventilation, and suppression methods. The impact has been seen with an increase in rapidly developing hostile fires that flashover more quickly than homes built prior to the mid 1980s. The increase in flashover events required investigation, and in recent years, research and experiments conducted by the National Institute of Standards and Technology and Underwriters Laboratory has provided a better understanding of the factors contributing to these events. The result of the research and experiments related to room-and-contents fires concluded that fires in modern homes with modern furnishings can enter a new phase in fire development where the fire becomes dormant or enters the decay stage due to insufficient oxygen to sustain continued grow. This stage results in a ventilation-limited fire that typically has no visible flame or active burning. However, there is still significant fuel from the heat and room content that continues to pyrolize, producing fire gases and smoke. Once outside oxygen is introduced into a decay stage compartment, the result will be a rapid ventilation-induced flashover. As homes and furnishings continue to be built in this manner, more and more homes will have the potential for ventilation-induced flashover fires that will lead to revaluation of the traditional fire growth curve.

- Smoke-stained windows (an indication of a significant fire)
- Turbulent smoke
- Ugly yellowish smoke (containing sulfur compounds)

Characteristics of Liquid-Fuel Fires

Fires involving liquid or gaseous fuels have some different characteristics from fires that involve solid fuels. To understand the behavior of these fires, you need to understand the characteristics of the combustion of liquid and gaseous fuels.

Recall that solid fuels do not burn in the solid state, but instead must become heated and converted to a vapor combine with sufficient oxygen before they will undergo combustion. Liquids share the same characteristic: they must be converted to a vapor and be mixed in the proper concentration with air before they will burn. Three conditions must be present for a vapor and air mixture to ignite:

- The fuel and air must be present at a concentration within a flammable range.
- There must be an ignition source with enough energy to start ignition.
- The ignition source and the fuel mixture must make contact for long enough to transfer the energy to the air–fuel mixture.

As liquids are heated, the molecules in the material become more active and the speed of vaporization of the molecules increases. Most liquids will eventually reach their boiling points during a fire. As the boiling point is reached, the amount of flammable vapor generated increases significantly **Figure 24**.

Chapter 4 Fire Growth

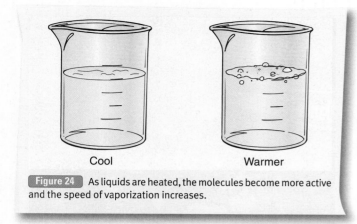

Figure 24 As liquids are heated, the molecules become more active and the speed of vaporization increases.

Because most liquid fuels are a mixture of compounds—for example, gasoline contains approximately 100 different compounds—the fuel does not have a single boiling point. Instead, the compound with the lowest ignition temperature determines the flammability of the mixture.

The amount of liquid that will be vaporized is also related to the <u>volatility</u> of the liquid. The higher the temperature, the more liquid that will evaporate. Liquids that have a lower molecular weight will tend to vaporize more readily than liquids with a higher molecular weight. As more of the liquid vaporizes, the mixture may reach a point where enough vapor is present in the air to create a flammable vapor–air mixture.

Two additional terms are used to describe the flammability of liquids: flash point and flame point **Table 1**. The <u>flash point</u> is the lowest temperature at which a liquid produces a flammable vapor. It is measured by determining the lowest temperature at which a liquid will produce enough vapor to support a small flame for a short period of time (the flame may go out quickly). The <u>flame point</u> (or <u>fire point</u>) is the lowest temperature at which a liquid produces enough vapor to sustain a continuous fire.

Characteristics of Gas-Fuel Fires

By learning about the characteristics of flammable gas fuels, fire inspectors can help to prevent injuries or deaths by identifying potential problems during a fire inspection. Two terms are used to describe the characteristics of flammable vapors: vapor density and flammability limits. For example, during an inspection, an odor of natural gas or other flammable gas or liquid requires investigation to assure there are no leaks, that proper venting exists, and that vapor limits are acceptable to prevent any fire from occurring.

Table 2 Vapor Density of Common Gases

Gaseous Substance	Vapor Density
Gasoline	>3.0
Ethanol	1.6
Methane	0.6
Propane	1.6

▮ Vapor Density

<u>Vapor density</u> refers to the weight of a gas fuel and measures the weight of the gas compared to air **Table 2**. The weight of air is assigned the value of 1. A gas with a vapor density of less than 1 will rise to the top of a confined space or rise in the atmosphere. For example, hydrogen gas, which has a vapor density of 0.07, is a very light gas. A gas with a vapor density greater than 1 is heavier than air and will settle close to the ground. For example, propane gas has a vapor density of 1.51 and will settle to the ground. By comparison, carbon monoxide has a vapor density of 0.97—almost the same as that of air—so it mixes readily with all layers of the air.

▮ Flammability Limits

Mixtures of flammable gases and air will burn only when they are mixed in certain concentrations. If too much fuel is present in the mixture, there will not be enough oxygen to support the combustion process; if too little fuel is present in the mixture, there will not be enough fuel to support the combustion process. The range of gas–air mixtures that will burn varies from one fuel to another. Carbon monoxide will burn when mixed with air in concentrations between 12.5 percent and 74.0 percent. By contrast, natural gas will burn only when it is mixed with air in concentrations between 4.5 percent and 15.0 percent.

<u>Flammability limits</u> (or <u>explosive limits</u>) are the highest and lowest concentrations for a particular gas in air to be ignitable. These terms are used interchangeably because under most conditions, if the flammable gas–air mixture will not explode, it will not ignite. The <u>lower explosive limit (LEL)</u> refers to the minimum amount of gaseous fuel that must be present in a gas–air mixture for the mixture to be flammable or explosive. In the case of carbon monoxide, the lower explosive limit is 12.5 percent. The <u>upper explosive limit (UEL)</u> of carbon monoxide is 74 percent. Test instruments are available to measure the percentage of fuels in gas–air mixtures and to determine when an emergency scene is safe.

▮ Energy Required for Ignition of Vapors

The principal sources of ignition of flammable liquids include flames; electrical, static, or frictional sparks; and hot surfaces. Flames must be capable of heating the vapor to its ignition temperature in the presence of air in order to be a source of ignition. Electrical sparks from commercial electrical supply installations may be hot enough to ignite flammable mixtures. Hot surfaces

Table 1 Flash Point and Flame Point

Fuel	Flash Point (°F)	Flash Point (°C)
Gasoline	−45	−43
Ethanol	54	12
Diesel fuel	104–131	40 to 55
SAE N 10 motor oil	340	171

Source: Principles of Fire Protection Chemistry and Physics. NFPA, 1998, p. 115.

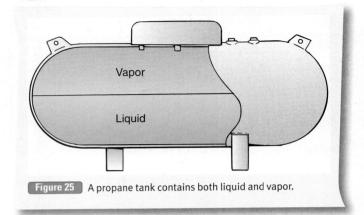

Figure 25 A propane tank contains both liquid and vapor.

Figure 26 Interior design is now regulated by the fire code.

can be a source of ignition if they are large enough and hot enough. The smaller the heated surface, the hotter it must be to ignite a mixture. The larger the heated surface in relation to the mixture, the more rapidly ignition will take place and the lower the temperature necessary for ignition.

Boiling Liquid, Expanding-Vapor Explosions

One potentially deathly set of circumstances involving liquid and gaseous fuels is the so-called <u>boiling-liquid, expanding-vapor explosion (BLEVE)</u>. A BLEVE occurs when a liquid fuel is stored in a vessel under pressure. The vessel is partly filled with the liquid, and the rest of the vessel is occupied by the same compound in the form of a vapor. A propane tank is an example Figure 25.

If this sealed container is subjected to heat from a fire, the pressure that builds up from the expansion of the liquid will prevent the liquid from evaporating. Normally evaporation cools the liquid and allows it to maintain its temperature. As heating continues, the temperature inside the vessel may exceed the boiling point of the liquid. The vessel can then fail, releasing all of the heated fuel in a massive explosion. The released fuel instantly becomes vaporized and ignited as a huge fireball.

The key to preventing a BLEVE is to cool the top of the tank, which contains the vapor. This action will prevent the fuel from building up enough pressure to cause a catastrophic rupture of the container.

Fire Growth

Fire growth and spread are greatly influenced by building construction. Whether or not a compartment goes to flashover, for example, is greatly influenced by the materials lining the walls and ceilings of the space. What follows is a description of these hazards and some of the fire protection methods employed to deal with them. It is important for you to be able to identify such potential situations and initiate appropriate actions.

Flame spread, or the more accurate description, fire growth, is a particularly hazardous fire phenomenon. On November 28, 1942, rapid fire growth was responsible for 492 deaths in the Cocoanut Grove nightclub fire in Boston Figure 26. At the time, flammable decorations were blamed for the rapid spread of the fire. Inadequate exits and overcrowding were blamed for the monstrous death toll. Pictures from the fire clearly show telltale globs of burned adhesives used to glue highly combustible acoustical tile to the ceiling. After other deadly 1940s fires with rapid flame spread in a soldier's hostel in Newfoundland and in Mercy Hospital in Iowa, fire officials began to understand that there was a problem with some building materials.

Building or Contents Hazard

Flame spread, or rapid fire growth, can be a problem caused by both the building itself and its contents. The fire growth building problem can be differentiated by a location characteristic:
- Hidden
- Exposed

The fire growth contents problem can result from:
- Furnishings
- Interior finish, including decorations
- Furnishings
- Building construction elements

In addition to fire growth, high flame-spread materials may contribute heavily to the fire load and to the generation of smoke and toxic products. These are additional threats to the life safety of both occupants and fire fighters.

Hidden Building Elements

A wide variety of materials and hidden building elements contribute greatly to rapid fire growth. For example, batt (or paneled) insulation laid in ceilings must be kept free of light fixtures because the heat from the fixture can ignite the paper vapor seal, igniting a fire inside of the ceiling. Any renovation may result in prior materials of construction being covered over by the new materials of construction. It is not uncommon for paneling or wall board to cover combustible insulation or dropped noncombustible ceiling materials and even exposed wiring. A vigilant fire inspector pays particular attention to these situations.

Combustible fiberboard is commonly used as insulating sheathing on wood frame buildings. It is also used as soundproofing. This material can support a fire hidden in the walls.

Foamed-plastic insulation is also used as sheathing, concealed in cavity walls, and glued to the interior surface of masonry wall panels. Foamed plastic applied to walls and ceilings for insulation has been involved in many disastrous fires. When such insulation is installed, it should be protected from exposure to flame by a 0.5″ (12.7 mm) gypsum board covering. This protects only against ignition from a small source. In a well-developed fire, as the gypsum board fails, the plastic will be involved, possibly suddenly and explosively as the gypsum fails or falls away.

Foamed plastic may be manufactured so that its flame spread is reduced, but it still can melt. It also lacks dimensional stability. When the plastic is used structurally, this may lead to disaster.

Air-duct insulation commonly installed years ago was usually made of a hair felt with a high flame spread. The presently used aluminum-faced foil (not aluminum-faced paper), glass-fiber insulation presents little flame spread problem.

Electrical insulation may be self-extinguishing. However, the tests of this material are conducted on wire not under load. When electrical wiring is operated at or above its rated capacity, the heat can break down the insulation and flammable gases can be emitted. The McCormick Place fire in Chicago is thought to have started in the flaming insulation of an overloaded extension cord. Large groups of electrical wires can support self-sustaining ignition.

■ Interior Finish

There are three ways in which interior finishes increase a fire hazard:
- They increase fire extension by surface flame spread.
- They generate smoke and toxic gases.
- They may add fuel to the fire, contributing to flashover.

Common interior finishes include such materials as wallboard, wallpaper, lay-in ceiling tile, vinyl wall covering, and interior floor finish items, such as carpeting, in addition to low-density fiberboard and combustible tiles.

In all cases, only approved ceiling, wall and floor finishes are installed in structures and documented to be approved product. In some cases, (e.g. healthcare occupancies and educational occupancies), state licensure requirements have prerequisites for interior finishes in buildings. Most materials have a flamespread rating, classification of index.

Ceilings

Both combustible and non-combustible ceiling materials can be found. Older style combustible ceilings that may still be found, as well as wooden ceilings and even wood paneling has been found in older structures. Concrete and drywall/gypsum board construction is prevalent today which provides good limitation of fire growth unless combustible materials are used as cover to the ceiling, similar to walls (e.g. carpeting, tapestry, certain paints and sprayed on products.)

Walls

Interior walls may be found of gypsum board, some type of wood, paneling, concrete (usually painted over) or even covered with flammable or combustible products, e.g. tapestry, paint, garments, etc. Most non-wood wall assemblies will not quickly contribute to flame spread or fire growth. However, those finishes which have an easily ignitable surface that can sustain fire growth present a problem and require action by the fire inspector.

Floors

Floor finishes may range from non-combustible products, such as concrete to surfaces to coverings that can sustain fire growth (such as floor tile) to carpeting which can support combustion (wood etc.) due to the use of combustible padding and glue under the carpet as well as the materials of the carpet itself.

Decorations

Present-day decorations and building contents present a major fire hazard. At one time, a popular decoration that provided a real flame spread or rapid fire growth problem was the Christmas tree. Although there are still tragedies during Christmas season, the incidence of such fires has been reduced. Regardless of precautions, the amount of decorations in places of assembly should be strictly limited to reduce the potential volume of fuel for a fire.

Decorations may range from combustible wall hangings to items hanging from automatic fire sprinklers or other ceiling attachments. Being observant is your greatest ally as a fire inspector. If something looks like it is out of place, it probably is, and an appropriate query and action is justified.

Decorations, usually hanging overhead and/or on the walls are very difficult to control from a fire prevention perspective. There are few applicable regulations. This is typically a generic provision in the code requiring limitations on decorative hangings. This leaves much discretion to the fire inspector and suggests the fire inspector act in a judicious and fair manner when finding decorations. Decorations often represent a serious problem because the hazard goes unrecognized even by people who consider themselves fire conscious.

> **Fire Inspector Tips**
>
> Materials used as interior finish must be assigned a flamespread rating. Ratings range from Class A (0 to 25), Class B (26 to 75), and Class C (76 to 200).

> **Fire Inspector Tips**
>
> The use of carpeting has changed in recent years. It is now also used on walls and ceilings. It is not always evident to designers or even to fire officials that the location where a material is installed may increase the rate at which a fire can grow. A spectacular 1980 fire in the Las Vegas Hilton Hotel spread from floor to floor outside the building because of flammable carpeting on the walls and ceilings of the elevator lobbies. The fire grew into one terrifying fire front extending many stories in height, and claimed eight lives.
>
> It has become commonplace for daycare centers to use carpeting on the walls. Unless the carpeting has achieved the appropriate interior finish fire rating, it must be removed.

Furnishings

Furnishings (drapes, curtains, decorations, chairs, couches, bedding, etc.) all may contribute to fire growth. In general, furnishings (what is in a room or structure) provide fuel for the fire. Their potential to contribute to fire growth is directly related to their materials of construction, their potential to interact with a heat source that can raise its temperature to an ignition point and sustain fire development. Almost any furnishing can burn so fire prevention is the foundation of limiting fire development and fire growth.

Building Construction Elements

Building construction elements may be fire resistive, non-combustible, semi-fire resistive, or combustible. Needless to say, if combustible they have the ability to contribute to the fire load. Depending upon the method of construction, the fire may extend quickly or be contained. Meeting code requirements is important to look for to assure construction limits fire growth or to advise the fire department that full involvement warrants fire department defensive operations.

Flame Spread and Smoke Development Ratings

Flame spread and related smoke development is the movement of fire beyond the burning area and igniting adjacent material, all the while developing products of combustion (smoke) as part of the process. Flame spread and smoke development ratings are based upon testing methodologies established by the American Society for Testing and Materials (ASTM), Test Method E-84. This is commonly known as the "tunnel test." The testing is typically conducted by Underwriters Laboratories or some similar third party testing facility. Each material specification sheet available from the manufacturer has the related rating information on it.

Wrap-Up

Chief Concepts

- Matter exists in three states: solid, liquid, and gas.
- Energy exists in many forms, including chemical, mechanical, and electrical.
- Conditions necessary for a fire include fuel, oxygen, heat, and a self-sustaining reaction.
- Fire may be spread by conduction, convection, and radiation.
- The four principal methods of fire extinguishment are cooling the fuel, excluding oxygen, removing the fuel, and interrupting the chemical reaction.
- Fires are categorized as Class A, Class B, Class C, Class D, and Class K. These classes reflect the type of fuel that is burning and the type of hazard that the fire represents.
- Solid-fuel fires develop through four phases: the ignition phase, the growth phase, the fully developed phase, and the decay phase.
- The growth of room-and-contents fires depends on the characteristics of the room and the contents of the room.
- Special considerations related to room-and-contents fires include flameovers, the thermal layering of gases, and backdrafts.
- Liquid-fuel fires require the proper mixture of fuel and air, an ignition source, and contact between the fuel mixture and the ignition.
- The characteristics of gas-fuel fires are different from the characteristics of other types of fires.
- Vapor density reflects the weight of a gas compared to air.
- Flammability limits vary widely for different fuels.
- A boiling-liquid, expanding-vapor explosion (BLEVE) is a catastrophic explosion in a vessel containing a boiling liquid and a vapor.
- Fire growth and spread are greatly influenced by building construction. Whether or not a compartment goes to flashover, for example, is greatly influenced by the materials lining the walls and ceilings of the space.

Hot Terms

Backdraft The sudden explosive ignition of fire gases when oxygen is introduced into a superheated space previously deprived of oxygen.

Boiling-liquid, expanding-vapor explosion (BLEVE) An explosion that occurs when a tank containing a volatile liquid is heated.

Chemical energy Energy that is created or released by the combination or decomposition of chemical compounds.

Class A fires Fires involving ordinary combustible materials, such as wood, cloth, paper, rubber, and many plastics.

Class B fires Fires involving flammable and combustible liquids, oils, greases, tars, oil-based paints, lacquers, and flammable gases.

Class C fires Fires that involve energized electrical equipment, where the electrical conductivity of the extinguishing media is of importance.

Class D fires Fires involving combustible metals such as magnesium, titanium, zirconium, sodium, and potassium.

Class K fires Fires involving combustible cooking media such as vegetable oils, animal oils, and fats.

Conduction Heat transfer to another body or within a body by direct contact.

Convection Heat transfer by circulation within a medium such as a gas or a liquid.

Decay phase The phase of fire development in which the fire has consumed either the available fuel or oxygen and is starting to die down.

Electrical energy Heat that is produced by electricity.

Endothermic Reactions that absorb heat or require heat to be added.

Exothermic Reactions that result in the release of energy in the form of heat.

Fire A rapid, persistent chemical reaction that releases both heat and light.

Wrap-Up

Fire tetrahedron A geometric shape used to depict the four components required for a fire to occur: fuel, oxygen, heat, and chemical chain reactions.

Fire triangle A geometric shape used to depict the three components of which a fire is composed: fuel, oxygen, and heat.

Flameover (rollover) A condition in which unburned products of combustion from a fire have accumulated in the ceiling layer of gas to a sufficient concentration (i.e., at or above the lower flammable limit) such that they ignite momentarily.

Flame point (fire point) The lowest temperature at which a substance releases enough vapors to ignite and sustain combustion.

Flammability limits (explosive limits) The upper and lower concentration limits (at a specified temperature and pressure) of a flammable gas or vapor in air that can be ignited, expressed as a percentage of the fuel by volume.

Flashover A condition in which all combustibles in a room or confined space have been heated to the point at which they release vapors that will support combustion, causing all combustibles to ignite simultaneously.

Flash point The minimum temperature at which a liquid or a solid releases sufficient vapor to form an ignitable mixture with the air.

Fuel All combustible materials. The actual material that is being consumed by a fire, allowing the fire to take place.

Fully developed phase The phase of fire development in which the fire is free-burning and consuming much of the fuel.

Gas One of the three phases of matter. A substance that will expand indefinitely and assume the shape of the container that holds it.

Growth phase The phase of fire development in which the fire is spreading beyond the point of origin and beginning to involve other fuels in the immediate area.

Hypoxia A state of inadequate oxygenation of the blood and tissue.

Ignition phase The phase of fire development in which the fire is limited to the immediate point of origin.

Ignition temperature The minimum temperature at which a fuel, when heated, will ignite in air and continue to burn.

Liquid One of the three phases of matter. A nongaseous substance that is composed of molecules that move and flow freely and that assumes the shape of the container that holds it.

Lower explosive limit (LEL) The minimum amount of gaseous fuel that must be present in the air mixture for the mixture to be flammable or explosive.

Matter Made up of atoms and molecules.

Mechanical energy Heat that is created by friction.

Plume The column of hot gases, flames, and smoke that rises above a fire. Also called a convection column, thermal updraft, or thermal column.

Radiation The combined process of emission, transmission, and absorption of energy traveling by electromagnetic wave propagation between a region of higher temperature and a region of lower temperature.

Smoke An airborne particulate product of incomplete combustion that is suspended in gases, vapors, or solid or liquid aerosols.

Solid One of the three phases of matter. A substance that has three dimensions and is firm in substance.

Thermal column A cylindrical area above a fire in which heated air and gases rise and travel upward.

Thermal layering The stratification (heat layers) that occurs in a room as a result of a fire.

Thermal radiation How heat transfers to other objects.

Upper explosive limit (UEL) The maximum amount of gaseous fuel that can be present in the air mixture for the mixture to be flammable or explosive.

Vapor density The weight of an airborne concentration (vapor or gas) as compared to an equal volume of dry air.

Volatility The ability of a substance to produce combustible vapors.

Fire Inspector *in Action*

You are partnering with the fire suppression team of the fire department to assist in creating a pre-fire plan by inspecting a freight terminal. This freight terminal handles cargo ranging from lumber to large animal feed to manufacturing chemicals. When you arrive at the freight terminal, you are handed a loose leaf book containing Safety Data Sheets (SDS) for all of the potentially hazardous materials that are shipped into and out of the terminal. The bills of laden and shipping labels are also made available to you.

1. There is a storage room where wooden pallets are stored. A fire in this room would be which class of fire?
    A. Class A fire
    B. Class C fire
    C. Class K fire
    D. Class B fire

2. You find an alcove where liquefied petroleum cylinders are being stored. This practice could present a hazard because:
    A. the tanks could be placed in front of an exit.
    B. if the tanks are involved in overheating, it could result in a BLEVE.
    C. they would be difficult to fill.
    D. the gas would be lighter than air.

3. You find batteries for material handling equipment being recharged in a small warehouse storage room. This causes generation of hydrogen gas which is explosive. What safety precautions need to be taken?
    A. Venting above the operation.
    B. Ventilation below the operation.
    C. None. The warehouse vents will take care of the gas.
    D. Fans should be installed to dissipate the gas in the warehouse.

4. Flashover in a compartment is influenced by which of the following?
    A. The total BTU content of the material involved
    B. The flamespread of the interior finish and contents
    C. The ambient temperature
    D. The ignition source of the fire

Performing an Inspection

CHAPTER 5

NFPA 1031 Standard

Fire Inspector I

4.2.4 Investigate common complaints, given a reported situation or condition, so that complaint information is recorded, the AHJ-approved process is initiated, and the complaint is resolved. (p 95)

(A) Requisite Knowledge. Applicable codes and standards adopted by the jurisdiction and policies of the jurisdiction. (p 95)

(B) Requisite Skills. The ability to apply codes and standards, communicate orally and in writing, recognize problems, and resolve complaints. (p 95)

4.3 Field Inspection. This duty involves fire safety inspections of new and existing structures and properties for construction, occupancy, fire protection, and exposures, according to the following job performance requirements. (pp 79–93)

4.3.1 Identify the occupancy classification of a single-use occupancy, given a description of the occupancy and its use, so that the classification is made according to the applicable codes and standards. (p 81)

(A) Requisite Knowledge. Occupancy classification types; applicable codes, regulations, and standards adopted by the jurisdiction; operational features; and fire hazards presented by various occupancies. (p 81)

(B) Requisite Skills. The ability to make observations and correct decisions. (p 81)

4.3.4 Verify the type of construction for an addition or remodeling project, given field observations or a description of the project and the materials being used, so that the construction type is identified and recorded in accordance with the applicable codes and standards and the policies of the jurisdiction. (pp 86–93)

(A) Requisite Knowledge. Applicable codes and standards adopted by the jurisdiction, types of construction, rated construction components, and accepted building construction methods and materials. (pp 86–93)

(B) Requisite Skills. The ability to read plans, make decisions, and apply codes and standards. (pp 86–93)

4.3.11 Inspect emergency access for an existing site, given field observations, so that the required access for emergency responders is maintained and deficiencies are identified, documented, and corrected in accordance with the applicable codes, standards, and policies of the jurisdiction. (p 87)

(A) Requisite Knowledge. Applicable codes and standards, the policies of the jurisdiction, and emergency access and accessibility requirements. (p 87)

(B) Requisite Skills. The ability to identify the emergency access requirements contained in the applicable codes and standards, observe, make decisions, and use measuring tools. (p 87)

4.3.14 Recognize a hazardous fire growth potential in a building or space, given field observations, so that the hazardous conditions are identified, documented, and reported in accordance with the applicable codes and standards and the policies of the jurisdiction. (pp 88–92)

(A) Requisite Knowledge. Basic fire behavior; flame spread and smoke development ratings of contents, interior finishes, building construction elements, decorations, decorative materials, and furnishings; and safe housekeeping practices.

(B) Requisite Skills. The ability to observe, communicate, apply codes and standards, recognize hazardous conditions, and make decisions. (pp 88–92)

4.3.15 Determine code compliance, given the codes, standards, and policies of the jurisdiction and a fire protection issue, so that the applicable codes, standards, and policies are identified and compliance is determined. (pp 79–95)

(A) Requisite Knowledge. Basic fire behavior; flame spread and smoke development ratings of contents, interior finishes, building construction elements, life safety systems, decorations, decorative materials, and furnishings; and safe housekeeping practices. (pp 79–95)

(B) Requisite Skills. The ability to observe, communicate, apply codes and standards, recognize hazardous conditions, and make decisions. (pp 79–95)

Fire Inspector II

5.2.3 Investigate complex complaints, given a reported situation or condition, so that complaint information is recorded, the investigation process is initiated, and the complaint is resolved in accordance with the applicable codes and standards and the policies of the jurisdiction. (pp 94–95)

(A) Requisite Knowledge. Applicable codes and standards adopted by the jurisdiction and policies of the jurisdiction. (pp 94–95)

(B) Requisite Skills. The ability to interpret codes and standards, recognize problems, and refer complaints to other agencies when required. (pp 94–95)

5.2.5 Recommend policies and procedures for the delivery of inspection services, given management objectives, so that inspections are conducted in accordance with the policies of the jurisdiction and due process of the law is followed. (p 95)

(A) Requisite Knowledge. Policies and procedures of the jurisdiction related to code enforcement as well as sources of detailed and technical information relating to fire protection and life safety. (p 95)

(B) Requisite Skills. The ability to identify approved construction methods and materials related to fire safety, read and interpret construction plans and specifications, educate, conduct research, make decisions, recognize problems, and resolve conflicts. (p 95)

5.3 Field Inspection. This duty involves code enforcement inspections and analyses of new and existing structures and properties for construction, occupancy, fire protection, and exposures, according to the following job performance requirements. (pp 79–93)

5.3.2 Identify the occupancy classifications of a mixed-use building, given a description of the uses, so that each area is classified in accordance with applicable codes and standards. (p 81)

(A) Requisite Knowledge. Occupancy classification, applicable codes and standards, operational features, and fire hazards presented by various occupancies. (p 81)

(B) Requisite Skills. The ability to interpret code requirements and recognize building uses that fall into each occupancy classification. (p 81)

5.3.10 Determine fire growth potential in a building or space, given field observations or plans, so that the contents, interior finish, and construction elements are evaluated for compliance, and deficiencies are identified, documented, and corrected in accordance with the applicable codes and standards and the policies of the jurisdiction. (pp 79–95)

(A) Requisite Knowledge. Basic fire behavior; flame spread and smoke development ratings of contents, interior finishes, building construction elements, decorations, decorative materials, and furnishings; and safe housekeeping practices. (pp 79–95)

(B) Requisite Skills. The ability to observe, communicate, interpret codes and standards, recognize hazardous conditions, and make decisions. (pp 79–95)

5.3.12 Verify code compliance of heating, ventilation, air conditioning, and other building service equipment and operations, given field observations, so that the systems and other equipment are maintained in accordance with applicable codes and standards and deficiencies are identified, documented, and reported in accordance with the policies of the jurisdiction. (p 91)

(A) Requisite Knowledge. Types, installation, maintenance, and use of building service equipment; operation of smoke and heat vents; installation of kitchen cooking equipment (including hoods and ducts), laundry chutes, elevators, and escalators; and applicable codes and standards adopted by the jurisdiction. (p 91)

(B) Requisite Skills. The ability to observe, recognize problems, interpret codes and standards, and write reports. (p 91)

Additional NFPA Standards

NFPA 13 *Standard for the Installation of Sprinkler Systems*
NFPA 101 *Life Safety Code*
NFPA 220 *Standard on Types of Building Construction*
NFPA 520 *Standard on Subterranean Space*
NFPA 555 *Guide on Methods for Evaluating Potential for Room Flashover*

FESHE Objectives

Principles of Code Enforcement

1. Describe the differences in how codes apply to new and existing structures. (p 92)
2. Identify appropriate codes and their relationship to other requirements for the built environment. (pp 79–95)

Knowledge Objectives

1. Describe the types of fire inspections.
2. Describe when the fire inspector should begin to inspect a building.
3. Describe the pre-inspection process.
4. Describe the fire inspection process.
5. Describe how to verify the proper installation and maintenance of heat, ventilation, and air conditioning systems; kitchen cooking equipment; laundry chutes; elevators; and escalators.
6. Describe how to verify that a new construction project meets all applicable codes, standards, and polices.
7. Describe how to verify that a remodeling project meets all applicable codes, standards, and polices.
8. Describe when and how to cite code violations.
9. Describe how to ensure that code violations are corrected.
10. Describe how to investigate a complaint against an occupancy and ensure that the complaint is resolved.
11. Describe the role of the Fire Inspector II in improving the inspection process.

Skills Objectives

1. Perform a fire inspection of a structure.

You Are the Fire Inspector

You receive a complaint from the manager of a jewelry store in your town. The business rents the property and the manager's complaint is that he recently observed the floor in the rear of the building to be sagging. You go to the building and observe three large safes in the rear of the store. The manager indicates that the safes have been in the business for the last ten years, and there have been no problems.

He also tells you that three months ago, he moved the safes from the side wall of the building to their present location in the rear of the business. Shortly after this move, he noticed what appeared to be sagging in the floor. The building is ordinary construction with the load-bearing walls on the left and right sides.

1. What is your legal and moral obligation now that you are aware of this hazardous condition?
2. What can you do immediately to protect life and property loss should the floor collapse?

Introduction

A fire inspection can reasonably ensure that a building will be safe for the occupants. The fire inspection is the primary objective of the fire inspector. The fire inspection process is performed to correct installation or construction problems. There are many points along the way that you, as the fire inspector, should be inspecting the building prior to its opening for business.

Before the building is constructed, many communities will have meetings where a potential developer will meet with the various departments of the municipality—such as building, fire, and zoning—to see if their project is even feasible. If it is, then some of the basic requirements for building a structure that meets the local codes and standards can be discussed. This type of informational meeting can lessen the chance that the building plans to be submitted will not meet local codes and standards.

Prior to the start of construction, many different building plans will have to be submitted and approved by the various departments of the municipality. The fire inspector and/or the plan reviewer will look at the plans submitted to that they submitted meet the local codes. Plan review is covered in detail in the chapter *Reading Plans*. While all of the information on the entire set of plans is important, some items of particular interest to you as a fire inspector are the occupant loads, travel distances to exits, number of exits, proper door swing, fire alarm protection, and fire sprinkler protection.

Types of Inspections

There are a number or basic or routine inspections that a fire inspector must perform. These include annual inspections, reinspections, complaint inspections, construction or final inspections, business license or change of occupancy inspections, and self-inspection. **Annual inspections** are inspections where you inspect the building because its turn has come up in the inspection cycle Figure 1 . Many agencies divide their jurisdiction into smaller inspection areas on a grid. Each area may be assigned a month, so when a month begins, the fire inspectors know that the buildings in that area must be inspected. Some states may also require biannual and quarterly inspections of certain occupancies.

Reinspections occur when code violations have been noted and you return to see if the owner is now compliant with the code. **Complaint inspections** occur when someone registers a concern of a possible code violation. You must then investigate the complaint to determine if it is valid. The immediacy of a complaint inspection is determined by the type of complaint. A locked exit door should be investigated and corrected immediately, while concerns about a fire extinguisher do not have the same urgency.

Construction or **final inspections** ensure compliance with fire code during and at the end of a building construction. These

Fire Inspector Tips

The fire inspector shall be involved with a building from pre-construction to as long as it is standing. Once the building is occupied the fire inspection process is really just a few steps—be prepared, be professional, be thorough, document, and follow up.

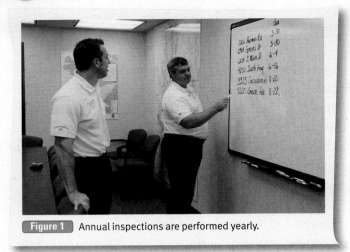

Figure 1 Annual inspections are performed yearly.

Figure 2 Construction or final inspections are conducted as a building is being constructed on specific building components including sprinkler systems.

Fire Inspector Tips

You should have a thorough knowledge of your agency's inspection schedule.

inspections evaluate elements including sprinkler systems, fire alarm systems, and fire pumps Figure 2 . **Business license** or **change of occupancy** inspections occur when the building department is notified of a new business requesting permission to open. Prior to issuing a business license or certificate of occupancy, the building department will work with the fire inspection agency to ensure that the occupancy is code compliant.

Finally there are **self-inspections**. These are not performed by a fire inspector but by the business owner. Self-inspections are usually performed on smaller buildings or when there are not enough fire inspectors to inspect all of the buildings in the jurisdiction. Many fire inspection agencies ask that a self-inspection form be submitted to the agency. Even if a self-inspection is performed annually, every a fire inspector must perform a professional inspection per the inspection schedule every few years.

When is the Best Time to Inspect?

You should become involved before the construction phase, long before the occupancy is scheduled to be opened. Inspections conducted during construction ensure that the portion(s) of the building being worked on are being constructed properly. For example, these inspections will verify that the sprinkler heads are actually connected to the sprinkler piping. Once the building is occupied, routine inspections are conducted to determine code compliance, as many circumstances can change after the occupancy is given final approval, such as locked exit doors, stock stored on the floor blocking the means of egress, or improper use of extension cords.

If the first inspection is conducted only days or hours before an opening of a new occupancy, considerable consternation between fire inspectors and building owners/occupants can occur, especially when violations are found that cannot be readily fixed. If a building owner or occupant is unable to comply with the code prior to the projected opening, the fire inspector has two options: one is to deny the opening until all violations are complied; the second is to open with a **conditional approval** which allows the business with minor, non-hazardous violations to open but requires that those violations be addressed. The issue of last minute violations can be avoided by performing various inspections as the building is being constructed. Even if an official inspection is not conducted during construction, walking through the job site just to become familiar with how the building is progressing is beneficial. You may be able to find and address future problems when they can be easily corrected.

Pre-Inspection Process

Before a fire inspection, you should review any previous fire inspection reports, correspondence, building modifications, and building plans and specifications, along with appropriate building department records and permits Figure 3 . Most of these items should be found in the building's inspection file and will give you an idea of what to expect prior to walking through the door. The building owner or the building department may be of assistance in providing as built diagrams or blue prints of the occupancy if they are not in the building's files. **As built diagrams** show how the final installation was actually completed. For example, it is common for a building owner to receive approval on a set of plans for a sprinkler system only to have a contractor make field adjustments during the installation process. When the contractor has completed his work, he should draw as built diagrams for the fire inspector showing the how the sprinkler system was actually installed.

You may also want to review the past occupancy records to develop a list of items that you should be especially aware of during your inspection. This may show a dangerous trend within the business, such as consistently locking the rear exit doors or blocking a means of egress with poor storage practices. Also perform a quick review of local codes and various reference books to refresh your memory on the specific occupancy-related hazards of the building you are going to inspect.

Figure 3 Before a fire inspection, you should review any previous fire inspection reports, correspondence, building modifications, and building plans and specifications, along with appropriate building department records and permits.

Fire Inspector Tips

Once the building has been approved for occupancy and is open for business, the building plans are filed in the event construction questions about the building come up during routine inspections that cannot be easily answered.

Classification of Construction and Occupancy

The type of building construction is critical in determining specific inspection requirements **Figure 4**. NFPA 220, *Standard on Types of Building Construction*, is one of the most widely used document to determine building construction. NFPA 220 lists five major categories of building construction. Each of these categories has hourly fire ratings given to various structural components such as exterior bearing walls, interior bearing walls, columns, beams, girders, trusses, floor-ceiling assemblies, roof-ceiling assemblies, and interior and exterior nonbearing walls. Each of these components has a fire rating between four hours to no rating needed. The chapter *Building Construction* discusses these categories in detail.

A Type I constructed building has structural members that are noncombustible or that contain limited combustible materials. Type II construction is also noncombustible. Type III construction allows for exterior member to be noncombustible, and interior members can contain combustible materials where allowed by the code. Type IV construction is typically known a heavy timber. This is where the lumber has dimensional sizes exceeding 4" (102 mm). Lastly Type V construction is ordinary wood frame—the typical single-family house.

Occupancies are easier to classify. Occupancy classification describes how the space is actually being used. Occupancy types are identified in the model codes, and the criteria for classifying an occupancy accompany the occupancy classifications. The occupancy identification of a particular building also may be included in previous fire inspection reports or citations.

Figure 4 Just looking at a building is no way to tell its construction materials or construction method. A building may have a fake brick façade.

However, if you rely on past fire inspection reports, make sure that the occupancy type has not changed since that report was written. Businesses within a building can change over years, changing the occupancy type. A greeting card store would be a mercantile occupancy, however if the space becomes a restaurant, then the occupancy becomes an assembly occupancy. Checking the building's reports will only tell you what occupancy type was last in the space.

Codes

It is critical to know which codes you can legally enforce in your jurisdiction. The mere fact that there are sets of model codes available does not give you the right to enforce those codes. The local jurisdiction must legally adopt a specific set of codes. This adoption is known as **enabling legislation**. It is the law that gives you the right to note any violations of the code and have them corrected.

Once a set of codes has been adopted, only those codes must be enforced. For example, if the 2006 edition of NFPA 1, *Fire Code*, was adopted by your jurisdiction, you must use that edition until the ordinance is changed. You would not be able to apply the 2009 edition of NFPA 1 simply because it is newer.

Fire Inspector Tips

Having a copy of the local codes in your office for reference and for a building owner to review is imperative.

Fire Inspector Tips

When checking on overcrowding in a place of assembly, the safety issue at hand trumps the owner's or occupant's wishes. The only way to verify overcrowding is during actual business operation. Advanced notice that there will be routine non-scheduled inspections for overcrowding and exiting related issues may go a long way to minimizing conflicts with owners.

When looking at the model codes, the first chapter is very important as it usually gives the scope of the code, lists the authority and responsibility of the fire inspector, discusses the appeals process, etc. Model codes are applied differently to existing buildings. Generally speaking, unless a building is renovated or modified, building codes are not applied retroactively.

Another item that should be considered is authority to inspect. Some states will require that to inspect public schools you must be certified by the state. There may also be state or local requirements that you must hold a minimum state or national certification to perform fire inspections. If that is the case, there are often requirements that your certification must be renewed every few years, which usually requires showing documentation of continuing education hours.

Tools and Equipment

In preparation for conducting a fire inspection, you should have a set of basic equipment. Additional specialty tools and equipment should be available for shared use by all fire inspectors in the agency. A list of the typical tools you should have access to includes:

- Writing tools—Pens, pencils, markers, and a clipboard to use as a writing surface and for form storage.
- Forms—A sufficient number of the various forms used should be carried, as you will never know what situation you may encounter. In recent years, more departments have been making a move away from paper reports and have been using handheld computers. These computers can allow the inspection to automatically sync up to the database at the department.
- Flashlight—A flashlight is used to check areas behind appliances, rooms where the lighting does not function, and spaces above ceilings Figure 5 .
- Measuring device—This can be used to determine room size, which is needed for determining proper occupant load. It will also be helpful to determine exiting concerns.
- Step ladder—Often the business will have shelving or machines which make observing all areas difficult. A small ladder may give you enough extra height to check these areas out.
- Pitot tube—A pitot tube is used on fire hydrants to determine the amount of water that is available for a building Figure 6 . This is a crucial piece of information fire sprinkler contractors need in order to design their systems. The local water department should be consulted prior to performing any of these tests. Many locales will state they must perform or witness the test, as shutting down a hydrant too quickly could rupture a water main.
- Calculator—This can be used to determine the size of the room, and to determine the occupant load.
- Camera—This tool is helpful to document some of the violations that are found. It verifies the violation, and documented examples can be used to train other fire inspectors. Additionally, many departments are taking pictures of all sides of the building and attaching them to the computer files. This is very helpful for the fire department during the preplanning phase. When computers are placed inside vehicles, pictures can be another resource for the incident commander. In addition, if the fire inspection report goes to court, pictures are proof of the condition of the building at the time the picture was taken.

Figure 5 A flashlight is used to check areas behind appliances, rooms where the lighting does not function, and spaces above ceilings.

Figure 6 A pitot tube is used on fire hydrants to determine the amount of water that is available for a building.

Fire Inspector Tips

Tools and equipment such as sound meters, light meters, and Pitot tubes should be professionally calibrated by trained technicians.

Fire Inspector Tips

Some tools are not necessary to carry on a daily basis. For example, a pitot tube is used to measure the water flow from a fire hydrant. This is not a test that will be performed at every inspection. Pitot tubes may be available in the office for use by several fire inspectors to share on an as needed basis.

Some tools, especially those that are more expensive to purchase or to maintain, such as gas monitors that require periodic calibration, are usually kept in the office for all to use. In most departments, fire inspectors return to the office on a daily basis making tool sharing quite feasible. However, in a state or county agency, fire inspectors may only return to a central office once or twice a week. This makes coordinating your inspection needs with the other fire inspectors in the office very important.

- Safety gear—Most work sites will require contractors to wear hard hats. Other sites may suggest gloves, goggles, or hearing protection. The fact that you are a fire inspector does not preclude you from following this common sense requirement.
- Electronic devices—Electronic tools may help you communicate, collect data, or reference codes or regulations.
- Coveralls—A layer of durable clothing will protect your work uniform from dust and dirt in areas such as commercial kitchens and woodworking facilities.

Standard Forms

The forms used during an inspection will vary from agency to agency. A set of standard forms should be used by all fire inspectors in the office. Some of the basics forms are:
- Inspection forms
- Final or construction inspection forms
- Complaint forms
- Stop work orders

An inspection form is used to note any violations found on the inspection or to note if no violations were found. The inspection form should consist of the following elements:
- A section showing the business name, address, and phone number.
- A section showing the date of the inspection.
- A section showing the area being inspected if the complex is large, if there are multiple buildings, or you are just inspecting one specific section.
- A section for listing any code violations with the specific code citations.
- A section for listing the reinspection date to remind the owner when you will be returning to check on code compliance.
- Areas for the signatures of the fire inspector and a representative of the owner. It is a good idea to also have a place where the names could be printed beneath the signatures.
- A section with a short legal statement stating the fire inspector's authority for the fire inspection is optional. The legal statement should be reviewed by your agency's legal staff.

There are two primary styles of inspection forms. One form is primarily blank lines on which you must write in a detailed description of any code violations. The second type of form is a checklist Figure 7 . The checklist is an organized way to verify compliance with certain codes and list violations, but this alone may not give the owner enough information to locate and correct problems. Ample space should be provided for you to provide additional details on the code violation. Another downside of a checklist is it encourages the tendency to only look for the items on the list, potentially missing other code violations.

A <u>complaint form</u> lists in detail any complaint that is lodged with the fire inspection agency and is investigated. This form should contain the date, time, and location of the business; signature spaces; and a record of the alleged violation. When possible, it should also state the name of person making the complaint and the person's contact information; however, the person may remain anonymous. There needs to be a section to indicate what was noted by the fire inspector when the complaint was investigated, as well as the date and time the inspection was performed. It should also provide a section to advise what corrective action was taken or recommended, and the date the complaint was closed.

<u>Final</u> or <u>construction inspection forms</u> are used when inspecting specialized systems such as fire alarm, sprinkler, hood and duct suppression systems, as well as for other types of construction phase inspections, such as a ceiling inspection or a final inspection. A standard inspection form can be used to note the needed details and whether or not the inspection passed; however, some agencies will utilize a form for these specialized inspections to denote that these inspections are different from a routine fire inspection. On the final or construction inspection forms there may be a series of check boxes to indicate the type of inspection being performed or there may be a specific inspection form for each type of inspection. Specific inspection forms will contain certain items that must be examined during the inspection, including the gauge pressures and gas valves functionality.

The <u>stop work order</u> is not used frequently and it must be used judicially. It should be used when contractors do not have the clearance for performing the work, or when the work is not correct and must be corrected prior to performing additional work. It is generally wise to consult with the building official and your supervisor about the stop work order, as the owner will not be happy and probably will seek to have it removed.

All of the forms should be in triplicate—one copy for the owner, one for the occupancy file, and the original for the

Inspection Checklist
Inspection Procedures

PREINSPECTION CHECKLIST

Equipment: _____

General
- ❏ Identification (photo ID)
- ❏ Business work hours

Clothing
- ❏ Coveralls
- ❏ Overshoes
- ❏ Boots

Personal Protective Equipment (PPE)
- ❏ Hard hat
- ❏ Safety shoes
- ❏ Safety glasses
- ❏ Gloves
- ❏ Ear protection
- ❏ Respiratory protection

Tools
- ❏ Flashlight
- ❏ Tape measure(s)
- ❏ Pad (graph paper) and pen or pencil
- ❏ Magnifying glass

Test gauges
- ❏ Combustible gas detector
- ❏ Pressure gauges
- ❏ Pitot tube or flow meter

Plans and Reports
- ❏ Previous reports
- ❏ Violation notices
- ❏ Previous surveys
- ❏ Applicable codes and standards

Notes: _____

SITE INSPECTION

Property Name: _____

Address: _____

Occupancy Classification
- ❏ Assembly
- ❏ Educational
- ❏ Day care
- ❏ Health care
- ❏ Ambulatory health care
- ❏ Detention and correctional
- ❏ One- and two-family dwelling
- ❏ Lodging and rooming
- ❏ Hotel/Motel/Dormitory
- ❏ Apartment
- ❏ Residential board and care
- ❏ Mercantile
- ❏ Business
- ❏ Industrial
- ❏ Storage
- ❏ Mixed

Copyright © 2002 National Fire Protection Association (Page 1 of 2)

Figure 7 A standard inspection form in the checklist format.

Hazard of Contents
- ❏ Light (low)
- ❏ Mixed
- ❏ Ordinary (moderate)
- ❏ Special hazards
- ❏ Extra (high)

Exterior Survey
- ❏ Housekeeping and maintenance

Building construction type
- ❏ Type I (fire resistive)
- ❏ Type IV (heavy timber)
- ❏ Type II (noncombustible)
- ❏ Type V (wood frame)
- ❏ Type III (ordinary)
- ❏ Mixed

Construction problems
Building height _____ feet _____ stories
- ❏ Potential exposures
- ❏ Outdoor storage
- ❏ Hydrants

Fire department connection
- ❏ Vehicle access
- ❏ Is it obstructed?
- ❏ Is it identified?
- ❏ Drainage (flammable liquid and contaminated runoff)
- ❏ Fire lanes marked

Building Facilities
- ❏ HVAC systems
- ❏ Gas distribution systems
- ❏ Conveyor systems
- ❏ Electrical systems
- ❏ Refuse handling systems
- ❏ Elevators

Fire Detection and Alarm Systems
See Form A-8.

Fire Suppresion Systems
See Form A-10.

Closing Interview
- ❏ Imminent fire safety hazards
- ❏ Housekeeping issues
- ❏ Maintenance issues
- ❏ Overall evaluation

Items to be researched:
- ❏ _____
- ❏ _____
- ❏ _____

Report
- ❏ Draft
- ❏ Review
- ❏ Final

Notes: _____

Figure 7 A standard inspection form in the checklist format (*Continued*).

> **Safety Tips**
>
> Fires can grow due to many hazardous conditions. While fires can start from many things, fire growth often times is related to housekeeping issues. Some of the more common issues include:
> - Storing stock too close to the sprinkler heads
> - Garbage cans overflowing
> - An accumulation of materials in furnace and water heater rooms
> - Storing boxes and mops too close to pilot lights on furnaces and water heaters
> - Storing flammable materials incorrectly

> **Fire Inspector Tips**
>
> Inspecting some occupancies during your normal business hours may not provide a true picture of what the business is like when fully operational. This is especially true of assembly occupancies such as nightclubs. Performing a full inspection of a nightclub at 11:00 pm might be difficult so it would be more prudent to conduct the full inspection during the day. Inspections should not disrupt business more than is absolutely necessary. This will allow for time to conduct a detailed inspection without disturbing the business operation; however, it may still be necessary to do a spot check during their normal business hours. Those spot checks should focus on occupant loads, visible exits, that the unlocked exit doors, and the clear means of egress.

> **Fire Inspector Tips**
>
> With few exceptions, inspection timetables are subject to each agency's policy. Some may state once a year, with target hazards being inspected twice a year. Local ordinances may specify the frequency of inspections. Additionally, occupancies such as schools, hospitals, and daycare centers may, through state law, require more frequent inspections. The best plan is to go through each building once a year. If that is not possible, begin with known life or high value target hazards in the community.

resinspection(s). The original is used to note the date on which the various code violations become compliant. It can also be used to document any conversations you had with the owner regarding the inspection and the related reinspections.

Scheduling and Introductions

There are two ways to begin the inspection—scheduled and unannounced. Each has its pluses and minuses. Scheduled inspections ensure that you will be able to fully conduct the fire inspection; however, it also allows the business to conduct its own pre-inspection and correct any usage violations prior to your arrival. The unannounced inspection allows you to get a "real" look at the business and understand how the business operates daily. A drawback to this inspection is the risk of being turned away due to an inability of the business to accommodate you at the specific time, wasting your trip.

While the codes give you the authority to conduct fire inspections, you are not allowed to do so without permission of the owner. Performing a fire inspection without permission is trespassing. The exception to this is an **exigent circumstance**, meaning an immediate life safety issue which requires immediate actions to be taken. Non-operating emergency lights or fire extinguishers with inspection tags out of date are not exigent circumstances. Locking emergency exits in a mall during the Christmas season is an exigent circumstance.

If permission to conduct the fire inspection is denied multiple times without valid explanation, then, as a last resort, you should request a court order to conduct the fire inspection. At times, working with the municipality provides the owner motivation to allow the fire inspection because the municipality often issues business licenses and often require compliance with the local codes.

Upon arrival, you must first seek permission from the owner, manager, or other individual with the authority to allow you to inspect the occupancy. A teenager working as a cashier may not have the authority to grant permission. Obtaining permission can be as simple as walking in to the business and asking to speak to the owner or manger. Often, the owner or manager will accompany you during the inspection. If not, he or she will typically assign another individual to assist you. It is always best to have a representative of the business along with you during the inspection. He or she can quickly answer any questions and can give you access to locked areas. Having a representative witness your inspection also lessens the possibility of you being falsely accused of an action.

Asking for permission, as opposed to demanding to make an inspection, will help develop good rapport with the owner. When permission is not granted, most often the occupancy will simply ask if you can come back at a time more convenient for them. This should be honored.

The Fire Inspection Process

The process of inspecting a building will vary with the occupancy. Occupancies vary in size and scope. Although some will be single-story buildings, others may have multiple buildings or multiple floors. Before the onsite inspection, it is important to verify that you have current information about the occupancy. Check your forms to see if there are any documents you will need to request of the owner. If your forms show a specific occupancy type, you should confirm that the occupancy class has not changed.

Presentation

First impressions count! Your first few minutes with the property or occupancy representative will set the stage for your entire professional relationship with that person. You will be judged on your appearance, your attitude toward your work, and the way you interact with others **Figure 8**. Wearing a uniform generally makes it easy to see what agency you are representing. In spite of the ease of recognition a uniform brings, many agencies require the fire inspector to have an ID card visible. You should never hesitate to show any form of identification when asked and should even compliment the person for asking. This shows

Chapter 5 Performing an Inspection

Figure 8 You will be judged first upon your appearance, your attitude toward your work, and towards those with whom you interact.

that you realize that others may misrepresent themselves for unscrupulous reasons. It also shows that you approach your job as a fire inspector seriously and professionally.

If your jurisdiction does not have uniforms, consider wearing appropriate professional attire. In many cases this could be a dress shirt, tie, badge, and nametag. During hot summer months, a short sleeve shirt with a collar may be appropriate. Having the name of the organization you represent on the shirt when possible, will connect you with an agency. When that is not possible, ID cards become much more important.

When working at job sites, safety concerns change dress code expectations. Boots and jeans may be appropriate. It should not be forgotten that when at a job site most workers are required to wear hard hats. You are not immune to hard hat or eye protection requirements.

Once you have arrived with the appropriate uniform, show the appropriate professional attitude and mannerisms. You can show professionalism by waiting your turn to speak and thanking people for their assistance. Do not smoke, chew gum, make loud noises, or walk around while waiting for the owner. Your mannerisms outwardly demonstrate your mental attitude. For instance, when first meeting with the property or occupancy representative, make eye contact and firmly shake hands. Introduce yourself by stating your name, title, and the reason you are there—to seek permission to inspect the property. Do not forget that you represent the fire inspection agency and the fire department.

In the event that the owner shows some hostility towards you, remain professional. Ask the owner if another fire inspector can perform the inspection. When you return to the office, make sure your supervisor is notified of the problem. If there has been a pattern of this, it may be appropriate to ask another fire inspector to accompany you on the next inspection.

Conducting the Exterior Inspection

Prior to actually meeting the building owner, certain observations can be made regarding the building Figure 9 . From the exterior you can observe vehicle access, fire lanes, fire hydrant access, caps missing on fire hydrants, sprinkler connection visibility and access, and exterior building issues. These include broken windows, tilting walls, and missing or falling bricks Figure 10 . Other things to note would be ponds, gates, and fences or other barriers. If this is a large complex it may be helpful if you have a site plan of the property.

Before conducting an exterior inspection, notify building personnel of your presence and intentions. This is more professional than explaining your presence and behavior to security guards or other personnel.

Figure 9 As you drive up, take note of the exterior of the building.

Figure 10 From the exterior you can observe vehicle access, fire lanes, fire hydrant access, or caps missing on fire hydrants.

Conducting the Interior Inspection

Adequate time should be allotted to perform a thorough inspection of the building's interior and complete required documentation. If this is your first time at the building, additional time should be allotted to learn the building layout. If you have inspected the building before, it may be a good idea to rotate other fire inspectors through the buildings. Different eyes see different things.

When conducting the fire inspection, follow a predetermined and structured order so that no areas of the building are missed. You may begin at the top floor and work down, or start at the front and work toward the rear, or start by going left, or right, and continuing that direction until you arrive back at the starting point. It is important that the inspection be systematic, thorough, and well-documented. It is important to look into every room.

In the cases of large, complex occupancies and properties, you may need to break the inspection into sections or buildings in order to make sure nothing is overlooked. This may entail conducting the fire inspection over multiple days. You must also remember that the person who is accompanying you has other job duties that are not being performed while he or she is escorting you.

Having the owner along helps you gain access to locked or restricted areas. Just because a door is locked does not mean that it is exempt from the inspection. There may be restricted areas that require signing into a log in order to access them. In rare cases there may be areas that contain trade secrets of the business and entry may be refused. Documenting those areas where access is not fully gained is important. Rapport with the owner is crucial. Try to determine information by asking the following:

- "How large is the area?"
- "Can I just step inside the door for a quick once over?"
- "Could someone from a safety committee provide some documentation of the hazards?"

Ask questions about the building during the inspection. The answers may give some insight to some possible hazards you have not previously noticed. Be thorough and do not rush. Often the owner will escort you to where he or she thinks you want to go. Do not be afraid to stop and look at areas along the way.

Code Violations

When documenting code violations, it helps to be specific about the problem and the needed steps toward code compliance. The owner will need to fix the violation, and you or another fire inspector must be able to relocate the problem area when it is time for reinspection. Rather than simply stating the issue, you should indicate what is necessary in order to reach compliance. For example, if a fire extinguisher is out of date, state that the fire extinguisher requires a current inspection tag. This provides the owner with direction, increasing the likelihood of compliance during reinspection.

On rare occasions, you may encounter a process or hazard that is extremely dangerous to the occupants. This is considered an exigent circumstance and may warrant closing a business. This action needs to be considered with extreme care. Some fire inspectors feel that have a right to close a business if violations are not corrected. Businesses can be closed, but unless it is an exigent circumstance, it must be done through the legal system. The fire inspector, in court, would need to present evidence that the business is too dangerous to stay open. It is entirely possible that the judge may feel that the list of code violations are not emergent enough to warrant closing the business. At that point, the judge can fine or impose other sanctions upon the business owner to advance compliance. Continued non-compliance would mean possible contempt of court with possible additional sanctions of fines, jail time, or ordering the business closed.

Fire Protection Features

Fire alarm systems and automatic fire sprinkler systems should be a top priority when evaluating the building's fire protection features **Figure 11**. In addition, fire hydrants on or close to the property should be routinely inspected to ensure working order. If the department or municipal water purveyor inspects the hydrants regularly, you may only want to check that the caps are in place and that they are not stuck or painted closed. If the hydrant is not maintained, then permission should be gained from the proper authorities to open the caps and flow the hydrant to see that there is good water supply and that all parts of the hydrant work properly.

Fire detection systems can detect the presence of smoke and fire, alert occupants, notify the fire department, activate fire suppression systems, close fire doors, open smoke vents, and control the building's heating, ventilation, and air conditioning (HVAC) systems. You should have a working knowledge of how these components perform. All of the features of a fire detection system must be inspected; however, it should not be your responsibility to do more than a visual inspection of the fire detection system components. Besides being very time consuming, you could be held liable if a piece breaks or the system cannot be reset. You should note on the inspection form that a recent full fire detection system inspection report must be submitted by the owner showing what devices were tested and if they passed inspection.

It is important to note which type of devices are installed. If heat detectors are installed where smoke detectors should be

Figure 11 Fire alarm systems and automatic fire sprinkler systems should be at the top of your list when taking into account the building's fire protection features.

Figure 12 A kitchen may call for a specialized fire suppression system.

installed per the code, this is a violation of the code. A smoke detector should be located in an area best suited to smoke detection, for example, in a foyer, lobby, or conference room. A smoke detector located in a machine shop or garage may not achieve the same intended goal and will cause unintended alarms.

The same logic holds true for a fire sprinkler system. While there are typically not many parts that need to be tested, a sprinkler inspection report should be provided by the owner. You are obligated to check for closed valves, proper pressures on various gauges, missing sprinkler coverage, and that heads are installed properly.

A kitchen may call for a specialized fire suppression system Figure 12 . As with other fire suppression systems, you should witness the initial installation to be certain that gas valves close, electricity and fans shut off, and all other components of the system function properly. Following the final inspection, a minimum of an annual inspection should occur. A visual inspection of the fire suppression system during the fire inspection would include looking for caps on the nozzles, accumulations of grease, missing nozzle coverage, and grease filters turned the wrong direction.

Hazard Recognition

The role of the fire inspector is to ensure a safe building for all occupants. During the fire inspection you should be on the lookout for various hazards. There are a few hazard violations that routinely appear, including:

- **Electrical**—Electrical cords cannot be spliced; circuit breakers must be identified; extension cords cannot be used in place of permanent wiring, openings are not allowed in electrical panels; clear access to the electric panel must be maintained; cover plates are needed for junction boxes, switches, and outlets; and no multi-plug adapters are allowed.
- **Exit/emergency lights**—Lights must function properly and must not be obstructed.
- **Exiting**—Exit doors must be operational and unobstructed, doors must close and latch but may not use deadbolt locks, storage is not allowed in halls or stairwells, exit doors must swing outward, and exit signage is required.
- **Fire extinguishers**—Fire extinguishers must have current inspection tags and a minimum of 2A10BC rating; extinguishers must be mounted properly, unobstructed, and operational, indicated by proper signage where appropriate; extinguishers should be properly spaced for minimal travel distance; they should be of the proper type for the hazard; and there should be enough extinguishers to comply with fire codes.
- **Fire detection and suppression systems**—This equipment must be accessible and kept in a normal status, fire department connections must be capped and accessible, storage cannot be too close to sprinklers, and rooms must be properly labeled. Specialized annual inspections are required on these systems.
- **General**—A key box containing proper keys is required, good housekeeping must be maintained, address numbers on the building must be visible, high pressure cylinders must be secured to the wall, gas meters must be protected, fire hydrants must be visible and accessible, no excessive amounts of flammable liquids may be stored, flammable materials must be kept in the proper containers, ashtrays should be provided in smoking areas, no smoking signs must be provided where smoking is not permitted, emergency vehicle access must be unobstructed, and fire lanes must be identified.
- **Heating appliances**—All combustibles must be kept 36″ (914 mm) away from a heat source. The heating appliance must be in good repair and easily accessible.
- **Openings**—All pipe chases must be filled, ceiling tiles must be intact and in place, any holes in drywall must be patched, any openings in fire walls must be repaired, and fire doors must work properly.

Contents

It makes no sense to inspect a building but not inspect its contents. On January 16, 1967, Chicago's McCormick Place, a large convention center thought to be fire proof due to its steel and concrete construction, burned to the ground. The fire protection needs to match the hazards within. The size or construction of a building is not the determining factor when classifying a content hazard. A 50,000 square foot (4645 m²) building with storage of aluminum canoes is less hazardous than a 5,000 square foot (465 m²) building storing plastic cups. This is because the amount of heat released with plastic cups is significantly greater than the canoe.

In the NFPA 101, *Life Safety Code*, content hazards are classified as low, ordinary, or high. <u>High hazard contents</u> are classified as those that are likely to burn with extreme rapidity and from which explosion is likely. High hazard contents include but are not limited to flammable liquids, grain dust, wood flour, plastic dust, aluminum or magnesium dust, hazardous chemicals, and explosives Figure 13 .

<u>Ordinary hazard contents</u> are classified as those that are likely to burn with moderate rapidity and give off a considerable volume of smoke Figure 14 . Most buildings contain ordinary hazard contents, such as wood and metal furnishings, paper and

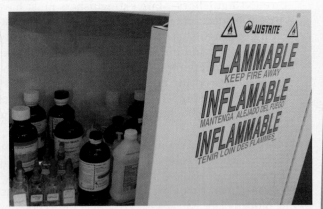

Figure 13 High hazard contents include but are not limited to flammable liquids, grain dust, wood flour, plastic dust, aluminum or magnesium dust, hazardous chemicals, and explosives.

Figure 15 Low hazard contents are those of such low combustibility that no self-propagating fire therein can occur.

Figure 14 Ordinary hazard contents are those that are likely to burn with moderate rapidity and give off a considerable volume of smoke.

office supplies, machine shop and mechanical equipment. **Low hazard contents** are classified as those of such low combustibility that no self-propagating fire therein can occur Figure 15. The storage of noncombustible materials is considered low hazard. Places originally classified as low hazard, more often than not become ordinary or even high hazard as business practices and storage of materials change. For instance, the storage of aluminum canoes is a low hazard. Removing the canoes and storing fiberglass or wooden canoes changes the hazard and the hazard designation.

Having the building and fire protection features meet the contents hazard is a must. When the building has an automatic fire sprinkler system, you must make sure the system is designed for the contents hazard present. NFPA 13, *Standard for the Installation of Sprinkler Systems*, has a number of different commodity or hazard classifications, each of those requiring a certain sprinkler design density. Any changes in contents, storage arrangement, or occupancy should be investigated to make sure the fire protection features still match the contents, storage arrangement, and occupancy.

All sprinkler systems are not created equal. An automatic fire sprinkler system designed for a warehouse storing aluminum canoes is not sufficient when the warehouse is converted to a plastics workshop. Determining the proper sprinkler coverage on an initial visit is extremely difficult. A hydraulic placard should be attached to the sprinkler riser. This placard will tell how the system was designed. By comparing the placard and the building contents, you may be able to determine the proper sprinkler protection. If there is any doubt, a design professional or sprinkler contractor should be brought in to evaluate the sprinkler design for the given hazard. If you are familiar with the building, a change in the contents may be noted. Through questioning the owner and comparing known current and past contents, a decision may be made about adequate fire protection.

When evaluating the contents, pay attention to the configuration. Contents left in hallways, on floors, and strewn throughout an occupancy indicates a disregard for occupancy safety. It also will have the potential to spread fire, should one occur. For example, rack storage is required to have a certain number of feet between racks. Having stock on the floor will increase the potential that a fire will quickly catch additional rack storage on fire. Contents stacked too high may impede proper sprinkler operation. If there are no sprinklers, space must be given to provide room for streams from a fire hose. These distances can be found in NFPA 13.

Building Features

Means of egress, fire doors, fire walls, staircase enclosures, smoke ventilation systems, hung ceiling systems, emergency lighting, and exit lighting should be inspected Figure 16. Look for the following items:

- Check the means of egress including stairway enclosures.
- Make certain that doors swing in the right direction.
- Travel distances and the common paths of travel are not exceeded.
- Doors close and latch.
- Exit doors are not locked.
- Fire doors should close and latch. If there are automatic devices used to operate and release the door, those pieces must function smoothly.

Figure 16 Means of egress must be inspected.

Figure 17 Buildings built long before the fire inspector was born may be renovated multiple times hiding a multitude of construction and fire protection problems. Checking for vertical openings and hidden shafts is a must in older buildings.

- Smoke evacuation systems and systems designed to pressurize a stairwell or floor are complicated and should be tested by a professional company.
- Ceiling tiles missing from the area around the sprinkler head are a deterrent to its operation.
- Exit lighting requires looking to see if the lights are illuminated.
- Emergency lighting will require some type of testing. In most cases this can be as easy as pushing the test button to see if the lights illuminate. It gets more difficult when those lights are more than 15′ (457 cm) off the floor. Often times those units will have a separate circuit breaker and the lights will come on simply by turning it off. Never do that yourself. Let the building owner do that, as you will have no idea what else may be tied into the breaker. There have been instances where a breaker was turned off and computers also shut down, causing the loss of many hours of work.
- More mundane features such as carpeting and interior finishes must also be checked for compliance. In certain public buildings, the furniture needs to have a tag indicating that it will not promote flame spread.

Many areas of a building are obscured from normal view and you may need to enter concealed spaces, voids, and areas above the hung ceiling and the true ceiling of an occupied space. Visual inspections ensure the existence and integrity of fire protection features Figure 17 . Since these are void areas, the prime reason to look behind these areas is to ensure that there are no breaches in the walls or ceilings to ensure that a fire cannot pass into another area. Look for adequate fire stopping of penetrations and that walls are tight to ceilings.

Heating, Ventilation, and Air Conditioning
You must evaluate HVAC, and other building service equipment and operations, to ensure the systems and other equipment are designed in accordance with the local codes and standards. Contractors install this equipment and should inspect it regularly. Ask the building owner for an HVAC inspection report from the contractor. One of the items to check during your inspection is the presence of smoke or fire dampers when fire walls are penetrated by HVAC components. Additionally, some HVAC units will need duct smoke detectors and carbon monoxide detectors.

Kitchen Cooking Equipment
Cooking equipment generally does not require much in regards to special inspections. The contractor will install and test the equipment. Your role will be to see that the device is secure, meaning will not fall or tip; proper gas and electric lines are in place; and fans work as designed. You will also ensure that the grease hoods are being cleaned and maintained properly. Many kitchens require a fire suppression system. These systems require specialized testing by a contractor. Ask the building owner for the current inspection report from the contractor.

Laundry/Garbage Chutes
Chutes penetrate multiple floors of a building and can act as a flue to spread smoke and fire. Chutes should be inspected to ensure the integrity of the chute is still intact. Where sprinkler heads are present, they should be inspected to see that nothing is blocking them from proper operation. At the bottom of the chute, a door is typically seen. The chute should have fire-rated doors with spring closures and held open by a fusible link. Nothing should impede its closing, and the fusible link must be in place; a common hazard is to see the door held open by a solid wire. Ensure that there is nothing blocking the chute. Chutes should also be in separated from corridors by a room that is non-combustible.

Elevators and Escalators

Elevators and escalators are intricate devices and you should not be involved in their actual testing. Most municipalities and states will require a licensed elevator and escalator inspector to test and certify these devices every six to twelve months. Your role is to see that the inspection certificate is current and posted.

Interior Finish

As a fire inspector, it is important to understand the impact various interior finishes have on the potential to help or hinder fire's movement. The interior finish is the exposed surface of the floor, walls, and ceiling. Anything attached to them will help to control the speed at which fire would spread. Drapery, curtains, and the like would not be counted as an interior finish unless it is attached to the wall. As building plans are submitted there should be listing of what the interior finishes are and rated for in the various areas within the building. The model code specify the interior flame spread rating for various buildings. Class A is the highest rating and has a flame spread of 0–25. The other classes are B and C, and the flame spread goes to 200. After that, there is no rating given.

As you conduct your inspection, evaluating the interior finish includes examining finishes on walls, floors, and ceilings. It is important to ensure that combustible interior finishes, (e.g. plywood paneling, cloth, or decorative wood) meet code requirements. This is determined initially through the plan review process and confirmed during inspections prior to occupancy of the building.

Some occupancies may have stage curtains or a foam type product on the wall. Copies of the certificates indicating that the product is fire rated and approved for its intended use are good to place in inspection files. These certificates may also list a flameproofing material that must be reapplied every number of years. Carpets placed on walls lose their flame rating unless they have been tested and listed for wall use.

■ Preplan Sketch

Often times a preplan sketch for the fire department will be drawn during the inspection process. This process takes time, so it may be easier to conduct the inspection and return at a later date for the preplan sketch **Figure 18** . If that is the case, when asking permission to walk the building for the preplan sketch, make it clear that you are not there to conduct a fire inspection, but just to draw a floor plan for the fire department's use in case of an emergency. Also consider taking pictures of the property and obtaining a building plan.

■ New Construction Considerations

During the plan review process, it is determined that the proper size, height, and occupancy of the building is acceptable. Typically, fire inspectors do not have much involvement with the physical construction of the building. The building department will require various inspections along the building process. Some of those inspections may be a footing or foundation inspection; various electric, plumbing, and carpentry inspections; and open ceiling inspections. Performing inspections with the other inspectors helps to ensure that construction is performed correctly. In

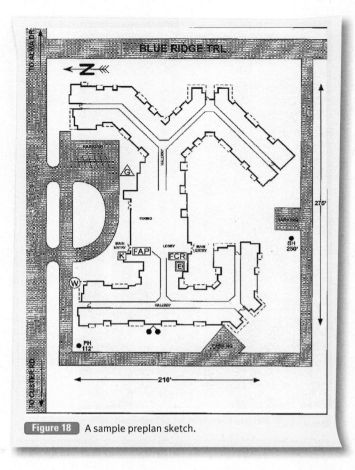

Figure 18 A sample preplan sketch.

addition, it gives you great insight to the roles of the other inspectors and makes you much more informed in how the various roles interact, resulting in more thorough future inspections.

When fire protection systems are installed such as fire alarms, sprinklers, fire pumps, and kitchen hood and duct suppression systems, you need to see the installation as it is progressing. Catching a mistake at this stage is much easier to correct than trying to fix the problem after the walls and ceiling have been installed and finished. When a system is completely installed, you need to witness a full acceptance test of each system conducted by the contractor. You should know what is required during the final inspection and should ask the contractor to perform those things if they have not been done. Those requirements can be found in specific codes and adopted fire protection system standards. Additionally, many reputable companies have check sheets that you can use as a guide during the test. A company's safety representative is often a wealth of knowledge you can access at no cost.

When conducting the final inspections of the various fire protection features, a copy of the plans that were approved should be taken to the job site. There you will compare types of devices approved, and their locations, against the actual installation.

■ Remodeling Considerations

A remodeling of a building is no different from an inspection viewpoint than a new building. Plans must be submitted and approved, and inspections are required at various times of the construction by many different inspectors. Any difficulties may come from the fact that the contractor is starting with an existing building, not a clean slate. Often remodeling an existing building will activate requirements that the area being worked on, or in some cases the entire building, must be brought to the current codes for new construction. This could mean such things a requiring a full sprinkler system.

If the building is not vacant, more attention is needed as there are general safety concerns to the occupants during the construction process. Additionally, at times the means of egress may be reduced or need to be temporarily relocated, requiring significant signage.

Post-Inspection Meeting

It is important to have a conversation with the owner regarding the findings of the inspection Figure 19. You must always take the high road. It is common that many building owners do not like to see the fire inspector show up. To them, the fire inspector may just be someone that costs them money. When talking with the owner about the inspection, be firm in your needs, while at the same time listen and be empathetic to his or her concerns. When there are disagreements, understanding the owner's viewpoint can often create a middle ground for compliance.

Hopefully the owner has accompanied you on the inspection and items noted were explained at that time. If not, you may need to take the owner to some locations to explain the code violations noted. It is important that you fully explain the reason behind any code violations. To just say that "because the code says so" will not help your cause. If the owner can be educated about the hazards, hopefully their reoccurrence will be lessened.

You should also take this time to note any other concerns that may not be actual code violations, but would be of interest to the owner such as fire extinguishers due for inspection in two months, making sure the sidewalks are kept clear of snow, etc. This can be noted on the inspection form for documentation. A copy of the inspection form should be left with the owner. If there are code violations, you should state when you will return to reinspect. As a final act, thank the building occupant or representative for their cooperation, even if it was lacking. This shows your professional attitude.

Figure 19 It is important to have a conversation with the owner regarding the findings of the inspection.

Fire Inspector Tips

You should be aware of your local appeal process and inform the owner of his or her right to appeal.

Documentation

Documenting the inspection and code violations in writing and in a prescribed format is the best way to attest to the findings of your inspection. Legal requirements and lawsuits have changed the way fire inspectors must document each and every element of their inspections. Inspection records could be recalled many years after an inspection has occurred. Two prime reasons for that would be a lawsuit or if there were a fire and people are looking to see what was noted in previous years' inspections.

The inspection is documented in the inspection report. The inspection report can be a check list, free form, or a combination of both. How to write an inspection report is covered in detail in the chapter *Writing Reports and Keeping Records*. When there are a great many issues or the issues are complex, it may be best to forgo the standard inspection form and to create a formal letter documenting the violations and the expectations. Any noncompliance or code violation noticed during the inspection must be documented in the written report of the inspection. A copy of the inspection form and a date for a return inspection should be left with the building owner. If necessary, mailing the form is acceptable; however, the only way to document if it was received is by certified mail.

Noting the date for the reinspection is important because by documenting a violation, you are stating you are aware of fire or life safety code violations. In Adams v. State of Alaska, October 1, 1976, the State undertook a fire inspection in a hotel and noted hazards, some of which were extreme life hazards. The State advised the hotel of the hazards, but did not take additional action. Several months later, a fire killed and injured many people. As a result of the inspection, the State owed a duty to exercise reasonable care to abate the hazards. By failing to return and have the hazards abated, the State breached their duty by their inaction.

Generally, the inspection report is best delivered in person. This way the owner may ask questions. A few extra minutes explaining why an item is a code violation provides guidance and understanding to the owner, and may lessen the chance of its reoccurrence. Issues that may seem clear to you may be unfamiliar to the owner, so extra time must be spent to explain those issues. When appropriate, ask if the owner would like a copy of the section of the code noting the violation. Upon returning to the office, simply copy that section of the code and fax or email it to the owner.

At no time should you become the agent of the owner. Only offer guidance based on the code. It is not your job to fix the code violation. Attempting to do so could lead to you assuming responsibility for the fix, which could become legally problematic.

When documenting an inspection, keep in mind that the information must be easily retrievable and easily readable by other fire inspectors and the occupancy owner. An inspection report stored on your portable computing device is great for you but is of no value to anyone else.

Records should be available in the office. While documentation is moving towards a paperless system, hard copies are readily available and not dependent upon a computer. The file should indicate if the inspection is ongoing or has been completed. If a new business moves in to an old space, the old records should be maintained as they show the history of the occupancy, regardless of the specific business. If, however, the building is destroyed, the files should still be maintained, although not necessarily in an active file. Those records could be moved to a dead file, in the event there are future questions about the property.

Destroying any records should only be done after proper approval has been given. Typically this is given at the state level. Fire inspection records are public documents and can be requested through the Freedom of Information Act. When a request is received, gather the documents requested and forward those to the appropriate authority in the organization for review so that any confidential information can be removed.

Code Violations

Code violations are not always black and white in nature. There is the letter of the code and the intent of the code, and you should consider both. Much of this wisdom comes with experience. For example, if the code states a fire extinguisher cannot be mounted more than five feet from the floor, and you inspect a well-kept building with an owner conscientious of safety concerns; however, you note the fire extinguisher is six feet from the ground and has been located there for a number of years. Is this a violation of the code or have they met the intent of the code? How about a small building where the only exit signage you find is a sign above the front door that is not internally illuminated? These types of issues are the judgment calls an inspector must routinely make. Would you be completely in the right if you chose to cite those as code violations? Absolutely.

All noted code violations must be corrected as soon as possible. Some items, such as locked exit doors, must be corrected immediately; others, such as unlit exit signs, do not have the same degree of urgency. If you note violations that have been overlooked for years, the owner may question the violation. This can be a tough situation. The only thing that can really be said is that for some reason this was missed on previous inspections but must now be changed to comply with the code.

When an owner or occupant receives an inspection report, it constitutes an important part of their business and business protection. The purpose of the inspection and the subsequent report is compliance with applicable codes. Follow-up dates are necessary when code violations have been found, providing the timetable for corrective action on the owner or occupant's part. Code violations should have reasonable timetables for compliance. Remember: what is reasonable to you may not be reasonable to the person receiving the report. When assigning the correction time to each violation, consider many factors including the seriousness of the violation, financial cost to correct the violation, and the ability to get the violation corrected. Replacing light bulbs in exit signs is not difficult or expensive, but adding a sprinkler system is much more involved, not to mention more costly.

Any code violations that are corrected during the inspection must also be noted in the final report. Document the fact that a code violation was actually noted during the inspection, by marking the violation with "complied on-site." Keep in mind that easy-to-fix code violations generally reappear once you leave.

When there are complex code violations, more research into the appropriate codes and standards may be necessary. In this situation, advise the owner that you are not certain about a specific item and you want to make certain of the code requirement before finalizing the report. You would not want to cite a violation when one does not exist. For example, if you encounter a chemical and are not certain how much can be stored, if it must be separated from other chemicals, or if it should be in its own room, referencing the code may be required.

When noting code violations, citing the code reference adds legitimacy to your assessment and allows the owner to look up the specific area themselves, should he or she choose. The vast majority of the time, the owner will not question the violation and will be more concerned about time frames or costs. For those owners who question a violation, you should be able to furnish the reference quickly if you elected to not cite code directly. That may mean having a master list of common violations with the associated code reference(s) or knowing where to look in the codes for the answer.

Reinspection dates are at your discretion and agency policy. Many fire inspectors routinely give 30 days prior to their return; however, some agencies have found that shorter time frames, such as two weeks, keep the violations more in the forefront of the owners' priorities. If items are severe a day or two may be appropriate. Large ticket items it may require months or years to complete, but more frequent follow-ups will ensure that the owner is taking some action, such as getting bids for the work.

The reinspection date is not when the code violations must be complied; instead, code violations must be corrected as soon as possible. The reinspection date is just an approximation of when you will return to check on code compliance. If you are allowing multiple dates for compliance, the inspection form should note that.

As a last resort, if there is non-compliance following repeated efforts, the municipality may need to be called in. They typically have the leverage of issuing fines and revoking business licenses when appropriate. When all else fails, going to court may be needed. It is unfortunate and time consuming, but it shows the seriousness of the inspection process. If after a legal proceeding, a violation is allowed to exist, the judge is the one that has allowed it, not the fire inspector.

> **Fire Inspector Tips**
>
> When returning to check on compliance, only look for those items in question.

> **Fire Inspector Tips**
>
> Save pertinent notes from an inspection as part of the file. Don't throw away something today that you feel is irrelevant. In the future, you may be wish you had those scribbled notes to clarify something you have put in your formal report.

> **Fire Inspector Tips**
>
> Make sure you know the occupancy classification of what you are inspecting, and that you are using the correct edition of the appropriate code as a reference. Just because a new code had been adopted, it does not automatically mean existing occupancies must meet the new requirements.

Investigating Complaints

When the fire inspection is the result of a complaint, you must have the information regarding the specific problem that needs to be addressed. A complaint that an occupancy is "dangerous" provides little direction to the inspector. Whenever possible, gather additional information about the nature of the risk, such as exit issues, overcrowding, or a specific dangerous situation. While a "fire trap" is not very specific, the complaint must be evaluated. How soon depends on the urgency of the complaint.

When responding to a complaint, the owner should not know that you plan to inspect because it is important that you see the condition as it exists, not after there has been an opportunity to repair it. Once the complaint is investigated, the building owner should be advised of the results. If there are code violations, a time frame should be given for compliance. Some items, such as locked exit doors, should be corrected prior to your leaving. Once on-site for the investigation, additional problems may be found, necessitating a complete fire inspection in the immediate future.

When documenting a complaint, the form needs to state the alleged violation. When possible, it should also state the name of person making the complaint. You need to state the date, time, and results of your investigation. If a violation exists, the form should state what corrective action was taken or recommended. The complaint stays open until the violation is repaired. You must return to see that the corrective action has in fact been taken. When repaired, the repair date should be noted. At that point the form is satisfied and can be filed.

If a complaint is received, and the investigation finds no violation, the complaint form should still be completed, indicating that the complaint was unfounded.

Improving the Inspection Process

Knowledge and experience must go hand in hand. You can have all of the knowledge in the world, but knowing when and how to apply that takes some time. Conversely, going and conducting inspections without knowing what to look for wastes everyone's time.

Knowledge can be gained from reading books and attending classes, but it cannot stop there. As you progress in the career of a fire inspector, new ideas, equipment, and trends emerge. It is your responsibility to stay abreast of the changes. The easiest way is to talk to contractors who should be experts in their field. If you show a willingness to learn, most are eager to help. You can also call the manufacturer. Tell them that you would like some additional information about certain equipment or processes.

One of the best ways to stay informed is to join a local fire inspectors association. Here, common problems are discussed, vendors talk about new products, and there is the opportunity to meet other fire inspectors. The ability to call another fire inspector to discuss a problem is invaluable.

Gaining experience takes time. The more you go into the same buildings, the more familiar you will become, and the quicker you can conduct the inspection. Additionally, the personnel at the building will be accustomed to seeing you and will know what to expect. While still being professional, the initial formality begins to become a little more informal and relaxed. Combining knowledge and experience will make you a more confident fire inspector. You will also be able to see the building and possible violations on a different level than just black and white.

Wrap-Up

Chief Concepts

- A fire inspection will reasonably ensure that a building will be safe for the occupants.
- There are a number or basic or routine inspections including annual inspections, reinspections, complaint inspections, construction or final inspections, business license or change of occupancy, and self-inspection.
- The fire inspection process begins before the construction phase, long before the occupancy is scheduled to be opened.
- It is critical to know which codes you can legally enforce in your jurisdiction. The mere fact that there are sets of model codes available does not give you the right to enforce those codes. The local jurisdiction must legally adopt a specific set of codes.
- The forms used during an inspection will vary from agency to agency. A set of standard forms should be used by all fire inspectors in the office. Some of the basics forms are:
  - Inspection form
  - Final or construction inspection form
  - Complaint form
  - Stop work order
- While the codes give you the authority to conduct fire inspections, you are not allowed to do so without permission of the owner. Performing a fire inspection without permission is trespassing. The exception to this is exigent circumstances.
- The process of inspecting a building will vary with the occupancy. Some occupancies may have multiple buildings, others will be multiple floors, others will be very small buildings, and another could be a massive one story building.
- The exterior inspection includes vehicle access, fire lanes, fire hydrant access, caps missing on fire hydrants, sprinkler connection visibility and access, and exterior building issues. These include broken windows, tilting walls, and missing or falling bricks.
- When conducting the interior inspection, follow a predetermined and structured order. You may begin at the top floor and work down, or start at the front and work toward the rear, or start by going left, or right, and continuing that direction until you arrive back at the starting point. It is important to look into every room.
- Often times a preplan sketch for the fire department will be drawn during the inspection process. This process takes time, so it may be easier to conduct the inspection and return at a later date for the preplan sketch.
- A post-inspection meeting should be the last step in the physical inspection of the building. When talking with the

Wrap-Up

owner about the inspection, be firm in your needs, while at the same time listen and be empathetic to his or her concerns. When there are disagreements, understanding the owner's viewpoint can allow for a middle ground for compliance.

- Documenting the inspection and code violations in writing and in a prescribed format is the best way to attest to the findings of your inspection. Inspection records could be recalled many years after an inspection has occurred. Two prime reasons for that would be a lawsuit or if there were a fire and people are looking to see what was noted in previous years' inspections.
- Code violations are not always black and white in nature. There is the letter of the code and the intent of the code, and you should consider both.
- When the fire inspection is the result of a complaint, you must have the information regarding the specific problem that needs to be addressed.

■ Hot Terms

Annual inspections Inspections performed as part of the regular inspection cycle

As built diagrams A set of drawings provided by a contractor showing how a system was actually installed, which may be different from the approved plans

Business license or **change of occupancy inspections** Inspections that occur when the building department is notified of a new business requesting permission to open

Complaint inspections Inspections that occur when someone registers a concern of a possible code violation

Complaint form Form that lists in detail any complaint that is lodged with the fire inspection agency and is investigated

Wrap-Up

Construction or **final inspections** Inspections that are conducted as a building is being constructed, including sprinkler systems, fire alarm systems, and fire pumps

Conditional approval Grants partial approval for work being done, often issued for the final certificate of occupancy when violations are minor and do not pose a hazard

Enabling legislation Legislation in which local jurisdiction adopt a specific set of codes

Exigent circumstance An immediate life safety issue which requires that immediate actions be taken

Final or **construction inspection form** A form used when inspecting specialized systems such as fire alarm, sprinkler, hood, and duct suppression systems, as well as for other types of construction phase inspections, such as a ceiling inspection or a final inspection prior to issuance of a certificate of occupancy

High hazard contents Contents that are likely to burn with extreme rapidity or from which explosions are likely (NFPA 520)

Low hazard contents Contents of such low combustibility that no self-propagating fire therein can occur (NFPA 520)

Ordinary hazard contents Those contents likely to burn with moderate rapidity and give off a considerable volume of smoke

Reinspection An inspection performed to determine if code violations have been corrected

Self inspections Inspection performed by the building owner or occupant

Stop work order A form used when contractors do not have the clearance for performing the work, or when work must be corrected prior to performing additional work

Fire Inspector *in Action*

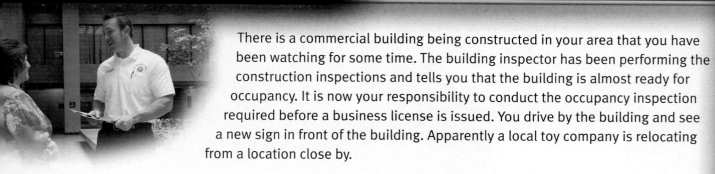

There is a commercial building being constructed in your area that you have been watching for some time. The building inspector has been performing the construction inspections and tells you that the building is almost ready for occupancy. It is now your responsibility to conduct the occupancy inspection required before a business license is issued. You drive by the building and see a new sign in front of the building. Apparently a local toy company is relocating from a location close by.

1. What do you plan to do in preparation for making the initial inspection?
   A. Visit the property and look around on your off hours.
   B. Look in the file of the toy company's current location for past history of violations and the degree of compliance.
   C. Go to the property and introduce yourself as soon as they take occupancy.
   D. Talk to the building inspector to make sure everything is code compliant so you don't have to walk the entire building.

2. Before you get a request to inspect the building, the owner invites you to visit the old location and meet with him to discuss code compliance in the new building.
   A. Don't go, this would be a warrantless search and therefore illegal as you do not have an inspection request nor court order.
   B. Meet with the owner and tour the building to find out what you will have to address when they move.
   C. This would be a wasted trip as the new building would be code compliant as it is brand new.
   D. Meet with the owner and immediately put him on notice that you will not tolerate violating your fire code.

3. You learn that manufacturing toys requires vast amounts of different kinds of chemicals, some of which react with each other. How do you feel is the best way to handle this?
   A. Take a chemistry course at the local community college.
   B. Talk to other inspectors and see what they know about chemistry.
   C. Learn who the chemical supplier is and ask for assistance.
   D. Allow the owner to do as he feels is right; he has been in business before.

4. It appears that this inspection is going to be very complex and confusing, so the best format to use for the inspection report would be:
   A. a check off sheet with items and check boxes.
   B. a verbal interview with the owner stating what is required with nothing in writing that can be used against the inspector.
   C. a detailed report in essay form with attachments received from suppliers.
   D. all that is needed is to refer the owner to comply with the code and set a date for compliance.

Reading Plans

CHAPTER 6

NFPA 1031 Standard

Fire Inspector I

4.2.3 Recognize the need for plan review, given a situation or condition, so that requirements for plan reviews are communicated in accordance with the applicable codes and standards and the policies of the jurisdiction. (p 108)

(A) Requisite Knowledge. Plan review policies of the jurisdiction and the rationale for the plan review. (p 108)

(B) Requisite Skills. The ability to communicate orally and in writing. (p 108)

4.4 Plans Review. There are no plan review job performance requirements for Fire Inspector I. (p 108)

Fire Inspector II

5.2.2 Process a plan review application, given a specific request, so that the application is evaluated and processed in accordance with the applicable codes and standards and the policies of the jurisdiction. (pp 108–113)

(A) Requisite Knowledge. Plan review application process, code requirements of the jurisdiction, and policies and procedures of the jurisdiction. (pp 108–113)

(B) Requisite Skills. The ability to communicate orally and in writing on matters related to code requirements, policies, and procedures of the jurisdiction. (pp 108–113)

5.3.3 Evaluate a building's area, height, occupancy classification, and construction type, given an approved set of plans and construction features, so that it is verified that the building is in accordance with applicable codes and standards. (pp 108–112)

(A) Requisite Knowledge. Building construction with emphasis on fire-rated construction, evaluation of methods of construction and assemblies for fire rating, analysis of test results, and manufacturer's specifications. (pp 108–112)

(B) Requisite Skills. The ability to identify characteristics of each type of building construction and occupancy classification. (pp 108–112)

5.3.11 Verify compliance with construction documents, given a performance-based design, so that life safety systems and building services equipment are installed, inspected, and tested to perform as described in the engineering documents and the operations and maintenance manual that accompanies the design, so that deficiencies are identified, documented, and reported in accordance with the applicable codes and standards and the policies of the jurisdiction. (pp 108–112)

(A) Requisite Knowledge. Applicable codes and standards for installation and testing of fire protection systems, means of egress, and building services equipment. (pp 108–112)

(B) Requisite Skills. The ability to witness and document tests of fire protection systems and building services equipment. (pp 108–112)

5.4 Plans Review. This duty involves field verification of shop drawings, plans, and construction documents to ensure that they meet the intent of applicable codes and standards for fire and life safety, according to the following job performance requirements. (pp 112–113)

5.4.1 Classify the occupancy, given a set of plans, specifications, and a description of a building, so that the classification is made in accordance with the applicable codes and standards and the policies of the jurisdiction. (pp 112–113)

(A) Requisite Knowledge. Occupancy classification, applicable codes and standards, regulations, operational features, and fire hazards presented by various occupancies. (pp 112–113)

(B) Requisite Skills. The ability to read plans. (pp 108–112)

5.4.3 Review the proposed installation of fire protection systems, given shop drawings and system specifications for a process or operation, so that the system is reviewed for code compliance and installed in accordance with the approved drawings, and deficiencies are identified, documented, and reported in accordance with the applicable codes and standards and the policies of the jurisdiction. (pp 110–111)

(A) Requisite Knowledge. Proper selection, distribution, location, and testing of portable fire extinguishers; methods used to evaluate the operational readiness of water supply systems used for fire protection; evaluation and testing of automatic sprinkler, water spray, and standpipe systems and fire pumps; evaluation and testing of fixed fire suppression systems; and evaluation and testing of automatic fire detection and alarm systems and devices. (pp 110–111)

(B) Requisite Skills. The ability to read basic floor plans or shop drawings and identify symbols used by the jurisdiction. (pp 110–111)

5.4.4 Review the installation of fire protection systems, given an installed system, shop drawings, and system specifications for a process or operation, so that the system is reviewed for code compliance and installed in accordance with the approved drawings, and deficiencies are identified, documented, and reported in accordance with the applicable codes and standards and the policies of the jurisdiction. (pp 110–111)

(A) Requisite Knowledge. Proper selection, distribution, location, and testing of portable fire extinguishers; methods used to evaluate the operational readiness of water supply systems used for fire protection; evaluation and testing of automatic sprinkler, water spray, and standpipe systems and fire pumps; evaluation and testing of fixed fire suppression systems; and evaluation and testing of automatic fire detection and alarm systems and devices. (pp 110–111)

(B) Requisite Skills. The ability to read basic floor plans or shop drawings. (pp 108–112)

5.4.6 Verify the construction type of a building or portion thereof, given a set of approved plans and specifications, so that the construction type complies with the approved plans and applicable codes and standards. (pp 108–112)

(A) Requisite Knowledge. Building construction with emphasis on fire-rated construction, evaluation of methods of construction and assemblies for fire rating, analysis of test results, and manufacturer's specifications. (pp 108–112)

(B) Requisite Skills. The ability to identify characteristics of each type of building construction. (pp 108–112)

Additional NFPA Standards

- **NFPA 1** *Fire Code*
- **NFPA 10** *Standard for Portable Fire Extinguishers*
- **NFPA 13** *Standard for the Installation of Sprinkler Systems*
- **NFPA 13R** *Standard for the Installation of Sprinkler Systems in Low-Rise Residential Occupancies*
- **NFPA 13D** *Standard for the Installation of Sprinkler Systems in One- and Two-Family Dwellings and Manufactured Homes*
- **NFPA 17** *Standard for Dry Chemical Extinguishing Systems*
- **NFPA 17A** *Standard for Wet Chemical Extinguishing Systems*
- **NFPA 70** *National Electric Code*
- **NFPA 72** *National Fire Alarm and Signaling Code*
- **NFPA 101** *Life Safety Code®*
- **NFPA 170** *Standard for Fire Safety and Emergency Symbols*
- **NFPA 220** *Standard on Types of Building Construction*
- **NFPA 251** *Standard Methods of Tests of Fire Resistance of Building Construction and Materials*
- **NFPA 704** *Standard System for the Identification of the Hazards of Materials for Emergency Response*
- **NFPA 720** *Standard for the Installation of Carbon Monoxide (CO) Detection and Warning Equipment*

FESHE Objectives

Fire Plans Review

1. Describe at least three reasons for performing plan checks, the objectives of a proposed plans review program, the impact of such a program, and how the program will enhance current fire prevention programs. (p 103)
2. Develop a graphic illustration of a model plans review system, identifying at least four components involved in the system including the use of plans review checklists. (pp 108–112)
3. List three methods to monitor and evaluate the effectiveness of code requirements according to applicable standards. (pp 108–112)
4. Determine fire department access, verify appropriate water supply, and review general building parameters. (pp 108–112)
5. Determine occupancy classification and construction type; calculate occupant load, height, and area of a building. (pp 108–112)
6. Determine the appropriateness of the three components of a building's egress system (exit access, exit, and exit discharge), verify building compartmentation and the proper enclosure of vertical openings. (pp 108–112)
7. Identify special hazards, verify interior finish and establish the proper locations for pre-engineered fire extinguishing systems. (pp 108–112)
8. Verify the compliance of a heating, ventilating, and air conditioning (HVAC) system; review sources requiring venting and combustion air; verify the proper location of fire dampers; and evaluate a stairwell pressurization system. (p 110)
9. Verify the proper illumination for exit access and the arrangement of exit lighting; perform a life safety evaluation of the egress arrangement of a building. (pp 108–112)
10. Verify the design of a fire alarm and detection system, and an offsite supervisory system for compliance with applicable standards. (pp 108–112)

Knowledge Objectives

After studying this chapter, you will be able to:

1. List the reasons why a plan review may be required.
2. Describe how to evaluate a plan review application.
3. Describe the types of drawings included in a plan.
4. List the types of views provided by the drawings in a set of plans.
5. Describe how to determine the occupancy of a structure using a set of plans.
6. Describe how to determine the building construction type using a set of plans.
7. Describe how to determine if adequate access for fire apparatus is provided in the plans for a structure.
8. Describe how to evaluate the building's egress system using a set of plans.
9. Describe how to evaluate the building's fire extinguishing systems using a set of plans.
10. Describe how to evaluate the building's fire alarm and detection system using a set of plans.
11. Describe how to evaluate a building's fire protection system using a set of plans.
12. Describe how to evaluate the building's life safety systems and building services equipment using a set of plans.

Skills Objectives

After studying this chapter, you will be able to:

1. Process a plan review application.
2. Perform a plan review to ensure that all applicable codes and standards are met.
3. Perform an in-the-field verification of shop drawings, plans, and construction documents to ensure proper compliance.

You Are the Fire Inspector

It's your first day of being assigned to review plans and an architect comes into your office with twelve large rolls of plans for review. The plans are for a new multi-million dollar hotel that will be built in your jurisdiction. The architect says that all code requirements have been met and that he would like to get started with construction right away.

1. What information needs to be gathered from the individual submitting the plans?
2. Where do you begin in a multiple set, large project that needs review?
3. What equipment will you need to perform a proper review?

Introduction

Plan review is an important and often overlooked part of fire prevention and code enforcement. Plan review should be thought of as the one time you will have to change the structure and layout of the building to meet the applicable codes that will not involve a major reconstruction. Plan review allows you to be involved in a building from the initial design phases all the way through completion of the project. As the fire inspector, you are afforded the chance to play a proactive roll in fire prevention within the community. The intent of plan review is to confirm that all applicable codes and standards have been met by the building designer. In order to perform a thorough plan review, you will need to be familiar with all applicable codes and standards as well as how to read and interpret construction documents.

The objective of plan review is to make certain that buildings are safe for the public to occupy and fire fighters to operate. As the fire inspector, you are also ensuring that fire protection systems will operate as designed and achieve their performance goals. The fire companies that will be responding to the completed building rely heavily on a proper plan review. Upon completion of a building and prior to an incident, the fire department should be made aware of fire lanes, access roads, water supplies, exterior doors to the structure, hazardous materials and processes, and fire protection systems installed in the building.

The fire prevention program of your department will be enhanced by your efforts in plan review. You will have the opportunity to educate those that are responsible for designing structures and systems in your jurisdiction. Developing good relationships with these design professionals will help you make sure that building in your area is done according to all applicable codes and standards. In addition, the community at large will benefit from buildings that are inherently safer.

Authority to Review

Most model codes include a section on plan review that gives the AHJ the authority to review and approve a project prior to construction. NFPA 1, *Fire Code*, states the requirement in Chapter 1, *Administration*. It is important to note that the first requirement in this section sets forth what type of drawings, construction documents, and shop drawings are reviewed. The purpose of this requirement is to keep you from spending time on conceptual designs that may never be built. As a rule, documents marked "not for construction" should not be reviewed by fire inspectors. The section also gives you the authority to require any field changes or modifications be submitted for approval.

Fire Inspector Tips

Based upon your state, there may be requirements that you obtain various certifications prior to being permitted to perform plan reviews. All plan reviewers must have the appropriate training and qualifications to properly perform the job.

Types of Plans

Plans, blueprints, construction documents, shop drawings, and **plan sets** are all terms used to refer to the documents that will be reviewed during a plan review. At first glance, a set of drawings in a plan set can be very intimidating, but you will find that information on the plans is fairly standardized and all necessary information can be found easily Figure 1.

Fire Inspector Tips

Plan reviews are performed to ensure code compliance for all new systems or structures being installed in the jurisdiction, so the fire inspector has input into the location and accessibility of features of fire protection used by responding fire fighters, and so the fire inspector may locate any specific hazards located in the building.

Figure 1 In time, the drawings in plan sets will become familiar.

Plans, blueprints, construction documents, shop drawings, and plan sets are created by design professionals. A design professional can refer to an architect or engineer. The exact education level and licensing of who can do what will depend on state or local regulations. For the most part, site plans will be created by civil engineers and bear their stamp and signature. Architects will complete building plans and layouts, and they will add their stamp and signature to these drawings. Engineers will complete mechanical, electrical, and fire protection drawings based on their area of expertise, and add their stamp and signature as well.

Most drawings in plan sets will contain a title block that is normally on the right side of the page. The title block contains basic information about the drawing and includes some or all of the following information:

- Date
- Drawing number
- Job number
- Revision number or date
- Project name and location
- Designer's name or initials
- Company information for the architectural or engineering firm
- Architect or engineer's stamp and signature
- Scale
- Legend

Most drawings in plan sets are to a certain scale which allows the designer to have a very large object (a building) drawn onto a relatively small sheet while still maintaining proportion. Smaller projects may not be drawn to scale and as long as there is sufficient dimensioning to determine that code requirements are met, this should not be an issue. The actual scale that is used on a particular drawing is determined by the designer who produces the drawings. This decision is based on the sheet size and the actual size of what needs to be shown. The scale is usually noted in the title block of a drawing but may be shown elsewhere. Different details on the same sheet may use different scales. In this case, the scale is shown under each detail.

There are two commonly used types of scales: architect's scale and engineer's scale. An architect's scale normally contains eleven different combinations ranging from 1/8″ = 1′ to 3″ = 1′. Engineer's scales are in increments of 10 and range from 1″ = 10′ to 1″ = 100′. This type of scale is used when a large area needs to be fit onto one sheet. These scales are commonly used on site plans which will be discussed in detail later in the chapter.

The drawings in the plan set may also have a legend to show the symbols and reference marks used in the plan set. The use of symbols has been standardized through NFPA 170, *Standard for Fire Safety and Emergency Symbols*. This document contains symbols for a multitude of uses and is broken down into chapters that group symbols by general use, fire service use, architectural and engineering drawings and insurance diagrams, pre-incident planning, and emergency management mapping. If the designer does not use the standardized symbols of NFPA 170, there should be a legend that you can consult to see what the symbols represent.

Code Analysis

A <u>code analysis</u> is a summary of the features of fire protection and building characteristics that will affect the plan review. A code analysis may be provided and could be located on the title page or on a sheet of its own. The code analysis should list the applicable building and fire codes for the project and based on those codes should show the following information:

- Occupancy classification—The adopted model building code and NFPA 101, *Life Safety Code*® may have different classifications for the occupancy and both should be shown in the plan set. If there are multiple occupancies within the building, they should all be listed. If the occupancies will be separated, the hourly rating of the fire-resistance rated separation should be included.
- Construction type—The construction type of the proposed building is based on NFPA 220, *Standard on Types of Building Construction*, or the adopted model building code classification. The chapter *Building Construction* discusses the five types of building construction in detail.
- Building area—This will show the height and area for the proposed building and should also show the height and area limitations from the building or fire codes for the occupancy type. The building code, *Life Safety Code*®, and local and state zoning laws may dictate the maximum area per floor and number of potentially occupied levels that a specific occupancy type can have.

Fire Inspector Tips

General notes will be provided sporadically throughout the plan set. They contain important information and should be read thoroughly.

- Occupant load—Each area of the building will have an occupant load based on the minimum number of occupants which the egress system will need to support. The total for the building will be shown in this portion of the code analysis. The occupant load factors come from NFPA 101, Table 7.3.1.2, and the adopted model building code. Occupant load factors and how to calculate the occupant load is discussed in detail in the chapter *Occupancy Safety and Evacuation Plans*.
- Fire protection systems—Most buildings built today will have some form of a fire protection system ranging from an elaborate smoke evacuation system in a stadium to smoke alarms in a single family residence. Sprinklers, suppression systems, fire alarm systems may all be required by applicable building and fire codes. The code analysis should indicate whether these systems will be included and also whether or not they are required.
- Egress—Often, the code analysis will include information about the egress arrangement. This may consist of travel distance, common path of travel, dead-end corridors, number of exits, remoteness, and accessibility of the means of egress. Again the code requirements and building information should be provided.

Type of Drawings in Plan Sets

There are several types of drawings that you will encounter in a plan set. It is important that you know what information is contained in which type of drawing. The following section gives a basic overview of the drawing types that you will encounter in a plan set.

Site Plans

A site plan is a drawing that shows the overview of the lot being reviewed Figure 2 . This drawing usually contains contour lines of the elevation on site, proposed and existing buildings, property lines, and utilities. This drawing should contain a North indicator, so you can determine the orientation of the lot and proposed building. The site plans are usually identified with a "C" as part of the drawing number in the title block as they are predominantly drawings from a civil engineer.

Structural Plans

The **structural plan** is a drawing showing the proposed building's load-bearing components. Structural plans will be the first look at a proposed structure and will allow you to determine the construction type if it is not listed in the notes or summary Figure 3 . The structural plans will contain information about the foundation, floor, and roof construction. All load-bearing elements will be shown on a structural plan drawing to include columns, beams, girders, trusses, and arches. Structural plans bear the letter "S" in their drawing number in the title block.

Architectural Plans

Architectural plans will show the floor plans, reflected ceiling plans, building sections, elevations, and the building details Figure 4 . The drawings in the floor plans will give you the layout of the building, showing walls, doors, windows, and staircases. The drawings in the reflected ceiling plans will show the layout of the ceiling, beams, joists, architectural features, step changes, and skylights. The drawings in the building sections are larger scale drawings that will show more detail than the floor plan. The drawings will also show the features of a specific component, for example stair details will show the tread rise and run, toe dimensions, and elements of the

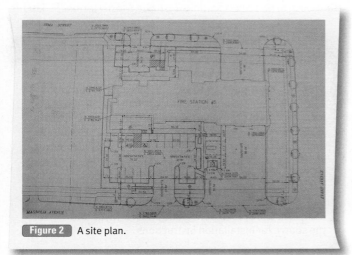

Figure 2   A site plan.

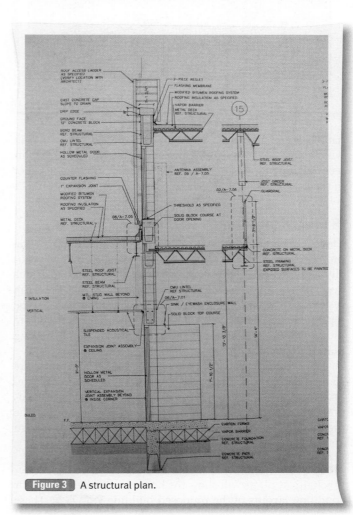

Figure 3   A structural plan.

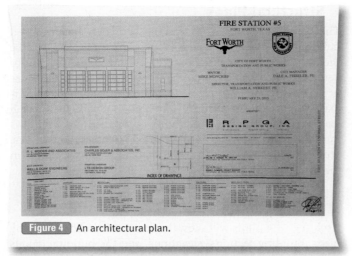

Figure 4 An architectural plan.

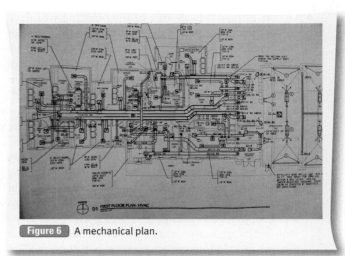

Figure 6 A mechanical plan.

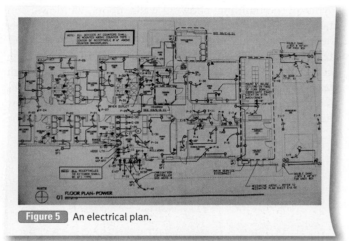

Figure 5 An electrical plan.

handrails and guards. Architectural plans are the drawings in which most of the life safety requirements are verified. Architectural drawings bear the letter "A" in their drawing number in the title block.

Electrical Plans

Electrical plans show circuits, outlets, and lighting within the proposed structure Figure 5. The electrical drawings will usually be based on a floor plan with the electrical annotations overlaid. The drawings will be labeled with an "E." The fire alarm system may be part of the electrical drawings but may also be separated and labeled with "FA" for fire alarm or "FP" for fire protection; these letters would all appear as part of the drawing number in the title block.

As a fire inspector, it is important to determine if the fire alarm drawings submitted are construction documents or a preliminary set. It is common for the architect to draw a "typical" fire alarm system. For the fire inspector, it is useful to review the preliminary drawing to see if the basics of the fire alarm system are correct.

Mechanical Plans

Mechanical plans will show the heating, ventilation, and air condition layouts of the proposed building Figure 6. These drawings will be marked with the letter "M" preceding the page number. The plumbing drawings may be part of the mechanical set or may be separate drawings marked with "P." Those drawings will show water and drain piping as well as fixtures. Often sprinkler system drawings are included as part of the mechanical set but could also be separated and labeled "FP." As with fire alarm systems, the architect may show a "typical" sprinkler system. This can give a basic overview of the system.

Specifications Book

In addition to the plan set, large projects may be accompanied by a specifications book. The specifications book is a collection of all of the information about a certain project, including the details of the components required in order to meet the project's design criteria. It also contains information specific to how items are to be stored prior to their installation on site. For example, the specifications book may require the contractor to store the interior doors in a clean and dry environment. Information in the specifications book may be helpful to you during site visits. For example, if fire-rated doors are stored improperly per the specifications book, this may shorten their life or lead to failure from corrosion down the road.

For the most part, the specifications book will follow the *Construction Specifications Institute Masterformat*™. This is

Fire Inspector Tips

There is a distinction between design drawings and shop drawings. Design drawings, usually designed and certified by the engineer or architect, are presented at the outset of a project. Shop drawings, however, continue to be edited in response to construction conditions. For this reason, shop drawings for a fire sprinkler system may not be completed until well into the construction process. This allows the designer of the system to review field conditions that may not have been noticeable during the initial design of the building. Shop drawings are required for specific systems such fire sprinklers, fire alarms, and fire protection hood systems. These drawings are the source for installation instructions.

Fire Inspector Tips

Reviewing drawings can be a long and arduous process, so it is important that you have the area you work in set up to be conducive to a plan review. The following is a list of supplies that should be found in an office or area for conducting reviews:
- Drafting table or large desk for drawings
- Adequate lighting
- Proper climate control
- Scales
- Compass
- Colored pencils/markers
- String
- Code books
- Checklists or worksheets
- Contact information for the designer

Fire Inspector Tips

Companies have created software programs to assist in plan reviews. Any potential software should be carefully evaluated by every member of the fire inspection team.

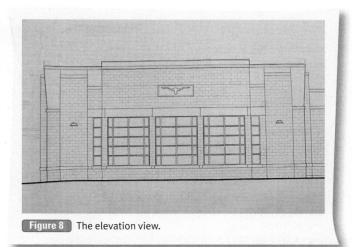

Figure 7 The plan view.

Figure 8 The elevation view.

essentially a way of standardizing the format for all the data that goes into the project. There are two versions that are commonly used. The first version was published in 1995 and contained 16 divisions of construction information. The second version was published in 2004 and contained 50 divisions. Both versions are still in use, but a 2014 Update version is now available. The choice on which version is used on a project will most likely come from the designer. Not all of the information contained in the specifications will be of use to you; however, the information presented in *Openings, Finishes, Furnishings, Fire Suppression,* and *Electrical* will be.

Type of Views

There are four basic types of views that will be encountered in a plan set. Each view is different, and the proper view will be chosen based on the type of information needed on a particular sheet:
- **Plan View**—A view where a horizontal slice is made in the building or area and everything above or below the slice is shown Figure 7. Examples of plan views include floor plans, site plans, and reflected ceiling plans.
- **Elevation View**—Elevations show the exterior of the building and will be labeled either using the direction the drawing is facing or by front, rear, left, and right Figure 8.
- **Sectional View**—A vertical slice of a building showing the internal view Figure 9. The section will be indicated on a floor plan by a cutting plane line and a symbol. Sections are also used in structural drawings to show the structural members as they are positioned vertically.

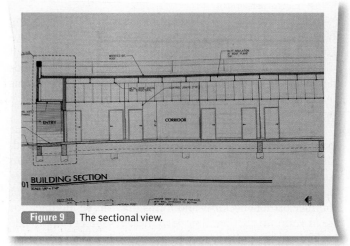

Figure 9 The sectional view.

- **Detail View**—Views of a specific element of construction or building feature in a larger scale providing more clarity Figure 10. It can be vertical or horizontal depending on what needs to be shown. In general, a detail view will be provided for a system or assembly that needs to be constructed on site to precise specification. The detail view affords you the ability

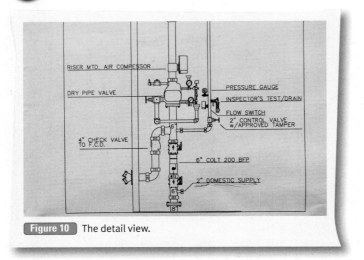

Figure 10 The detail view.

to accurately measure a feature that may not be possible on a smaller scale. An example would be trying to determine the distance between a handrail and wall on a 1/8 scale floor plan. The 2¼″ required by NFPA 101 would be impossible to determine at that scale. However, if a 1½″ scale detail was provided, this distance could be verified.

Plan Review Process

The plan review process can be broken down into the following phases:
- Application
- Review
- Approval
- Site visits
- Commissioning

The plan review process often begins with meetings during the conceptual phases of a project and continues through occupancy of the building. The meetings will most often include all of the stakeholders in the project. This may include the building owner, construction manager, architect, and engineer. This procedure can last for more than a year when dealing with large projects.

With one project having the potential to occupy a long time span, it is important for you to have a process that is streamlined. You should have standardized forms and checklists that you can use when reviewing drawings. Professional associations for fire marshals as well as other jurisdictions can be good resources for finding existing checklists and forms that can be modified for use in your jurisdiction. If none of these resources can provide a useful starting point, it may be your responsibility to develop checklists based on the codes and standards used in your jurisdiction. This can be accomplished by reading the codes and standards that you will be reviewing projects with and developing simple yes/no questions that can help you determine if the drawings meet these codes.

It is fairly common for the architect involved in a project to want to meet with the fire department and building department before final drawings are submitted for the review process.

A preliminary meeting helps to establish expectations and communications among various parties. During these meetings, the requirements for plan review should be explained. The architect should be made aware of current code editions, location of local code amendment resources, fees associated with the plan review, and an estimate of the time required for review. It is important to keep a log of all meetings that includes: attendees, topics discussed, and any tentative agreements made.

Application Phase

Every submittal made to your agency should follow a uniform route. This often begins with an application for plan review and payment of a fee. Make sure that a record is kept of when the plan set application is accepted, when the plan set is approved and/or disapproved, when the plan set is released, and when the plan set is returned. Having a log of that information is crucial to ensuring that the plan set is reviewed in a timely fashion. quickly accessible, and tracked as it moves through the review process.

Review Phase

During the review phase you will examine all of the plans contained in the plan set.

Site Plan Review

The site plan is the generally the first look at the proposed building in relation to lot lines and other structures. You should begin by looking at the access to the structure on these drawings. Proper access is important for fire apparatuses as well as fire fighters. Most model codes have minimum widths, heights, and location of fire department access roads. You should be familiar with the turning radius of the fire department's apparatus as well as the height of the tallest piece of apparatus. Other items that need to be considered with apparatus access are turnarounds, dead-ends, dry wells, and weight limits of any bridges on the site. Once the access roads are determined to be adequate, you should look at fire fighter access into the structure. The access roads must come close enough to a door into the structure so that fire fighters can enter the structure with hose lines that are connected to the fire pumper apparatus located on the access

> **Fire Inspector Tips**
>
> Meetings with architects should be encouraged because strong professional partnerships are key to a successful plan review. These meetings also build good public relations.

> **Fire Inspector Tips**
>
> The application should contain all of the important project information. This will be your reference throughout the entire project. Make sure that you have accurate contact information for all interested parties. This will help you later on if you need to discuss any issues that arise during the plan-review process.

road and still have enough hose to traverse the building interior. NFPA 1, for example, requires these access roads extend to within 50′ (1524 cm) of an exterior door that can be opened from the outside.

The site plan will also show any hydrants that will be installed on the site and you should use this to determine if hydrant spacing is appropriate for the structure. To do this, you will need to know the construction type and size of the building. That information will be used to establish the required fire flow as well as the hydrant spacing. Calculating the fire flow is covered in Chapter 9: *Fire Flow and Fire Suppression Systems*.

Another important and often overlooked item on the site plan is the landscaping. This may be included on the site plan or may be provided in separate landscape drawings. You should look at the locations of trees and bushes to make sure they do not have the potential to obstruct the fire department connection, hydrants, exit discharge, or fire access roads. Landscaping concerns not addressed in plan review can become obvious years later.

The site plans are an integral tool to be used for subsequent property inspections. This will be what future fire inspectors use to see if access roads have been obstructed. They also play a key role in preplanning. The incident commander will be able to determine ahead of time what roads can support the weight of fire apparatus, where the closest water supply is located, and where entry to the structure can be attained.

Structural Plans

The structural plans contain information regarding the building's load-bearing components. You will determine the construction type from the structural drawings. Most model building codes specify a maximum height and area. These height and area limitations are based on the occupancy of the building as well as the construction type. The *Life Safety Code®* also has construction limitations on some specific occupancy types such as assembly, day care, health care, and detention and correction centers.

You should be able to identify the key items on a structural drawing and should be familiar with the different types of loads that are exerted on a structure. Of particular importance to you is the use of lightweight wood trusses. These assemblies are becoming increasingly popular due to the reduction in cost of labor. Responding fire fighters need to be aware of buildings built using lightweight wood trusses because these building components have been shown to fail relatively early under fire conditions.

Architectural Plans

The architectural plans are where a majority of the information for compliance with the adopted codes will be located. The first thing that you should look for when reviewing the architectural plans is the occupancy classification. It is quite possible that the building will include more than one occupancy type. When two or more occupancies exist and are not separated by fire-resistance rated construction, the most restrictive code requirements of the two occupancies apply. Once the occupancy type has been determined, you should determine the occupant loads for all of the spaces in the building. Chapter 7: *Occupancy Safety and Evacuation Plans* discusses means of egress and occupant loads in detail.

> **Fire Inspector Tips**
>
> You should be aware of what to look for on the structural drawings; however, the review of the building specifics is usually left to the building department.

> **Fire Inspector Tips**
>
> Most jurisdictions have building departments that operate independently of the fire department and also look at projects for compliance with codes and standards. Building departments can be a resource to you in your work. They can be another set of eyes on a project or offer an area of expertise that you do not have. It is a good rule to try and collaborate with these people and foster good working relationships.

The next step in reviewing the architectural plan is to determine if the plan set meets the specific egress requirements of the occupancy. You will need to determine that the means of egress is appropriate for the following: types of components, egress capacity, number of exits, arrangement of means of egress, travel distance to exits, discharge from exits. This begins with determining the number of occupants that will likely be in the structure due to the use of the space and the occupant load factor. Once you determine the number of people that will be in the area, you need to make sure that they have enough room to evacuate. For example, NFPA 101 attributes a per person/per inch factor for movement over level areas and stairs. It also takes into consideration occupants who may have decreased speed of travel in an emergency or it also considers high hazard occupancies where fire spread may be rapid. You will have to measure the distances traveled for occupants to get into a protected exit to make sure that is within the allowable limits for the occupancy classification.

Once these tasks are accomplished, you will continue examining all fire protection systems. This will include vertical openings, hazardous areas, interior finishes, fire alarms, extinguishing requirements, stairwell pressurization systems, fire dampers, and separation spaces. From there you will examine the building features, utilities, HVAC, elevators, and laundry and trash chutes.

Vertical openings are created in almost all buildings and include stairs, elevators, escalators, ventilation ducts, and other utilities or building features. Vertical openings will also be shown on as pipe chases and ducts. These openings can create a path for smoke and fire to travel through the building. Vertical shafts in a structure will normally be required to have a fire resistance rating based on the number of floors that they penetrate. It is important to ensure that when a fire-resistance-rated assembly is penetrated with a duct for ventilation that that duct is provided with a damper that prevents the spread of fire into that shaft.

Important parts of fire protection are containment of smoke and fire to one area and preservation of the means of egress. Smoke creates a significant amount of pressure and when not properly contained, can travel through a building at incredible

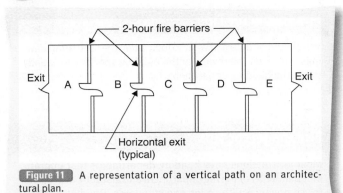

Figure 11 A representation of a vertical path on an architectural plan.

Fire Inspector Tips

The level of fire-resistance rating for a particular vertical opening will be derived from the building code and Life Safety Code. NFPA 251 is used to create a standardized testing process for these materials. A third-party testing agency such as Underwriters Laboratory (UL) will test certain assemblies such as walls and partitions and assign them an hourly fire-resistance rating. Assemblies that pass the tests will be given assigned a number and will be included in the Fire Resistance Directory published by UL.

Fire Inspector Tips

You may need to consult specific codes and standards that deal with a certain type of conveying system or utility to make sure that those requirements are met in the plan set.

Fire Inspector Tips

There are fire and life safety concerns specific to hazardous occupancies. If the occupancy has several unique concerns, then there may be an individual standard dedicated to that type of occupancy. For example when there is a fire alarm activation at a prison, the occupants are not permitted evacuate outside with the option to return later. The same is true for a hospital, but for different reasons. Consult the *Uniform Fire Code* for a full list of common hazardous occupancies, and to find references to the appropriate standards for common hazardous occupancies. Given the broad scope of structures and hazards, it is more important for you to know where to find the answers than to try and know all the answers.

speed. Containing the smoke and fire involves the use of fire-resistance rated construction, opening protection, and fire stop systems.

When an architectural plan shows an hourly rated wall, you need to verify that it is an assembly that has been tested and is in accordance with NFPA 251, *Standard Methods of Tests of Fire Resistance of Building Construction and Materials* (or other adopted test standard), and will match the tested assembly once built

Figure 11. Any doors in a rated wall need to be listed fire doors; penetrations for ducts will need to have dampers; and penetrations for cable, wires, pipes, and similar items will need to be protected by a firestop system. Much like the rated wall assembly, the firestop system is an assembly is specific to the penetrating item, the wall type being penetrated and the size of the opening. During the site visit phase it is important that you verify that the firestop system is appropriate and, during subsequent fire inspections, that the field-built assembly matches the architectural plan.

Electrical and Mechanical Plans

The electrical and mechanical plans will show information about the building utilities and HVAC systems. Electrical plans will show the wiring diagrams and lighting plan for the building. These will help you to determine locations of emergency lighting and exit signs as well as if lighting is provided in the means of egress. It will be difficult if not impossible to determine if the level of illumination of the means of egress will be adequate. Upon approval of the plans, you should advise the designer that the illumination level will be verified during a site visit. Exit signs should also be field verified for visibility.

In a residential occupancy, the electrical drawings may also show the location of any smoke and carbon monoxide detectors. Local and state codes in the jurisdiction will determine when and where these devices are required. When they are required, the codes usually direct the installation to follow the requirements of the NFPA 72, *National Fire Alarm and Signaling Code*, and NFPA 720, *Standard for the Installation of Carbon Monoxide (CO) Detection and Warning Equipment*.

Working with the mechanical plans, you will be able to determine what type of heating appliances (furnaces, water heaters, etc.) are being used and where they are located. This is important because the building and fire codes may require separation of heating appliances with fire-resistance rated construction depending on the size of the appliance and the occupancy type. In addition, the locations of fuel burning appliances in a residential occupancy will usually drive requirements for carbon monoxide detection.

Building utilities will be provided throughout the building and because of that, wires, pipes, and ducts will create vertical openings and penetrations in fire-resistance rated walls and ceilings. You will need to verify that the proper fire-resistance rating is provided around pipe chases and shafts for ventilations systems. Ducts may require smoke and/or fire dampers when they exit fire-resistance-rated shafts, the building code and fire code will determine what rating is required. In addition, building utilities are responsible for a large number of penetrations in fire-resistance-rated assemblies. The Life Safety Code as well as other model building codes specify when firestop systems are required and what constitutes a proper firestop system. Not all penetrations in fire-resistance rated walls and ceilings will require a fire stop system, so consulting the codes is necessary to determine requirement standards.

Fire Protection System Plans

Fire protection systems are often required in new buildings. A common example is an automatic sprinkler system. The adopted model building code and Life Safety Code will be the driving force in the requirements for sprinklers. This may be a specific requirement or an alternative offered by the designer when something in the building cannot meet the prescriptive nature of the code. In the building codes, the installation of a sprinkler system allows the building to be built taller and with larger area than a building of the same construction type not protected by a sprinkler system.

There are three standards that govern the installation of sprinkler systems; NFPA 13, NFPA 13R, and NFPA 13D. The first, NFPA 13, *Standard for the Installation of Sprinkler Systems*, can be used in most buildings and is intended to protect life and property within the building. The next two, NFPA 13R, *Standard for the Installation of Sprinkler Systems in Low-Rise Residential Occupancies*, and NFPA 13D, *Standard for the Installation of Sprinkler Systems in One- and Two-Family Dwellings and Manufactured Homes*, are primarily concerned with life safety and are not intended to control or suppress a fire, although that is often the result. With any of the systems mentioned above, the review process will essentially be the same. When the following steps are adhered to, a thorough evaluation of sprinkler plans and calculations will be the result. The chapter *Fire Suppression Systems* discusses these systems and how to thoroughly evaluate these systems in detail.

The evaluation begins with determining if the occupancy or commodity classification chosen by the designer is correct. For example, offices would have a classification of light hazard. This is referring to the expected heat release and speed of fire spread based on the anticipated fire loading due to the usage of the space.

Fire Detection Systems Plans

Many buildings also require the installation of a fire detection system. NFPA 72, *National Fire Alarm and Signaling Code*, sets forth the minimum installation requirements for these systems, NFPA 70, *National Electrical Code*, also regulates fire detection installation. The first thing to look at with a fire detection plan is the intent of the system. The building code or fire code will dictate where fire detection systems need to be installed and will go into specific locations for initiating and notification devices. In addition, the Americans with Disabilities Act (ADA) sets requirements for where audible and visible notification devices are required.

Initiating devices can be installed for a multitude of reasons, including occupant notification and evacuation, elevator recall, door closing, door unlocking, and activation of extinguishing systems. You should determine that initiating devices are listed and that the installation proposed is in accordance with the device's listing and NFPA 72. The maximum spacing of heat and smoke detectors will be determined by a NFPA 72 standards or the detector's listed spacing, which will be included on the cut sheet of the device published by the manufacturer and include by the system designer in the plan submittal. Other factors that will reduce the spacing of detectors include ceiling slope, height, and construction (beams, joists, and ceiling pockets).

Local or state codes may also mandate the installation of manual fire alarm boxes within the building. If these devices are required, NFPA 72 will state that the operable part of these devices need to be mounted not less than 3½' and not more than 4½' above floor level. They should be installed where they are conspicuous, unobstructed, and accessible. The travel distance to the manual fire alarm box should not exceed 200' (6096 cm) measured horizontally and on the same floor.

Notification appliance location will depend upon the intent of the system and the building's fire emergency plan. As with initiating devices, the notification appliances need to be listed and installed in accordance with their listing. NFPA 72 sets the maximum sound pressure level of the ambient sound and notification appliance at the minimum hearing distance of 110 dBA. You will not be able to determine this from the plans, but upon system acceptance, this should be verified. The devices need be 15 dBA above the average ambient sound levels which will depend on the use of the building.

Other systems in the building such as sprinklers, fire pumps, water tank levels, and suppression systems may be tied in to the fire-detection system. It is important to identify how the alarm will be activated. Will water flowing from the sprinkler system send an alarm signal to the fire alarm control panel and notify the occupants to evacuate the building? Will the closing of a sprinkler valve be sent as a supervisory signal to the fire alarm? Will a signal be transmitted from there to an attended location, off site to a central station, or to the fire department?

Secondary Power

NFPA 72 specifies how long a fire alarm system is required to function and alarm on secondary power. This depends on the type of fire-alarm system and if a standby generator is available. You should verify that the number and type of devices shown on the plan set matches what is included on the battery calculations.

Chemical Suppression System Plans

A common fire protection system is the chemical suppression system. These systems are usually chemical or gaseous systems designed to protect a specific hazard or area. A common location for a chemical suppression system is the commercial kitchen. A commercial kitchen will have a dry chemical or wet chemical extinguishing system depending on the type of cooking medium and the age of the system. Those types of suppression systems are governed by NFPA 17, *Standard for Dry Chemical Extinguishing Systems*, and NFPA 17A, *Standard for Wet-Chemical Extinguishing Systems*, respectively.

Gaseous suppression systems are found in computer rooms or areas housing sensitive electrical or communications equipment. It's important for you to make sure that the system is adequate for the hazard that is being protected. The specific standard for the installation being proposed will have to be consulted to ensure that the system meets the requirements. The fire alarm code in the jurisdiction may also dictate that these systems tie in to the fire-detection system. In these instances the fire alarm code should be consulted to determine the specific suppression-system signals that need to be reported to the fire-detection system.

Deficiencies and Variances

When deficiencies are found during a plan review, they need to be brought to the attention of the submitter and, depending on the severity, may require correction before the permit is issued. This notification should be done in writing and should include the deficiency, its location within the plan set, and the applicable code violation. If several deficiencies are found or if the problems are complex in nature, a meeting with the submitter may be appropriate. During these meetings it is important that you do not redesign the building. Your job is to simply determine if the designer's solutions meet the applicable code requirements. You should be helpful and explain the deficiencies but should remember that it is outside the scope of your authority to redesign a building.

Unfortunately, not all designers are adept in ensuring that their designs meet the applicable fire codes. When it becomes apparent that a designer is not familiar with the fire codes, you may want to suggest that the designer hire a consultant who knows the codes before the plans are submitted for review. In the end this will save both you and the designer valuable time, effort, and fees associated with repeat plan reviews.

There may come a time when a project cannot conform to a section of code. This usually occurs when an older or historic building is renovated or changes use. When a situation such as this occurs, it may be necessary to seek a **variance** (or appeal) from the adopted code. A variance is relief from a particular code requirement. Section 1.10 of NFPA 1 discusses general requirements of the **Board of Appeals**. You will usually be required to appear before the Board of Appeals when a project you have reviewed comes for a hearing. You should be well prepared on the subject being discussed and be prepared to answer questions relative to the review or the deficiencies found. A review of the project and all notes relating to it are an important part of a successful hearing. The fire inspector is often asked to comment on the project, alternatives proposed, and whether the fire inspector supports the project moving forward with the appeal. It's important to remember that codes represent the minimum amount of safeguards and that new construction should find a way to meet these minimums without needing an appeal or a variance.

Once all of the deficiencies are corrected and a permit has been issued, you will still need to work with the project. During the building process, it is inevitable that changes will need to be made in the field and these changes will usually include deviation from the approved drawings. It is within your authority to require that those changes be submitted for review and approval. The scope of the changes proposed should help you determine if a submission of changes is necessary.

■ Approval

If all of the plans in the plan set meet your local codes and standards, then the plan set may be approved. You will notify the interested parties that the plan set is approved and note the approval of the plan set in the record on the future building. Once the plan set is approved, the next step is to ensure that it is built according to plan.

■ Site Visit

A successful building design is not complete until it is built according to the approved plan set. The process of verifying this approved design requires site visits, also called in-the-field inspections, during construction. Periodically, you should visit the construction site to ensure that the approved plan set is being used for construction and that no unauthorized changes have been made.

> **Safety Tips**
>
> Visiting a construction site can become hazardous if you are not properly prepared. There are many injury hazards and as with every other aspect of the fire department the proper PPE must be worn. For visiting a building under construction, this includes a hard hat, safety-glasses, steel-toed shoes, and possibly a dust mask or respirator.

> **Fire Inspector Tips**
>
> Occupied buildings will often have some additional concerns that were not apparent on the drawings. It's important to these issues during the plan review process so that the final inspection and occupancy of the building will go smoothly. Almost every occupancy has storage of some quantities of hazardous materials. If the specific materials, quantities, and locations can be identified during the plan-review process you can do a review under NFPA 1, *Fire Code,* and applicable building code to determine if they are within the allowable limits and what safeguards need to be in place. Depending on the amount and type of storage, the building or area with hazardous materials may need to be identified with an NFPA 704 diamond. This symbol helps responding fire fighters to quickly identify the extent of the chemical hazards located in the building.

> **Fire Inspector Tips**
>
> Portable fire extinguishers are sometimes shown on the floor plans but are not easily located in the field. NFPA 1 contains a table showing which occupancies require portable fire extinguishers. Once it is determined that portable extinguishers are required, NFPA 1 has extracted text from NFPA 10, *Standard for Portable Fire Extinguishers*, stating maximum travel distance, size, and units of extinguishing agent required. The class of extinguisher required will depend on how the contents of the building are classified. It's important to verify in the field that the extinguishers are either mounted and visible, or signed and easily accessible.

You should review the content of approved plan set and the local procedures on performing a site visit prior to visiting the site. Field inspections must be coordinated with the construction superintendent to make the best use of time. When you arrive on site, you should contact the project manager or other person in charge of the site and explain your intentions to conduct an in-the-field inspection. Being that this is an active construction site, proper personal protective equipment (PPE) such as a hard hat, steel-toed boots, eye protection, hearing protection, and a respirator or dust mask may be required depending upon the conditions you will encounter.

If the approved plan set is not being used, you should halt construction until approval is obtained. This can be accomplished by following the procedures set up by your agency. Ordinances, laws, or other regulations will have set up the process for stopping work if the work has been misrepresented or is not being conducted in accordance with the approved plan set. This may seem excessive and will usually not be well received by the project manager, but it is a necessary step in ensuring that what is being built meets standards. You should explain that it is far easier to halt operations temporarily than it is to rebuild on the project.

Commissioning

The final step in the plan review process consists of a final inspection and **commissioning**, or testing, of all of the building's fire protection systems prior to occupancy. Testing all of the fire protection systems installed ensures that they react in accordance with their basis of design. The specifics of how a system is properly commissioned will be contained in the specific installation standard. You should consult those standards and be familiar with the testing procedures prior to arriving on site. In addition to system acceptance, you should work with the responding fire fighters to identify important features of the site as well as the building.

Performance Based Design

Historically the fire codes have been prescriptive in nature. This means that there were specific requirements that needed to be met. An alternative that is now available to designers is **performance-based design**, in which fire safety solutions are designed to achieve a specified goal for a specified use or application. Under this option, a qualified person will prepare an evaluation of a building to include the performance objectives and applicable scenario. This includes any calculations that are performed to demonstrate the proposed design's fire and life safety performance. NFPA 1 lays out design scenarios that may be used by the person completing the performance-based design. You have the final say in the approval of the design as well as the parameters used by the designer. This may seem like an intimidating task for a fire inspector when the design may have been performed by a **fire protection engineer**, a person with special training in protection of life and property from fire events.

Technical Assistance

When a proposed design, operation, process, or new technology exceeds your capabilities, NFPA 1 allows for the use of technical assistance. This technical assistance involves the review of the matter in question by an independent third party. The owner or designer is usually responsible for the cost of the third party review, so it's important that you quickly inform the submitter when a third party review will be mandated. The evaluation performed by the third party review will be submitted to you for final approval. If you need some direction in the technical assistance process, consult *Guidelines for Peer Review in the Fire Protection Design Process* published by the Society of Fire Engineers (SFPE).

Equivalencies and Alternatives

Most of the model codes and standards have provisions for equivalencies and alternatives. **Equivalencies** set forth the permissibility of a method, process, system or device of superior quality, strength, durability, etc. It is important to note that the submitter is required to provide technical documentation to determine the equivalency and to show that the equivalency is approved for the intended purpose. If you are installing a sprinkler system with a pipe that is not of the types listed in NFPA 13, as long as there is technical documentation showing its durability and suitability for sprinkler system service it would be allowed under this clause, subject to approval of the AHJ.

An **alternative clause** allows for the code provisions to be altered to an alternative that would not reduce the level of safety within the building. In many cases the use of alternatives in existing buildings may actually result in a safer building than was originally being occupied.

Wrap-Up

■ Chief Concepts

- The objective of a plan review is to make certain that buildings are safe for the public to occupy and fire fighters to operate in.
- Plans, blueprints, construction documents, shop drawings, and plan sets are all terms used to refer to the documents that will be reviewed during a plan review.
- The code analysis should list the applicable building and fire codes for the project and based on those codes should show the following information.
- The types of drawings in plan sets include site plans, structural plans, architectural plans, electrical plans, and mechanical plans.
- Large projects may be accompanied by a specifications book.
- The four basic types of plan set views are plan view, elevation view, sectional view, and detail view.
- The plan review process consists of the following phases: application, review, approval, site visits, and commissioning.
- With performance-based design review, a qualified person will prepare an evaluation of a building to include the performance objectives and applicable scenario. This includes any calculations that are performed to demonstrate the proposed design's fire and life safety performance.

■ Hot Terms

Alternative clause This clause allows for the code provisions to be altered and an alternative offered that would not reduce the level of safety within the building.

Architectural plan A drawing showing floor plans, elevation drawings, and features of a proposed building's layout and construction.

Board of Appeals A group of persons appointed by the governing body of the jurisdiction adopting the code for the purpose of hearing and adjudicating differences of opinion between the authority having jurisdiction and the citizenry in the interpretation, application, and enforcement of the code. [NFPA 1]

Code analysis A summary of the features of fire protection and building characteristics in a plan set.

Commissioning The time period of plant testing and operation between initial operation and commercial operation.

Detail view A view on a drawing of a specific element of construction or building feature in a larger scale to provide further clarity.

Electrical plans Design documents in a plan set showing the power layout and lighting plan of a proposed building.

Elevation view A view in a drawing showing the exterior of the building.

Equivalencies The use of systems, methods, or devices of equivalent or superior quality, strength, fire resistance, effectiveness, durability, to those prescribed by a code or standard.

Fire protection engineer An engineer with specialized training in protection of life and property from fire events.

Mechanical plans Drawings in a plan set showing the proposed plumbing, HVAC, or other mechanical systems for a building.

Performance-based design A design process whose fire safety solutions are designed to achieve a specified goal for a specified use or application. [NFPA 914]

Plan set Created by design professionals, plan sets include a series of drawings detailing how a proposed building will be built. Also known as plans, blueprints, construction documents, or shop drawings.

Plan view A view on a drawing where a horizontal slice is made in the building or area and everything above or below the slice is shown.

Sectional view On a drawing in a plan set, a vertical slice of a building showing the internal view of the building.

Site plan A drawing showing the building and surrounding area, including items such as roads, driveways, and hydrants.

Specifications book A collection of all of the information about a project that may be provided to the fire inspector in addition to the plan set during a plan review.

Structural plan A drawing showing the proposed building's load-bearing components.

Variance A waiver allowing a condition that does not meet a recognized code or standard to continue to exist legally.

Fire Inspector *in Action*

A developer is planning to build a new hotel and restaurant in the center of your town. The town planners are very excited about the developer's plans and hope that it will be the first step in revitalizing the downtown area. As a Fire Inspector II, you attend the initial meetings with the developer and his architect to discuss how the project can meet their vision and meet the codes. After a year of meeting, you are handed a set of plans to review.

1. When you open the plan set, you should find:
   A. site plans.
   B. structural plans.
   C. electrical plans.
   D. all of the above.

2. The mechanical plans will indicate:
   A. the location for all electronic devices for the structure.
   B. heating, air conditioning, and ventilation layouts.
   C. the location of the boiler room.
   D. the location of all machines in the hotel and restaurant.

3. On the mechanical plan, you see a proposed solar panel for the roof. This is the first time you have encountered such a device, so you:
   A. inform the developer that you first have to take an online course on solar energy before you can complete the review.
   B. bring the plans over to your brother-in-law, who is a mechanical engineer.
   C. contact an independent third party with technical expertise.
   D. tell the developer that solar panels are not approved in your jurisdiction.

4. The final step of the plan review process is:
   A. commissioning.
   B. site visit.
   C. approval.
   D. a letter to the developer.

Occupancy Safety and Evacuation Plans

CHAPTER 7

NFPA 1031 Standard

Fire Inspector I

4.3.2 Compute the allowable occupant load of a single-use occupancy or portion thereof, given a detailed description of the occupancy, so that the calculated allowable occupant load is established in accordance with applicable codes and standards. (pp 119–122)

(A) Requisite Knowledge. Occupancy classification; applicable codes, regulations, and standards adopted by the jurisdiction; operational features; fire hazards presented by various occupancies; and occupant load factors. (pp 119–122)

(B) Requisite Skills. The ability to calculate occupant loads, identify occupancy factors related to various occupancy classifications, use measuring tools, and make field sketches. (pp 119–122)

4.3.3 Inspect means of egress elements, given observations made during a field inspection of an existing building, so that means of egress elements are maintained in compliance with applicable codes and standards and deficiencies are identified, documented, and reported in accordance with the applicable codes and standards and the policies of the jurisdiction. (pp 124–140)

(A) Requisite Knowledge. Applicable codes and standards adopted by the jurisdiction related to means of egress elements, maintenance requirements of egress elements, types of construction, occupancy egress requirements, and the relationship of fixed fire protection systems to egress requirements and to approved means of egress elements, including, but not limited to, doors, hardware, and lights. (pp 124–140)

(B) Requisite Skills. The ability to observe and recognize problems, calculate, make basic decisions related to means of egress, use measuring tools, and make field sketches. (pp 124–140)

4.3.10 Verify that emergency planning and preparedness measures are in place and have been practiced, given field observations, copies of emergency plans, and records of exercises, so that plans are prepared and exercises have been performed in accordance with applicable codes and standards and deficiencies are identified, documented, and reported in accordance with the applicable codes and standards and the policies of the jurisdiction. (pp 140–143)

(A) Requisite Knowledge. Requirements relative to emergency evacuation drills that are required within the jurisdiction, ways to conduct and/or evaluate fire drills in various occupancies, and human behavior during fires and other emergencies. (pp 140–143)

(B) Requisite Skills. The ability to identify the emergency evacuation requirements contained in the applicable codes and standards and interpret plans and reports. (pp 140–143)

4.3.11 Inspect emergency access for an existing site, given field observations, so that the required access for emergency responders is maintained and deficiencies are identified, documented, and corrected in accordance with the applicable codes, standards, and policies of the jurisdiction. (pp 124–140)

(A) Requisite Knowledge. Applicable codes and standards, the policies of the jurisdiction, and emergency access and accessibility requirements. (pp 124–140)

(B) Requisite Skills. The ability to identify the emergency access requirements contained in the applicable codes and standards, observe, make decisions, and use measuring tools. (pp 124–140)

Fire Inspector II

5.3.1 Compute the maximum allowable occupant load of a multi-use building, given field observations or a description of its uses, so that the maximum allowable occupant load calculation is in accordance with applicable codes and standards. (pp 119–123)

(A) Requisite Knowledge. How to calculate occupant loads for an occupancy and for building use; code requirements, regulations, operational features, and fire hazards presented by various occupancies. (pp 119–123)

(B) Requisite Skills. The ability to calculate occupant loads, identify occupancy factors related to various occupancy classifications, use measuring tools, read plans, and use a calculator. (pp 119–123)

5.3.5 Analyze the egress elements of a building or portion of a building, given observations made during a field inspection, so that means-of-egress elements are provided and located in accordance with applicable codes and standards and deficiencies are identified, documented, and reported in accordance with the policies of the jurisdiction. (pp 124–140)

(A) Requisite Knowledge. Acceptable means-of-egress devices. (pp 124–140)

(B) Requisite Skills. The ability to calculate egress requirements, read plans, and make decisions related to the adequacy of egress. (pp 124–140)

5.3.7 Evaluate emergency planning and preparedness procedures, given existing or proposed plans and procedures and applicable codes and standards, so that compliance is determined. (pp 140–143)

(A) Requisite Knowledge. Occupancy requirements for emergency evacuation plans, fire safety programs for crowd control, roles of agencies and individuals in implementation and development of emergency plans. (pp 140–143)

(B) Requisite Skills. The ability to compare submitted plans and procedures with applicable codes and standards adopted by the jurisdiction. (pp 140–143)

5.4.2 Compute the maximum allowable occupant load, given a floor plan of a building or portion of the building, so that the calculated occupant load is in accordance with the applicable codes and standards and the policies of the jurisdiction. (pp 119–123)

(A) Requisite Knowledge. How to calculate occupant loads for an occupancy and building use, code requirements, regulations, operational features such as fixed seating, and fire hazards presented by various occupancies. (pp 119–123)

(B) Requisite Skills. The ability to calculate accurate occupant loads, identify occupancy factors related to various occupancy classifications, use measuring tools, read plans, and use a calculator. (pp 119–123)

5.4.5 Verify that means of egress elements are provided, given a floor plan of a building or portion of a building, so that all elements are identified and checked against applicable codes and standards and so that deficiencies are discovered and communicated in accordance with the policies of the jurisdiction. (pp 124–140)

(A) Requisite Knowledge. Applicable codes and standards adopted by the jurisdiction, the identification of standard symbols used in plans, and field verification practices. (pp 124–140)

(B) Requisite Skills. The ability to read plans and research codes and standards. (pp 124–140)

Additional NFPA Standards

- **NFPA 1** *Fire Code*
- **NFPA 13** *Standard for the Installation of Sprinkler Systems*
- **NFPA 70** *National Electric Code*
- **NFPA 72** *National Fire Alarm and Signaling Code*
- **NFPA 101** *Life Safety Code®*
- **NFPA 5000** *Building Construction and Safety Code*

FESHE Objectives

Fire Plans Review

1. Determine occupancy classification, construction type: calculate occupant load and, the height and area of a building. (pp 119–123)
2. Determine the appropriateness of the three components of a building's egress system (exit access, exit, and exit discharge), verify building compartmentation and the proper enclosure of vertical openings. (pp 124–140)
3. Verify the proper illumination for exit access, the arrangement of exit lighting and perform a life safety evaluation of the egress arrangement of a building. (pp 124–140)

Knowledge Objectives

After studying this chapter, you will be able to:

1. Discuss the importance of calculating the occupant load in determining the egress requirements of the building.
2. Describe how to ensure that the means of egress meet all applicable codes and standards.
3. Describe how to ensure that emergency access for an existing site meets all applicable codes and standards.
4. Describe how to evaluate emergency evacuation plans to ensure they meet all applicable codes and standards.
5. Describe how to evaluate evacuation drills in various occupancies.

Skills Objectives

After studying this chapter, you will be able to:

1. Compute the occupant load for a single-use occupancy.
2. Compute the occupant load of a multi-use building.
3. Compute a maximum allowable occupant load.

You Are the Fire Inspector

As a newly assigned fire inspector following a brief stint shadowing more senior personnel, you are given your first assignment. Upon visiting a local tavern, the owner advises that he plans to subscribe to the sports service, NFL Sunday Ticket. This requires that owner to pay a fee based the maximum occupant load of the tavern. The owner asks that you establish the occupant load keeping in mind that the lower the number of seats, the lower the fee that must be paid.

1. What observations must you make before beginning the process of establishing the occupant load?
2. Does the type of seating have an impact on the maximum number of occupants?
3. What effect will the employee service areas have on the computations?

Introduction

While there is no one universal set of building, fire, or life safety codes in use across the United States, the National Fire Protection Association (NFPA), and the International Code Council (ICC) are the two most commonly used building and fire model codes. Each state either adopts a model building or fire code, or creates their own codes to meet their specific building and fire safety needs. Some cities, towns, and counties further modify the model codes to meet their particular needs. For example, model building and fire codes require fire sprinklers in most buildings at 12,000 square feet (1115 m^2). However, many communities across the county have modified that requirement require fire sprinklers at 5,000 square feet (465 m^2) or even for every new building built, regardless of size. Some communities have required sprinklers in single family homes for over 20 years. However, it is only recently that the controversial subject of sprinklers in single family homes has made its way into the model codes.

NFPA 101, *Life Safety Code* is used in this text as the code of reference for determining the means of egress requirements. The *Life Safety Code* was first developed in 1913 by the Committee on Safety to Life. In 1927 it was named the *Building Exits Code*, and later named the *Life Safety Code*. It is neither a building nor fire code. There are some building and fire code requirements, but they are related to safely exiting a building, not the construction or protection of the building, as such. The *Life Safety Code* forms the basis for most of the egress requirements contained in the model building and fire codes. The *Life Safety Code* also addresses the construction, protection, and occupancy features necessary to minimize dangers to life from the effects of fire, including smoke, heat, and toxic gases. It also establishes minimum criteria for the design of egress facilities to allow for the prompt escape of occupants from buildings or to safe areas within buildings.

Occupant Load

The <u>occupant load</u> is the number of people who might occupy a given area. The occupant load reflects the maximum number of people anticipated to occupy the building space(s) at any given time. The occupant load must not be based only on normal occupancy, because the greatest hazard can occur when an unusually large crowd is present, which is a condition often difficult for the fire inspector to control.

While creating the occupant load figure is not too difficult, the occupant load figure can be fluid. For example, a large

Fire Inspector Tips

Navigating the *Life Safety Code* can be confusing at times, especially to new fire inspectors. To assist the fire inspector, the *Life Safety Code* has an annex that contains explanatory material. This annex offers great insight to what the authors of the *Life Safety Code* intended. Areas in the *Life Safety Code* that have annex material are marked with an asterisk (*). By referring to this material found in the back of the *Life Safety Code*, you may be able to make more informed decisions. A solid bar in the margin indicates a change from the previous edition of the *Life Safety Code*. It will not tell you what the change was, only that a change occurred from one edition to another.

Fire Inspector Tips

You must have access to and be fully knowledgeable of the codes, standards, and regulations in effect in your jurisdiction.

open room will have one occupant load Figure 1. When it is filled with table and chairs, such as in a banquet, it has another occupant load figure Figure 2. When posting occupant loads, some fire departments indicate two numbers. The first is for a table and chair configuration and the second for an open area.

Ideally the design professional for the building should determine the occupant load figure based upon the uses within the building, and include that on the building plans that are submitted to the authority having jurisdiction (AHJ). This information is critical to determine the number, distribution, and travel distances to those exits. It is also incumbent upon you during the plan review process to review those numbers for accuracy. Mistakes do happen and some unscrupulous designers may try to increase the occupant load due to pressures from the building owner.

Occupant Load Factors

Standard occupant load factors are listed in Table 7.3.1.2 in the *Life Safety Code* and are shown here in Table 1. The table shows two types of occupant load factors: gross and net. Most of the factors shown in the table are gross factors. The gross factor is calculated for the building as a whole and is measured from wall to wall. The net factor is only the space that can be occupied. It

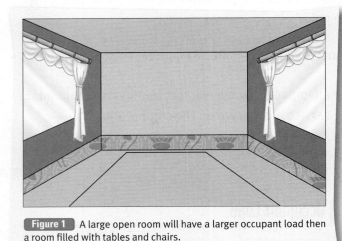

Figure 1 A large open room will have a larger occupant load then a room filled with tables and chairs.

Figure 2 Chairs and tables take up floor space and reduce the occupant load in a room.

Table 1 Occupant Load Factors

Table 7.3.1.2 Occupant Load Factor

Use	(ft² per person)ª	(m² per person)ª
Assembly Use		
Concentrated use, without fixed seating	7 net	0.65 net
Less concentrated use, without fixed seating	15 net	1.4 net
Bench-type seating	1 person/18 linear in	1 person/455 linear mm
Fixed seating	Number of fixed seats	Number of fixed seats
Waiting spaces	See 12.1.7.2 and 13.1.7.2.	See 12.1.7.2 and 13.1.7.2.
Kitchens	100	9.3
Library stack areas	100	9.3
Library reading rooms	50 net	4.6 net
Swimming pools	50 (water surface)	4.6 (water surface)
Swimming pool decks	30	2.8
Exercise rooms with equipment	50	4.6
Exercise rooms without equipment	15	1.4
Stages	15 net	1.4 net
Lighting and access catwalks, galleries, gridirons	100 net	9.3 net
Casinos and similar gaming areas	11	1
Skating rinks	50	4.6

(Continues)

Table 1 Occupant Load Factors *(Continued)*

Use	(ft² per person)[a]	(m² per person)[a]
Educational Use		
Classrooms	20 net	1.9 net
Shops, laboratories, vocational rooms	50 net	4.6 net
Day-Care Use	35 net	3.3 net
Health Care Use		
Inpatient treatment departments	240	22.3
Sleeping departments	120	11.1
Ambulatory health care	150	13
Detention and Correctional Use	120	11.1
Residential Use		
Hotels and dormitories	200	18.6
Apartment buildings	200	18.6
Board and care, large	200	18.6
Industrial Use		
General and high-hazard industrial	100	9.3
Special-purpose industrial	NA	NA
Business Use (other than below)	100	9.3
Concentrated business use[f]	50	4.6
Air traffic control tower observation levels	40	3.7
Storage Use		
In storage occupancies	NA	NA
In mercantile occupancies	300	27.9
In other than storage and mercantile occupancies	500	46.5
Mercantile Use		
Sales area on street floor[b,c]	30	2.8
Sales area on two or more street floors[c]	40	3.7
Sales area on floor below street floor[c]	30	2.8
Sales area on floors above street floor[c]	60	5.6
Floors or portions of floors used only for offices	See business use.	See business use.
Floors or portions of floors used only for storage receiving, and shipping, and not open to general public	300	27.9
Mall buildings[d]	Per factors applicable to use of space[e]	Per factors applicable to use of space[e]

NA: Not applicable. The occupant load is the maximum probable number of occupants present at any time.

[a]All factors are expressed in gross area unless marked "net."

[b]For the purpose of determining occupant load in mercantile occupancies where, due to differences in grade of streets on different sides, two or more floors directly accessible from streets (not including alleys or similar back streets) exist, each such floor is permitted to be considered a street floor. The occupant load factor is one person for each 40 ft² (3.7 m²) of gross floor area of sales space.

[c]For the purpose of determining occupant load in mercantile occupancies with no street floor, as defined in 3.3.271, but with access directly from the street by stairs or escalators, the floor at the point of entrance to the mercantile occupancy is considered the street floor.

[d]For any food court or other assembly use areas located in the mall that are not included as a portion of the gross leasable area of the mall building, the occupant load is calculated based on the occupant load factor for that use as specified in Table 7.3.1.2. The remaining mall area is not required to be assigned an occupant load.

[e]The portions of the mall that are considered a pedestrian way and not used as gross leasable area are not required to be assessed an occupant load based on Table 7.3.1.2. However, means of egress from a mall pedestrian way are required to be provided for an occupant load determined by dividing the gross leasable area of the mall building (not including anchor stores) by the appropriate lowest whole number occupant load factor from Figure 7.3.1.2(a) or Figure 7.3.1.2(b). Each individual tenant space is required to have means of egress to the outside or to the mall based on occupant loads calculated by using the appropriate occupant load factor from Table 7.3.1.2. Each individual anchor store is required to have means of egress independent of the mall.

[f]See A. 7.3.1.2.

Fire Inspector Tips

When multiple uses exist in a building where both gross and net factors are used, calculate each area independently using the appropriate factors, and then add them together for the building's total occupant load.

Fire Inspector Tips

Occupant load is based on use of a space, not the occupancy. A school is an educational occupancy, but it has spaces of classrooms, kitchens, libraries, laboratories, swimming pools, etc. Occupant load is determined by the nature of the use of a building or space and the amount of space available for that use. Since different generic uses are characterized by different occupant densities, the Life Safety Code has established occupant load factors for each use.

is the gross figure minus any tables, columns, or other unusable space. Net factors are used in occupancies where there will be higher concentrations of people.

When determining occupant loads with net factors, only consider the areas of the building that will actually be occupied. For example, if there is a room that is 30′ × 30′ (9.1 m × 9.1 m), and has four 12″ × 12″ (305 mm × 305 mm) columns, you have a gross square footage of 900 (83.61 m²); however, if the chart calls for using net factors, then you would subtract the 4 square feet (38 cm²) that the columns occupy and then end up with a figure of 896 net square feet (83.24 m²).

Calculate the Occupant Load for a Single-Use Occupancy

Let's walk through an example to learn how to calculate the occupant load for a single-use occupancy. A one-story free standing greeting card business has 900 square feet (83.6 m²) in the retail area and a 300 square foot (27.9 m²) back room used for storage. The occupancy use is mercantile. You note that for mercantile use for sales on a street floor, Table 7.3.1.2 shows a factor of 30, with each factor representing the square feet per person. Divide the factor into the area you are calculating and you get an occupant load of 30 occupants **Figure 3**.

When consulting Table 7.3.1.2 in the Life Safety Code, be sure to note if there are any footnotes attached to the occupant load factor. Any footnotes must be read to see if they impact your calculations. In this example, the 300 square foot (27.9 m²) back room is used for storage and there is a factor of 300 (27.9) tied to storage use. Therefore the backroom has an occupant load of 1, giving the entire building an occupant load of 31 **Figure 4**.

Now let's assume the store has two levels, with a sales area located on the second floor of the building. Looking at Table 7.3.1.2, you see that an occupant load factor of 40 (3.7) is used, therefore the occupant load in the retail area is 22.5 people. As the calculations do not allow for half a person, 22 is the occupant load.

Calculate the Occupant Load for a Multiple-Use Occupancy

Let's walk through an example to learn how to calculate the occupant load for a multiple-use occupancy. Let's calculate the occupant load for a school. While the occupancy is educational, remember that occupant loads are based on use of the area. Schools are one occupancy type with many different uses. In this example, there are six classrooms, one library, one exercise room with equipment, and a gymnasium with bleachers. Each classroom is 20′ × 40′ (6.1 m × 12.2 m). There are 25 student desks each measuring 4 square feet (37 cm²) and a teacher's desk measuring 32 square feet (3 m²). The library is 60′ × 60′ (18.3 m × 18.3 m). The exercise room is 40′ × 40′ (12.2 m × 12.2 m). The gymnasium has bleachers along both walls. Each wall of bleachers has 5 rows that are 25′ long (7.6 m).

Beginning with the classrooms, you see by Table 7.3.1.2 that a factor of 20 (1.9) net must be used. Determine the square footage of the classroom by multiplying 20′ (6.1 m) by 40′ (12.2 m). This gives you 800 square feet (74 m²). Now determine how much square footage the students' desks take up by 4 square feet (37 cm²) by 25. This equals 100 square feet (9.3 m²). Now deduct the 100 square feet (9.3 m²) from the total square footage, which leaves you with 700 square feet (65 m²). The teacher's desk is 32 square feet (3 m²), so subtract that from the total. This leaves you with an area of 668 net square feet (62 m²). Using the occupant load factor of 20 (1.9), divide 20 (1.9) into the 668 net square feet (62m²) and come up with 33.4, or 33 students for each of the six classrooms **Figure 5**.

$$\frac{1 \text{ occupant}}{300 \text{ occupancy load factor} \sqrt{300 \text{ square feet}}}$$

Figure 4 When consulting Table 7.3.1.2 in the Life Safety Code, be sure to note if there are any footnotes attached to the occupant load factor. Any footnotes must be read to see if they impact your calculations.

$$\frac{30 \text{ occupants}}{30 \text{ occupancy load factor} \sqrt{900 \text{ square feet}}}$$

Figure 3 To determine the occupant load for a single-use occupancy, divide the occupant load factor into the area.

$$\frac{33.4 = 33 \text{ students}}{20 \text{ occupant load factor} \sqrt{668 \text{ square feet}}}$$

Figure 5 Calculating the occupant load for a multiple-use occupancy.

Now apply the same series of steps to the other rooms in the school. The library is 3,600 square feet (334 m²). Using an occupant load factor of 100 (9.3), the occupant load for the stack room area of the library is 36. The exercise room is 1,600 square feet (149 m²). Table 7.3.2.1 allows for an occupant load factor of 50 (4.6), as the room has equipment. So 1600 (149) divided by 50 (4.6) equals an occupant load of 32.

The gymnasium has bleachers that when pulled out will allow for bench seating. Each wall has 25′ (7.6 m) of bleachers, each with 5 rows. According to Table 7.3.2.1, bench type seating is one person per 18″ (1 person per 455 mm). To continue in your calculations, convert the 25′ of bleachers into inches (meters into millimeters). Multiplying 25′ × 12 gives 300″ (7600 mm) per row. Divide the 300″ (7600) by 18″ (455 mm) to get 16.6, or 16 occupants per row. There are five rows per bleacher, so 16 × 5 equals 80 occupants. The other wall has an identical set of bleachers, so double the 80 and the total occupant load for the bleachers becomes 160 occupants.

When everything is added together, you have six classrooms each with a capacity of 33 for a total classroom count of 198, the library is 36, the exercise room is 32, and the gymnasium bleachers are 160. This gives our school a total occupant load of 426 **Figure 6**. Any additional space is calculated in the same fashion and added to the grand total.

■ Calculate Occupant Load Increases

The *Life Safety Code* will allow for the designed occupant load to be increased with the approval of the AHJ. Increases are allowed as long as other parts of the means of egress are sufficient. The means of egress is a continuous path of travel from any point in the building to a public way, and will be discussed in detail later in the chapter. Let's return to a school classroom as an example. Using Table 7.3.2.1, you calculate that the occupant load of the classroom is only 33. However, the classroom door

> **Fire Inspector Tips**
>
> The *Life Safety Code* also states that the AHJ is allowed to request diagrams of furniture arrangements for approval and that those diagrams be posted at the building. This may be particularly important for banquet halls or restaurants that to fit as many people as possible into the space.

must be a minimum width, and since you can determine what the exit capacity of a door is, it is possible that the size of the door might allow an egress capacity of 180 people. Technically, you could increase the capacity of the room, as long as you do not go beyond 180 people. Obviously that is not practical in a typical elementary school classroom Table 7.3.2.1 does not give any figure smaller than 7 square feet (0.65 m²) per person and for that reason it would be wise to never lower the capacity of an area beyond 7 square feet (0.65 m²) per person.

■ Sufficient Capacity for Occupant Load

Now that you know the amount of occupants the room is designed to hold, that still does not tell you if that number is acceptable. Those occupants now have to get to the outside of the building, so you need to look at numbers of exits, travel distances to those exists, widths of corridors and stairs, etc. The means of egress must be sufficient for the occupant loads that were just determined. The means of egress must be sized to accommodate all people occupying the area it serves. Sizing means making certain that the halls, doors, stairways are wide enough to accommodate the occupant loads entering those components. Sizing is accomplished by matching the occupant load against the calculated egress capacity. The sizing criteria of any component does not ensure that all occupants can leave immediately, but it does provide for timely exiting without any unacceptable queuing, such

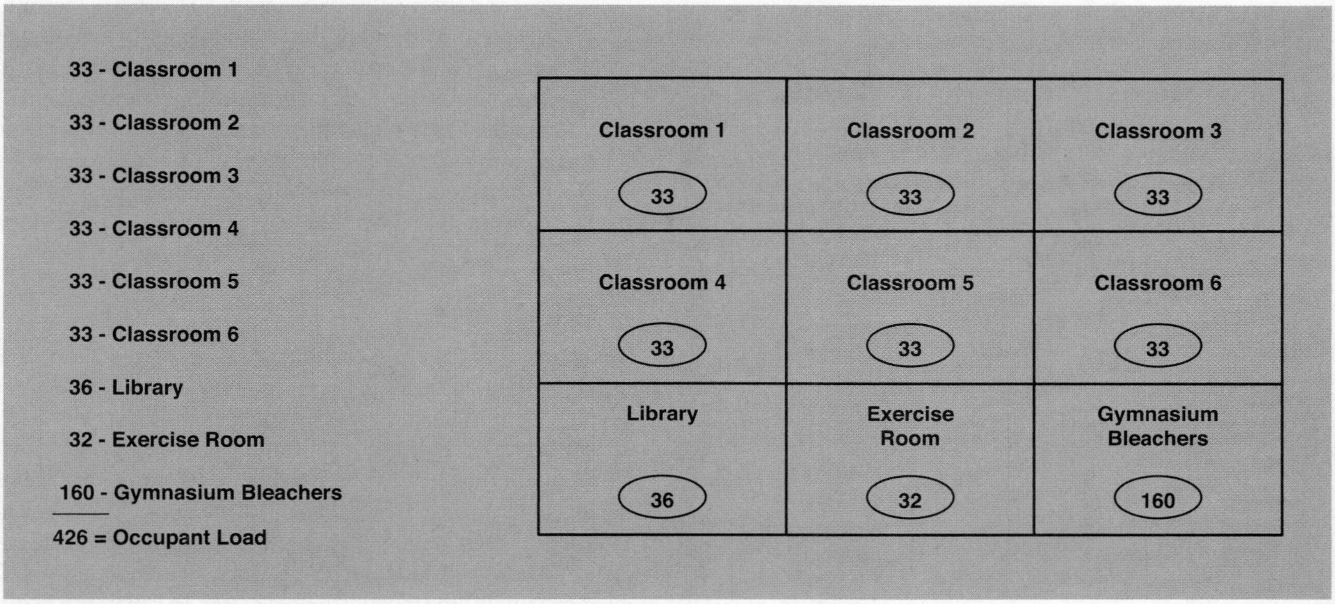

33 - Classroom 1
33 - Classroom 2
33 - Classroom 3
33 - Classroom 4
33 - Classroom 5
33 - Classroom 6
36 - Library
32 - Exercise Room
160 - Gymnasium Bleachers
―――
426 = Occupant Load

Classroom 1	Classroom 2	Classroom 3
33	33	33
Classroom 4	Classroom 5	Classroom 6
33	33	33
Library	Exercise Room	Gymnasium Bleachers
36	32	160

Figure 6 Add the occupant loads from each room together to find the grand total.

as occupants having to wait in line to pass through a door to enter an exit. The geometry of a building, its occupancy, occupant load, and the travel distance to exits determine the location of exits, the number of exits, and the capacity of exits.

Once the occupant load has been calculated, the means of egress system must provide egress capacity for a minimum of the calculated occupant load. Let's assume you have to walk out of a room, down a hall, down one flight of steps, and then through a door to get outside. Each of those points, the door, corridor, stairs, and outside door, are each a component of the means of egress. You would apply occupant load factors to each of those points to determine if they could handle the occupant load of the space. Consult the *Life Safety Codes* to determine the egress widths that must be provided per occupant per component **Table 2**.

For example, a room has an occupant load of 400 and has two exits. But that may still not be enough. If two 36" (914 mm) doors were provided, you would have a problem. The combined occupant load for those two doors is only 360 people. So, what are your options? A third exit could be added or the door width could be made wider to 42" (1067 mm) for a combined occupant load for 420. Then the two doors would properly handle the calculated 400 people in the room. This concept is explained in the section titled, *Evaluation of the Means of Egress*.

A means of egress system is only as good as its most constricting component. For example, if you have good egress capacity for 300 people until you get to a set of stairs that are calculated to 200, those stairs must either be widened, an additional staircase must be added, or the occupancy load of the areas feeding to that staircase must be reduced to the capacity the staircase will handle.

Multiple exits are required to provide alternate routes in case one exit becomes obstructed by fire and becomes unavailable. Although the exit with the greatest share of the total required egress capacity might be the exit closest to the majority of the occupants, it might also be the exit that is lost in a fire. Consequently, a disproportionate amount of the total egress capacity might be lost.

Means of Egress

The *Life Safety Code* and the model building codes use the term **means of egress**. A means of egress is a continuous path of travel from any point in a building to a public way that is safely away from the building. A means of egress consists of three separate and distinct parts:

- **Exit access**—That portion of a means of egress that leads to the entrance of an exit. In other words, the exit access is the travel anywhere within the building to the exit from the building. This may include travel within corridors, on stairs, or traversing open floor areas. There are limits an occupant can travel within the exit access to actually reach an exit, which is called the "Maximum Travel Distance."
- **Exit**—That portion of a means of egress that is between the exit access and the exit discharge. An exit may be comprised of vertical and horizontal means of travel. For example, entering an exit stairway on the fourth floor would end the exit access for that floor and begin the exit even though there are many floors of stairs to walk until the exit discharge is reached.
- **Exit discharge**—That portion of a means of egress between the termination of the exit and a public way. In other words, the exit discharge is the area between the exit and the nearest public way. All exits must terminate at a public way. While the requirements for exit discharge are vaguely defined, the entire distance must be identifiable, reasonably direct, and essentially unimpeded. This could include such things as removing the accumulation of snow and ice during the winter, making certain the terrain is even, and removing obstacles that might hinder movement to the public way. Most designs will exit onto a walkway that leads to the public way.

Exit Access

The exit access may be a corridor, aisle, balcony, gallery, room, porch, or roof. Basically it is anything that takes you to an exit. The length of the exit access establishes the travel distance to an exit, an extremely important feature of a means of egress, since an occupant might be exposed to fire or smoke during the time it takes to reach an exit **Figure 7**. The maximum travel distance to an exit is regulated by the *Life Safety Code* and is the distance an occupant is allowed to travel until an exit is encountered. The average maximum travel distance is 200' (61 meters), but this distance varies with the occupancy.

Fire Inspector Tips

Once occupant loads are established, it is not uncommon that they are exceeded. Much of controlling this rests with the business manager or, in the case of larger public assembly situations, a crowd manger. It is difficult for a fire inspector to routinely check the occupant load of a bar or nightclub at 1 AM. In locations where there are many late evening assembly occupancies or where there are documented instances of overcrowding, fire departments often will send fire inspectors to conduct basic inspections during those peak hours to check for overcrowding or other fire violations.

Table 2 Egress Widths

Area	Stairways (width per person)		Level Components and Ramps (width per person)	
	Inch	mm	Inch	mm
Board and care	0.4	10	0.2	5
Health care, sprinklered	0.3	7.6	0.2	5
Health care, nonsprinklered	0.6	15	0.5	13
High hazard contents	0.7	18	0.4	10
All others	0.3	7.6	0.2	5

Table 7.3.3.1 Capacity Factors from NFPA 101, Life Safety Code

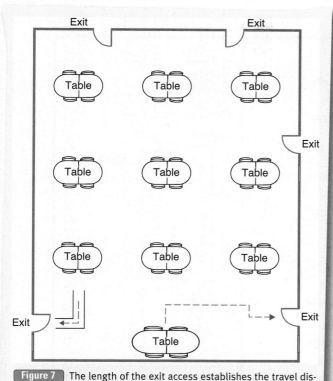

Figure 7 The length of the exit access establishes the travel distance to an exit, an extremely important feature of a means of egress, since an occupant might be exposed to fire or smoke during the time it takes to reach an exit.

The travel distance must be measured from the most remote point in a room or floor area to an exit. In most cases, the travel distance can be increased up to 50 percent if the building is completely protected with an approved supervised automatic sprinkler system. This is something that must be noted during the plan review stage of construction. The need to shorten excessive travel distances must be made at this point because it could be almost impossible to make changes once the building is completed. If, in the course of a fire inspection in a completed building, the travel distance seems too far, it should be measured for compliance. If there is a problem, corrections should be made. Compliance may be just a matter of reconfiguring a furniture layout. If immediate corrections are not possible, then a plan for compliance when remodeling or building additions should be agreed upon.

The width of an exit access should be at least sufficient for the number of persons it must accommodate: the occupant load of the room(s). These widths are regulated by the *Life Safety Code* and have minimum dimensions. For example, if you have an area with an occupant load of 100 people, you could not put in a corridor of 20″ (508 mm). The model building and fire codes, as well as the *Life Safety Code*, will dictate these minimum requirements.

In some occupancies, the width of the access is determined by the activity in the occupancy. One example is a new hospital, where patients may need to be moved in beds. The corridors in the patient areas of the hospital must be 8′ (2.4 meters) wide to allow for a bed to be wheeled out of a room and turned 90 degrees.

As these occupancies and widths have been determined for you by the *Life Safety Code*, it is only necessary that you understand that the dimensions change based on the occupancy and you must know where to look for the proper widths in the *Life Safety Code*. As these items relate to exiting, the specifics will be found in chapters dealing with means of egress requirements.

A fundamental principle of exit access is a free and unobstructed way to the exits. If the access passes through a room that can be locked or through an area containing a fire hazard more severe than is typical of the occupancy, the principles of free and unobstructed exit access are violated. A good example is a restaurant. If you are in the seating area and the main exit is through the front door, you would not be allowed to pass through the kitchen to get to the second exit. Obviously, a kitchen has a higher hazard than sitting at tables, and thus the exit would not be allowed for patrons Figure 8. However, the kitchen exit is a viable exit for those working in the kitchen, as they are not moving through a higher hazard.

The floor of an exit access should be level. If this is not possible, differences in elevation may be overcome by a ramps or stairs. Where only one or two steps are necessary to overcome differences in level, in an exit access, a ramp may be preferred, because people may trip in a crowded corridor and fall on the stairs if they do not see the steps or notice that those in front of them have stepped up or down.

■ Exit

Examples of exits are doors leading directly outside at ground level, such as the front door of a business, or through a protected passageway to the outside at ground level. Examples of protected passageways include smokeproof towers, protected interior and outside stairs, exit passageways, enclosed ramps, and enclosed escalators or moving walkways in existing buildings. The actual exit may be as narrow as the width of a door, as in the case of a door from a retail area that leads directly to the outside, or it could be many, many floors, in the case of multi-floored buildings. In the case of multi-floored buildings, enclosed stairwells must meet certain requirements. As soon as you enter one of those stairwells, the exit access ends, you are in the exit stage, regardless of how many floors it may take to reach outside safety. Elevators are not accepted as exits, except in very special circumstances.

The specific placement of exits is a matter of design judgment, given the specifications of travel distance, allowable dead ends, common path of travel, and exit capacity. Exits must be remote from each other, providing separate means of egress so that occupants can have a choice in the exit they wish to use Figure 9. This concept is important when it is necessary for occupants to leave a fire or smoke-filled area and move toward an exit. If occupants have no choice but to enter the fire area to reach an exit, it is doubtful whether they would be willing to do so.

■ Exit Discharge

Ideally, all exits in a building should discharge directly to the outside or through a fire-rated passageway to the outside of the building. A maximum of 50 percent of the exit stairs may

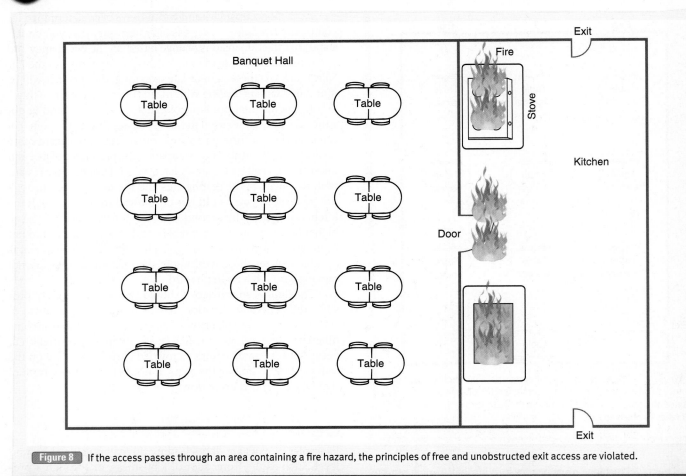

Figure 8 If the access passes through an area containing a fire hazard, the principles of free and unobstructed exit access are violated.

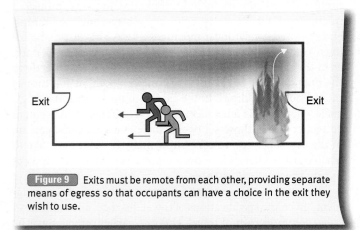

Figure 9 Exits must be remote from each other, providing separate means of egress so that occupants can have a choice in the exit they wish to use.

discharge onto the street floor of the building. The obvious disadvantage of this arrangement is that if a fire occurs on the street level floor, it is possible for people using the exit stairs discharging to that floor to be discharged into the fire area. If any exits do discharge to the street floor the following conditions must be satisfied: the exits must discharge to a free and unobstructed public way outside of the building, the street floor must be protected by automatic sprinklers, and the street floor must be separated from any floors below by construction having a 2-hour fire resistance rating.

A free and unobstructed route to the public way is key. You cannot direct occupants to areas where they will be trapped by fencing. A common example of this would be an apartment complex that has a recreation building with an outside swimming pool. Many times the occupant load of the building will be greater than 50 thus requiring two exits. Often one exit may lead into the pool area. The difficulty with this arrangement is that because the owners do not want people entering the pool through a gate in the fence, none will be provided. This is unsafe because occupants who have been directed to the pool area cannot safely exit the premises.

Evaluation of the Means of Egress

The *Life Safety Code* requires that the number of exits from any occupancy be consistent with the requirements listed in Table 3. Throughout the 1970s and early 1980s, egress capacity of doors was computed based on units of exit width. A unit of exit width was equal to 22″ (559 mm) and was recognized for an egress capacity of 100 persons. A distance equal to 12 to 21″ (304 to 533 mm) was recognized as a half unit capable of safely exiting 50 persons. Therefore, a 32″ (813 mm) exit door was credited with a 100 person egress capacity and a 34″ (864 mm) door was credited for 150 person egress capacity.

This former method of computing egress capacity gave way to a factor-based system based upon the entire width of the door,

Chapter 7 Occupancy Safety and Evacuation Plans

Table 3 Number of Exits per Occupancy

Occupant Load	Minimum Number Exits
0 to 499	2 (with limited exceptions for less than 50)
500 to 999	3
1000 or more	A minimum of 4

Exits Serving More than One Story

Where an exit serves more than one story, the occupant load of each story is considered individually when computing the required capacity of the exit at that story, provided that the required egress capacity of the exit (stairs) is not decreased (narrowed) in the direction of egress travel. Once a maximum required egress capacity is determined, such required capacity must be maintained—in the direction of egress travel—for the remainder of the egress system **Figure 12**.

Required stair width is determined by the required egress capacity of each floor the stair serves, considered independently. It is not necessary to accumulate occupant loads from floor to floor to determine stair width. Each story or floor is considered separately when calculating the occupant load to be served by the means of egress from that floor. The size or width of the stair need only accommodate the portion of the floor's occupant load assigned to that stair. However, in a multistory building, the floor requiring the

or egress component, on a per occupant basis. Presently, for a level means of egress, (doors and hallways) the factor is 0.2″ (5 mm) per occupant. Therefore, a 36″ (914 mm) door has egress capacities of 180 and a 34″ (864 mm) door has egress capacities of 170 occupants **Figure 10**.

For stairs, the factor used is 0.3″ (7.6 mm) per occupant. Occupants can walk more quickly on a flat surface than a flight of stairs, so a higher factor allows for fewer occupants. For example, a 36″ (914 mm) door has the capacity of 180 occupants. If you have a 36″ (914 mm) stair, you would only be able to allow 120 occupants on the stairs. To accommodate the same number of occupants entering through the door, the stairs must be widened. In this example, the stairs would need to be 54″ (1372 mm) wide to accommodate 180 occupants **Figure 11**.

The *Life Safety Code* also adds specific conditions to different occupancies based on the type of occupants served. For example, in a mercantile occupancy, there must be a main aisle capable of exiting 50 percent of the occupants without having to travel through checkout stands. This ensures that during an emergency, occupants are not required to work their way around cash registers and other fixtures designed to support the checkout process.

History shows that over 65 percent of occupants will make an attempt to exit a building the same way that they entered. Because of that fact, the *Life Safety Code* requires that assembly occupancies must have a main exit capable of exiting 50 percent of its capacity. If that is a problem, the capacity of a building must be reduced to reflect these limitations, or the exit capacity must be increased to meet the maximum capacity of the building.

$$\text{Means of Egless Factor 0.2″} \sqrt{\frac{180 \text{ occupants}}{36 \text{ inches - Door Width}}}$$

Figure 10 To calculate the egress capabilities of a door, divide the door width by the factor.

$$\frac{180 \text{ - occupants}}{\underset{54″}{\times 0.3″}} \text{ - Means of Egless Factor}$$

Figure 11 To calculate the width of the stairs, multiply the occupants by the factor 0.3.

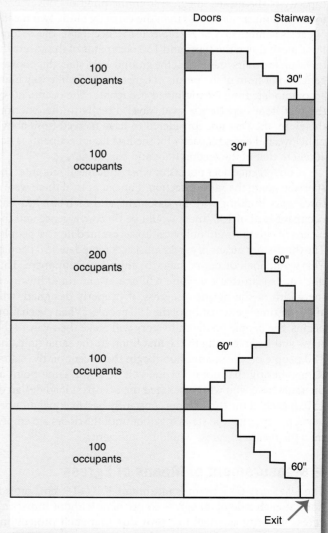

Figure 12 The floor requiring the greatest egress capacity dictates the minimum stair width from that floor down to the exit discharge.

greatest egress capacity dictates the minimum stair width from that floor down to the exit discharge, in the direction of egress travel. It is not permissible to reduce such stair width along the remainder of the stairs encountered in traveling to the level of exit discharge; that is, stairs encountered in the direction of egress travel. Exits serving floors above the floor of greatest egress capacity are permitted to use egress components sized to handle the largest demand created by any floor served by that section of stair run.

For example, consider a five story building. The first, second, fourth, and fifth floors each have a capacity of 100, and the third floor has a capacity of 200. The stairway for fourth and fifth floors must accommodate 100 people; however, the third floor the occupant load is 200 and the stairs must continue that width (amount of people) to the exit—you are not allowed to narrow it back to 100 at the second floor.

Egress Capacity from a Point of Convergence

Where means of egress from a story above and a story below meet together, it is known as convergence. At that point, the capacity of the means of egress from the point of convergence to the exit must not be less than the capacity of the two means of egress combined. For example, if 150 occupants of a second floor move down the stairs and 150 occupants of the basement travel up the stairs, meeting at the ground-level stair, this convergence point from of the means of egress to the public way must accommodate the 300 combined occupants. The occupants of the first floor experience level travel directly to the exterior; therefore, they are not considered to have merged from above or below, and egress capacity for the first floor occupants is not added to that of the second floor and basement.

Convergence does not occur when multiple floors enter the stairwell going the same direction. For example, if there were a three story building with no basement, and each story had an occupant load of 150, there would be no convergence and the means of egress would only need to accommodate 150 people. The third-story means of egress must accommodate 150 people. This would include doors, stairs, or any other component. This does not mean that everyone will exit at one time; however, the width of the means of egress, if properly designed, will allow for timely evacuation of the 150 people. When evacuation begins, the people on the third story will make their way to the stairs and begin exiting to the first floor. At the same time, the 150 people on the second floor begin the decent on the stairs. Remembering that not everyone will be at the same point at the same time, you will not exceed the occupant load design of 150. If each of the three floors constantly has an equal number of people entering the stairs together until the floors are empty, then the means of egress would be sufficient.

Measurement of Means of Egress

The *Life Safety Code* permits maximum 4½″ (114 mm) projections at each side of an egress component without impacting egress capacity calculations, provided that such projections occur at a height of not more than 38″ (965 mm), which is the maximum mounting height for handrails. Similarly, a 4½″ (114 mm) encroachment is permitted along stairs, corridors, passageways, and other components of the means of egress for purposes of calculating egress capacity, both without negatively impacting the width of the means of egress. This is because the human body is normally widest at shoulder level. Also, the body sway associated with walking, particularly on stairs, is greater at shoulder height than it is at waist height. Other projections, however, might constitute obstructions and cause impediments to the free flow of pedestrian travel, and must be factored in when determining the width of the means of egress. Only those projections specified in the *Life Safety Code* are permitted to encroach on the required width without having to subtract the encroaching space from the overall width before performing the egress width calculation.

Corridor Capacity

To determine the required capacity of a corridor, divide the occupant load that utilizes the corridor for exit access by the required number of exits to which the corridor connects. The corridor capacity cannot be less than the required capacity of the exit to which the corridor leads. For example, a corridor system serves a floor with a 660 occupant load and has three required means of egress (for between 500 and 999 occupants). The corridor must be wide enough to accommodate the portion of the floor's occupant load that it serves. Typical design practice is to divide the floor occupant load by the number of means of egress provided. In this three exit example, each exit could lead into a separate corridor; therefore, each corridor needs a capacity for only 220 people **Figure 13**. The required corridor width is determined to be 44″ (1120 mm) wide, unless the *Life Safety Code* dictates a wider corridor, such as in health care occupancy.

The required corridor capacity cannot be less than the required capacity of the exit to which it leads. In other words, the corridor is not permitted to create a bottleneck that is too narrow and impedes the flow of occupants to the exit door.

Minimum Width

The width of any means of egress must be not less than that required for a given egress component specified by the *Life Safety Code*, and usually not less than 36″ (915 mm). One of the unique features of the *Life Safety Code* is that each occupancy has provisions for new and existing buildings. The determination of

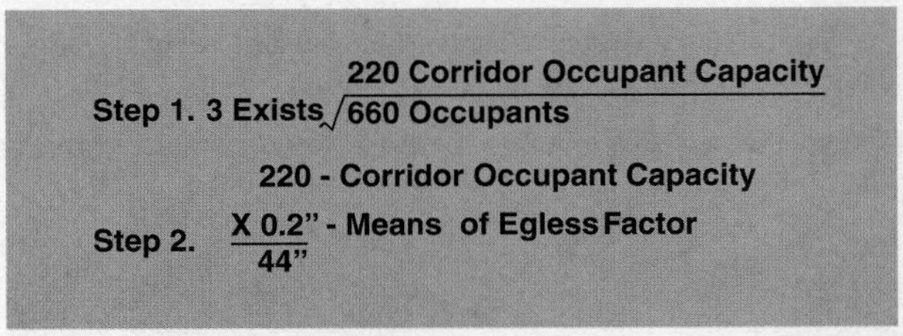

Step 1. 3 Exists / 660 Occupants = 220 Corridor Occupant Capacity

Step 2. 220 Corridor Occupant Capacity × 0.2″ Means of Egress Factor / 44″

Figure 13 To determine the corridor capacity, first divide the occupant load for the floor by the number of means of egress. This is the corridor occupant capacity. Next, determine the width of the corridor by multiplying the corridor occupant capacity by the means of egress factor.

new and existing is at the time the *Life Safety Code* is adopted. While a new building would require a minimum width, there may be occasions where the *Life Safety Code* will reduce that requirement for existing buildings.

If any egress width in an existing building is greater than required for new buildings, you would not be allowed to reduce it to the requirement of an existing building. For example, if an existing building has a corridor 50″ (1270 mm) wide. The *Life Safety Code* only requires 44″ (1118 mm) in new buildings and allows 36″ (915 mm) in existing buildings; however, the owner could not take the existing 50″ (1270 mm) and reduce it to 36″ (915 mm) simply because it is an existing building.

■ Number of Means of Egress

The number of means of egress from any area is at least two, unless specifically allowed by the *Life Safety Code*. The first chapters of the *Life Safety Code* list the general requirements on the number of means of egress. Chapters in the *Life Safety Code* listing specific requirements for occupancies take precedence over general requirements. If nothing specific is mentioned, then the general requirement prevails. This is why it is important to not only refer to the general requirements, but also to the specific occupancy, new or existing.

Most of the occupancy chapters in the *Life Safety Code* provide redundancy with respect to the number of means of egress by requiring at least two means of egress. Some occupancies identify specific arrangements under which only a single means of egress is permitted. Where large numbers of occupants are to be present on any floor or portion of a floor, then more than two means of egress must be provided.

If the occupant load is more than 500 but not more than 1000, then no less than three means of egress are required. If the occupant load is more than 1000, then no less than four means of egress are required. Several occupancies establish not only the minimum number of means of egress, but also the minimum number of actual exits that must be provided on each floor.

In most occupancies, meeting the requirements for egress capacities and travel distances means the required minimum number of means of egress will automatically be met. However, in occupancies characterized by high occupant loads, such as assembly and mercantile occupancies, or with unusual geometry that increases distance to exits, such as many buildings in Las Vegas, compliance with requirements for more than two exits per floor might require specific attention.

Multi-story Buildings

Similar to the procedure for determining required egress capacity, the number of required exits is based on a floor-by-floor consideration, rather than the accumulation of the occupant loads of all floors. For example, if the fourth floor of a building has an occupant load of 700, it requires three exits. If the third floor of the same building has an occupant load of 400, it would require only two exits. Regardless of the fact that the two floors together have an occupant load in excess of 1000, four exits are not required. However, the number of exits cannot decrease as an occupant proceeds along the egress path. The three exits required from the fourth floor in this example cannot be merged into two exits on the third floor, even though the third floor requires only two exits. The number of required exits must be carried out down to the ground floor.

■ Remoteness of Exits

The *Life Safety Code* also requires that whenever multiple exits are required, they must be separated by a distance equivalent to one half of the diagonal of the space being served **Figure 14**. This distance is reduced to one-third of the diagonal if the building is equipped with an automatic fire sprinkler system. In other words, if the area has a diagonal distance of 200′ (60 m), then the exits must be a minimum of 100′ (30 m) apart if the building is not sprinklered. If the building is sprinklered, then the exits only need to be 66′ (20 m) apart. This is done to minimize the chance that a single fire will block both exits. This determination needs to be made at the plan review stage. In existing buildings, it may be impossible to move exits away from each other.

■ Measurement of Travel Distance to Exits

Table 4 is a compilation of the requirements of individual occupancies for permissible distances for common path of travel, dead-end corridors, and travel distance to not less than one required exit. Travel distance to exits is the maximum distance that occupants are permitted to travel from their position in a building to the nearest exit. There is no formula by which this

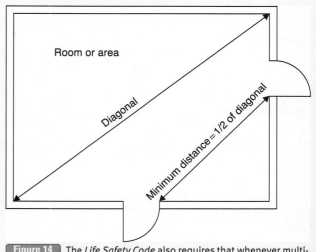

Figure 14 The *Life Safety Code* also requires that whenever multiple exits are required, they must be separated by a distance equivalent to one half of the diagonal of the space being served.

Fire Inspector Tips

The occupant load of each story considered individually shall be required to be used in computing the width of the means of egress and the number of exits at each story, provided that the required number of exits and means of egress widths is not decreased in the direction of egress travel.

distance can be established. Maximum allowed travel distances are based on factors that include the following:

1. Type and number of obstructions (for example, display cases, seating, heavy machinery) around which occupants must travel
2. The presence or absence of automatic fire sprinkler systems

Allowable travel distances will vary with the type and size of occupancy and the degree of hazard present. As shown in **Table 4** maximum travel distances can vary from 75′ (23 m) in nonsprinklered high hazard storage occupancies to 400′ (122 m) in sprinklered mall buildings or sprinklered special-purpose industrial occupancies meeting additional criteria. For most occupancies, the allowable travel distance is permitted to be increased if the building is protected throughout by automatic sprinkler systems.

There are no formulas or exact criteria for determining maximum permitted travel distances. Many factors have been considered and weighed in establishing these distances; they are the result of observing people who are in motion, consensus judgment, and many years of studying the results of fires in which the pre-fire conditions of a building were known.

Excessive travel distances can be a factor in large losses of life in fires because they increase the time required to reach the safety of an exit, whether the exit is a door directly to the outside or into a properly enclosed exit stair on an upper floor of a building. There is evidence that excessive travel distances played a role in a number of the fatalities on the casino floor at the

> **Fire Inspector Tips**
>
> You should have a working understanding of common terms and how to assess the correct distances. It would not be uncommon that during an inspection a building owner would ask your opinion about some remodeling being contemplated. Knowledge of the terms may raise red flags as the proposed plan is told to you and this could aid the owner's planning. A **dead end** exists when an occupant enters a corridor and, finding no exit, is forced to retrace the path traveled to reach a choice of egress travel paths. Although relatively short dead ends are permitted by the *Life Safety Code*, it is better practice to eliminate them wherever possible, as they increase the danger of persons being trapped during a fire.

Table 4 Travel Distances to Exits

Type of Occupancy	Common Path Limit				Dead-End Limit				Travel Distance Limit			
	Unsprinklered		Sprinklered		Unsprinklered		Sprinklered		Unsprinklered		Sprinklered	
	ft	m	ft	m	ft	m	ft	m	ft	m	ft	m
Assembly												
New	20/75	6.1/23[a]	20/75	6.1/23[a]	20	6.1[b]	20	6.1[b]	200	61[c]	250	76[c]
Existing	20/75	6.1/23[a]	20/75	6.1/23[a]	20	6.1[b]	20	6.1[b]	200	61[c]	250	76[c]
Educational												
New	75	23	100	30	20	6.1	50	15	150	45	200	61
Existing	75	23	100	30	20	6.1	50	15	150	45	200	61
Day Care												
New	75	23	100	30	20	6.1	50	15	150	45[d]	200	61[d]
Existing	75	23	100	30	20	6.1	50	15	150	45[d]	200	61[d]
Health Care												
New	100	30	100	30	30	9.1	30	9.1	NA	NA	200	61[d]
Existing	NR	NR	NR	NR	30	9.1	30	9.1	150	45[d]	200	61[d]
Ambulatory Health Care												
New	75	23[c]	100	30[c]	20	6.1	50	15	150	45[d]	200	61[d]
Existing	75	23[c]	100	30[c]	50	15	50	15	150	45[d]	200	61[d]
Detention and Correctional												
New — Use Condition II, III, IV	100	30	100	30	50	15	50	15	150	45[d]	200	61[d]
New — Use Condition V	100	30	100	30	20	6.1	20	6.1	150	45[d]	200	61[d]

Table 4 Travel Distances to Exits *(Continued)*

Type of Occupancy	Common Path Limit				Dead-End Limit				Travel Distance Limit			
	Unsprinklered		Sprinklered		Unsprinklered		Sprinklered		Unsprinklered		Sprinklered	
	ft	m	ft	m	ft	m	ft	m	ft	m	ft	m
Existing — Use Condition II, III, IV, V	50	15[f]	100	30[f]	NR	NR	NR	NR	150	45[d]	200	61[d]
Residential												
One- and two-family dwellings	NR	NR	NR	NR	NR	NR	NR	NR	NR	NR	NR	NR
Lodging or rooming houses	NR	NR	NR	NR	NR	NR	NR	NR	NR	NR	NR	NR
Hotels and dormitories												
New	35	10.7[g, h]	50	15[g, h]	35	10.7	50	15	175	53[d, i]	325	99[d, i]
Existing	35	10.7[g]	50	15[g]	35	10.7	50	15	175	53[d, i]	325	99[d, i]
Apartments												
New	35	10.7[g]	50	15[g]	50	15	50	15	175	53[d, i]	325	99[d, h]
Existing	35	10.7[g]	50	15[g]	50	15	50	15	175	53[d, h]	325	99[d, h]
Board and care												
Large, new	NA	NA	75	23	NA	NA	30	9.1	NA	NA	250	76
Large, existing	110	33.5	160	48.8	50	15	50	15	175	53[d, h]	325	99[d, h]
Mercantile												
Class A, B, C												
New	75	23	100	30	20	6.1	50	15	150	45	250	76
Existing	75	23	100	30	50	15	50	15	150	45	250	76
Business												
New	75	23[k]	100	30[k]	20	6.1	50	15	200	61	300	91
Existing	75	23[k]	100	30[k]	50	15	50	15	200	61	300	91
Industrial												
General	50	15	100	30	50	15	50	15	200	61[l]	250	76[m]
Special purpose	50	15	100	30	50	15	50	15	300	91	400	122
High hazard	0	0	0	0	0	0	0	0	0	0	75	23

(Continues)

Table 4 Travel Distances to Exits (Continued)

Type of Occupancy	Common Path Limit Unsprinklered ft	m	Common Path Limit Sprinklered ft	m	Dead-End Limit Unsprinklered ft	m	Dead-End Limit Sprinklered ft	m	Travel Distance Limit Unsprinklered ft	m	Travel Distance Limit Sprinklered ft	m
Aircraft servicing hangars, finished ground level floor	50	15[n]	100	30[n]	50	15[n]	50	15[n]	Note[l]	Note[l]	Note[l]	Note[l]
Aircraft servicing hangars, mezzanine floor	50	15[n]	75	23[n]	50	15[n]	50	15[n]	75	23	75	23
Storage												
Low hazard	NR	NR	NR	NR	NR	NR	NR	NR	NR	NR	NR	NR
Ordinary hazard	50	15	100	30	50	15	100	30	200	61	400	122
High hazard	0	0	0	0	0	0	0	0	75	23	100	30
Parking structures, open[o]	50	15	50	15	50	15	50	15	300	91	400	122
Parking structures, enclosed	50	15	50	15	50	15	50	15	150	45	200	60
Aircraft storage hangars, finished ground level floor	50	15[n]	100	30[n]	50	15[n]	50	15[n]	Note[l]	Note[l]	Note[l]	Note[l]
Aircraft servicing hangars, mezzanine floor	50	15[n]	75	23[n]	50	15[n]	50	15[n]	75	23	75	23
Underground spaces in grain elevators	50	15[n]	100	30[n]	50	15[n]	100	30[n]	200	61	400	122

NR: No requirement. NA: Not applicable.

[a]For common path serving >50 persons, 20 ft (6.1 m); for common path serving ≤50 persons, 75 ft (23 m).
[b]Dead-end corridors of 20 ft (6.1 m) permitted; dead-end aisles of 20 ft (6.1 m) permitted.
[c]See Chapters 12 and 13 for special considerations for smoke-protected assembly seating in arenas and stadia.
[d]This dimension is for the total travel distance, assuming incremental portions have fully utilized their permitted maximums. For travel distance within the room, and from the room exit access door to the exit, see the appropriate occupancy chapter.
[e]See business occupancies, Chapters 38 and 39.
[f]See Chapter 23 for special considerations for existing common paths.
[g]This dimension is from the room/corridor or suite/corridor exit access door to the exit; thus, it applies to corridor common path.
[h]See the appropriate occupancy chapter for requirements for second exit access based on room area.
[i]See the appropriate occupancy chapter for special travel distance considerations for exterior ways of exit access.
[j]See 36.4.4 and 37.4.4 for special travel distance considerations in covered malls considered to be pedestrian ways.
[k]See Chapters 38 and 39 for special common path considerations for single-tenant spaces.
[l]See Chapters 40 and 42 for special requirements on spacing of doors in aircraft hangars.
[m]See Chapter 40 for industrial occupancy special travel distance considerations.
[n]See Chapters 40 and 42 for special requirements if high hazard conditions exist.
[o]See 42.8.2.6.2 for special travel distance considerations in open parking structures.
Table A.7.6 from NFPA 101, *Life Safety Code*

MGM Grand Hotel fire in Las Vegas in 1980. Of the 85 fatalities, 18 victims were located on the casino level, and some apparently were overrun by the flames.

■ Common Path of Travel

A <u>common path of travel</u> is the distance an occupant must walk until there is a decision of which means of egress to use. Consider an office with one door opening into a corridor with exits at each end. The distance it takes to get through the office to the corridor is the common path of travel. Once the corridor is reached, a choice of what exit to use can be made.

■ Measuring Travel Distance

The natural exit access (path of travel) is influenced by the contents and occupancy of the building. **Figure 15** illustrates the path along which travel distance to an exit is measured. In (a), the stair is not appropriately enclosed to qualify as an exit; second-floor travel distance measurement continues to the first floor at the exit door to the outside. In (b), the stair is properly enclosed and constitutes an exit; travel distance measurement ends on the second floor at the entrance door to the exit stair enclosure. The travel paths marked as 1 through 6 shows that travel distance is measured as follows:

1. Starting at the most remote point subject to occupancy
2. On the floor or other walking surface
3. Along the centerline of the natural path of travel
4. Around corners and obstructions with a clearance of 12″ (305 mm)
5. Over open exit access ramps and open exit access stairs in the plane of tread nosings
6. Ending where the exit begins

Travel distance is that length of travel to an exterior exit door [as shown in (a)], an enclosed exit stair [as shown in (b)], an exit passageway, or a horizontal exit. It includes all travel within the occupied space until an occupant reaches that level of protection afforded by the nearest exit. Therefore, where stairs form part of an exit access rather than an exit, the travel over such stairs must be included in the travel distance measurement [as shown in (a)] **Figure 16**. The measurement of travel distance along stairs is to be made in the plane of the tread nosings, not along each riser and tread.

In reviewing plan sets for compliance with the travel distance limitations established for any occupancy, it is important to know the natural path of travel and the obstacles that will be present. In **Figure 17** (a) and (b) depict the same building. In (a), points X and Y are located at the same physical distance from the exit door. Without further information related to the layout of furniture and partitions, it isn't clear whether the occupant will be able to travel in a straight line, as shown from point Y to the exit door, or will need to follow a longer travel

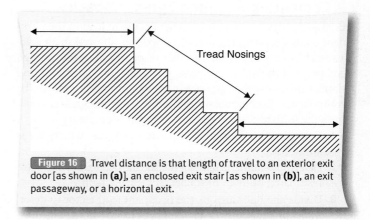

Figure 16 Travel distance is that length of travel to an exterior exit door [as shown in **(a)**], an enclosed exit stair [as shown in **(b)**], an exit passageway, or a horizontal exit.

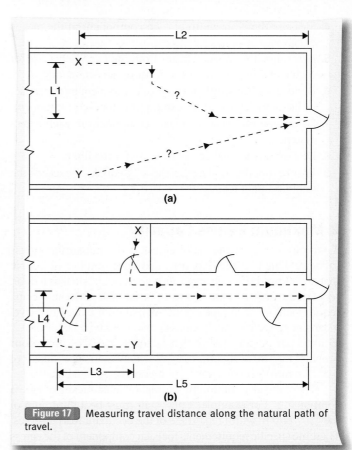

Figure 17 Measuring travel distance along the natural path of travel.

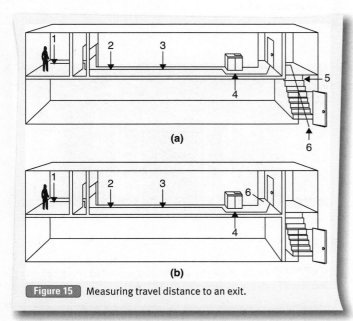

Figure 15 Measuring travel distance to an exit.

path that zigzags around obstacles, as shown from point X to the exit door.

In (b), the placement of partitions appears on the plan. An occupant is unable to travel in a straight "beeline" path to the exit door from either point X or point Y; the partitions preclude this. The occupant at point Y must first move in a direction opposite from that of the building's exit door to reach the room door before turning and traveling in a direction toward the exit door. In this case, the travel distance is calculated by adding together travel path segments L3, L4, and L5.

The maximum permitted travel distance is that length of travel that must not be exceeded to reach the nearest exit. Although more than one exit might be required, only the travel distance to the closet exit is regulated.

■ Computing Required Egress Capacity

Egress capacity is different from the occupant load in the occupant loads deal with the amount of people allowed in an area. The egress capacity is the number of people allowed through any point of the means of egress, such as doors, corridors, and stairs. Consulting the *Life Safety Code* will provide a code for the components. To compute the minimum required egress capacity from the individual floors of a building, it is necessary to:

1. Calculate the floor area, either net or gross, whichever is applicable.
2. Determine from the *Life Safety Code* the occupant load factor [estimated number of square feet (m²)/person.].
3. Divide the floor area by the occupant load factor to determine the minimum number of people for whom exits must be provided for that floor (occupant load).
4. Measure the clear width of each component in the means of egress.
5. Determine the capacity factor from the *Life Safety Code* for each exit component for the appropriate occupancy.
6. Divide the clear width of each exit component by the capacity factor to determine the exit capacity for each component.
7. Determine the most restrictive component in each egress system.
8. Determine the total egress capacity for the floor.
9. Ensure that the total egress capacity equals or exceeds the total occupant load.

■ Maximum Egress Capacity

A limitation on the means of egress is the maximum egress capacity. This is defined as the maximum number of people that are allowed through a specific egress point, such as a door, set of stairs, or a corridor. You must be familiar with the specific requirement for each type of exit. An exit door placed in a level means of egress can safely egress persons using a factor of 0.2″ (5 mm) per person. Using that factor, a 36″ (914 mm) door provides for 180 occupants to egress safely. Stairs use a factor of 0.3″ (7.6 mm) per occupant, a reduction of 50 percent.

There are also minimum widths for most means of egress components. For example, a new stair must be a minimum of 42″ (107 cm) in width regardless of the occupant load it serves. There are specific requirements for each component of the egress system including the handrails, guard rails, treads, and risers of stairs and corridors. A thorough review of the means-of-egress chapter of the model fire code will provide the parameters that must be considered when evaluating these components.

Means of Egress Elements and Arrangements

Means of egress elements are the components of the means of egress including exit access, exit enclosures, exit discharges, stairways, ramps, doors, hardware, exit markings, illumination, etc.

■ Doors

All exits must have doors; however, not every door is an exit. When a door is marked as an exit, certain requirements in the *Life Safety Code* take effect such as identification, clear access, locking, widths, etc. Other doors, not classified as exits do not fall into these requirements. Take a small shopping strip with three small stores as an example. Each store has it front and rear exit. One business decides to take over the other two stores, and have one large, open store. The size of the store might still only require two remote exits. While individually each store required the rear exit, when combined, all three are not needed. However, if the doors continue to have exit signs, denoting them as exits, they must be maintained as such as people will be directed to those doors in an emergency. If they were to remove the exit signage from two of the doors, those two doors simply become convenience doors, and the *Life Safety Code* requirements for exit doors do not apply.

Doors should be side-hinged or pivoted swinging type and should swing in the direction of exit travel, except in small rooms when allowed by the *Life Safety Code* Figure 18.

> **Fire Inspector Tips**
>
> Where two or more occupancy requirements apply and are in conflict, such as in a mixed occupancy, apply the most restrictive requirements.

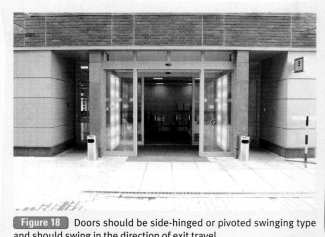

Figure 18 Doors should be side-hinged or pivoted swinging type and should swing in the direction of exit travel.

Horizontal sliding, vertical, or rolling doors can be used as means of egress in certain occupancies. In assembly occupancies and in schools, panic hardware equipped with latches must be installed on all egress doors that serve rooms with occupant loads of 100 or more.

When doors protect exit facilities, as in stairway enclosures and horizontal exits, they normally must be kept closed to limit the spread of smoke. If open, they must be closed immediately in case of fire. Although ordinary, fusible-link operated devices to close doors in case of fire are designed to close in time to stop the spread of fire, they do not operate fast enough to stop the spread of smoke and are therefore not permitted by the *Life Safety Code*.

Sometimes, people keep self-closing doors open with hooks or with wedges under the door. Doors also can be blocked open to provide ventilation, for the convenience of building maintenance personnel, or to avoid the accident hazard of swinging doors. The following measures have been provided in the *Life Safety Code* to alleviate these unallowable situations:

1. Doors that are normally kept open can be equipped with door closers and automatic hold-open devices that release the doors and allow them to close upon activation of the fire alarm system.
2. Doors that are normally closed can be equipped to open electrically or pneumatically when a person approaches the door, as long as precautions are used to prevent the door from automatically opening when there is smoke in the area.
3. Doors that normally are closed can be opened and held open manually by monitors, as in schools.
4. Smokeproof towers that protect against smoke can be used, even if the doors are open.

In the event of electrical failure, the door must, however, close and remain closed unless it is opened manually for egress purposes.

Another major maintenance difficulty with exit doors are exterior doors that are locked to prevent unauthorized access. The *Life Safety Code* specifies that when the building is occupied, all doors must be kept unlocked from the side from which egress is made. If a door is locked it must be opened without the use of any special knowledge, and with only one action. In other words, if a door is deadbolted closed, a person inside should be able to unlock the door by simply opening the handle. This setup is very common on hotel doors. Other measures to prevent unauthorized use of exit doors include:

- An automatic alarm that rings when the door is opened
- Visual supervision such as wired-glass panels, closed circuit television, and mirrors, which may be used where appropriate
- Automatic photographic devices to provide pictures of users

A single door in a doorway should not be less than 32″ (813 mm) wide in new buildings and 28″ (711 mm) in existing buildings. To prevent tripping, the floor on both sides of the door should have the same elevation for the full swing of the door.

Panic Hardware

Egress doors in assembly and educational occupancies, such as schools or movie theaters, must be equipped with panic hardware when the occupant load exceeds 100 Figure 19. Panic hardware is defined as a door-latching assembly incorporating a device that releases the latch upon the application of force in the direction of egress travel. Panic hardware devices are designed to facilitate the release of the latching device on the door when a pressure of not more than 15 lb (6.8 kg) is applied in the direction of exit travel. Even though the door may be locked, pressing on this hardware will open the door. Panic hardware that has been tested and listed for use on fire-protection-rated doors is called "fire exit hardware." The panic hardware is the fire rated item. If panic hardware is needed on fire-protection-rated doors, only fire exit hardware can be used.

Horizontal Exits

A horizontal exit is a means of egress from one area to an area of refuge in another area on approximately the same level Figure 20. Typically these means of egress are through 2-hour fire barriers. With a horizontal exit, space must be provided in the area of refuge for the people entering the refuge area in addition to the normal occupant load. The *Life Safety Code* recommends 3 square feet (0.28 m^2) of space per person,

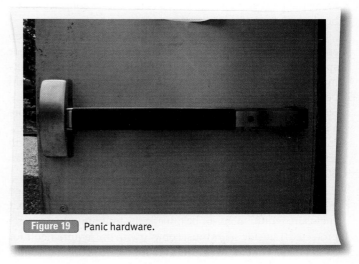
Figure 19 Panic hardware.

Figure 20 A horizontal exit.

with the exception of healthcare and detention and correctional occupancies, where 6 to 30 square feet (0.56 to 2.79 m²) of space is recommended. This is information that can be researched in the *Life Safety Code* should there be any question regarding how many people can occupy a horizontal exit.

Horizontal exits cannot comprise more than 50 percent of the total required exit capacity, except in healthcare facilities, where horizontal exits may comprise 66 percent of the total required exit capacity, or in detention and correctional facilities, where horizontal exits can comprise 100 percent of the total exit capacity. Horizontal exits are used in many healthcare facilities where the evacuation of patients over stairs is difficult, if not impossible, especially when considering patients in a hospital are confined to beds, many with ventilators or other machines. A horizontal exit arrangement within a single building and between two buildings is illustrated in Figure 21.

■ Stairs

Exit stairs are designed to minimize the danger of falling, because a person falling on a stairway could result in the exit being blocked. Stair width must be based on calculated occupant loads. There should be no decrease in the width of the stairs along the path of travel, since this could create congestion or a bottle neck.

Steep stairs are dangerous and stair treads must be deep enough to give good footing. For new stairs, the *Life Safety Code* specifies a maximum 7″ (178 mm) height (rise) and a minimum 11″ (279 mm) tread (run) (7–11 stairs). Landings should be provided to break up any excessively long individual flight. Stairs can only be a maximum of 12′ (366 cm) in length before a landing is required. Continuous railings are required for new stairs. These railings must also start before the stairs begin and terminate after the stairs end Figure 22. New stairs more than 60″ (1.5 m) wide need to have one or more center rails.

Stairs can serve as exit access, exit, or exit discharge. When used as an exit, they must be in an enclosure that meets exit enclosure requirements or be outside the building and properly protected. Outside stairs must comply with the requirements for exterior stairs and be arranged so that persons who fear heights will not be reluctant to use them. They should not be exposed to fire conditions originating in the building, and, where necessary, should be shielded from snow and ice. Exterior stairs should not be confused with fire escape stairs Figure 23.

Stair enclosures involve the principles that are designed to limit fire and smoke spread. Doors or openings from each story are necessary to prevent the stairway from serving as a flue for products of combustion. In general, stairway enclosures should include not only the stairs, but also the path of travel from the bottom of the stairs to the exit discharge, so that occupants have a protected, enclosed passageway all the way to the public area. The stair enclosure should be of 1-hour construction when connecting three or fewer floors and of 2-hour construction when connecting four or more floors.

■ Smokeproof Enclosure

Smokeproof enclosures are stair enclosures designed to limit the spread of the products of combustion from a fire. Smokeproof enclosures provide the highest protected type of stair enclosure recommended by the *Life Safety Code*. The stair tower is only accessed through balconies open to the outside air, vented vestibules, or mechanically pressurized vestibules, so that smoke, heat, and flame will not spread readily into the tower even if the doors are accidentally left open.

■ Ramps

Ramps, enclosed and otherwise, arranged like stairways, are sometimes used instead of stairways where there are large crowds and to provide both access and egress for nonambulatory persons. To be considered safe, exits ramps must have a very gradual slope. An example of a very common ramp would be at large

Figure 22 Continuous railings are required for new stairs.

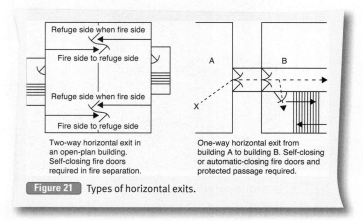

Figure 21 Types of horizontal exits.

Figure 23 Exterior stairs should not be confused with fire escape stairs. **A.** Exterior stairs. **B.** Fire escape stairs.

scale professional sporting events, where many thousands of people are leaving the venue at one time **Figure 24**.

Exit Passageways

A hallway, corridor, passage, tunnel, or underfloor or overhead passageway may be used as an exit passageway, providing it is separated and arranged according to the requirements for exits. The use of a hallway or corridor as an exit passageway requires close review. The *Life Safety Code* specifies that an exit enclosure should not be used for any purpose that could interfere with its value as an exit. For example, in an industrial occupancy, the use of a gasoline-powered forklift in a corridor designated as an exit passageway would violate the intent of the *Life Safety Code*. Also, penetration of the enclosure by ductwork and other utilities is typically not allowed by the *Life Safety Code*.

In addition, multiple doors in a corridor used as an exit increases the likelihood that a door could fail to close and latch, resulting in fire contamination. The door openings in exit enclosures should be limited to those necessary for access to the enclosure from normally occupied spaces. Therefore, doors and other openings to spaces such as boiler rooms, storage areas, trash rooms, and maintenance closets are not allowed in an exit passageway.

An exit passageway should not be confused with an exit access corridor. Exit access corridors do not have as stringent construction protection requirements as do exit passageways, since they provide access to an exit rather than being an actual component of the exit.

Fire Escape Stairs

Fire escapes should be stairs, not ladders. Fire escapes are, at best, a poor substitute for standard interior or exterior stairs. The *Life Safety Code* only permits existing fire escapes and only in existing buildings. Fire escapes can create a severe fire exposure to people if flames in lower levels come through windows, blocking the path beneath them **Figure 25**. The best location for fire escapes is on exterior masonry walls without exposing windows, with access to fire escape balconies by exterior fire doors. Where window openings expose fire escapes, shatter-resistant wired-glass in metal sashes should be used to provide maximum protection for the fire escape. When the building has an automatic sprinkler system the fire exposure hazard to personnel on fire escapes is minimized. An obvious problem with fire escape in northern climates is that outdoor fire escapes may be obstructed by snow and ice.

Figure 24 Exit ramp.

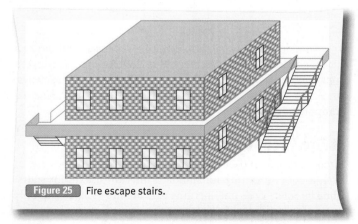

Figure 25 Fire escape stairs.

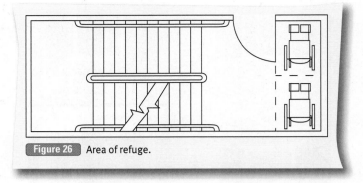

Figure 26 Area of refuge.

Elevators are not generally recognized as exits. However, elevators are permitted to be used, under limited conditions, to serve areas of refuge for the mobility impaired. The *Life Safety Code* also recognizes elevators, under very limited conditions, as the second exit from limited access towers such as Federal Aviation Administration (FAA) control towers.

Areas of Refuge

An <u>area of refuge</u> is a protected area in a building designed for use by people with mobility impairments Figure 26 . It is a staging area that provides relative safety to its occupants until they can be completely evacuated. All new buildings must address the issue of "accessible means of egress," meaning that the area of refuge must meet the requirements for the handicapped. While not common, the *Life Safety Code* states that areas of refuge may be in a different building, accessed using corridors, balconies, etc. It further states that any floor with an approved, supervised sprinkler system is an area of refuge.

Ropes and Ladders

Ropes and ladders are not recognized as a substitute for standard exits from a building. There only possible use is in existing one- and two-family dwellings where it is impractical to add a secondary means of escape. In this case, a suitable rope or chain ladder, or a folding metal ladder may be suitable rather than having to leap from a window. However, it should be recognized that aged, infirm, very young, and physically handicapped persons cannot use ladders, and that if the ladder passes near or over a window in a lower floor, flames from the window can prevent the use of the ladder.

In the situations where these items are used, education of the public is key. It needs to be reinforced that the only time to leave through a window is as a last resort—when smoke or fire is in the room, posing an immediate threat. It is important to stress that a closed door, even a bedroom door, can keep fire out for about 10 minutes. Remember, just because you cannot exit down the hallway to get out, does not mean you need to leap out of a second floor window. Open the window, prepare to leave, but do so *only* when actually physically threatened by the fire.

Windows

Windows are not exits. They may be used as access to fire escapes in existing buildings if they meet certain criteria concerning the size of the window opening and the height of the sill from the

Escalators, Moving Walkways, and Elevators

Escalators are not recognized as an acceptable means-of-egress component in new construction. In some select cases they may be allowed in existing buildings, but the *Life Safety Code* should be referenced for specific details. Moving walkways also may be used as means of egress if they conform to the general requirements for ramps, if inclined, and for passageways, if level.

floor. Windows may also be considered a means of escape from certain residential occupancies, if they meet minimum requirements specified in the *Life Safety Code*.

Windows are required in school rooms subject to student occupancy, unless the building is equipped with a standard automatic sprinkler system, and in bedrooms in one- and two-family dwellings that do not have two separate means of escape. These windows are for rescue and ventilation and must meet the criteria for size of opening, method of operation, and height from the floor.

Exit Lighting

In buildings where artificial lighting is provided for normal use, the illumination of the means of egress is required to ensure that occupants can see to evacuate the building quickly Figure 27 . The intensity of the illumination should be not less than 1 footcandle (10.77 lu/m^2) measured at the floor. It is desirable that such floor illumination be provided by lights recessed in the wall and located approximately 1 foot (30.5 cm) above the floor because such lights are then unlikely to be obscured by the smoke that might occur during a fire. In auditoriums and other places of public assembly where movies or other projections are shown, the *Life Safety Code* permits a reduction in this illumination for the period of the projection.

Emergency Lighting

The *Life Safety Code* requires emergency power for illuminating the means of egress in most occupancies. Some examples of where emergency lighting is required are assembly occupancies; most educational buildings; healthcare facilities; most hotels and apartment buildings; Class A and B mercantiles; and business buildings based on occupant load and number of stories.

Figure 27 Exit lighting.

Well-designed emergency lighting using a source of power independent from that of the normal building power automatically provides the necessary illumination in the event of an interruption of power to normal lighting. The failure of the public power supply or the opening of a circuit breaker or fuse should result in the automatic operation of the emergency lighting system.

Reliability of the exit illumination is important. NFPA 70, the *National Electrical Code®*, details requirements for the installation of emergency lighting equipment. Battery-operated electric lights and portable lights normally are not used for primary exit illumination, but they may be used as an emergency source under the restrictions imposed by the *Life Safety Code*. Luminescent, fluorescent, or other reflective materials are not a substitute for required illumination, since they are not normally sufficiently intense to justify recognition as exit floor illumination.

Where electric battery-operated emergency lights are used, suitable facilities are needed to keep the batteries properly charged. Automobile-type lead storage batteries are not suitable because of their relatively short life when not subject to frequent recharge. Likewise, dry batteries have a limited life, and there is a danger that they may not be replaced when they have deteriorated. If normal building lighting fails, well-arranged emergency lighting provides necessary floor illumination automatically, with no appreciable interruption of illumination during the changeover.

The emergency lighting is not designed to take the place of regular building lighting, but to provide a means to illuminate the means of egress. Where a generator is provided, a delay of up to 10 seconds is considered tolerable. In the cases of a generator being used as emergency power, it typically will power the normal building lighting. In that case, separate emergency lighting packs are not needed. The normal requirement is to provide such emergency lighting for a minimum period of 1½ hours. In the case of a long power failure, it is common to receive calls from residents in apartment buildings stating the hallways are black. As mentioned above, the lighting lasts only for 90 minutes so people have time to safely exit the building. They are not designed for the duration of a power failure.

Exit Signs

All required exits and access ways must be identified by readily visible signs when the exit or the way to reach it is not immediately visible to the occupants Figure 28 . Directional "EXIT" signs are required in locations where the direction of travel to the nearest exit is not immediately apparent. The character of the occupancy will determine the actual need for such signs. In assembly occupancies, hotels, department stores, and other buildings with transient populations, the need for signs will be greater than in a building with permanent or semi-permanent populations because those permanent or semi-permanent populations are more familiar with the various exit routes. Even in permanent residential-occupancy buildings, signs are needed to identify exit facilities, such as stairs, that are not used regularly during the normal occupancy of the building. It is just as important that doors, passageways, or stairs that are

Figure 28 Exit sign.

> **Fire Inspector Tips**
>
> While not required by the *Life Safety Code* yet, photoluminescent technology exists as a means to identify the means of egress in a power failure and occupancies may choose to utilize this growing technology.

not exits but are so located that they may be mistaken for exits be identified by signs with the words "NO EXIT", with "no" larger than "exit".

Exit signs should be readily visible: prominent, well-located, and unobscured. Decorations, furnishings, or other building equipment should not obscure the visibility of these signs. The *Life Safety Code* does not make any specific requirement for sign color but requires that signs be of a distinctive color. The *Life Safety Code* specifies the size of the sign, the dimensions of the letters, and the levels of illumination for both externally and internally illuminated signs. The *Life Safety Code* also requires in specific occupancies that exit signs are needed at the on the wall close to the floor.

Maintenance of the Means of Egress

The provision of a standard means of egress with adequate capacity does not guarantee the safety of the occupants in the event of an evacuation if the means of egress are not properly maintained. Many building owners and property managers do not assign someone to be responsible for the maintenance of the means of egress. As a result you may find stairways used as storage for materials during peak sales or manufacturing periods. In apartment buildings, rubbish, bicycles, baby carriages, and other obstructions are often found in stairway enclosures. Exit doors may be found locked or with hardware in need of repair. Doors blocked open or removed from openings into stairway enclosures may permit rapid spread of smoke or hot gases throughout the building. Loose handrails and loose or slippery stair treads may cause persons evacuating a building to fall in the path of others seeking escape. Maintaining the means of egress in safe operating condition at all times is as important as the proper design of the egress system itself.

Evacuation and Emergency Plans

The *Life Safety Code*, as well as the model fire codes, have areas that deal with emergency planning and evacuation drills. All of those sources need to be consulted as there may be varying requirements for what should be included in the evacuation plan, the frequency of evacuation drills, when evacuation drills are to be held, and the training of personnel. An **evacuation plan** allows for the safe and orderly removal of people from a building in the event of a fire or other emergency and is one part of a larger emergency plan. **Emergency plans** are created to deal with various emergencies that might affect a building other than fire, such loss of power, flooding, chemical spills, etc. The evacuation plan takes effect when there is a need to have all occupants leave the building or area. Emergency plans can be small, such as in a small store, or they may be more involved, as in a high rise office building. Regardless of the size of the building, every evacuation plan has the same basic rules that can be expanded as the building, and need, gets larger.

Developing the emergency plan is ultimately the responsibility of the business. The fire inspector is a resource to offer input in its development. The emergency plan should include all people in the building, including those with disabilities. Some business may not have a person with a permanent disability working there, but they probably have had one there occasionally. Take the example of a person with a broken leg using crutches or a wheel chair on a short term basis. During this short time, these individuals have the same needs of those who are permanently handicapped.

The emergency plan has the same basic components as a home escape plan that children are instructed about in school fire safety classes. Those points include Plan, Alert, Get Out, and Meeting Place. The emergency plan is a basic outline and should easily evolve into a more sophisticated plan based upon the building. A small business's plan may assign specific safety tasks to only two or three key people, while a high rise office building may have 20 or more.

Plan Requirements

Every business needs an evacuation plan that includes two ways out of all areas. It should include the location of fire extinguishers and fire alarm pull stations **Figure 29**. The evacuation plan is not useful unless it is explained and practiced regularly. It also must be tailored to the specific needs of the business. A good plan for one business may not work for another. It should also cover all hours that people are occupying the business. Once the evacuation plan is agreed upon, it must then be taught to every employee. When that has been accomplished, drills, announced and unannounced, must be conducted, reviewed, and modified, as necessary. Evacuation drills must continue to be carried out on a regular basis.

Alert

Occupants should be taught to report all signs of fire or smoke, regardless of size, even if they are already extinguished. The fire department needs to investigate the cause of each fire for documentation or, when appropriate, to aid in possible national product recalls. Fire officials will also make certain there are no

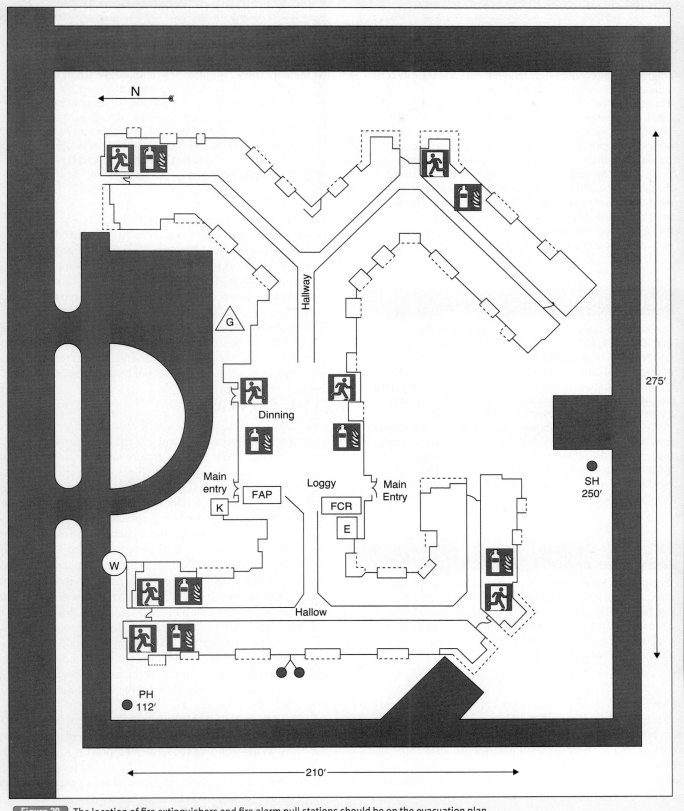

Figure 29 The location of fire extinguishers and fire alarm pull stations should be on the evacuation plan.

Safety Tips

Emergency planning is an emerging science that is influenced by many factors and events of the current world. The Columbine, Colorado, school incident has exposed a new paradigm in the mix of fire safety requirements and security considerations. Should students and teachers immediately leave a building when a fire alarm sounds, without any verification that a fire emergency exists? If not, at what point is the "Go/No Go" decision made, and by whom? Where is the balance security and safety? While most facilities would evacuate a building for a fire alarm activation, the question changes drastically when the school is in lockdown. In these situations, if the fire alarm sounds, are occupants automatically going to do what they have been taught and practiced—to evacuate the building? These decisions must be made jointly with police, fire, and school officials, and should include a response to each of the possible situations the building could encounter.

Fire Inspector Tips

One scenario that can be debated regarding evacuations is that if a new five-story residence with maximum fire protection features were constructed, would it be safe to protect in place, or should all the occupants evacuate? Occupants should evacuate whenever possible. If you are protecting people in place because the building is new, then when does it become old and require evacuation? If people are told to stay in their units, will they disregard evacuative fire procedures in other buildings? Fire safety education programs stress that that in a fire situation the only safe place is outside the building. If, for any reason, you are unable to safely exit the building, the next safest place to be is inside your unit with the door closed.

Fire Inspector Tips

When evaluating emergency plans, ensure that the emergency plan accounts for the occupant load of the area, areas of refuge for mobility impaired people, the hours the business is open, occupants have a clear path to a public way, that the means of egress clearly marked, and the provisions in case the main exit is not available.

Fire Inspector Tips

The development of emergency plans is no longer simply left to a fire or police official. Such plans must involve an array of experts to fully address the potential hazards and establish the necessary contingencies to address each of the many risks. The shelter-in-place concepts widely used in health care facilities are now becoming a model for other occupancies. Many county and state emergency management agencies, along with the Federal Emergency Management Agency provide excellent resources for assistance in developing emergency plans. Direct building owners and managers to these resources for developing an emergency plan.

hidden fires. In order to alert other employees and visitors in the building, all employees should know the locations of the fire alarm pull stations and how to operate them. They also need to know that anytime they suspect a fire they can activate the alarm without repercussion. If the building is not equipped with an alarm system, occupants need to be alerted using intercoms or face-to-face communication.

▮ Get Out and Stay Out

Research has shown that most occupants will not willingly evacuate a building without visible signs of emergency. People should evacuate immediately, and should not take time to finish projects or gather personal belongings. A floor sweeper should be assigned to make certain everyone has exited and possibly to escort some occupants. The only exception to this rule, during evacuation drills, is for those positions that are mission critical to the organization. That determination should be made by the building management staff. Those individuals, even though they are not leaving, should be questioned by you to what they would do during a true emergency situation, such as knowing the locations of exits, what to do once outside, etc. Once the area is evacuated and secured, people must not re-enter for any reason. People may need to be posted at various points to prevent re-entry.

▮ Meeting Place

Once outside, occupants should be moved to a predetermined area away from the building and any arriving emergency equipment. Parking lot drive lanes are not appropriate evacuation areas because positioning people there will interfere with fire department's needs for that space to reach the building or to stage equipment. It is also important to keep people together in order to account for the safety of each person. Once outside, some people may begin walking around, or getting into cars and leaving. As fire units begin to arrive, the actions of people leaving or milling around will certainly impede fire department operations.

Employees should be accounted for, which in schools and larger buildings may mean to have a "go box" with names and phone numbers of all employees or students. Additionally, if possible, a daily roster should be included to account for those off sick, out to meetings, etc. An easy way to identify that everyone has been accounted for, without the use of radios, is with a red or green card. Once outside, if everyone has been accounted for, a green card is displayed. If people are not accounted for, then a red card is displayed.

Since fires and other emergencies can occur at any time, every business needs a well-rehearsed emergency plan. Life safety may very well depend on the newest employee. At the Beverly Hills Country Club fire, in Southgate, Kentucky, it was a teenage busboy that stopped a performance and made an announcement about the fire. His action is credited for saving 1,500 lives.

▮ Evacuation Leaders

Evacuation leaders assist visitors and employees in exiting the area Figure 30 . They should, during regular work hours, make sure that exits remain visible and unobstructed. Upon hearing

Chapter 7 Occupancy Safety and Evacuation Plans

Figure 30 Evacuation leaders are used to assist visitors and employees in exiting the area.

■ Evacuation Plan Success

The evacuation plan will not work without practice. All employees must be made aware of, and involved in, the emergency plans of the business. Once the evacuation plans have been formalized, all employees must review them regularly. Announced evacuation drills will then need to take place and be critiqued. Following that, unannounced evacuation drills must occur. Announced evacuation drills are good in that they consider the needs of the occupants; however, if everyone is aware of the evacuation drill, it is not a true picture of how people will respond to a real fire alarm. Unannounced evacuation drills provide this opportunity.

Questions about the evacuation plan should be encouraged and answers promptly provided. All employees with specific duties must be trained and drilled in those duties, as well as the overall emergency plan. New employees should, as part of their orientation, be trained and fully understand the evacuation plans of the building, and the specific needs of their assigned work area. Additionally, any employees who have been re-assigned to a different job or area need to be trained in the emergency procedures for the new job and area.

■ Evacuation Drills

Evacuation drills evaluate overcrowding points in the means of egress, overcrowding of a certain exit while another exit is available, occupants not evacuating, use of designated areas once outside, and doors left open.

In a small building, you may evacuate the entire building. As buildings get taller this may not be practical, or needed, so the fire alarm may only sound on select floors. In this situation, your observations should be limited to those floors. Evacuating selected floors is referred to as staged evacuation. While quick evacuation is important, it is more important that a safe and orderly evacuation is performed.

You also want to see if occupants have any problem locating an exit. Most of us are routine in what we do, such as always driving to work the same way or going into work using the same door. For that reason, it may be beneficial to block an exit to see if occupants know where their second exit is located. Putting a sign at the door stating "fire" can accomplish this, but blocking the door with equipment or preventing the exit door from opening will test how much effort and time occupants will delay before going to a different exit.

Following an evacuation drill, key personnel should be gathered to discuss the results. Additionally, it may be wise to poll the occupants to gather their feedback.

notification of an emergency, the evacuation leaders should ensure that people in their immediate work area are moved to a safe area, preferably outside of the building. When exiting, do not use any elevators, as they could stop at the fire floor and trap the occupants. Prior to exiting, evacuation leaders should make a final sweep of the area, if possible, checking washrooms, walk-in coolers, spray booths, etc. The last person out of an area should close the door when leaving, and leave it unlocked so that the fire department can access the area. If possible, one employee should remain in a safe area near those exits to prevent unauthorized personnel from re-entering the building.

Wrap-Up

Chief Concepts

- The National Fire Protection Association (NFPA) and the International Code Council (ICC) are the two most commonly used building and fire model codes.
- Each state either adopts a model building or fire code, or creates their own codes to meet their specific building and fire safety needs.
- The occupant load reflects the maximum number of people anticipated to occupy the building space(s) at any given time.
- While coming up with the occupant load figure is not too difficult, the occupant load figure can be fluid. A large open room will have one occupant load. When it is filled with table and chairs, as in a banquet hall, it has smaller occupant load figure. When posting occupant loads, some fire departments indicate the two potential occupant load figures.
- The means of egress will affect the occupant load. The means of egress must be sized to accommodate the occupant load.
- A means of egress consists of three separate parts: exit access, exit, and exit discharge.
- The exit access may be a corridor, aisle, balcony, gallery, room, porch, or roof. The length of the exit access establishes the travel distance to an exit, an extremely important feature of a means of egress, since an occupant might be exposed to fire or smoke during the time it takes to reach an exit.

Wrap-Up

- Examples of exits are doors leading directly outside at ground level, such as the front door of a business, or through a protected passageway to the outside at ground level. The latter includes smokeproof towers, protected interior and outside stairs, exit passageways, enclosed ramps, and enclosed escalators or moving walkways in existing buildings.
- Ideally, all exits in a building should discharge directly to the outside or through a fire-rated passageway to the outside of the building.
- At least two means of egress must exist in any area, unless specifically allowed by the *Life Safety Code*. The first chapters of the *Life Safety Code* list the general requirements on the number of means of egress.
- Means-of-egress elements are the components of the means of egress including exit access, exit enclosures, exit discharges, stairways, ramps, doors, hardware, exit markings, illumination, etc.
- In buildings where artificial lighting is provided for normal use, illumination of the means of egress is required to ensure that occupants can see to evacuate the building quickly.
- Maintaining the means of egress in safe operating condition at all times is as important as the proper design of the egress system itself.
- An evacuation plan allows for the safe and orderly removal of people from a building in the event of a fire or other emergency, and it is one part of a larger emergency plan. The evacuation plan takes effect when there is a need to have all occupants leave the building or area.
- Emergency plans are created to deal with various emergencies that might affect a building besides fire, such loss of power, flooding, chemical spills, etc.
- An evacuation plan should include:
  - A specific plan tailored to the building
  - An alert system
  - "Get out and stay out"
  - A meeting place

Wrap-Up

■ Hot Terms

Area of refuge An area that is either (1) a story in a building where the building is protected throughout by an approved, supervised automatic-sprinkler system and has not less than two accessible rooms or spaces separated from each other by smoke-resisting partitions; or (2) a space located in a path of travel leading to a public way that is protected from the effects of fire, either by means of separation from other spaces in the same building or by virtue of location, thereby permitting a delay in egress travel from any level. (NFPA 101)

Common path of travel The portion of exit access that must be traversed before two separate and distinct paths of travel to two exits are available. (NFPA 101)

Dead end corridor A passageway from which there is only one means of egress. (NFPA 301)

Emergency plan A document that outlines procedures for occupants to deal with all types of building-related emergency situations.

Evacuation plan A document that is part of an emergency plan and outlines procedures to vacate building occupants in a safe, orderly, and efficient manner.

Exit That portion of a means of egress that is separated from all other spaces of a building or structure by construction or equipment as required to provide a protected way of travel to the exit discharge. (NFPA 101)

Exit access That portion of a means of egress that leads to an exit. (NFPA 101)

Exit discharge That portion of a means of egress between the termination of an exit and a public way. (NFPA 101)

Horizontal exit An exit between adjacent areas on the same deck that passes through an A-60 Class boundary that is contiguous from side shell to side shell or to other A-60 Class boundaries. (NFPA 301)

Means of egress A continuous and unobstructed way of exit travel from any point in a building or structure to a public way, consisting of three separate and distinct parts: (a) the exit access, (b) the exit, and (c) the exit discharge. A means of egress comprises the vertical and horizontal travel and includes intervening room spaces, doorways, hallways, corridors, passageways, balconies, ramps, stairs, enclosures, lobbies, escalators, horizontal exits, courts, and yards. (NFPA 101)

Occupant load The number of people who might occupy a given area.

Panic hardware A door-latching assembly incorporating a device that releases the latch upon the application of a force in the direction of egress travel. (NFPA 101B)

Smokeproof enclosures A stair enclosure designed to limit the movement of products of combustion produced by a fire. (NFPA 101)

The net seating area of a ground floor ballroom measures 140 feet (42.6 m) by 110 feet (33.5 m). The room has been set for a large dinner with the seating arrangement consisting of table and chairs requiring a net area of 15 square feet (1.4 m²) per person. The facility has four exit doors. The main entrance/exit door is comprised of a double 36" (914 mm) door and three additional doors each measuring 36" (914 mm) in width. All egress doors are measured at 0.2" (5.1 mm) of width per person and all four of the exits are marked with exit signage.

1. What is the maximum occupant load of the room based only on floor area?

2. What is the maximum occupant load of the room based only on total egress capacity?

3. An assembly occupancy must provide a main exit that will accommodate 50 percent of the total occupant load of the room. Does this requirement change the maximum allowable occupant load of the room?

Fire Alarm and Detection Systems

CHAPTER 8

NFPA Objectives

Fire Inspector I

4.3.6 Determine the operational readiness of existing fire detection and alarm systems, given test documentation and field observations, so that the systems are in an operational state; maintenance is documented; and deficiencies are identified, documented, and reported in accordance with the policies of the jurisdiction. (pp 150–163)

(A) Requisite Knowledge. A basic understanding of the components and operation of fire detection and alarm systems and devices and applicable codes and standards. (pp 150–163)

(B) Requisite Skills. The ability to observe, make decisions, recognize problems, and read reports. (pp 150–163)

Additional NFPA Standards

NFPA 72 *National Fire Alarm and Signaling Code*

FESHE Objectives

Fire Plans Review

1. Verify the design of a fire alarm and detection system and an offsite supervisory system for compliance with applicable standards. (pp 150–163)

Knowledge Objectives

After studying this chapter, you will be able to:

1. Describe the basic components and functions of a fire alarm system.
2. Describe the basic types of fire alarm initiation devices and indicate where each type is most suitable.
3. Describe the fire inspector's role in inspection fire detection systems.

Skills Objectives

After studying this chapter, you will be able to:

1. Test a fire detection system.

You Are the Fire Inspector

You are a new fire inspector, this is only your fourth month on the job when the phone rings. It is a fire alarm contractor who is calling requesting a functional test of a new fire alarm system, which is being installed in a popular department store. The contractor is behind on the project and is anxious for the inspection. While performing the functional test you realize the alarm technician has installed an end of line resistor in the fire alarm control panel. You know this device should be at the end of the line, on the last detector.

1. How should you remedy this situation?
2. How can you prevent this situation in the future?

Introduction

Fire alarm systems are critical for communicating vital information to the occupants and the outside world. Fire alarm systems can protect the occupants and the structure by providing early warning to both the occupants and fire fighters which can minimize both loss of life and property damage during fires. This chapter will provide an overview of the types and features of fire alarm systems. It's also important to ensure that the fire alarm system was installed properly and complies with all local codes and standards.

Fire Alarm and Detection Systems

Many new construction projects, including single family homes, require some sort of fire alarm and detection system. Fire alarm and detection components are integrated in a single system that can perform a myriad of different functions, using multiple devices. A fire detection system recognizes when a fire is occurring and activates the fire alarm system, which then alerts the building occupants and activates building functions such as recalling elevators, shutting down HVAC units, unlocking doors, releasing door hold-open devices, and, in most cases, notifying the fire department. Some fire detection systems also automatically activate fire suppression systems to control the fire.

Fire alarm and detection systems range from simple, such as a small one-zone system in a coffee shop, to complex fire detection and control systems for high-rise buildings Figure 1 . Many fire alarm and detection systems in large buildings also control other systems to help protect occupants and control the spread of fire and smoke. Although these

Figure 1 Fire alarm and detection systems range from simple one zoned system to complex fire detection and control systems. As a fire inspector, you need to understand the components for all types of systems.

systems can be complex, they generally include the same basic components.

Plan Review

An inspection of a fire alarm and detection system begins in the plan review stage. Reviewing a plan of this component is no different than an electrical, plumbing, or HVAC proposal. The fire alarm contractor must submit a proper set of documents for approval prior to work beginning. This is the point where you or the plan reviewer must confirm that the plan includes proper type of devices in the correct locations, the proper number of devices, proper spacing, and adequate power.

A good set of submitted plans would include a scaled drawing of the area showing what initiating devices and notification appliances are going to be installed. It should also show the locations of the control panel, annunciator if present, and any remote power supplies. The plans should also indicate what wiring will be used for the initiating devices and notification appliances. Some locations will require a sample of each wire to be submitted with the plans.

There should also be copies of the manufacturer's specification sheets for the products that will be used in the fire alarm and detection system. From these, you or the plan reviewer can determine if the particular equipment being proposed for installation is acceptable. For example, if an addressable control panel is being provided, conventional detectors cannot be used. In this case, addressable detectors are needed or special modules for the conventional detectors are required.

Other information that should be included in the plans are voltage drops and battery calculations. The voltage drops will show the amount of voltage initially supplied in the notification circuit, and then a deduction for each device on that circuit. If there is not enough voltage to power the last device, larger wiring or an additional notification circuit will be needed. The battery calculations will show the amount of power needed for the system in standby and full alarm conditions, calculate that need for 24 or 60 hours according to local requirements, and then indicate the appropriate batteries to be used. Battery standby is used to provide secondary power in the event a long power failure.

Fire Alarm and Detection System Components

A fire alarm and detection system has three basic components: an alarm initiation device, an alarm notification device, and a control panel. The **alarm initiation device** is either an automatic or manually operated device that, when activated, causes the system to indicate an alarm. The **alarm notification appliance** is generally an audible device, more recently accompanied by a visual indication, that alerts the building occupants once the system is activated. The control panel links the initiation device to the notification appliance and performs other essential functions as required.

Fire Alarm System Control Panels

Most fire alarm and detection systems have several alarm initiation devices in different areas and use both audible and visible devices to notify the occupants of an alarm. The **fire alarm control panel** serves as the "brain" of the system, linking the initiating devices to the notification devices.

The control panel manages and monitors the proper operation of the system. It can indicate the source of an alarm, so that responding fire personnel will know what activated the alarm and where the initial activation occurred. The control panel also manages the primary power supply and provides a backup power supply for the system. It may perform additional functions, such as notifying the fire department or central station monitoring company when the alarm system is activated, and may interface with other systems and facilities such as recalling elevators, unlocking stairwell doors to allow people in the stairs to re-enter the floors, releasing doors held open by magnetic hold open devices, and shutting down ventilation systems.

Control panels vary greatly depending on the age of the system and the manufacturer. For example, an older system may simply indicate that an alarm has been activated, whereas a newer system may indicate that alarm occurred in a specific zone within the building **Figure 2**. The most modern panels actually specify the exact location of the activated initiation device. These panels are known as addressable panels.

Fire alarm control panels are used to silence the alarm and reset the system. These panels should always be locked, or in a room that is locked from the public. Many newer systems require the use of a password or key before the alarm can be silenced or reset. A problem with having a password is that the fire fighters will need to know the passwords to potentially hundreds of alarm systems. It is much easier to have the control panel access gained through the use of a key, which can be kept in the key box at the building. Alarms should not be reset until the activation source has been found and checked

Fire Inspector Tips

It is also very helpful to have an alarm matrix such the one listed in NFPA 72, *National Fire Alarm and Signaling Code*. An **alarm matrix** is a chart showing what will happen with the fire alarm and detection system when an initiating device is activated. Here, all the various types of devices are listed, including the inputs to the control panel and the possible outputs (functions) the panel can perform.

Fire Inspector Tips

While the smoke alarms and smoke detectors seem to be one in the same there are slight differences between the terms. Smoke detectors are devices that are placed in a building and connected to a control panel. Smoke alarms refer to the single station units found in almost all homes.

Figure 2 Most modern fire alarm control panels indicate the zone where an alarm was initiated.

by fire fighters to ensure that the situation is under control. If the system is reset prior to identifying the problem, fire fighters have no good way to determine the problem, and the activation may reoccur.

Some types of alarms require the activated devices to be reset prior to the control panel being reset. An example would be for activated manual pull stations. Other alarms, such as trouble or supervisory alarms, will reset the control panel automatically after the problem is resolved. A common example of this would be a power failure.

Many buildings have an additional display panel in a separate location, usually near the front door of the building. This panel, which is called a **remote annunciator**, enables fire fighters to ascertain the type and location of the activated alarm device as they enter the building eliminating the need for fire fighters to hunt down the control panel to determine the problem Figure 3 . There are two types of annunciators. The older style simply indicates the location of the alarm. The newer annunciators are known as functional annunciators. This type of annuciator will provide as much detail about the problem as the control panel. In addition, it has the capability of acknowledging, silencing, and resetting the alarm. This allows the main control panel to be put in any location of the building, as full control of the alarm system is performed at the functional annunciator. These typically are found just inside the main entrance of the building.

The fire alarm control panel should also monitor the condition of the entire alarm system to detect any faults. Faults within the system are indicated by a trouble alarm, which shows that a component of the system is not operating properly and requires attention. It also monitors the integrity of the control panel's circuits. Trouble alarms do not activate the building's fire alarm but will make an audible sound and illuminate a light at the alarm control panel. They will also transmit a notification to a remote service location, such as a central station monitoring company.

Another type of alarm that the control panel will monitor is a supervisory alarm. Supervisory alarms sound when something changes from the normal ready condition within the system. For example, tamper switches are put on control valves of the fire sprinkler system, and when a control valve is closed—shutting down the sprinkler system—an alarm is sounded at the control panel and transmitted to the monitoring location. As with trouble alarms, this activation will not sound the general building's alarms.

A fire alarm system is usually powered by 110-volts, even though the system's appliances may use a lower voltage. Some extremely old alarm systems, however, require 110 volts for all components.

In addition to the normal power supply, the codes and standards require a backup power supply for all alarm systems Figure 4 . In most systems, a battery in the fire alarm control panel is activated automatically when the external power is interrupted. The model building and fire codes and their referenced standards will specify how long the system must be able to function on the battery backup. Typically 24 or 60 hours of backup power is provided. Many agencies use the 60-hour systems, which would cover the power loss in a building for an entire weekend. In large buildings, the backup power supply could be an emergency generator. If either the main power supply or the battery backup power fails, the trouble alarm should sound.

Depending on the building's size and floor plan, an activated alarm may sound throughout the building or only in particular areas. In high-rise buildings, fire alarm systems are often programmed to alert only the occupants on the same floor as the activated alarm as well as those on the floors immediately above and below the affected floor. If an initiating device is activated on any other floor, the alarms in that those areas would then activate as well. Some systems have a public address feature, enabling the fire alarm panel

Figure 3 A remote annunciator allows fire fighters to quickly determine the type and location of the activated alarm device.

Figure 4 Model building and fire codes and their referenced standards require that alarm systems have a backup power supply, which is activated automatically when the normal electrical power is interrupted.

to play a prerecorded message, or for a fire department officer to provide specific instructions or information for occupants.

■ Residential Smoke Alarms

Current building and fire prevention codes require the installation of fire detectors with an alarm component in all residential dwelling units. <u>Single-station smoke alarms</u> are most commonly used Figure 5 . Millions of single-station smoke alarms have been installed in private dwellings and apartments, and countless numbers of lives have been saved due to them.

Smoke alarms can be either battery-powered or hard-wired to a 110-volt electrical system. Most building codes require hard-wired, AC-powered smoke alarms with battery backup in all newly constructed dwellings; battery-powered units are popular for existing residencies. The major concern with a battery-powered smoke alarm is ensuring that the battery is replaced on a regular basis. The International Association of Fire Chiefs (IAFC) has a campaign to get home owners to change the batteries in their smoke alarms. The Change Your Clock, Change Your Battery campaign is designed to remind people to replace the batteries in the spring and fall when daylight saving time occurs.

Newer battery-powered smoke alarms are available with lithium batteries that will last for 10 years. These units have the battery sealed inside the detector to prevent it from being used somewhere else. As smoke alarms should be replaced after 10 years, this poses no problem.

The original codes stated that the smoke alarm should be placed in the hallway outside of the sleeping rooms. That concept would work well if bedroom doors were left open and smoke was allowed to work into the hallway and activate the smoke alarm. Since bedroom doors are closed a good portion of the time, the most up-to-date codes also require new homes to have a smoke alarm in every bedroom and on every floor level. They must be also interconnected so that, if the basement smoke alarm is activated, alarms will sound on all levels of the home to alert the occupants. This arrangement is called multiple-station smoke alarms.

Many homes have added smoke and heat detectors as part of their home security systems. These devices operate just like the devices on an approved fire alarm system, except they are connected to a burglar alarm panel. These systems will require a passcode to set or reset the system and may be monitored by a central station. As home alarms are generally a burglar alarm system, it is unlikely that the fire department would have a password or code to reset the system.

Ionization versus Photoelectric Smoke Detectors

Two types of fire detection devices may be used in a smoke alarm to detect combustion:

- Ionization detectors are triggered by the invisible products of combustion.
- Photoelectric detectors are triggered by the visible products of combustion.

<u>Ionization smoke detectors</u> work on the principle that burning materials release many different products of combustion, including electrically charged microscopic particles. An ionization detector senses the presence of these invisible charged particles (ions).

An ionization smoke detector contains a very small amount of radioactive material inside its inner chamber. The radioactive material releases charged particles into the chamber, and a small electric current flows between two plates Figure 6 . When smoke particles enter the chamber, they neutralize the charged particles and interrupt the current flow. The detector then senses this interruption and activates the alarm.

<u>Photoelectric smoke detectors</u> use a light beam and a photocell to detect larger visible particles of smoke Figure 7 . They operate by reflecting the light beam either away from or onto the photocell, depending on the design of the device. When visible particles of smoke pass through the light beam, they interfere with the light beam, thereby activating the alarm.

Figure 5 The most common residential fire alarm system today is a single-station smoke alarm.

Figure 6 An ionization smoke detector.

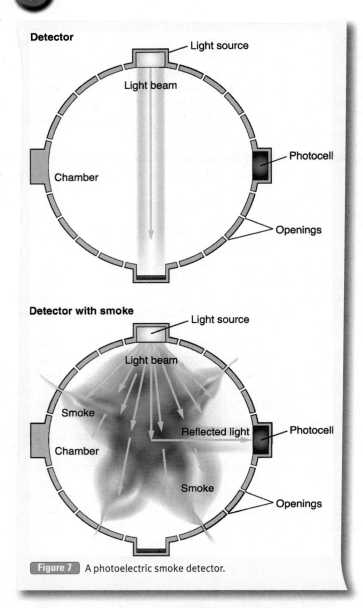

Figure 7   A photoelectric smoke detector.

Figure 8   Ionization smoke detectors react more quickly than photoelectric smoke detectors to a fast-burning fire, such as a fire in a wastepaper basket, which may produce little visible smoke.

Ionization smoke detectors are more common and less expensive than photoelectric smoke detectors. They react more quickly than photoelectric smoke detectors to fast-burning fires, such as a fire in a wastepaper basket, which may produce little visible smoke Figure 8 . On the downside, fumes and steam from common activities such as cooking and showering may trigger unwanted alarms.

Photoelectric smoke detectors are more responsive to slow-burning or smoldering fires, such as a fire caused by a cigarette caught in a couch, which usually produces a large quantity of visible smoke. They are less prone to false alarms from regular cooking fumes and shower steam than are ionization smoke detectors. Most photoelectric smoke alarms require more current to operate, however, so they are typically connected to a 110-volt power source.

Both ionization and photoelectric smoke detectors are acceptable life-safety devices. Indeed, in most cases, they are considered interchangeable. Some fire codes require that photoelectric smoke detectors be used within a certain distance of cooking appliances or bathrooms, to help prevent unwanted alarms.

Table 1   Types of Smoke Detectors

Type of Detector	Features	Application
Ionization	Uses radioactive material within the device to detect invisible products of combustion.	Used to detect fires that do not produce large quantities of smoke in their early states. React quickly to fast-burning fires. Inappropriate for use near cooking appliances or showers.
Photoelectric	Uses a light beam to detect the presence of visible particles of smoke.	Used to detect fires that produce visible smoke. React quickly to slowburning, smoldering fires.

Combination ionization/photoelectric smoke alarms are also available. These alarms will quickly react to both fast-burning and smoldering fires. They are not suitable for use near kitchens or bathrooms, because they are prone to the same nuisance alarms as regular ionization smoke detectors. Table 1 indicates the general features and recommended applications for each type of smoke detector.

Most ionization and photoelectric smoke detectors look very similar to each other. The only sure way to identify the type of alarm is to read the label, which is often found on the back of the case. An ionization alarm must have a label or engraving stating that it contains radioactive material.

From a fire inspector's viewpoint, the best option is to recommend to home owners to install an alarm that has both ionization and photoelectric sensors. It should also be noted that the most common type of alarm, the ionization, has performed well for many years and has saved many lives. Any smoke alarm is better than no smoke alarm. As a result of the press that some studies have evoked, some jurisdictions have made a requirement that only photoelectric alarms are allowed to be used in homes. Always state your local jurisdiction's policies when making recommendations.

■ Alarm Initiation Devices

Alarm initiation devices begin the fire alarm process either manually or automatically. Manual initiation devices require human action; automatic devices function without human intervention. Manual fire alarm boxes are the most common type of alarm initiation devices that require human action. Automatic initiation devices include various types of heat and smoke detectors and other devices that automatically recognize the evidence of a fire.

Manual Initiation Devices

Manual initiation devices are designed so that building occupants can activate the fire alarm system on their own if they discover a fire in the building. Many older alarm systems could be activated only manually. The primary manual initiation device is the manual fire alarm box, or **manual pull-station** Figure 9. Such a station has a switch that either opens or closes an electrical circuit to activate the alarm.

Pull-stations come in a variety of sizes and designs, depending on the manufacturer. They can be either single-action or double-action devices. **Single-action pull-stations** require a person to pull down a lever, toggle, or handle to activate the alarm. The alarm sounds as soon as the pull-station is activated. **Double-action pull-stations** require a person to perform two steps before the alarm will activate. They are designed to reduce the number of false alarms caused by accidental or intentional pulling of the alarm. The person must move a flap, lift a cover, or break a piece of glass to reach the alarm activation device. Designs that use glass are no longer in favor, because the glass must be replaced each time the alarm is activated and because the broken glass poses a risk of injury.

Once activated, a manual pull-station should stay in the "activated" position until it is reset. This enables responding fire fighters to determine which pull-station initiated the alarm. Resetting the pull-station requires a special key, a small screwdriver, or an Allen wrench. The pull-station must be reset before the building alarm can be reset at the alarm control panel. A keyed pull station may be preferred as it can be kept in the building's locked key box. Small screwdrivers or Allen wrenches tend to get lost easily.

A variation on the double-action pull-station, designed to prevent malicious false alarms, is a single station pull box covered with a piece of clear plastic Figure 10. These covers are often used in areas where malicious false alarms occur frequently, such as high schools and college dormitories. The plastic cover must be opened before the pull-station can be activated. Lifting the cover triggers a loud tamper alarm at that specific location, but does not activate the fire alarm system. Snapping the cover back into place resets the tamper alarm. The intent is that a person planning to initiate a false alarm will drop the cover and run when the tamper alarm sounds. In most cases, the pull-station tamper alarm is not connected to the fire alarm system.

Figure 9 Several types of manual fire alarm boxes (also known as manual pull-stations) are available.

Even automatic fire alarm systems can be manually activated by pressing a button or flipping a switch on the alarm system control panel. This is more useful for testing the system, as it would be impractical to expect someone to go to the control panel and activate the alarm manually during a fire.

Automatic Initiation Devices

Automatic initiation devices are designed to function without human intervention and will activate the alarm system when they detect evidence of smoke or fire. These systems can be programmed to transmit the alarm to the fire department or an on- or off-site monitoring facility, even if the building is unoccupied, and to perform other functions when a detector is activated.

Automatic initiation devices can use any of several types of detectors. Detectors are activated by smoke, heat, the light produced by actual flame, or specific gases created by fires.

Troubleshooting Smoke Alarms

Most problems with smoke alarms are caused by a lack of power, dirt in the sensing chamber, or defective alarms.

Power Problems

Smoke alarms require power from a battery or from a hard-wired 110-volt power source. Some smoke alarms that are hard-wired to a 110-volt power source are equipped with a battery that serves as a backup power source.

The biggest problem with battery-powered alarms is a dead or missing battery. Some people do not change batteries when recommended, resulting in inoperable smoke alarms. People may also remove smoke alarm batteries to use the battery for another purpose or to prevent nuisance alarm activations. The solution in this situation is to install a new battery in the alarm.

Hard-wired smoke detectors become inoperable if someone turns the power off at the circuit breaker. The solution in this situation is to turn the circuit breaker back on.

Most smoke alarms containing batteries will signal a low-battery condition by emitting a chirp every few seconds. This chirp indicates that the battery needs to be replaced to keep the smoke alarm functional. Many hard-wired smoke alarms that are equipped with battery backups will chirp to indicate that the backup battery is low, even when the 110-volt power source is operational. This feature assures that the smoke alarm will always have an operational backup system.

Dirt Problems

A second problem with smoke alarms is the increased sensitivity that results when dust or an insect becomes lodged in the photoelectric or ionization chamber. Such an obstruction causes the alarm to activate when a small quantity of water vapor or smoke enters the chamber, leading to unnecessary alarms. The solution in this situation is to remove the cover of the chamber and gently vacuum out the chamber. Follow the manufacturer's instructions for this process.

Alarm Problems

The third common problem you may encounter with smoke alarms is a worn-out detector. It is recommended that smoke alarms be replaced every 10 years because the sensitivity of the alarm can change. Some worn-out alarms may become overly sensitive and emit false alarms. Others may develop decreased sensitivity and fail to emit an alarm in the event of a fire. The date of manufacture should be stamped on each smoke alarm. If a detector is more than 10 years old, it should be replaced with a new alarm.

Understanding the basic functions and troubleshooting of simple smoke alarms enables you to respond to citizens who call your fire department when they encounter problems with their smoke alarms. Most importantly, it enables you to keep smoke alarms operational. Remember—only *operational* smoke alarms can save lives.

Smoke Detectors

A <u>smoke detector</u> is designed to sense the presence of smoke and refers to a sensing device that is part of a fire alarm system. This type of device is commonly found in school, hospital, business, and commercial occupancies that are equipped with fire alarm systems Figure 11.

Figure 10 A variation on the double-action pull-station, designed to prevent malicious false alarms, has a clear plastic cover and a separate tamper alarm.

Figure 11 Commercial ionization smoke detector.

Smoke detectors come in a variety of designs and styles geared toward different applications. The most common smoke detectors are ionization and photoelectric models, which operate in the same way that residential smoke alarms do. However, the smoke detectors used in commercial fire alarm systems are much more sophisticated and more expensive than residential smoke alarms. In commercial fire alarm systems, photoelectric detectors are typically used, as ionization detectors have a tendency to trigger false alarms more frequently.

Each detection device is rated to protect a certain floor area, so in large areas the detectors are often placed in a grid pattern. Newer smoke detectors also have a visual indicator, such as a steady or flashing light, that indicates when the device has power or has been activated.

A <u>beam detector</u> is a type of photoelectric smoke detector used to protect large spans such as churches, auditoriums, airport terminals, and indoor sports arenas. In these facilities, it would be difficult or costly to install large numbers of individual smoke detectors, but a single beam detector can be used for the entire length Figure 12.

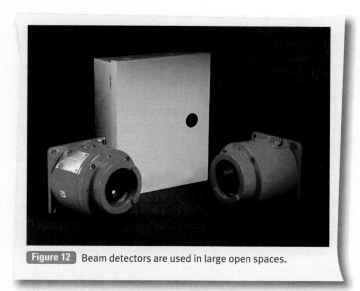

Figure 12 Beam detectors are used in large open spaces.

A typical beam detector has two components: a sending unit, which projects a narrow beam of light across the open area, and a receiving unit, which measures the intensity of the light when the beam strikes the receiver. When smoke interrupts the light beam, the receiver detects a drop in the light intensity and activates the fire alarm system. Most photoelectric beam detectors are set to respond to a certain **obscuration rate**, or percentage of light blocked. If the light is completely blocked, as when a solid object is moved across the beam, the trouble alarm will sound, but the fire alarm will not be activated.

Smoke detectors are usually powered by a low-voltage circuit and send a signal to the fire alarm control panel when they are activated. Both ionization and photoelectric smoke detectors are self-restoring. After the smoke condition clears, the alarm system can be reset at the control panel.

Heat Detectors

Heat detectors are also commonly used as automatic alarm initiation devices. Heat detectors can provide property protection, but cannot provide reliable life-safety protection because they do not react quickly enough to incipient fires. Because of this, the detectors have written on them "not a life safety device." They are generally used in situations where smoke alarms cannot be used, such as dusty environments and areas that experience extreme cold or heat. These detectors are often installed in unheated areas, such as attics and storage rooms, as well as in boiler rooms and manufacturing areas.

Heat detectors are generally very reliable and less prone to false alarms than are smoke detectors. You may come across heat detectors that were installed 30 or more years ago and are still in service; however, some older units have no visual trigger that tells which device was activated, so tracking down the cause of an alarm may be very difficult. Newer models have an indicator light that shows which device was activated.

Several types of heat detectors are available, each of which is designed for specific situations and applications. **Spot detectors** are individual units that can be spaced throughout an occupancy, so that each detector covers a specific floor area. The detectors may be in individual rooms or spaced at intervals along the ceiling in larger areas.

Heat detectors can be designed to operate at a fixed temperature or to react to a rapid increase in temperature. Either fixed-temperature or rate-of-rise devices can be configured as spot or line detectors.

Fixed-Temperature Heat Detectors

Fixed-temperature heat detectors, as the name implies, are designed to operate at a preset temperature. A typical temperature for a light-hazard occupancy, such as an office building, would typically be 135°F (57°C); however, they can be provided in other activation temperatures, depending on the locations where these will be installed. Fixed-temperature detectors typically include a metal alloy that will melt at the preset temperature. The melting alloy releases a lever-and-spring mechanism, to open or close a switch. Most fixed-temperature heat detectors must be replaced after they have been activated, even if the activation was accidental.

Rate-of-Rise Heat Detectors

Rate-of-rise heat detectors will activate if the temperature of the surrounding air rises more than a set amount in a given period of time. A typical rating might be "greater than 12°F (6.7°C) in 1 minute." If the temperature increase occurs more slowly than this rate, the rate-of-rise heat detector will not activate. By contrast, a temperature increase at a pace greater than this rate will activate the detector and set off the fire alarm. Rate-of-rise heat detectors should not be located in areas that normally experience rapid changes in temperature, such as near garage doors in heated parking areas, heating registers, or a commercial kitchen's cooking line.

Some rate-of-rise heat detectors have a **bimetallic strip** made of two metals that respond differently to heat: A rapid increase in temperature causes the strip to bend unevenly, which opens or closes a switch. Another type of rate-of-rise heat detector uses an air chamber and diaphragm mechanism: As air in the chamber heats up, the pressure increases. Gradual increases in pressure are released through a small hole, but a rapid increase in pressure will press upon the diaphragm and activate the alarm. Most rate-of-rise heat detectors are self-restoring, so they do not need to be replaced after an activation unless they were directly exposed to a fire.

Rate-of-rise heat detectors generally respond more rapidly to most fires than do fixed-temperature heat detectors. However, a slow-burning fire, such as a smoldering couch, may not activate a rate-of-rise heat detector until the fire is well established. Combination rate-of-rise and fixed-temperature heat detectors are available. These devices balance the faster response of the rate-of-rise detector with the reliability of the fixed-temperature heat detector.

Line Heat Detectors

Line detectors use wire or tubing strung along the ceiling of large open areas to detect an increase in heat. An increase in temperature anywhere along the line will activate the detector. Line detectors are found in churches, warehouses, and industrial or manufacturing applications.

One wire-type model has two wires inside, separated by an insulating material. When heat melts the insulation, the wires short out and activate the alarm. The damaged section of insulation must be replaced with a new piece after activation.

Another wire-type model measures changes in the electrical resistance of a single wire as it heats up. This device is self-restoring and does not need to be replaced after activation unless it is directly exposed to a fire.

The tube-type line heat detector contains a sealed metal tube filled with air or a nonflammable gas. When the tube is heated, the internal pressure increases and activates the alarm. Like the single-wire line heat detector, this device is self-restoring and does not need to be replaced after activation unless it is directly exposed to a fire.

Flame Detectors

Flame detectors are specialized devices that detect the electromagnetic light waves produced by a flame Figure 13. These devices can recognize a very small fire extremely quickly—as soon as a match is struck—and prior to the flame stabilizing on the match.

Typically flame detectors are found in places such as aircraft hangars or specialized industrial settings in which early detection and rapid reaction to a fire are critical. Flame detectors are also used in explosion suppression systems, where they detect and suppress an explosion as it is occurring.

Flame detectors are complicated and expensive. Another disadvantage of these models is that other infrared or ultraviolet sources, such as the sun or a welding operation, can set off an unwanted alarm. Flame detectors that combine infrared and ultraviolet sensors are sometimes used to lessen the chances of a false alarm.

Gas Detectors

Gas detectors are calibrated to detect the presence of a specific gas that is created by combustion or that is used in the facility. Depending on the system, a gas detector may be programmed to activate either the building's fire alarm system or a separate alarm. These specialized instruments need regular calibration if they are to operate properly. Gas detectors are usually found only in specific commercial or industrial applications.

Air Sampling Detectors

Air sampling detectors continuously capture air samples and measure the concentrations of specific gases or products of combustion. These devices draw in air samples through a sampling unit and analyze them using an ionization or photoelectric smoke detector. When air sampling detectors are installed in the return air ducts of large buildings, they are known as **duct detectors**. They will sound an alarm and shut down the air-handling system if they detect smoke.

More complex systems are sometimes installed in special hazard areas to draw air samples from rooms, enclosed spaces, or equipment cabinets Figure 14. The samples pass through gas analyzers that can identify smoke particles, products of combustion, and concentrations of other gases associated with a dangerous condition. Air sampling detectors are most often used in areas that hold valuable contents or sensitive equipment and, therefore, where it is important to detect problems early.

Alarm Initiation by Fire Suppression Systems

Other fire protection systems in a building may activate the fire alarm system. Automatic sprinkler systems are usually connected to the fire alarm system through a water flow paddle, and will activate the alarm if a water flow occurs Figure 15. Dry pipe sprinkler systems use pressure switches to activate the fire alarm system. Such a system not only alerts the building occupants and the fire department to a possible fire, but also ensures that someone is made aware that water is flowing, in case of an accidental discharge. Any other fire-extinguishing systems in a building, such as those found in kitchens or containing halogenated or carbon dioxide agents, should also be tied into the building's fire alarm.

> **Fire Inspector Tips**
>
> Duct detectors are best powered by the control panel. Some duct detectors are powered by the air handling unit as they are often integral to the HVAC unit. The problem with this is when a maintenance person comes to work on the air handling unit they will shut the unit down. Doing so sends the fire alarm into trouble, as power to the detector has been lost.

Figure 13 Flame detectors are specialized devices that detect the electromagnetic light waves produced by a flame.

Figure 14 Air sampling detector air intake cone.

Chapter 8 Fire Alarm and Detection Systems

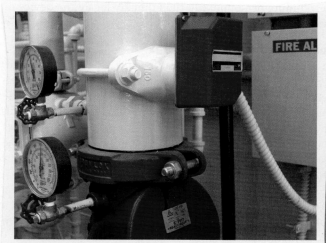

Figure 15 Automatic sprinkler systems use an electric flow switch to activate the building's fire alarm system.

In addition to the water flow alarm, valves that can shut down the sprinkler system must have tamper switches on them. This is to indicate that someone is shutting down a valve that will affect the sprinkler system. In sprinklered buildings, this is a prime reason why trouble alarms are investigated by trained personnel.

■ Alarm Notification Appliances

Audible and visual alarm notification appliances such as bells, horns, and electronic speakers produce an audible signal when the fire alarm is activated. Most newer systems also incorporate visual alerting devices. These audible and visual alarms alert occupants of a building to a fire.

Older systems used a variety of sounds as notification devices and a building could have many different alarms ringing throughout the day. This inconsistency often led to confusion over whether the sound was actually an alarm. More-recent model building and fire prevention codes have adopted a standardized audio pattern, called the **temporal-3 pattern**, that must be produced by any audio device used as a fire alarm. The temporal 3 is three short sounds from the audible device then a longer pause, and then it repeats itself until the alarm is silenced. Even single-station smoke alarms designed for residential occupancies now use this sound pattern. As a consequence, people can recognize a fire alarm immediately.

Some public buildings also play a recorded evacuation announcement in conjunction with the temporal-3 pattern. The recorded message is played through the fire alarm speakers and provides safe evacuation instructions. In facilities such as airport terminals, this announcement is recorded in multiple languages. This kind of system may also include a public address feature that fire department or building security personnel can use to provide specific instructions, relay information about the situation, or give notice when the alarm condition is terminated.

Many new fire alarm systems incorporate visual notification devices such as high-intensity strobe lights as well as audio devices **Figure 16**. Visual devices alert hearing-impaired occupants to a fire alarm and are very useful in noisy environments where an audible alarm might not be heard.

Figure 16 This alarm notification device has both a loud horn and a high-intensity strobe light.

■ Other Fire Alarm Functions

In addition to alerting occupants and summoning the fire department, fire alarm systems may control other building functions, such as air-handling systems, fire doors, and elevators. To control smoke movement through the building, the system may shut down or start up air handling systems. Fire doors that are normally held open by electromagnets may be released to compartmentalize the building and confine the fire to a specific area. Doors allowing reentry from exit stairways into occupied areas may be unlocked. Elevators will be summoned to a predetermined floor, usually the main lobby, so they can be used by fire crews.

Responding fire personnel must understand which building functions are being controlled by the fire alarm, for both safety and fire suppression reasons. This information should be gathered during preincident planning surveys and should be available in printed form or on a graphic display at the control panel location.

■ Fire Alarm Systems

Some fire alarm systems give very little, or no, information at the alarm control panel; others specify exactly which initiation device activated the fire alarm. The systems can be further subdivided based on whether they are zoned or coded systems.

In a **zoned system**, the alarm control panel will indicate where in the building the alarm was activated. Almost all alarm systems are now zoned to some extent. Only the most rudimentary alarm systems give no information at the alarm control panel about where the alarm was initiated. In a coded system, the zone is identified not only at the alarm control panel but also through throughout the building using the audio notification device.

Table 2 shows how, using these two variables, systems can be broken down into four categories: noncoded alarm, zoned noncoded alarm, zoned coded alarm, and master-coded alarm.

Noncoded Alarm System

In a noncoded alarm system, the control panel has no information indicating where in the building the fire alarm was activated. The alarm typically sounds a bell or horn. Fire department personnel must then search the entire building to find which initiation device was activated. This type of system is generally found only in older, small buildings.

Zoned Noncoded Alarm System

The zoned noncoded alarm system is the most common type of system, particularly in smaller buildings. With this model, the building is divided into multiple zones, often by floor or by wing. The alarm control panel indicates in which zone the activated device is located and, sometimes, which type of device was activated. Responding personnel can go directly to that part of the building to search for the problem and check the activated device.

Many zoned noncoded alarm systems have an individual indicator light for each zone. When a device in that area is activated, the indicator light goes on.

Computerized alarm systems, known as addressable system, may use addressable devices. This is a step up from the zoned systems. In these systems, each individual initiation device—whether it is a smoke detector, heat detector, or pull-station—has its own unique identifier. When the device is activated, the display on the control panel will indicate a specific point. When programmed properly, the display will provide information such as, "smoke detector, room 101."

Responding personnel can then quickly locate the device or devices that have been activated.

Zoned Coded Alarm

In addition to having all the features of a zoned alarm system, a zoned coded alarm system indicates which zone has been activated over the announcement system. Hospitals often use this type of system, because it is not possible to evacuate all staff and patients for every fire alarm. The audible notification devices sound in a sequence that provides a code to indicate which zone was activated. A code list tells building personnel which zone is in an alarm condition.

More modern systems in these occupancies use speakers as their alarm notification devices. This approach means that a voice message indicating the nature and location of the alarm can accompany the audible alarm signal.

Master-Coded Alarm

In a master-coded alarm system, the audible notification devices for fire alarms also are used for other purposes. For example, a school may use the same bell to announce a change in classes, to signal a fire alarm, to summon the janitor, or to make other notifications. Most of these systems have been replaced by modern speaker systems that use the temporal-3 pattern fire alarm signal and have public address capabilities. This type of system is rarely installed in new buildings.

■ Fire Department Notification

The fire department should always be notified when a fire alarm system is activated. In some cases, a person must make a telephone call to the fire department or an emergency communications center. In other cases, the fire alarm system may be connected directly to the fire department or to a remote location where someone on duty receives a signal from the fire alarm system and calls the fire department.

There are two primary ways fire alarm systems transmit their signals to the fire department: phone lines and radio signal. Phone lines are the traditional method. When using a dedicated phone line—one that is only for monitoring the fire alarm—the monthly costs can be high. Recently there has been a move to switch from phone lines to radio to transmit the alarm signal. Radios have far less related trouble signals than when using phone lines. Additionally if a radio goes bad, it could be replaced in about an hour. If the phone line is dug up, it may be many days before it is repaired, leaving the building without automatic fire department notification.

As shown in **Table 3**, fire alarm systems can be classified in five categories, based on how the fire department is notified of an alarm.

Protected Premises Fire Alarm System

A protected premises fire alarm system does not notify the fire department. Instead, the alarm sounds only in the building to notify the occupants. Buildings with this type of system should have notices posted requesting occupants to call the fire department and report the alarm after they exit **Figure 17**. It is common that after a period of a local alarm ringing, the occupants will

Table 2 Categories of Alarm Annunciation Systems

Category	Description
Noncoded alarm	No information is given on what device was activated or where it is located.
Zoned noncoded alarm	Alarm system control panel indicates the zone in the building that was the source of the alarm. It may also indicate the specific device that was activated.
Zoned coded alarm	The system indicates over the audible warning device which zone has been activated. This type of system is often used in hospitals, where it is not feasible to evacuate the entire facility.
Master-coded alarm	The system is zoned and coded. The audible warning devices are also used for other emergency-related functions.

Table 3 Fire Department Notification Systems

Type of System	Description
Protected Premises	The fire alarm system sounds an alarm only in the building where it was activated. No signal is sent out of the building. Someone must call the fire department to respond.
Remote Supervising Station	The fire alarm system sounds an alarm in the building and transmits a signal to a remote location. The signal may go directly to the fire department or to another location where someone is responsible for calling the fire department.
Auxiliary System	The fire alarm system sounds an alarm in the building and transmits a signal to the fire department via a public alarm box system.
Proprietary Supervising System	The fire alarm system sounds an alarm in the building and transmits a signal to a monitoring location owned and operated by the facility's owner. Depending upon the nature of the alarm and arrangements with the local fire department, facility personnel may respond and investigate, or the alarm may be immediately retransmitted to the fire department. These facilities are monitored 24 hours a day.
Central Station	The fire alarm system sounds an alarm in the building and transmits a signal to an off-premises alarm monitoring facility. The offpremises monitoring facility is then responsible for notifying the fire department to respond, and they must send a person to investigate the signals and provide silence or reset services.

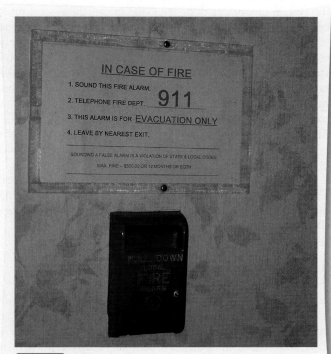

Figure 17 Buildings with a local alarm system should post notices requesting occupants to call the fire department and report the alarm after they exit.

call the fire department and ask if they are responding. This may be the first indication to the fire department that the alarm has activated. Most people assume that when a fire alarm activates, it is only minutes until the arrival of the fire department. For the safety of occupants, it is highly suggested that these alarms transmit their signals to a monitoring station.

Remote Supervising Station Systems

A remote supervising station system sends a signal directly to the fire department or to another monitoring location via a telephone line or a radio signal. This type of direct notification can only be installed in jurisdictions where the fire department is equipped to handle direct alarms. If the signal goes to a monitoring location, that site must be continually staffed by someone who will call the fire department.

Auxiliary Systems

Auxiliary systems can be used in jurisdictions with a public fire alarm box system. A building's fire alarm system is tied into a master alarm box located outside the building. When the alarm activates, it trips the master box, which transmits the alarm directly to the fire department communications center. This would only be seen in large older cities, as most cities have phased out public fire alarm boxes.

Proprietary Supervising Systems

In a proprietary supervising system, the building's fire alarms are connected directly to a monitoring site that is owned and operated by the building's owner. Proprietary systems are often installed at facilities where multiple buildings belong to the same owner, such as universities or industrial complexes. Each building is connected to a monitoring site on the premises (usually the security center), which is staffed at all times Figure 18. When an alarm sounds, the staff at the monitoring site reports the alarm to the fire department, usually by telephone or a direct line. The code does not state that the buildings need to be on the same property. Take, for example, a large box retailer who owns buildings throughout the country. Because they are all owned by the same owner, they fall into the proprietary category and can have all building monitored in one location, which may be on the opposite side of the country from your jurisdiction.

Central Stations

A central station is a third-party, off-site monitoring facility that monitors multiple alarm systems. Individual building owners contract and pay the central station to monitor their facilities Figure 19. When an activated alarm at a covered building transmits a signal to the central station by telephone or radio, personnel at the central station notify the appropriate fire department of the fire alarm. The central station facility may be located in the same city as the facility or in a different part of the country. One of the requirements of a central station is to provide a "runner." This is someone who would respond to the building to investigate and repair the alarm if necessary.

Usually, building alarms are connected to the central station through leased or standard telephone lines; however, the use of either cellular telephone frequencies or radio frequencies is becoming more common. Cellular or radio connections may be used to back up regular telephone lines in case they fail; in

A

B

Figure 18 In a proprietary system, fire alarms from several buildings are connected to a single monitoring site owned and operated by the buildings' owner.

remote areas without telephone lines, they may be the primary transmission method.

Wiring Concerns

Generally speaking, correct wiring begins at one device and goes to the next, and the next, until that last device is connected on that circuit. This type of wiring is done to ensure the integrity of the wiring, and the devices are part of that integrity. When a device is removed from the circuit, or wiring is cut the panel sees a problem and reports a trouble alarm. Until that problem is corrected, some of the devices will not work. Everything up to the missing detector will function as the wiring circuit is still

Figure 19 A central station monitors alarm systems at many locations.

intact. The devices past the missing detector will not operate as the wiring has been "opened" because of the missing detector.

People not familiar with wiring fire alarms will add devices by **T tapping** into the existing wiring. Instead of properly wiring from one device into the next, they take a couple of wires and join them to the existing circuit. While the device will function if activated, the problem is that it is not **supervised**. That means that if the device is removed from the wiring, the control panel will not know, and no trouble alarm will be transmitted.

Fire Alarm and Detection System Maintenance

Fire alarm systems must be kept in proper working order for the system to work as designed. That means the system must be maintained properly and when a device is broken it must be repaired immediately. In addition, there should be a monthly visual inspection looking for things such as visibility of pull stations, access to the pull stations, smoke detectors with the caps still in place, damage to any devices, and anything else that might prohibit smoke or heat detectors from operating such as painted heat detectors and bags on smoke detectors Figure 20. Also, yearly system tests should be conducted to determine that the system is operating as required. NFPA 72 specifies the requirements of inspections and maintenance.

One thing that is not visible is the strength of the batteries. They should be replaced about every four years. Load testing during an inspection can determine exactly when the batteries should be replaced.

Fire Alarm and Detection System Testing

As a fire inspector, other than at the initial acceptance test to verify monitoring and alarm signals, you should not conduct any tests on fire detection systems, but rather call for the owner to have the system inspected by an alarm company yearly. You will then study the inspection results to determine if action needs to be taken. You need to read this report closely. Often deficiencies are noted and recommended for correction. Unless you follow up with the owner, there is no way to know if repairs have actually been made to return the alarm to full operational status.

Chapter 8 Fire Alarm and Detection Systems

be shut off. This is the only way to tell of the audio visual devices have enough power to operate properly. Each device component of the system needs to be activated to see that it functions as designed. If auxiliary functions such as releasing magnetically held doors, or shutting down an HVAC unit are programmed, those must be checked as well to see that those functions occurred.

If there is an interruption in the wiring, such as a missing detector, a trouble alarm should occur. The best way to check this is to actually take some devices off of each circuit and see if a trouble alarm sounds.

When inspecting audio visual devices, it should be noted that for those people who might be susceptible to seizures the visual devices located within the space must flash together. They cannot each flash independently of each other.

During the pre-occupancy inspection, you should ensure that proper signals go to the monitoring station. This is also a time to verify that the proper alarm sequence was received by the monitoring system. Each of the three types of alarms—trouble, supervisory, and fire—are increasingly more important. A properly functioning alarm system will have each type of alarm override the lower level alarm.

During the pre-occupancy inspection, the circuit breaker needs to be identified and locked out at the breaker box to prevent an accidently shutting down of the fire alarm power. It is a good idea to also put the circuit number in the control panel for future use. The batteries in the control panel, and any power supplies in the field, must be marked with the year that they were manufactured.

Once you are satisfied with the results of the final pre-occupancy inspection, approval can be given. From this point on, unless the system is modified, you do not need to be this physically involved in the annual inspection process.

Documentation and Reporting of Issues

When conducting the pre-occupancy inspection, any deficiencies need to be noted. There should be a copy of the results for you, the alarm company, and the owner. If you have to return, then only those items need to be inspected again. If the system meets your approval, then simply indicate on the paperwork that the system is approved. Often the alarm company will have an inspection report that they use and will want you to sign. This paperwork should be kept in the occupancy file.

Annual inspection paperwork should be submitted to the owner by an alarm company indicating the status of the system. If deficiencies are noted, continued follow up must occur until the proper repairs a made. The most recent inspection paperwork needs to be kept on site at the control panel. Some alarm companies are installing tubes or boxes to house that paperwork.

Figure 20 There should be a monthly visual inspection of fire detection systems looking for things such as visibility of pull stations, access to the pull stations, smoke detectors with the caps still in place, or damage to any devices.

Figure 21 When the fire detection system is first installed, you must observe the testing of the system by two alarm technicians.

When a system is first installed, every part of the system needs to be tested. It is the only way you can sign off that all parts of the system worked when installed. On large systems this could take hours or days. Two alarm technicians with communication equipment should be present. One needs to be in the field actually testing the devices with you **Figure 21**. The second needs to silence or reset the system and report what the control panel displays. The alarm company should perform an inspection themselves prior to your arrival to determine if everything works properly.

The fire alarm and detection system should be designed to work on battery power, so for the test the 110 volt power should

Wrap-Up

Chief Concepts

- Fire alarm and detection systems can protect the occupants and the structure by providing early warning to both the occupants and fire fighters, which can minimize both loss of life and property damage during fires.
- A fire alarm and detection system recognizes when a fire is occurring and activates the fire alarm system, which then alerts the building occupants, activates building functions such as recalling elevators, shutting down HVAC units, unlocking doors, releasing door hold open devices, and in most cases, notifying the fire department.
- An inspection of a fire alarm and detection system begins in the plan review stage. The fire alarm contractor must submit a proper set of documents for approval prior to work beginning. You or the plan reviewer must confirm that the proper type of devices are in the correct locations, that the proper number of devices are installed, that spacing concerns are addressed properly, and that there is enough power to supply the fire alarm and detection system.
- A fire detection or alarm system has three basic components: an alarm initiation device, an alarm notification device, and a control panel.
- The alarm notification appliance is generally an audible device, more recently accompanied by a visual indication, that alerts the building occupants once the system is activated.
- The control panel links the initiation device to the notification appliance and performs other essential functions as required.
- Two types of fire detection devices may be used in a smoke alarm to detect combustion:
  - Ionization detectors are triggered by the invisible products of combustion.
  - Photoelectric detectors are triggered by the visible products of combustion.
- Alarm initiation devices begin the fire alarm process either manually or automatically. Manual initiation devices require human action; automatic devices function without human intervention.
- Automatic initiation devices can use any of several types of detectors. Some detectors are activated by smoke; others react to heat, others to the light produced by actual flame, and still others to specific gases.
- Audible and visual alarm notification devices such as bells, horns, and electronic speakers produce an audible signal when the fire alarm is activated. Most new systems also incorporate visual alerting devices.
- In addition to alerting occupants and summoning the fire department, fire alarm systems may control other building functions, such as air-handling systems, fire doors, and elevators.
- Some fire alarm systems give very little, or no, information at the alarm control panel; others specify exactly which initiation device activated the fire alarm. The systems can be further subdivided based on whether they are zoned or coded systems.
- There are two ways fire alarm systems transmit their signals to the fire department—phone lines and radio signal.
- Correct wiring begins at one device and goes to the next, and the next, until that last device is connected on that circuit. This type of wiring is done to ensure the integrity of the wiring, and the devices are part of that integrity. When a device is removed from the circuit, or wiring is cut the panel sees a problem and reports a trouble alarm.
- Fire alarm systems must be kept in proper working order for the system to work as designed. That means the system must be maintained properly and when a device is broken it must be repaired immediately. In addition, there should be a monthly visual inspection.
- As a fire inspector, other than at the initial acceptance test to verify monitoring and alarm signals, you should not conduct any tests on fire detection systems, but rather call for the owner to have the system inspected by an alarm company yearly. You will then study the inspection results to determine if action needs to be taken.

Wrap-Up

Hot Terms

Air sampling detector A system that captures a sample of air from a room or enclosed space and passes it through a smoke detection or gas analysis device.

Alarm initiation device An automatic or manually operated device in a fire alarm system that, when activated, causes the system to indicate an alarm condition.

Alarm matrix A chart showing what will happen with the fire alarm system when an initiating device is activated.

Alarm notification appliance An audible and/or visual device in a fire alarm system that makes occupants or other persons aware of an alarm condition.

Auxiliary system A fire alarm system that sounds an alarm in the building and transmits a signal to the fire department via a public alarm box system.

Beam detector A smoke detection device that projects a narrow beam of light across a large open area from a sending unit to a receiving unit. When the beam is interrupted by smoke, the receiver detects a reduction in light transmission and activates the fire alarm.

Bimetallic strip A device with components made from two distinct metals that respond differently to heat. When heated, the metals will bend or change shape.

Central station An off-premises facility that monitors alarm systems and is responsible for notifying the fire department of an alarm. These facilities may be geographically located some distance from the protected building(s).

Coded system A fire alarm system design that divides a building or facility into zones and has audible notification devices that can be used to identify the area where an alarm originated.

Double-action pull-station A manual fire alarm activation device that requires two steps to activate the alarm. The person must push in a flap, lift a cover, or break a piece of glass before activating the alarm.

Duct detector A smoke detector inside an air handling unit that will sound an alarm and shut down the air handling unit when smoke is detected.

Fire alarm control panel The component in a fire alarm system that controls the functions of the entire system.

Fixed-temperature heat detector A sensing device that responds when its operating element is heated to a predetermined temperature.

Flame detector A sensing device that detects the radiant energy emitted by a flame.

Gas detector A device that detects and/or measures the concentration of dangerous gases.

Heat detector A fire alarm device that detects abnormally high temperature, an abnormally high rate-of-rise in temperature, or both.

Ionization smoke detector A device containing a small amount of radioactive material that ionizes the air between two charged electrodes to sense the presence of smoke particles.

Line detector Wire or tubing that can be strung along the ceiling of large open areas to detect an increase in heat.

Manual pull-station A device with a switch that either opens or closes a circuit, activating the fire alarm.

Master-coded alarm An alarm system in which audible notification devices can be used for multiple purposes, not just for the fire alarm.

Noncoded alarm An alarm system that provides no information at the alarm control panel indicating where the activated alarm is located.

Obscuration rate A measure of the percentage of light transmission that is blocked between a sender and a receiver unit.

Wrap-Up

Photoelectric smoke detector A device to detect visible products of combustion using a light source and a photosensitive sensor.

Proprietary supervising system A fire alarm system that transmits a signal to a monitoring location owned and operated by the facility's owner.

Protected premises fire alarm system A fire alarm system that sounds an alarm only in the building where it was activated. No signal is sent out of the building.

Rate-of-rise heat detector A fire detection device that responds when the temperature rises at a rate that exceeds a predetermined value.

Remote annunciator A secondary fire alarm control panel in a different location than the main alarm panel; it is usually located near the front door of a building.

Remote supervising station system A fire alarm system that sounds an alarm in the building and transmits a signal to the fire department or an off-premises monitoring location.

Single-action pull-station A manual fire alarm activation device that takes a single step—such as moving a lever, toggle, or handle—to activate the alarm.

Single-station smoke alarm A single device usually found in homes that detects visible and invisible products of combustion and sounds an alarm.

Smoke detector A device that detects smoke and sends a signal to a fire alarm control panel.

Spot detector A single heat-detector device; these devices are often spaced throughout an area.

Supervised Electronically monitoring the alarm system wiring for an open circuit

T tapping Improper wiring of an initiating device so that it is not supervised.

Temporal-3 pattern A standard fire alarm audible signal for alerting occupants of a building.

Zoned coded alarm A fire alarm system that indicates which zone was activated both on the alarm control panel and through a coded audio signal.

Zoned noncoded alarm A fire alarm system that indicates the activated zone on the alarm control panel.

Zoned system A fire alarm system design that divides a building or facility into zones so that the area where an alarm originated can be identified.

Fire Inspector *in Action*

You are inspecting an eight-story mixed-use occupancy. The building contains commercial occupancies on the first floor and residential condominiums on the upper floors. The building is about 40 years old. It is equipped with multiple smoke detectors in each residential unit. The first floor and the hallways and lobbies of the upper floors are protected by sprinklers. The fire alarm system is divided into multiple zones. The alarm system is monitored by a central station, which is responsible for contacting the fire department in the event of an alarm.

1. How would you classify the fire alarm system in this building?
   A. A noncoded alarm system
   B. A zoned noncoded alarm system
   C. A zoned coded alarm system
   D. A master-coded alarm system

2. Which type of notification does this alarm system use?
   A. Local alarm system
   B. Remote station system
   C. Auxiliary system
   D. Central station

3. If this building were new, which test would you perform on this system?
   A. The central station alarm test
   B. The noncoded alarm activation test
   C. The auxiliary system test
   D. None

4. As part of the maintenance plan, fire detection systems should be tested:
   A. weekly.
   B. monthly.
   C. yearly.
   D. daily.

Fire Flow and Fire Suppression Systems

CHAPTER 9

NFPA Objectives

Fire Inspector I

4.3.5 Determine the operational readiness of existing fixed fire suppression systems, given test documentation and field observations, so that the systems are in an operational state, maintenance is documented, and deficiencies are identified, documented, and reported in accordance with the applicable codes and standards and the policies of the jurisdiction. (pp 194–196)

(A) Requisite Knowledge. A basic understanding of the components and operation of fixed fire suppression systems and applicable codes and standards. (pp 184–196)

(B) Requisite Skills. The ability to observe, make decisions, recognize problems, and read reports. (pp 184–196)

4.3.9 Compare an approved plan to an existing fire protection system, given approved plans and field observations, so that any modifications to the system are identified, documented, and reported in accordance with the applicable codes and standards and the policies of the jurisdiction. (pp 171, 184–196)

(A) Requisite Knowledge. Fire protection symbols and terminology. (pp 171, 184–196)

(B) Requisite Skills. The ability to read and comprehend plans for fire protection systems, observe, communicate, apply codes and standards, recognize problems, and make decisions. (pp 171, 184–196)

4.3.16 Verify fire flows for a site, given fire flow test results and water supply data, so that required fire flows are in accordance with applicable codes and standards and deficiencies are identified, documented, and reported in accordance with the applicable codes and standards and the policies of the jurisdiction. (pp 176–179)

(A) Requisite Knowledge. Types of water distribution systems and other water sources in the local community, water distribution system testing, characteristics of public and private water supply systems, and flow testing procedures. (pp 177–179)

(B) Requisite Skills. The ability to use Pitot tubes, gauges, and other data gathering devices as well as calculate and graph fire flow results. (pp 176–179)

Fire Inspector II

5.3.4 Evaluate fire protection systems and equipment provided for life safety and property protection, given field observations of the facility and documentation, the hazards protected, and the system specifications, so that the fire protection systems provided are approved for the occupancy or hazard being protected. (pp 184–196)

(A) Requisite Knowledge. Applicable codes and standards for fire protection systems, basic physical science as it relates to fire behavior and fire suppression, implications and hazards associated with system operation, installation techniques and acceptance inspection, testing and reports of maintenance of completed installations, and use and function of various systems. (pp 171, 184–196)

(B) Requisite Skills. The ability to recognize problems, use codes and standards, and read reports, plans, and specifications. (pp 171, 184–196)

5.3.11 Verify compliance with construction documents, given a performance-based design, so that life safety systems and building services equipment are installed, inspected, and tested to perform as described in the engineering documents and the operations and maintenance manual that accompanies the design, so that deficiencies are identified, documented, and reported in accordance with the applicable codes and standards and the policies of the jurisdiction. (pp 171, 184–196)

(A) Requisite Knowledge. Applicable codes and standards for installation and testing of fire protection systems, means of egress, and building services equipment. (p 171)

(B) Requisite Skills. The ability to witness and document tests of fire protection systems and building services equipment. (pp 194–196)

5.4.3 Review the proposed installation of fire protection systems, given shop drawings and system specifications for a process or operation, so that the system is reviewed for code compliance and installed in accordance with the approved drawings, and deficiencies are identified, documented, and reported in accordance with the applicable codes and standards and the policies of the jurisdiction. (pp 171, 194–196)

(A) Requisite Knowledge. Proper selection, distribution, location, and testing of portable fire extinguishers; methods used to evaluate the operational readiness of water supply systems used for fire protection; evaluation and testing of automatic sprinkler, water spray, and standpipe systems and fire pumps; evaluation and testing of fixed fire suppression systems; and evaluation and testing of automatic fire detection and alarm systems and devices. (pp 176–179, 194–196)

(B) Requisite Skills. The ability to read basic floor plans or shop drawings and identify symbols used by the jurisdiction. (p 171)

5.4.4 Review the installation of fire protection systems, given an installed system, shop drawings, and system specifications for a process or operation, so that the system is reviewed for code compliance and installed in accordance with the approved drawings, and deficiencies are identified, documented, and reported in accordance with the applicable codes and standards and the policies of the jurisdiction. (pp 171, 184–196)

(A) Requisite Knowledge. Proper selection, distribution, location, and testing of portable fire extinguishers; methods used to evaluate the operational readiness of water supply systems used for fire protection; evaluation and testing of automatic sprinkler, water spray, and standpipe systems and fire pumps; evaluation and testing of fixed fire suppression systems; and evaluation and testing of automatic fire detection and alarm systems and devices. (pp 171–196)

(B) Requisite Skills. The ability to read basic floor plans or shop drawings. (p 171)

Additional NFPA Standards

- **NFPA 12** *Standard on Carbon Dioxide Extinguishing Systems*
- **NFPA 12A** *Standard on Halon 1301 Fire Extinguishing Systems*
- **NFPA 13** *Standard for the Installation of Sprinkler Systems*
- **NFPA 13D** *Standard for the Installation of Sprinkler Systems in One and Two Family Dwellings and Manufactured Homes*
- **NFPA 13R** *Standard for the Installation of Sprinkler Systems in Low-Rise Residential Occupancies*
- **NFPA 14** *Standard for the Installation of Standpipes and Hose Systems*
- **NFPA 15** *Standard for Water Spray Fixed Systems for Fire Protection*
- **NFPA 16** *Standard for the Installation of Foam-Water Sprinkler and Foam-Water Spray Systems*
- **NFPA 17** *Standard for Dry Chemical Extinguishing Systems*
- **NFPA 17A** *Standard for Wet Chemical Extinguishing Systems*
- **NFPA 20** *Standard for the Installation of Stationary Pumps for Fire Protection*
- **NFPA 22** *Standard for Water Tanks for Private Fire Protection*
- **NFPA 24** *Standard for the Installation of Private Fire Service Mains and Their Appurtenances*
- **NFPA 25** *Standard for the Inspection, Testing, and Maintenance of Water-Based Fire Protection Systems*
- **NFPA 750** *Standard on Water Mist Fire Protection Systems*
- **NFPA 1141** *Standard for Fire Protection Infrastructure for Land Development in Wildland, Rural, and Suburban Areas*
- **NFPA 1963** *Standard for Fire Hose Connections*
- **NFPA 2001** *Standard on Clean Agent Fire Extinguishing Systems*

FESHE Objectives

Fire Plans Review

8. Determine fire department access, verify appropriate water supply, and review general building parameters. (pp 176–179)
11. Identify special hazards, verify interior finish and establish the proper locations for pre-engineered fire extinguishing systems. (pp 184–196)

Knowledge Objectives

After studying this chapter, you will be able to:

1. Describe the elements of a plan review of a fire suppression system.
2. Describe the two types of water distribution systems.
3. Describe the sources of water for a municipal water supply system.
4. Describe the major features of a municipal water distribution system.
5. Describe dry-barrel and wet-barrel hydrants.
6. Define static pressure, residual pressure, and flow pressure.
7. Describe the hydrant testing procedure.
8. Describe the factors to be considered during the design phase of an automatic sprinkler system.
9. Identify the six types of sprinkler systems.
10. Identify the types of sprinkler heads.
11. Identify the three types of standpipes.
12. Describe how to test the readiness of fire suppression systems.

Skills Objectives

After studying this chapter, you will be able to:

1. Determine fire flow.
2. Perform the underground flush test on a fire suppression system.
3. Perform the hydrostatic test on a fire suppression system.
4. Perform an air test on a fire suppression system.
5. Perform a main drain test on a fire suppression system.

You Are the Fire Inspector

As a new fire inspector, you are stationed in a mostly residential area and have only a few office buildings with automatic sprinkler systems. During your training, you gained a basic familiarity of the operation of fire suppression systems but you feel far from being an expert. Technical information on fire suppression systems is readily available and you also find that it is continually changing.

1. What can you do to gain real-life experience in these systems?
2. Would it be practical and possible for you to contact a local area large fire department, a fire service fire prevention association, a state fire training school, a specialized fire insurance company, or take courses through the NFPA to learn more about sprinkler systems?
3. Is there a technical group that would allow you to sit in on meetings or training sessions?

Introduction

Fire suppression systems include automatic sprinkler systems, standpipe systems, and specialized extinguishing systems such as dry chemical systems. Most newly constructed commercial buildings incorporate at least one of these systems, and increasing numbers of residential dwellings are being built with residential sprinkler systems as well. Some of the codes you will need to consult when working with the systems discussed in this chapter are:

- NFPA 12, *Standard on Carbon Dioxide Extinguishing Systems*
- NFPA 13, *Standard for the Installation of Sprinkler Systems*
- NFPA 13D, *Standard for the Installation of Sprinkler Systems in One and Two Family Dwellings and Manufactured Homes*
- NFPA 13R, *Standard for the Installation of Sprinkler Systems in Low and Mid-Rise Residential Occupancies*
- NFPA 14, *Standard for the Installation of Standpipes and Hose Systems*
- NFPA 15, *Standard for Water Spray Fixed Systems for Fire Protection*
- NFPA 16, *Standard for the Installation of Foam-Water Sprinkler and Foam-Water Spray Systems*
- NFPA 17, *Standard for Dry Chemical Extinguishing Systems*
- NFPA 17A, *Standard for Wet Chemical Extinguishing Systems*
- NFPA 20, *Standard for the Installation of Stationary Pumps for Fire Protection*
- NFPA 22, *Standard for Water Tanks for Private Fire Protection*
- NFPA 24, *Standard for the Installation of Private Fire Service Mains and Their Appurtenances*
- NFPA 25, *Standard for the Inspection, Testing, and Maintenance of Water-Based Fire Protection Systems*

Plan Review of Fire Suppression Systems

The evaluation of a fire suppression system begins in the plan review stage. The most common fire suppression system is the fire sprinkler system. Prior to any work beginning, the appropriate licensed sprinkler contractor must submit documents for review. Those documents would include at a minimum:

- Building construction data
- The system design criteria in accordance with the appropriate NFPA sprinkler standard, including occupancy class and commodity classification, current available water supply testing data, a plan of the sprinkler system layout, manufacturer cut sheets of the materials and equipment being used in the project, and hydraulic calculations of the system

During the plan review, the plan reviewer will ensure that there is appropriate sprinkler coverage throughout the building, that the proper type of sprinkler heads will be used, that the proper size and length of pipe will be used based on hydraulic calculations, and that the water supply test data is current. Once the submitted plans are reviewed, the sprinkler contractor needs to be contacted and advised if the plans are approved, or if the plans are not approved, the deficiencies that need to be corrected.

Fire Inspector Tips

In some cases, due to the complexity of sprinkler systems, the plan review is sent out to a private fire protection engineering company for review and approval.

Water Supply

The importance of a dependable and adequate water supply for fire-suppression systems and fire department operations is critical. There are two basic types of water supply distribution systems: public and private. In public water supplies, the system is designed to provide water for fire protection and normal consumptive demands. There are also water distribution systems on private property which provide water for fire protection and sanitary purposes. In some large manufacturing facilities, extensive water distribution systems exist to provide water for manufacturing processes and to provide water for fire protection.

Municipal water systems, and in some cases private water systems contracted by local governments to furnish water for fire protection, do so by providing water under pressure through fire hydrants Figure 1 or by static water sources such as lakes and streams. Static water sources also serve as drafting sites for fire department apparatus to obtain and deliver water to the fire scene Figure 2 .

Figure 2 Static water sources serve as drafting sites for fire department apparatus to obtain and deliver water to the fire scene.

Municipal Water Systems

Municipal water systems make clean water available to people in populated areas and provide water for fire protection. As the name suggests, most municipal water systems are owned and operated by a local government agency, such as a city, county, or special water district. Some municipal water systems are privately owned; however, the basic design and operation of both private and government systems are very similar.

Municipal water is supplied to homes, commercial establishments, and industries. Hydrants make the same water supply available to the fire department. In addition, most automatic sprinkler systems and many standpipe systems are connected directly to a municipal water source. A municipal water system has three major components: the water source, the treatment plant, and the distribution system.

Water Sources

Municipal water systems can draw water from wells, rivers, streams, lakes, or human-made storage facilities. The source will depend on the geographic and hydrologic features of the area. Many municipal water systems draw water from several sources to ensure a sufficient supply. Underground pipelines or open canals supply some cities with water from sources that are many miles away.

The water source for a municipal water system needs to be large enough to meet the total demands of the service area. Most of these systems include large storage facilities such as reservoirs, thereby ensuring that they will be able to meet the community's water supply demands if the primary water source is interrupted. The backup supply for some systems can provide water for several months or years. By contrast, in some other systems, the supply may last only a few days.

Water Treatment Facilities

Municipal water systems also include a water treatment facility, where impurities are removed from the water Figure 3 . The nature of the treatment system depends on the quality of the untreated source water. Water that is clean and clear from the source requires little treatment. Other systems must use extensive filtration to remove impurities and foreign substances. Some treatment facilities use chemicals to remove impurities and improve the water's taste. All of the water in the system must be suitable for drinking.

Figure 1 The water that comes from a hydrant is provided by a municipal or private water system.

Figure 3 Impurities are removed at the water treatment facility.

pressure-control/reduction devices are needed to keep parts of the system from being subjected to excessive pressures.

Most municipal water supply systems use both pumps and gravity to deliver water. Pumps may be used to deliver the water from the treatment plant to **elevated water storage towers**, which are above-ground water storage tanks designed to maintain pressure on a water distribution system, or to reservoirs located on hills or high ground. These elevated storage facilities maintain the desired water pressure in the distribution system, ensuring that water can be delivered under pressure even if the pumps are not operating Figure 5 . When the elevated storage facilities need refilling, large supply pumps are used. Additional

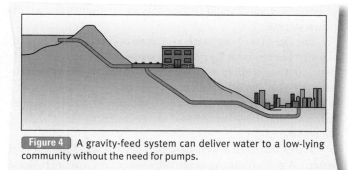

Figure 4 A gravity-feed system can deliver water to a low-lying community without the need for pumps.

Chemicals also are used to kill bacteria and harmful organisms and to keep the water pure as it moves through the distribution system to individual homes or businesses. After the water has been treated, it enters the distribution system.

Water Distribution System

Most water distribution systems use a combination of direct pumping systems and gravity systems. The distribution system delivers water from the treatment facility to the end users and fire hydrants through a complex network of underground pipes, known as **water mains**. In most cases, the distribution system also includes pumps, storage tanks, reservoirs, and other necessary components to ensure that the required volume of water can be delivered, where and when it is needed, at the required pressure.

Water pressure requirements differ, depending on how the water will be used. Generally, water pressure ranges from 20 pounds per square inch (psi) to 80 psi (138–552 kPa) at the delivery point. The recommended minimum pressure for water coming from a fire hydrant is 20 psi (138 kPa), but it is possible to operate with lower hydrant pressures under some circumstances.

Most water distribution systems rely on an arrangement of pumps to provide the required pressure, either directly or indirectly. In systems that use pumps to supply direct pressure, if the pumps stop operating, the pressure will be lost and the system will be unable to deliver adequate water to the end users or to hydrants. Most municipal systems have multiple pumps and backup power supplies to reduce the risk of a service interruption due to a pump failure. The extra pumps can sometimes be used to boost the flow for a major fire or a high-demand period.

In a pure **gravity-feed system**, the water source, treatment plant, and storage facilities are located on high ground while the end users live in lower-lying areas, such as a community in a valley Figure 4 . This type of system may not require any pumps, because gravity, through the elevation differentials, provides the necessary pressure to deliver the water. In some systems, the elevation pressure is so high that

Figure 5 Water that is stored in an elevated tank can be delivered to the end users under pressure.

pumps may be installed to increase the pressure in particular areas, such as a booster pump that provides extra pressure for a hilltop neighborhood.

A combination pump-and-gravity-feed system must maintain enough water in the elevated storage tanks and reservoirs to meet anticipated demands. If more water is being used than the pumps can supply, or if the pumps are out of service, some systems will be able to operate for several days by relying solely on their elevated storage reserves. Others may be able to function for only a few hours.

The underground water mains that deliver water to the end users come in several different sizes. Large mains, known as <u>primary feeders</u>, carry large quantities of water to a section of the town or city. Smaller mains, called <u>secondary feeders</u>, distribute water to a smaller area. The smallest pipes, called <u>distributors</u>, carry water to the users and hydrants along individual streets.

The size of the water mains required depends on the amount of water needed both for normal consumption and for fire protection in that location. Most jurisdictions specify the minimum-size main that can be installed in a new municipal water system to ensure an adequate flow. Some municipal water systems, however, may have undersized water mains in older areas of the community.

Water mains in a well-designed system will follow a grid pattern. A grid arrangement provides water flow to a fire hydrant from two or more directions and establishes multiple paths from the source to each area. This kind of organization helps to ensure an adequate flow of water for firefighting. The grid design also helps to minimize downtime for the other portions of the system if a water main breaks or needs maintenance work. With a grid pattern, the water flow can be diverted around the affected section.

Older water distribution systems may have dead-end water mains, which supply water from only one direction. Such dead-end water mains may still be found in the outer reaches of a municipal system. Hydrants on a dead-end water main will have a limited water supply. If two or more hydrants on the same dead-end main are used to fight a fire, the upstream hydrant will have more water and water pressure than the downstream hydrants.

Control valves installed at intervals throughout a water distribution system allow different sections to be turned off or isolated. These valves are used when a water main breaks or when work must be performed on a section of the system.

<u>Shut-off valves</u> are located at the connection points where the underground mains meet the distributor pipes. These valves control the flow of water to individual customers or to individual fire hydrants Figure 6 . If the water system in a building or to a fire hydrant becomes damaged, the shut-off valves can be closed to prevent further water flow.

Types of Fire Hydrants

The fire department accesses the public or private water supply via fire hydrants. Most fire hydrants consist of an upright steel casing (barrel) attached to the underground water distribution

Figure 6 A shut-off valve controls the water supply to an individual user or fire hydrant.

Fire Inspector Tips

On the discharge outlet, the number of threads per inch (or millimeter) and outside diameter of the male threads are regulated by NFPA 1963, *Standard for Fire Hose Connections*. The male threads on all fire hydrant discharge outlets must conform to the local fire department's hose threads. The discharge outlets are considered to be standard if they meet the standards of the American Water Works Association and Underwriters Laboratories (UL 246).

system. The two major types of fire hydrants in use today are the dry-barrel hydrant and the wet-barrel hydrant. Hydrants are equipped with one or more valves to control the flow of water through the hydrant. One or more outlets are provided to connect fire department hoses to the hydrant. These outlet nozzles are sized to fit the 2½" (63.5 mm) or larger fire hoses used by the local fire department.

Wet-Barrel Hydrants

<u>Wet-barrel hydrants</u> are used in locations where temperatures do not drop below freezing. These hydrants always have water in the barrel and do not have to be drained after each use. Wet-barrel hydrants usually have separate valves that control the flow to each individual outlet Figure 7 .

Dry-Barrel Hydrants

<u>Dry-barrel hydrants</u> are used in climates where temperatures can be expected to fall below freezing. The valve that controls the flow of water into the barrel of the hydrant is located at the base, below the frost line, to keep the hydrant from freezing Figure 8 . The length of the barrel depends on the climate and the depth of the valve. Water enters the barrel of the hydrant

Figure 7 A wet-barrel hydrant has a separate valve for each outlet nozzle.

Table 1 Fire Hydrant Colors

NFPA 24, *Standard for the Installation of Private Fire Service Mains and Their Appurtenances,* recommends that fire hydrants be color-coded to indicate the water flow available from each hydrant at 20 psi (138 kPa). It is recommended that the top bonnet and the hydrant caps be painted according to the following system.

Class	Flow Available at 20 psi	Color
Class C	Less than 500 gpm (1893 L/min)	Red
Class B	500–999 gpm (1893–3784 L/min)	Orange
Class A	1000–1499 gpm (3785–5677 L/min)	Green
Class AA	1500 gpm and higher (5678 L/min)	Light blue

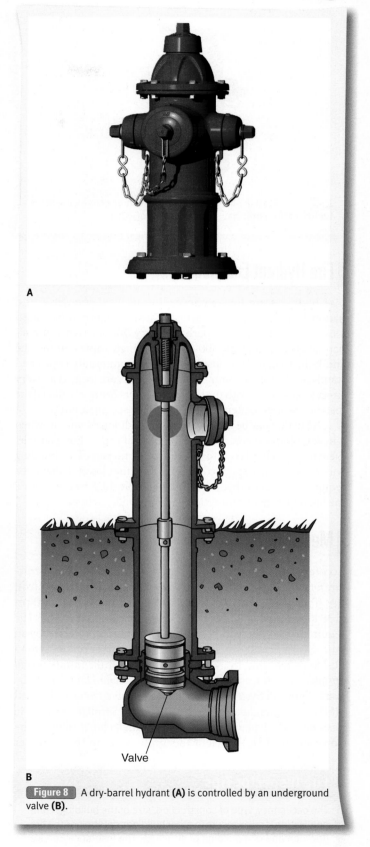

Figure 8 A dry-barrel hydrant (**A**) is controlled by an underground valve (**B**).

only when it is needed. Turning the nut on the top of the hydrant rotates the operating stem, which opens the valve so that water flows up into the barrel of the hydrant.

Whenever this type of hydrant is not in use, the barrel must remain dry. If the barrel contains standing water, it will freeze in cold weather and render the hydrant inoperable. After each use, the water drains out through an opening at the bottom of the barrel. This drain is fully open when the hydrant valve is fully shut. When the hydrant valve is opened, the drain closes, thereby preventing water from being forced out of the drain when the hydrant is under pressure.

A partially opened valve means that the drain is also partially open, so pressurized water can flow out. This leakage can erode (undermine) the soil around the base of the hydrant and may damage the hydrant. For this reason, a hydrant should always be either fully opened or fully closed. Most dry-barrel hydrants contain only one large valve that controls the flow of water **Figure 9**. Each outlet nozzle must be connected to a hose or an outlet valve, or have a hydrant cap firmly in place before the valve is turned on.

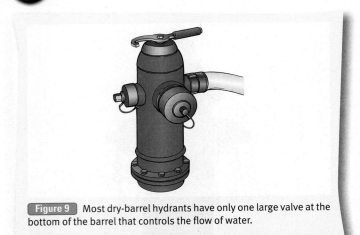

Figure 9 Most dry-barrel hydrants have only one large valve at the bottom of the barrel that controls the flow of water.

Fire Hydrant Locations

Fire hydrants are located according to local standards and nationally recommended practices. Fire hydrants may be placed a certain distance apart, perhaps every 500′ (152 m) in residential areas and every 300′ (91 m) in high-value commercial and industrial areas. Consult your local community codes and standards. In many communities, hydrants are located at every street intersection, with mid-block hydrants being installed if the distance between intersections exceeds a specified limit.

A builder may be required to install additional hydrants when a new building is constructed due to local codes. For example, there may be a local code that requires that no part of the building will be more than a specified distance from the closest hydrant or a requirement that the hydrant cannot be more the X feet away from the fire department connection to the sprinkler system.

Measuring Flow Rates

According to NFPA 1141, *Standard for Fire Protection Infrastructure for Land Development in Wildland, Rural, and Suburban Areas*, fire flow is the flow rate of a water supply, measured at 20 psi (138 kPa) residual pressure, that is available for firefighting. The required water demand flow rate for properties protected by an automatic sprinkler system is based upon the sprinkler system design as required by NFPA 13, *Standard for the Installation of Sprinkler Systems*, as well as NFPA 13R, *Standard for the Installation of Sprinkler Systems in Low-Rise Residential Occupancies*, and NFPA 13D, *Standard for the Installation of Sprinkler Systems in One- and Two-Family Dwellings and Manufactured Homes*. The flow rate must be sufficient to supply water to the sprinkler

Fire Inspector Tips

The occupancy, type of construction, size of the building, communication, and exposure to other buildings determines the fire flow demand of a building or structure. The Insurance Service Office (ISO) has a procedure that was developed for insurance rate purposes. This procedure has been in effect for many years and is of value in arriving at hydrant spacing recommendations.

system plus the additional equipment of the fire department, such as attack hose lines. This is determined by flowing water from a hydrant. Therefore, a test of a fire hydrant's available water supply is commonly referred to as a hydrant flow test.

Fire pumper companies are sometimes assigned to test the flow from fire hydrants in their districts or assist fire insurance companies and municipal and private water departments. As a fire inspector, you can use this information to determine the available fire flow. Records should be kept on all fire hydrant flow tests. The procedure for testing fire hydrants and determining fire flow are relatively simple, but a careful attention to detail is required.

Flow and Pressure

Static pressure is the pressure in a water supply system when the water is not moving. Static pressure is potential energy, because it would cause the water to move if there were some place the water could go. This kind of pressure causes the water to flow out of an opened fire hydrant. If there were no static pressure, nothing would happen when fire fighters opened a hydrant.

Static pressure is generally created by elevation pressure and/or pump pressure. An elevated storage tank creates elevation pressure in the water mains. Gravity also creates elevation pressure (sometimes referred to as head pressure) in a water system as the water flows from a hilltop reservoir to the water mains in the valley below. Pumps create pressure by bringing the energy from an external source into the system.

Static pressure in a water distribution system can be measured by placing a pressure gauge on a hydrant port and opening the hydrant valve. No water can be flowing out of the hydrant when static pressure is measured.

When measured in this way, the static pressure reading assumes that there is no flow in the system. Of course, because municipal water systems deliver water to hundreds or thousands of users, there is almost always water flowing within the system. Thus, in most cases, a static pressure reading actually measures the normal operating pressure of the system.

Residual pressure is the amount of pressure that remains in the system when water is flowing. When fire fighters open a hydrant and start to draw large quantities of water out of the system, some of the potential energy of still water is converted to the kinetic energy of moving water. However, not all of the potential energy turns into kinetic energy; some of it is used to overcome friction in the pipes. The pressure remaining while the water is flowing constitutes the residual pressure.

Residual pressure is important because it provides the best indication of how much more water is available in the system. The more water flowing, the less residual pressure. In theory, when the maximum amount of water is flowing, the residual pressure is zero, and there is no more potential energy to push more water through the system. In reality, 20 psi (138 kPa) is considered the minimum usable residual pressure, necessary to reduce the risk of damage to underground water mains or pumps.

Flow pressure measures the quantity of water flowing through an opening during a hydrant test. When a stream of water flows out through an opening (known as an orifice), all of the pressure is converted to kinetic energy. To calculate the volume of water flowing, measure the pressure at the center of the

Chapter 9 Fire Flow and Fire Suppression Systems

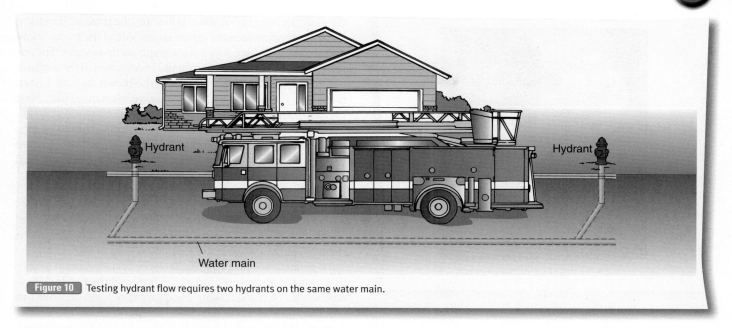

Figure 10 Testing hydrant flow requires two hydrants on the same water main.

Fire Inspector Tips

It is always a good idea to obtain approval with the local water system owner prior to doing any testing. They may prefer you to not flow test the hydrants without them. They might also be able to advise you of any work being done on the system that might influence your results.

water stream as it passes through the opening, and then factor in the size and flow characteristics of the orifice. A **Pitot gauge** is used to measure flow pressure in psi (or kilopascals), and to calculate the flow in gallons (or liters) per minute.

Obtaining a Static and Residual Pressure Readings

Testing fire hydrant flows requires two adjacent hydrants, a Pitot gauge, and an outlet cap with a pressure gauge. A Pitot gauge is used to measure flow pressure in psi (kPa), which is then used to calculate the fire flow in gallons (liters) per minute. As part of testing, you will measure the static pressure and residual pressure at one hydrant, and slowly open the other hydrant to obtain the hydrant flow data. The two hydrants should be connected to the same water main and preferably at about the same elevation Figure 10.

Hydrant Testing Procedure

The 2 ½″ (63.5 mm) hydrant cap is removed and a cap gauge with a bleeder valve is placed on one of the outlet nozzles of the first hydrant (the test hydrant) located nearest to the building in question Figure 11. The hydrant valve is then slowly and fully opened to allow water to fill the hydrant barrel. The bleeder valve on the cap gauge should be opened only until water comes out. Then the cap gauge should be closed to prevent inaccurate water pressure readings. The initial pressure reading on the cap gauge is recorded as the static pressure. When water at the first hydrant is

Figure 11 Place a cap gauge with a bleeder valve on one of the outlets of the first hydrant.

flowing from a discharge nozzle, the cap gauge reading is referred to as the residual reading.

At the second hydrant, remove one of the discharge nozzle caps and open the hydrant. The second hydrant should be downstream of the first hydrant. Obtain a Pitot pressure reading, by following these instructions

1. Remove the cap from a hydrant nozzle, preferably the 2½″ (63.5 mm) discharge nozzle. Using the larger nozzle injects more air and the reading must be slightly adjusted.
2. Record the size of the discharge nozzle being used Figure 12.
3. Fully open the hydrant and allow water to flow.
4. Hold the Pitot tube into the center of the flow and place it parallel to the discharge opening at a distance one-half of the inside diameter of the discharge nozzle Figure 13.
5. Record the pressure reading.
6. Slowly close the hydrant valve fully and replace the cap.

At the same time as the Pitot pressure reading is being taken, fire fighters at the first hydrant record the residual pressure

Figure 12 Measure the size of the discharge nozzle.

Figure 13 Hold the Pitot tube into the center of the flow and place it parallel to the discharge opening at a distance one-half of the inside diameter of the discharge nozzle.

reading. If the reading is a minimum of 25 percent different than the static reading, then it is considered a good reading. If it is less than 25 percent, additional nozzles or hydrants must be opened until the minimum 25 percent is reached.

A Pitot pressure reading must be taken at each nozzle of the hydrant. When calculating the results, the flow results must be combined. For example, if the first nozzle flowed 750 gallons (2839 L) and a second nozzle flowed 825 gallons (3123 L), the total flow would be 1,575 gallons (5962 L).

Determining Hydrant Flow Test Results

The easiest way to determine the flow of a fire hydrant is to enter the static pressure, residual pressure, and Pitot pressure readings into a computer program. Within seconds the available fire flow at 20 psi (138 kPa) will be given. If a computer program is not available, then the results can be recorded on semi-logarithmic graph paper especially made for hydrant flow testing. First, record the static pressure. As no water is flowing, the pressure is marked at the appropriate pressure on the far left side of the graph, above the zero mark. At the bottom of the graph are three scales. The one being used to show the results should be indicated. If the available water at 20 psi (138 kPa) is not able to be shown on scale A, then move to B, if it is still not able to be shown, then move to C. Next the Pitot pressure and residual pressure must be shown Figure 14 .

The Pitot pressure reading must be converted to gallons per minute (gpm) to determine the fire flow. The formula used to produce the fire flow is based on the discharge diameter and Pitot pressure:

$$Q = 29.83\, cd^2 \sqrt{p}$$

Q is the amount of water available (gpm), c is the coefficient of the hydrant (0.7, 0.8, or 0.9, see your local standards), d is the diameter of the discharge opening (2.5″ or 4.5″) and p is the Pitot pressure reading Figure 15 . Once that number has been determined, the intersection of the flow and the residual pressure should be marked using the appropriate scale. To make this calculation easier in the field, there are pocket cards available to help in calculating the flow. Once the static and residual pressures are plotted, a line is drawn between the two points until the line passes the 20 psi (140 kPa) line. The amount of water that is available at 20 psi (140 kPa) and the measured flow should be noted Figure 16 .

Automatic Sprinkler Systems

The most common type of fire suppression system is the automatic sprinkler system. Automatic sprinklers are reliable and effective, with a history of more than 100 years of

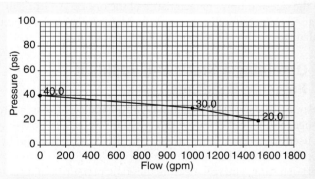

Figure 14 The results can be recorded on a special graph paper especially made for hydrant flow testing.

Q = 29.83(0.9)(2.5 2.5)(square root 25)
Q = 29.83(0.9)(6.25)(5)
Q = 838.96 of water flowing from the port
Available water, at 20 psi, of approximately 980 GPM

Figure 15 Q is the amount of water needed (gpm), c is the coefficient of the hydrant (0.7, 0.8, or 0.9), d is the diameter of the discharge opening (2.5″ or 4.5″) and p is the Pitot pressure reading.

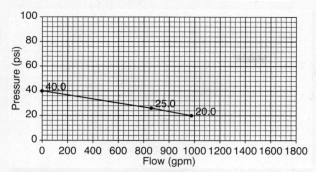

Figure 16 A line is drawn between the two points until the line passes the 20 psi (138 kPa) line. The amount of water that is available at 20 psi (138 kPa) and the measured flow should be noted.

sprinkler system piping controls the water supply to the sprinkler system. The piping brings the water to the sprinkler heads **Figure 17**.

Automatic sprinkler systems are designed to discharge water in sufficient density to control or extinguish a fire in the incipient stage. To accomplish this, the following factors must be considered during the design phase:

- Occupancy hazards
- Water supply

Fire Inspector Tips

It is essential that the sprinkler system matches the hazard. A sprinkler system designed to control and extinguish fire in an office with a small amount of combustibles will not be able to be effective in protecting a storage area where the fire load is severe, as in a tire storage warehouse.

successfully controlling fires. Unfortunately, few members of the general public have an accurate understanding of how automatic sprinklers work. In movies and on television, when one sprinkler head is activated, the entire system begins to discharge water. This inaccurate portrayal of how automatic sprinkler systems operate has made people hesitant to install automatic sprinklers, fearing that they will cause unnecessary water damage.

The reality is quite different. In almost all types of automatic sprinkler systems, the sprinkler heads open as each one is heated to its operating temperature. Usually, only one or two sprinkler heads open before the fire is controlled. There are some exceptions to this rule based upon sprinkler system design for a specific hazard.

A sprinkler system is a network of pipes that run underground and overhead (above ground). The underground pipes are connected to the building's water supply and supply water to the sprinkler system automatically. A valve in the sprinkler system riser or at the junction of the piping to the exterior water supply and the interior

Occupancy Hazards

A building's use is an essential consideration in designing a sprinkler system that is adequate to protect against the hazards in a particular type of occupancy. NFPA 13 uses an occupancy approach to sprinkler system design. All buildings that fall within the scope of NFPA 13 fall into specific occupancy hazards. Each of the hazards has specific requirements for the spacing of sprinkler heads, the sprinkler head discharge densities, and water supply requirements.

- Light-hazard class—This class includes occupancies in which the quantity and/or combustibility of materials is low and fires with relatively low rates of heat release are expected. Examples in this class include apartment buildings, churches, dwellings, hotels, public buildings, office buildings, and schools.
- Ordinary-hazard class—This class is subdivided into two subgroups. In general this class includes ordinary mercantile, manufacturing, and industry properties.
  - Group 1—This group includes properties where combustibility is low, the quantity of combustibles is moderate, stockpiles of combustibles are no higher than 8′, and fires with moderate rates of heat release are expected. Examples in this group include canneries, laundries, and electronics plants.
  - Group 2—This group includes properties where the quantity and combustibility of the contents is moderate to high, stockpiles of material do not exceed 12′, and fires with moderate rate of heat release are expected. Examples in this group include cereal mills, distilleries, and machine shops.

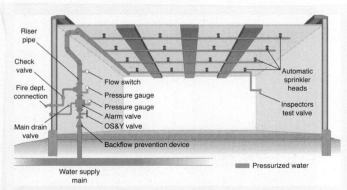

Figure 17 The basic components of an automatic sprinkler system include sprinkler heads, piping, control valves, and a water supply.

- Extra Hazard Occupancy Class—This class is subdivided into two groups and may produce severe fires.
  - Group 1—This group includes occupancies with little or no flammable or combustible liquids, but with significant quantities of very highly combustible materials that will have a high heat release rate and might include lint, dust, or other materials that could contribute to rapidly developing fires. Some examples are die casting, metal extruding, rubber production operations, sawmills, and upholstering operations using plastics foams.
  - Group 2—This group includes occupancies with moderately substantial amounts of flammable or combustible liquids or occupancies in which shielding of combustibles is extensive. Some examples are asphalt saturating, flammable liquid spraying, open oil quenching, solvent cleaning, and varnish and paint dipping.
- Special Occupancy Conditions—Occupancies involving high-piled combustible stocks, flammable and combustible liquids, combustible dusts and fibers, large quantities of light or loose combustible material, chemicals and explosives can permit rapid spread of fire, and often cause the opening of excessive number of sprinkler heads with disastrous results. Complete automatic sprinkler protection with strong water supplies usually will control fires in occupancies containing these hazardous conditions, provided the severity of the hazards is recognized plainly, and the sprinkler system is designed appropriately for the hazards.

Water Supply

Every automatic sprinkler system must have a water supply of adequate volume, pressure, and reliability. The types of hazards being protected, occupancy classification, and fuel loading conditions determine the minimum water flow required for the system. A water supply must be able to deliver the required volume of water to the highest or most remote sprinkler in a structure while maintaining a minimum residual pressure in the sprinkler system. Sprinkler systems must have a primary water supply and may be required to have a secondary source.

A fire department connection (FDC), which is a fire hose connection through which the fire department can pump water into a sprinkler system or standpipe system, is also a necessary part of an automatic sprinkler system in order to ensure that the presence of the primary water supply is maintained Figure 18. A fire department pumper connected to the public water supply by way of a fire hydrant can pump water into the sprinkler system vertical pipe section (riser) through the fire department connection.

Fire Inspector Tips

Areas of an occupancy may fall into different classifications. For example, a hotel is a light hazard occupancy; however, areas such as a kitchen offer more hazards than the guests' rooms. The sprinkler system protection for the kitchen must be increased as a kitchen belongs in the ordinary hazard class.

A

B

Figure 18 A fire department connection.

Water Distribution Pipes

An automatic sprinkler system consists of an arrangement of pipes in different sizes. The system starts with an underground water supply water main (public or private). The underground supply water main contains a **check valve**, a valve that allows flow in one direction only, to prevent sprinkler system water from backflowing into the public water system and possibly contaminating the public drinking water supply. Water quality protection laws in some states may further require the installation of a back flow preventer to prevent backflowing.

Sprinkler System Risers

Sprinkler system **risers** are vertical sections of pipe that connect the underground supply to the rest of the piping in the system. The riser also has the system water flow control valve and associated hardware that is used for testing, alarm activation, system isolation, and maintenance. Risers supply the cross mains that directly serve a number of branch lines. Sprinkler heads are installed on the branch lines with nipple risers (short vertical pipes). Hangers, rings, clamps support the entire system which may be pitched or sloped to help facilitate drainage.

Pipe Arrangement

The arrangement of pipes is based on one of two methods: pipe schedule or hydraulic calculations. **Pipe schedule system** is a sprinkler system in which the pipe sizing is selected from a schedule that is determined by the occupancy classification and in which a given number of sprinklers may be supplied from specific sizes of pipe. This is the traditional method for designing sprinkler systems and has been in use for over 100 years. The design is based on tables in NFPA 13 that designate the maximum number of sprinklers that a given size of pipe can supply, the proper sprinkler spacing, and the proper occupancy classification. Pipe schedule designs are limited to light hazard and ordinary hazard occupancy classifications, and addition and modifications to existing extra hazard systems.

Some sprinkler contractors may still use this method but it is not as accurate as the hydraulic calculation method. Hydraulically designed systems are based upon the type of occupancy to be protected, the type of hazard, the required density (quantity of water to be discharged) and the minimum pressure for each operating sprinkler. The layout and size of the distribution pipes are determined from these design requirements.

Valves

In order to control the water supply, a sprinkler system includes several different valves, such as the main water supply control valve, the alarm valve, and other smaller valves used for testing and service. These smaller valves include check valves, drain valves, globe valves, and alarm valves. Many large systems have zone valves, which enable the water supply to different areas to be shut down without turning off the entire system. All of the valves play a critical role in the design and function of the system.

The type of main sprinkler system valve used depends on the type of sprinkler system installed. Options include an **alarm valve**, a **dry-pipe valve**, or a **deluge valve**. These valves are usually installed on the main riser, above the water supply control valve.

The primary functions of an alarm valve are to signal an alarm when a sprinkler head is activated and to prevent nuisance alarms caused by pressure variations and surges in the water supply to the system. The alarm valve has a clapper mechanism that remains in the closed position until a sprinkler head or inspector's test valve opens. The closed clapper prevents water from flowing out of the system and back into the public water mains when water pressure drops.

When a sprinkler head is activated, the clapper opens fully and allows water to flow freely through the system. The open clapper also allows water to flow to the **water-motor gong**, a mechanical alarm notification device that is powered by water moving through the sprinkler system. When water reaches the water-motor gong, an alarm sounds. In some installations, an electrical flow switch activates internal and external alarm systems.

Without a properly functioning alarm valve, sprinkler system flow alarms would occur frequently. The normal pressure changes and surges in a public water supply system will not normally open the clapper valve. The clapper valve prevents water from flowing to the water-motor gong or tripping the electrical water flow switches as a result of the pressure surges and changes in the public water supply system. In addition, some sprinkler systems may require the installation of a retard chamber due to severe water pressure changes in order to prevent the accidental tripping of the alarm valve.

Many of the newer wet pipe sprinkler systems do not use an alarm valve. Instead, they use a paddle-type water flow switch device that detects the movement of water and sounds the alarm. These paddle type devices use a dial device that can be adjusted from seconds to minutes to prevent a momentary surge of water pressure from activating the alarm.

In dry-pipe and deluge systems, the main valve functions both as an alarm valve and as a dam, holding back the water until the sprinkler system is activated. When the system is activated, the valve opens fully so that water can enter the **sprinkler piping**, which is the network of piping in a sprinkler system that delivers water to the sprinkler heads. Dry-pipe and deluge systems are described later in this chapter.

Main Water Supply Control Valves

Every sprinkler system must have at least one main control valve that allows water to enter the system. This water supply control valve must be of the "indicating" type, meaning that the position of the valve itself indicates whether it is open or closed. Examples are the **outside stem and yoke (OS&Y) valve**, the **post indicator valve (PIV)**, the **wall post indicator valve (WPIV)**, and the **butterfly valve**).

The OS&Y valve has a stem that moves in and out as the valve is opened or closed Figure 19 . If the stem is out, the valve is open. OS&Y valves are often found in a mechanical room in

> **Fire Inspector Tips**
>
> Sprinkler contractors may also use a grid design or looped system design. A grid system design allows water to access the sprinkler heads from multiple directions. The looped system also allows water to come from multiple directions; however, the branch lines are not interconnected like they are in a grid system.

Figure 19 An outside stem and yoke (OS&Y) valve is often used to control the flow of water into a sprinkler system.

Figure 21 A WPIV controls the flow of water from an underground pipe into a sprinkler system.

Figure 20 A post indicator valve (PIV) is used to open or close an underground valve.

Figure 22 A tamper switch activates an alarm if someone attempts to close a valve that should remain open.

is a type of indicating valve, and when the butterfly valve is turned, there is a flat piece of metal inside the pipe that is rotated. When open, it is in line with the water flow and when closed, it is perpendicular to the flow. Determining if the butterfly valve is open or closed is easy, as there is an external indicator. When open, the indicator is in line with the piping and when closed, it is at a right angle to the piping.

It is critically important that the sprinkler system is always charged with water and ready to operate if needed. To ensure that the water supply to the sprinkler system is not deliberately or accidentally shut off, all control valves should be chained and locked in the open position, or electronically monitored with a **tamper switch** Figure 22 . Tamper switches electronically monitor the position of the valve. If someone opens or closes the valve, the tamper switch sends a supervisory signal to the fire alarm control panel, indicating a change in the valve position. This signal only sounds at the fire alarm control panel, and at the monitoring station if the alarms are sent off-site. This alarm does not sound

the building, where water to supply the sprinkler system enters the building. In warmer climates, they may be found outside.

The PIV has an indicator that reads either open or shut depending on its position Figure 20 . A PIV is usually located in an open area outside the building and controls an underground valve. Opening or closing a PIV requires a wrench, which is usually attached to the side of the valve.

A WPIV is similar to a PIV but is designed to be mounted on the outside wall of a building Figure 21 . The butterfly valve all of the alarms in the building. If the change has not been authorized, the cause of the signal can be investigated and the problem can be corrected.

Sprinkler System Zoning

In large facilities, a sprinkler system may be divided into zones, where a specific valve or sprinkler riser controls the flow of water to a particular zone in an area of the building. The area of a building that a single sprinkler system riser can cover is regulated by NFPA 13, for example the maximum area for a single zone is 52,000 square feet (4831 m^2) for light and ordinary hazard occupancies. This is one reason that in a building's sprinkler room it would not be uncommon to see multiple risers next to each other, each protecting a particular area or zone of the building. This design in accordance with NFPA 13 also makes maintenance easy and can prove extremely valuable during firefighting operations.

After the fire is extinguished, sprinkler system water flow to the fire affected area can be shut off so that the activated sprinkler heads can be replaced. Fire protection in the rest of the building, however, is unaffected by the shutdown of the sprinklers in the fire damaged area.

Fire pumps are needed when the water comes from a static source. They may also be deployed to boost the pressure in some sprinkler systems, particularly for tall buildings Figure 23 . Because most municipal water supply systems do not provide enough pressure to control a fire on the upper floors of a high-rise building, fire pumps will turn on automatically when the sprinkler system activates or when the pressure drops to a preset level. In high-rise buildings, a series of fire pumps on upper floors may be needed to provide adequate pressure.

A large industrial complex could have more than one water source, such as a municipal system and a backup storage tank Figure 24 . Multiple fire pumps can provide water to the sprinkler and standpipe systems in different areas through underground pipes. Private hydrants may also be connected to the same underground system.

Water Flow Alarms

All sprinkler systems should be equipped with a method for sounding an alarm whenever water begins flowing in the pipes. This type of warning is important both in case of an actual fire and in case of an accidental activation. Without these alarms, the occupants or the fire department might not be aware of the sprinkler activation. If a building is unoccupied, the sprinkler system could continue to discharge water long after a fire is extinguished, leading to extensive water damage.

Most systems incorporate a mechanical flow alarm called a water-motor gong Figure 25 . When the sprinkler system is activated and the main alarm valve opens, some water is fed through a pipe to a water-powered gong located on the outside of the building. This gong alerts people outside the building that there is water flowing. This type of alarm will function even if there is no electricity.

Accidental soundings of water-motor gongs are rare, but sometimes, a water surge will cause a momentary alarm. If a water-motor gong is sounding, water is probably flowing from the sprinkler system somewhere in the building. Fire companies that arrive and hear the distinctive sound of a water-motor gong know that there is a fire or that something else is causing the sprinkler system to flow water.

Most modern sprinkler systems also are connected to the building's fire alarm system by either an electric **flow switch** or

Figure 23 In tall buildings, a fire pump may be needed to maintain appropriate pressure in the sprinkler system.

Fire Inspector Tips

Remember, fire pumps make pressure, not water. Some people think that when the water supply is insufficient for a building, they will just put in a fire pump. A pump will only provide more pressure to the already insufficient water supply.

Figure 24 An elevated storage tank ensures that there will be sufficient water and adequate pressure to fight a fire.

Figure 25 A water-motor gong sounds when water is flowing in a sprinkler system.

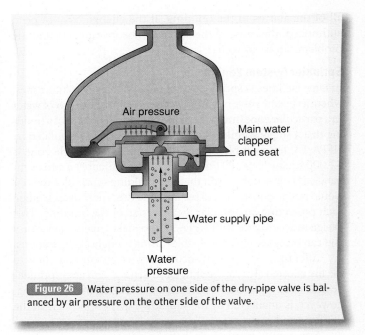

Figure 26 Water pressure on one side of the dry-pipe valve is balanced by air pressure on the other side of the valve.

a pressure switch. This connection will trigger the alarm to alert the building's occupants; a monitored system will notify the fire department or third-party monitoring company as well. Like water-motor gongs, flow and pressure switches can be accidentally triggered by water pressure surges in the system. To reduce the risk of accidental activations, these devices usually have a time delay before they will sound an alarm.

Types of Sprinkler Systems

There are four major classifications of automatic sprinkler systems. Each type of system includes a piping system that contains water from a source of supply to the sprinklers in the area under protection. The four major classifications are:

- Wet-pipe systems
- Dry-pipe systems
- Preaction systems
- Deluge systems

Although many buildings may use the same type of system to protect the entire facility, it is not uncommon to see two or three systems combined in one building. Some facilities use a wet sprinkler system to protect most of the structure, but implement a dry sprinkler or preaction system in a specific area. For example, the office area of a building will have a wet system installed, but the attic, which is unheated, will have a dry system, or even an anti-freeze system. Dry systems will also be common in large areas where the temperatures are below freezing, such as a cold storage warehouse.

Wet-Pipe Sprinkler Systems

A wet-pipe sprinkler system is the most common and least expensive type of automatic sprinkler system. As its name implies, the piping in a wet system is always filled with water. When a sprinkler head activates, water is immediately discharged onto the fire. The major drawback to a wet sprinkler system is that it cannot be used in areas where temperatures drop below freezing. Water will also begin to flow in such a system if a sprinkler head is accidentally opened or a leak occurs in the piping.

Larger unheated areas, such as a loading dock, can be protected with an antifreeze loop. An antifreeze loop is a small section of the wet sprinkler system that is filled with glycol or glycerin instead of water. A check valve separates the antifreeze loop from the rest of the sprinkler system. When a sprinkler head in the unheated area is activated, an antifreeze mixture sprays out first, followed by 100% water.

Dry-Pipe Sprinkler Systems

A dry-pipe sprinkler system operates much like a wet sprinkler system, except that the pipes are filled with pressurized air or nitrogen instead of water. A dry-pipe valve keeps water from entering the pipes until the air or nitrogen pressure drops below a threshold pressure. Dry systems are used in facilities, or areas, that may experience below-freezing temperatures, such as unheated warehouses, garages, or an unheated attic space.

The air or nitrogen pressure is set just enough to hold a clapper, which acts like a check valve, inside the dry-pipe valve in the closed position Figure 26 . When a sprinkler head opens, the air escapes. As the air pressure drops, the water pressure on the bottom side of the clapper forces the clapper open and water begins to flow into the pipes. When the water reaches the open sprinkler head, it is discharged onto the fire.

Dry sprinkler systems do not eliminate the risk of water damage from accidental activation. If a sprinkler head breaks, the air pressure will drop and water will flow, just as in a wet sprinkler system.

Dry sprinkler systems should have a high/low pressure alarm to alert building personnel if the system pressure changes. The activation of this alarm could mean one of two things: The compressor is not working or there is an air leak in the system. If the air pressure in the system is too low, the clapper will open and the system will fill with water. At that point, the system would essentially function as a wet sprinkler system, which could cause it to freeze in low temperatures. In this scenario, the system would have to be drained and reset to prevent the pipes from freezing.

Dry sprinkler systems must be drained after every activation so the dry-pipe valve can be reset. The clapper also must be reset, and the air pressure must be restored before the water is turned back on.

Accelerators and Exhausters

One problem encountered in dry-pipe sprinkler systems is the delay between the activation of a sprinkler head and the actual flow of water out of the head. The pressurized air that fills the system must escape through the open head before the water can flow. For personal safety and property protection reasons, any delay longer than 90 seconds is unacceptable. Large systems, however, can take several minutes to empty out the air and refill the pipes with water. To compensate for this problem, two additional devices are used: accelerators and exhausters.

An <u>accelerator</u> is installed at the dry-pipe valve **Figure 27**. The rapid drop in air pressure caused by an open sprinkler head triggers the accelerator, which allows air pressure to flow to the supply side of the clapper valve. The air is used to assist in opening the clapper. This quickly eliminates the pressure differential, opening the dry-pipe valve and allowing the water pressure to force the remaining air out of the piping.

An <u>exhauster</u> is installed on the system side of the dry-pipe valve, often at a remote location in the building. Like an accelerator, the exhauster monitors the air pressure in the piping. If it detects a drop in pressure, it opens a large-diameter portal, allowing the air in the pipes to escape to the atmosphere. The exhauster closes when it detects water, diverting the flow to the open sprinkler heads. Large systems may have multiple exhausters located in different sections of the piping.

■ Preaction Sprinkler Systems

A <u>preaction sprinkler system</u> is also known as an interlock system, and is similar to a dry sprinkler system with one key difference: In a preaction sprinkler system, a secondary device—such as a smoke detector or a manual-pull alarm—must be activated before water is released into the sprinkler piping. This type of a system is known a single interlock system. A double interlock system would require the activation of a detection device *and* the opening of a sprinkler head prior to filling the sprinkler pipes. Once the system is filled with water, it functions as a wet sprinkler system.

A preaction system uses a deluge valve instead of a dry-pipe valve. The deluge valve will not open until it receives a signal that an initiating device, such as a smoke detector or pull station, has been activated. Because a detection system usually will activate more quickly than a sprinkler system does, water in a preaction system will generally reach the sprinklers before a head is activated. The fused head will then cause the water to flow from the head.

The primary advantage of a preaction sprinkler system is its ability to prevent accidental water discharges. If a sprinkler head is accidentally broken or the pipe is damaged, the deluge valve will prevent water from entering the system. These systems are very expensive and complicated, and usually reserved for areas where accidental water from a sprinkler system would be devastating to the room's contents, such as libraries and museums.

■ Combined Dry-Pipe and Preaction Systems

A combined dry-pipe and precaution system includes the essential features of both types of systems. The piping contains air or nitrogen under pressure. A supplementary heat-detecting device opens the water control valve and an air exhauster at the end of the unheated water feed main. The system then fills with water and operates as a wet-pipe system. If the supplementary heat-detecting should fail, then the system will operate as a conventional dry-pipe system.

■ Deluge Sprinkler Systems

A <u>deluge sprinkler system</u> is a type of dry sprinkler system in which water flows from all the sprinkler heads as soon as the system is activated **Figure 28**. A deluge system does not have closed heads that open individually at the activation temperature; instead, all of the heads in a deluge system are always open.

Deluge systems can be activated in three ways:

1. A detection system can release the deluge valve when a detector is activated.
2. The deluge system can be connected to a separate pilot system of air-filled pipes with closed sprinkler heads.

Figure 27 An accelerator is installed at the dry-pipe valve.

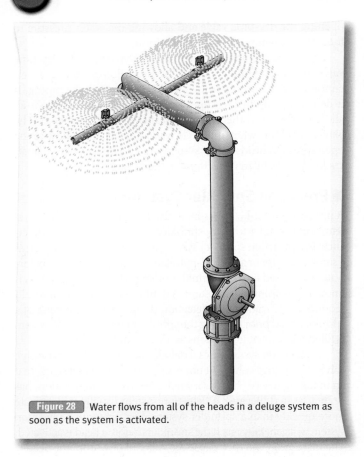

Figure 28 Water flows from all of the heads in a deluge system as soon as the system is activated.

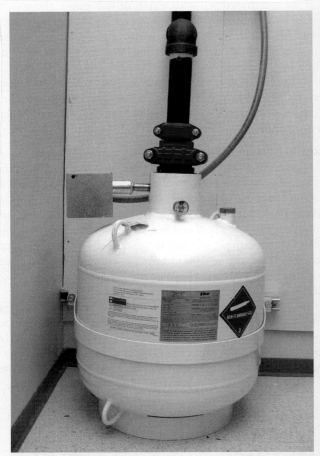

Figure 29 Special extinguishing systems are used in areas where water would not be effective or desirable.

A water-filled system is known as a hydraulic, or wet, pilot system. When a head on the pilot line is activated, the air pressure drops, opening the deluge valve.

3. Most deluge valves can be released manually.

Deluge systems are used in special applications such as aircraft hangars or industrial processes, where rapid fire suppression is critical. In some cases, foam concentrate is added to the water, so that the system will discharge a foam blanket over the hazard. Deluge systems are also used for special hazard applications, such as liquid propane gas loading stations. In these situations, a heavy deluge of water is needed to prevent a large, rapidly-developing fire.

■ Special Types

Automatic sprinkler systems are used to protect whole buildings, or at least major sections of buildings. Nevertheless, in certain situations, more specialized extinguishing systems are needed. Specialized extinguishing systems are often used in areas where water would not be an acceptable extinguishing agent Figure 29 . For example, water may not be the preferred agent of choice for areas containing sensitive electronic equipment or contents such as computers, valuable books, or documents. Water is also incompatible with materials such as flammable liquids or water-reactive chemicals. Areas where these materials are present may require a specialized extinguishing system.

Dry Chemical and Wet Chemical Extinguishing Systems

Dry chemical and wet chemical extinguishing systems are the most common specialized agent systems. Used in commercial

Fire Inspector Tips

Dry chemical extinguishing systems are not used as the agent of choice in commercial kitchens. When restaurants began to cook with vegetable oil, tests found that the standard dry chemical product would not extinguish the fire. While it would knock the fire down temporarily, the fire would quickly flash back to life. As a result, the NFPA created an additional fire classification, Class K, to address this specific hazard. Class K fires require the use of specialized extinguishing agents to extinguish a fire with cooking oils that have an extremely high combustion temperature. Chapter 10: Portable Fire Extinguishers discusses this topic in more detail. The only commercial kitchens that still allow for dry chemical extinguishing systems are those cooking systems that use animal cooking oils, not vegetable.

kitchens, they protect the cooking areas and exhaust systems. In addition, some gas stations have dry chemical systems that protect the dispensing areas. These systems are also installed inside buildings to protect areas where flammable liquids are stored or used. Both dry chemical and wet chemical extinguishing systems are similar in basic design and arrangement.

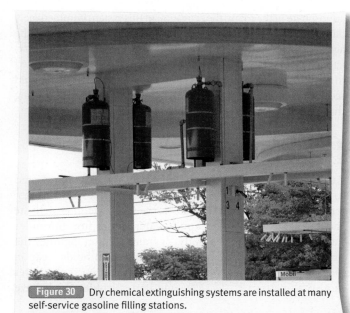

Figure 30 Dry chemical extinguishing systems are installed at many self-service gasoline filling stations.

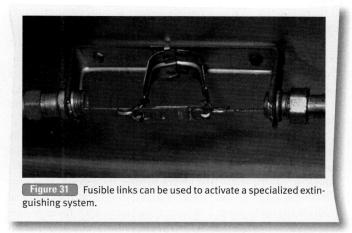

Figure 31 Fusible links can be used to activate a specialized extinguishing system.

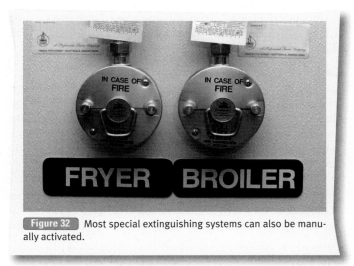

Figure 32 Most special extinguishing systems can also be manually activated.

Dry chemical extinguishing systems use the same types of finely powdered agents as dry chemical fire extinguishers Figure 30. The agent is kept in self-pressurized tanks or in tanks with an external cartridge of carbon dioxide or nitrogen that provides pressure when the system is activated.

Five compounds that are used as the primary dry chemical extinguishing agents are:
- Sodium Bicarbonate—rated for class B and C fires only
- Potassium bicarbonate—rated for class B and C fires only
- Urea-based potassium bicarbonate—rated for class B and C fires only
- Potassium Chloride—rated for class B and C fires only
- Ammonium phosphate—rated for Class A, B, and C fires

Wet chemical extinguishing systems discharge a proprietary liquid extinguishing agent. It is important to note that wet chemical extinguishing agents are not compatible with normal all-purpose dry chemical extinguishing agents. Only wet agents such as Class K, or B:C-rated dry chemical extinguishing agents should be used where these systems are installed.

All dry chemical extinguishing systems must meet the requirements of NFPA 17, *Standard for Dry Chemical Extinguishing Systems*. All wet chemical extinguishing systems must meet the requirements of NFPA 17A, *Standard for Wet Chemical Extinguishing Systems*. With both dry chemical and wet extinguishing agent systems, fusible-link or other automatic initiation devices are placed above the target hazard to activate the system Figure 31. A manual discharge station is also provided so that workers can activate the system if they discover a fire Figure 32. When the system is activated, the extinguishing agent flows out of all the nozzles. Nozzles are located over the target areas to discharge the agent directly onto a fire.

Many kitchen systems discharge agent into the ductwork above the exhaust hood as well as onto the cooking surface. This approach helps prevent a fire from igniting any grease build-up inside the ductwork and spreading throughout the system. Although the ductwork should be cleaned regularly, it is not unusual for a kitchen fire to extend into the exhaust system.

Dry and wet-chemical extinguishing systems should be tied into the building's fire alarm system. Kitchen extinguishing systems should also shut down gas or electricity to the cooking appliances and exhaust fans.

Clean Agent Extinguishing Systems (Halogenated Agents)

Clean agent extinguishing systems are often installed in areas where computers or sensitive electronic equipment are used or where valuable documents are stored. The agents used in these systems are nonconductive and leave no residue. Halogenated agents or carbon dioxide are generally used for this purpose because they will extinguish a fire without causing significant damage to the contents.

Clean agent systems operate by discharging a gaseous agent into the atmosphere at a concentration that will extinguish a fire. Smoke detectors or heat detectors installed in these areas activate the system, although a manual discharge station is also provided with most installations. Discharge is usually delayed 30 to 60 seconds after the detector is activated to allow workers to evacuate the area.

During this delay (the pre-alarm period), an abort switch can be used to stop the discharge. In some systems, the abort button must be pressed until the detection system is reset; releasing the abort button too soon causes the system to discharge.

If there is a fire, the clean agent system should be completely discharged before fire fighters arrive.

Clean agent extinguishing systems should be tied to the building's fire alarm system and indicated as a zone on the control panel. This notification scheme alerts fire fighters that they are responding to a situation where a clean agent has discharged. If the system has a preprogrammed delay, the pre-alarm should activate the building's fire alarm system.

Until the 1990s, Halon 1301 was the agent of choice for protecting areas such as computer rooms, telecommunications rooms, and other sensitive areas. Halon 1301 is a nontoxic, odorless, colorless gas that leaves behind no residue. It is very effective at extinguishing fires because it interrupts the chemical reaction of combustion. Unfortunately, Halon 1301 damages the environment—a fact that led to a manufacturing and importation ban. Alternative agents have been developed for use in new types of systems that are also replacing Halon 1301 systems. A stockpile of Halon 1301 is still available to recharge existing systems, because the new clean agents are not compatible with the Halon system equipment.

Carbon Dioxide Extinguishing Systems

Carbon dioxide extinguishing systems are similar in design to clean agent systems. The primary difference is that carbon dioxide extinguishes a fire by displacing the oxygen in the room and smothering the fire. Large quantities of carbon dioxide are required for this purpose, because the area must be totally flooded to extinguish a fire Figure 33.

Carbon dioxide systems may be designed to protect either a single room or a series of rooms. They usually have the same series of pre-alarms and abort buttons as are found in Halon 1301 systems. Because the carbon dioxide discharge creates an oxygen-deficient atmosphere in the room, the activation of this system is immediately dangerous to life. Any occupant who is still in the room when the agent is discharged is likely to be rendered unconscious and asphyxiated. Carbon dioxide extinguishing systems should be connected to the building's fire alarm system. Responding fire fighters should see that a carbon dioxide system discharge has been activated. Using this knowledge, they can deal with the situation safely.

Automatic Sprinkler Heads

Automatic sprinkler heads, which are commonly referred to as sprinkler heads, are the working ends of a sprinkler system. In most systems, the heads serve two functions: activate the sprinkler system and apply water to the fire. Sprinkler heads are composed of a body, or frame, which includes the orifice (opening); a release mechanism, which holds a cap in place over the orifice; and a deflector, which directs the water in a spray pattern Figure 34. Standard sprinkler heads have

Figure 34 Automatic sprinkler heads have a body with an opening, a release mechanism, and a water deflector.

Figure 33 Carbon dioxide extinguishes a fire by displacing the oxygen in the room and smothering the fire.

Fire Inspector Tips

NFPA 12A, *Standard on Halon 1301 Fire Extinguishing Systems*, and NFPA 2001, *Standard on Clean Agent Fire Extinguishing Systems*, provide the design and installation requirements for clean agent fire extinguishing systems.

Fire Inspector Tips

NFPA Standard 12, *Standard on Carbon Dioxide Extinguishing Systems*, provides the design and installation requirements for carbon dioxide extinguishing systems.

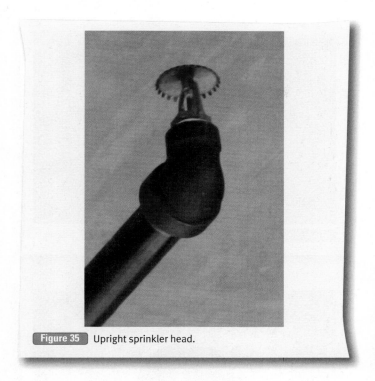

Figure 35 Upright sprinkler head.

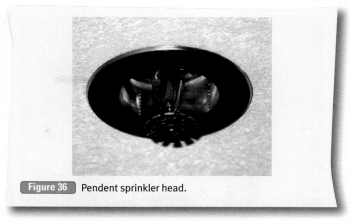

Figure 36 Pendent sprinkler head.

Figure 37 Sidewall sprinkler head.

a ½" (12.7 mm) orifice, but several other sizes are available for special applications.

Types of Sprinkler Heads

The upright, the pendent, and the sidewall sprinkler head are basic sprinkler heads. Sprinkler heads with different mounting positions are not interchangeable, because each mounting position has deflectors specifically designed to produce an effective water pattern down or out toward the fire. Each automatic sprinkler head is designed to be mounted in one of three positions. Additional sprinkler heads are variations of three basic sprinklers modified to address specific needs.

The <u>upright sprinkler head</u> are designed to be installed so that the water spray is directed upwards against the deflector Figure 35. Upright sprinkler heads are designed to be mounted on top of the supply piping, as their name suggests. Upright heads are usually marked SSU, for "standard spray upright."

<u>Pendent sprinkler heads</u> are designed to be installed so that the water stream is directed downward against the deflector Figure 36. Pendant sprinkler heads are designed to be mounted on the underside of the sprinkler piping, hanging down toward the room. These heads are commonly marked SSP, which stands for "standard spray pendant."

<u>Sidewall sprinklers</u> have special deflectors which are designed to discharge most of the water away from the nearby wall in a pattern resembling one quarter of a sphere, with a small portion of the discharge directed at the wall behind the sprinkler Figure 37. Sidewall sprinkler heads are designed for horizontal mounting, projecting out from a wall. Extended coverage sidewall sprinklers offer special extended, directional, and discharge water patterns.

<u>Dry sprinkler heads</u> provide protection in an area where 40 degrees Fahrenheit cannot be maintained. They are constructed to provide isolation between the head and the water supply by use of a cylinder that extends from the head to the threads. The threads reside in a heated area or are installed on a dry sprinkler system that is designed with pendent heads.

<u>Open sprinklers</u> are sprinklers from which the actuating elements, which trigger sprinkler activation, have been removed Figure 38. Open sprinkler heads are used with deluge systems. <u>Corrosion-resistant sprinklers</u> have special coating or platings such as wax or lead and are used in atmospheres which could corrode an uncoated sprinkler. <u>Nozzle sprinklers</u> are used in sprinkler applications requiring special discharge patterns, directional spray, fine spray, or other unusual discharge characteristics Figure 39. <u>Ornamental sprinklers</u> are sprinklers which have been painted or plated by the manufacturer. <u>Flush sprinklers</u> are sprinklers in which all or part of the body, including the shank thread, is mounted above the lower plane of the ceiling Figure 40. <u>Recessed sprinklers</u> are sprinklers in which all or part of the body, other than the shank thread, is mounted within a recessed housing Figure 41. <u>Intermediate level sprinklers</u> are sprinklers equipped with integral shields to protect their operating elements from the discharge of sprinklers installed at higher elevations Figure 42.

<u>Residential sprinklers</u> are sprinklers which have been specifically designed and listed for use in residential occupancies

Figure 38 Open sprinkler head.

Figure 40 Flush sprinkler head.

Figure 41 Recessed sprinkler head.

Figure 39 Nozzle sprinkler head.

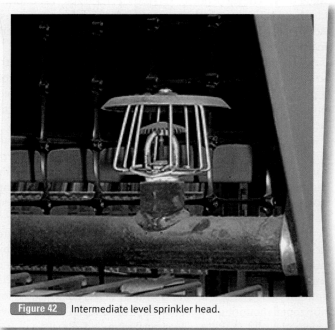

Figure 42 Intermediate level sprinkler head.

Figure 43. Sprinkler heads that are intended for residential occupancies are manufactured with a release mechanism that provides a faster response. Residential heads usually have smaller orifices and release a limited flow of water, because they are used in small rooms with a limited fire load. The spray pattern of the water is also different than a typical head in that it must hit higher on the wall. Because of this, residential heads and standard heads are not interchangeable.

An <u>early-suppression fast-response (ESFR) sprinkler head</u> has improved heat collectors that speed up the response and

Figure 43 Special sprinkler heads are used in residential systems. They open more quickly than commercial heads and discharge less water.

Old style sprinklers direct only 40 to 60 percent of the total water initially in a downward direction and are designed to be installed with the deflector either in the upright or pendent position. The primary difference between the old style sprinkler, which is designed for installations in either the upright or pendent position, and the current upright and pendent sprinklers is the design of the deflector.

■ Release Temperature

Sprinkler heads are also rated according to their release temperature. The release mechanisms hold the cap in place until the release temperature is reached. At that point, the mechanism is released, and the water pushes the cap out of the way as it discharges water onto the fire **Figure 45**.

<u>Fusible-link sprinkler heads</u> use a metal alloy, known as solder, that melts at a specific temperature **Figure 46**. This alloy links two pieces of metal that keep the cap in place. When the designated operating temperature is reached, the solder melts and the link breaks, releasing the cap. Fusible-link sprinkler heads come in a wide range of styles and temperature ratings.

<u>Frangible-bulb sprinkler heads</u> use a glass bulb filled with glycerin or alcohol to hold the cap in place **Figure 47**. The bulb also contains a small air bubble. As the bulb is heated, the liquid absorbs the air bubble and expands until it breaks the glass,

Fire Inspector Tips

In automatic sprinklers manufactured before the mid-1950s, the deflectors in both pendant and upright sprinkler heads directed part of the water stream up toward the ceiling. It was believed that this action helped cool the area and extinguish the fire. Sprinkler heads with this design are called old-style sprinklers, and many of them remain in service today.

Automatic sprinklers manufactured after the mid-1950s deflect the entire water stream down toward the fire. These types of heads are referred to as new-style heads or standard spray heads. New-style heads can replace old-style heads, but the reverse is not true. Due to different coverage patterns, old-style heads should not be used to replace any new-style heads.

Figure 44 The ESFR sprinkler head is larger than a standard sprinkler head and discharges about four to five times the amount of water.

ensure rapid release of water onto the fire **Figure 44**. They are used in large warehouses and distribution facilities where early fire suppression is important. These heads often have large orifices to discharge large volumes of water.

A listed <u>large-drop sprinkler head</u> generates large drops of water of such size and velocity as to enable effective penetration of a high-velocity fire plume. The large-drop sprinkler head has a proven ability to meet prescribed penetration, cooling, and distribution criteria prescribed in the large drop sprinkler examination requirements.

Figure 45 Automatic sprinkler heads are designed to activate at a wide range of temperatures. How quickly they react depends on their temperature rating.

Figure 46 In fusible-link sprinkler heads, two pieces of metal are linked together by an alloy such as solder.

Figure 47 Frangible-bulb sprinkler heads activate when the liquid in the bulb expands and breaks the glass.

Figure 48 Chemical-pellet sprinkler heads have a plunger mechanism that holds the cap in place.

Table 2 **Temperature Rating Coded by Color of Sprinkler Head**

Maximum Ceiling Temperature (°F)	Temperature Rating (°F)	Color Code	Glass Bulb Colors
100	135 to 170	uncolored or black	orange or red
150	175 to 225	white	yellow or green
225	250 to 300	blue	blue
300	325 to 375	red	purple
375	400 to 475	green	black
475	500 to 575	orange	black
625	650	orange	black

The chemical pellet will liquefy when the temperature reaches a preset point. When the pellet melts, the liquid compresses the plunger, releasing the cap and allowing water to flow.

A typical rating for sprinkler heads in buildings such as an office building is 165°F (73.9°C). The actual temperature of the head is determined by it location. For example, heads in an office might be 165°F (73.9 °C), but heads in an attic space, where ambient temperatures are higher, might be 212°F (100°C). If heads are located near space heaters in a parking garage, it would not be surprising to see some set at 285°F (140.6°C). Sprinkler heads use a color-coding system in the liquid to identify the temperature rating **Table 2**.

The temperature rating on a sprinkler head must match the anticipated ambient air temperatures. If the rating is too low for the ambient air temperature, accidental activations may occur, which would cause considerable water damage. Conversely, if the rating is too high, the system will be slow to react to a fire, and the fire may be able to establish itself and grow before the system activates.

Standpipe Systems

A **standpipe system** consists of a network of inlets, pipes, and outlets for fire hoses that are built into a structure to provide water for firefighting purposes. It includes one or more inlets supplied by a municipal water supply or fire department hoses using the fire department connection, piping to carry the water closer to the fire, and one or more outlets equipped with valves to which fire hoses can be connected **Figure 49**. Standpipe systems are required in many high-rise buildings, and they are found in many other structures as well. Standpipes are also installed to carry water to large bridges and to supply water to limited-access highways that are not equipped with fire hydrants.

Standpipes providing access to the water supply are found in buildings both with and without sprinkler systems. In many newer buildings, sprinklers and standpipes are combined into a single system. In older buildings, the sprinkler and standpipe systems are typically separate entities. Three categories of standpipes—Class I, Class II, and Class III—are distinguished based on their intended use.

releasing the cap. The volume and composition of the liquid and the size of the air bubble determine the temperature at which the head activates as well as the speed with which it responds.

Chemical-pellet sprinkler heads use a plunger mechanism and a small chemical pellet to hold the cap in place **Figure 48**.

Figure 49 Standpipe outlets allow fire hoses to be connected inside a building.

Figure 50 A Class I standpipe provides water for fire department hose lines.

Figure 51 A Class II standpipe is intended to be used by building occupants to attack incipient-stage fires.

Class I Standpipe

A <u>Class I standpipe</u> is designed for use by fire department personnel only. Each outlet has a 2½″ (63.5 mm) male coupling and a valve to open the water supply after the attack line is connected Figure 50. Responding fire fighters carry a fire hose, fire nozzle, and other equipment into the building with them, usually in some sort of roll, bag, or backpack. A Class I standpipe system must be able to supply an adequate volume of water with sufficient pressure to operate fire department attack hose lines.

Class II Standpipes

A <u>Class II standpipe</u> is designed for use by the building occupants. The outlets are generally equipped with a length of 1½″ (38.1 mm) single-jacket hose preconnected to the system Figure 51. These systems are intended to enable occupants to attack a fire before the fire department arrives, but their safety and effectiveness are questionable. After all, most building occupants are not trained to attack fires safely. If a fire cannot be controlled with a regular fire extinguisher, it is usually safer for the occupants to simply evacuate the building and call the fire department. Class II standpipes may be useful at facilities such as refineries and military bases, where workers are trained as an in-house fire brigade.

> **Fire Inspector Tips**
>
> On February 23, 1991 a fire occurred on the 22nd floor of the Meridian Plaza high rise office building in Philadelphia, Pennsylvania. One of the many factors that contributed to this out-of-control fire was the incorrect installation of the pressure-reducing hose valves.

> **Fire Inspector Tips**
>
> It is not uncommon to hear a contractor say, "But I already flushed the sprinkler system!" Oftentimes, the contractor is referring to the flush that is performed by the water department for chlorination of the sprinkler system. This flush does not substitute for an underground flush.

Class III Standpipes

A Class III standpipe has the features of both Class I and Class II standpipes in a single system. This kind of system has 2½" (63.5 mm) outlets for fire department use as well as smaller outlets with attached hoses for occupant use. As in Class II systems, the occupant hoses may have been removed—either intentionally or by vandalism—so the system may basically become a Class I system.

Water Flow in Standpipe Systems

Standpipes are designed to deliver a minimum amount of water at a particular pressure to each floor. The design requirements depend on the code requirements in effect when the building was constructed. The actual flow also depends on the water supply, as well as on the condition of the piping system and fire pumps. NFPA 14, *Standard for the Installation of Standpipes and Hose Systems*, establishes the design and installation criteria for all standpipe systems.

Pressure-reducing valves **Figure 52** are often installed at the outlets to limit the water pressure. A vertical column of water, such as the water in a standpipe riser, exerts a backpressure (also called head pressure). In a tall building, this backpressure can amount to hundreds of psi (kPa) at lower floor levels. If a hose line is connected to an outlet without a flow restrictor or a pressure-reducing valve, the water pressure could rupture the hose, and the excessive nozzle pressure could make the line difficult or dangerous to handle. Building and fire codes limit the height of a single riser and may also require the installation of pressure-reducing valves on lower floors.

Installation and maintenance of these devices is critical, as these devices can cause problems for fire fighters. An improperly adjusted pressure-reducing valve, for example, could severely restrict the flow to a hose line. One cause of this is when the wrong pressure-reducing valve has been installed on a floor.

Water Supplies

Both standpipe systems and sprinkler systems are supplied with water in essentially the same way. That is, many wet standpipe systems in modern buildings are connected to a public water supply, often with an electric or diesel fire pump to provide additional pressure when necessary. Some of these systems also have a water storage tank that serves as a backup supply. In these systems, the FDC on the outside of the building can be used to increase the flow, boost the pressure, or obtain water from an alternative source.

Dry standpipe systems are found in many older buildings. If freezing weather is a problem, such as in open parking structures, bridges, and tunnels, dry standpipe systems are still acceptable. Most dry standpipe systems do not have a permanent connection to a water supply, so the FDC must be used to pump water into the system. If a fire occurs in a building with dry standpipes, connecting the hose lines to the FDC and charging the system with water are high priorities. When a building is being remodeled, code officials might seek the installation of a dry standpipe, which is better than having no standpipe at all.

Some dry standpipe systems are connected to a water supply through a dry-pipe or deluge valve, similar to a sprinkler system. In such systems, opening an outlet valve or tripping a switch next to the outlet releases water into the standpipes. High-rise buildings often incorporate complex systems of risers, storage tanks, and fire pumps at various floors to deliver the needed flows to upper floors.

Fire Suppression System Testing

Underground Flush Test

As the fire inspector, you are not there to conduct the test but rather to observe the test and record the results. A fire sprinkler system must be inspected in stages. First there is

Figure 52 A pressure-reducing valve on a standpipe outlet may be necessary on lower floors to avoid problems caused by backpressure.

Chapter 9   Fire Flow and Fire Suppression Systems

> **Fire Inspector Tips**
>
> During testing, you should be looking to see that the sprinkler heads are located as submitted on the plans and that all areas of the building are protected by sprinklers. This is also the time to trace every pipe back to the water supply to make certain that every pipe is correctly connected.

the underground flush test. The underground flush test is performed prior to connecting the building's sprinkler pipes to the underground feed. It is performed to flush out any foreign material that may be in the supply line that could occlude the sprinkler lines. To perform the test, the supply valve line is completely opened. The flush should be continued until the water is clear and there is no evidence of foreign material coming out of the pipe. NFPA standards will dictate the amount of water that must be flowed through the underground feed during the test.

■ Hydrostatic Test

The next test is a hydrostatic test. This is a two-hour test conducted to make certain that the system components in the sprinkler system will not leak. The hydrostatic test involves filling the pipes with water until the pressure is 200 psi (1379 kPa) for two hours. Prior to pressurizing the pipe, the OS&Y for the sprinkler system must be closed. The idea is to pressurize the sprinkler pipes, not the city water main. In order for you to ensure that there are no leaks of any kind in the pipes, all of the pipes must be exposed. You should be looking at piping and on the floor, for any evidence of water leaking from pipes, fittings, or heads.

A common problem occurs when contractors install a drop ceiling or drywall that covers the piping. A drop ceiling can be temporarily removed with little difficulty. A greater challenge is a drywall ceiling. It is impossible for you to state that there were no leaks if the piping was not visible, so the drywall must be opened for a visual inspection.

■ Air Test

In a dry-pipe system, there is one additional test, an air test. Because the pipes in the system contain air or nitrogen, you must be sure the piping will hold pressure. To perform the air test, air pressure is pumped up to 40 psi (276 kPa) in the piping for 24 hours. If at the end of 24 hours, if the air pressure is relatively unchanged, then the system passed.

■ Main Drain Test

One last test to be conducted is a main drain test. The purpose of this test is to determine if the water supply to the sprinkler system is compromised. This test will only take a minute or two. First, water supply valves to the system must be opened. The static pressure shown on a riser gauge is recorded. Next the main drain valve is opened fully. When the pressure gauge stabilizes, you will record the residual pressure that the contractor gives you, and then the drain valve is slowly closed. You will then compare static pressure to the residual pressure. The difference between the two pressures should be minimal.

If this is the first time the main drain test is performed, the difference in pressures will serve as a benchmark for future tests. The sprinkler system will need an annual inspection, and will include a main drain test. When those results are received, a comparison to the previous year's results will indicate if the residual pressure has changed. If the residual pressure is significantly lower, the cause of the drop is worthy of investigation.

There have been instances where the valves to the building are only open a half turn. Although there is static pressure to the sprinkler system, when the drain is opened to supply water, the almost-closed valve does not allow much water to pass, and the pressure drops to almost zero. In this situation, there would be no water for the sprinklers in the event of a fire. The cause of the pressure drop needs to be investigated and corrected. It is almost always a closed or partially opened street valve; however, another cause is a backflow device not in proper working order, preventing water from entering the sprinkler system. In new construction testing, it is important that the municipal or private water department verify by actual valve testing that all street water valves are fully operational.

■ Hood and Duct System Testing

When testing a hood and duct system, you must make certain that the proper nozzles are installed at the proper locations above the cooking line and that there are nozzles in the hood and in the duct work. There should also be a new fusible element above every cooking appliance.

At this point, a tank of sample testing gas, usually nitrogen, is installed in place of the wet chemical product. All components of the kitchen cooking equipment and hood and duct system should be fully operational. The suppression system is then manually activated. A check is then made to assure that any gas and electric supply to the equipment shuts off, that the building's fire alarm sounds, and that every nozzle has gas coming out of it. When checking for the gas line to shut down, it is not enough to just see the gas valve close. There have been tests where the gas valve physically closes but did not shut down the gas due to a defective valve. Light a gas burner to test that the gas supply has been shut down. When the gas valve closes, the flame on the gas burner should extinguish.

There are two ways to activate the system and both should be checked, although once you are satisfied that the nozzles are getting "product," only the mechanical connections need to be inspected. The first is through the manual pull station. The second is through the fusible link to ensure that if a fire occurs that it will actually activate the system. The contractor can locate the most remote fusible element and remove it from the detection line, thereby activating the system.

■ Documentation and Reporting of Issues

Documentation is crucial as you are stating that every test was performed in the proper order at the time of inspection. While most contractors will have forms that they will ask you to sign off on, it is still a good idea that you have your own paperwork detailing the results of the inspection **Figure 53**. Request a copy of the contractor's testing forms.

Inspection Checklist
Sprinkler Systems

Building: _____
Address: _____
Inspector: _____ Date: _____
Date of Last Inspection: _____ Outstanding Violations: ❏ Yes   ❏ No

General
Date sprinklers installed: _____

Were building alterations/renovations made since last inspection?	❏ Yes	❏ No
Was new sprinkler system added since last inspection?	❏ Yes	❏ No
Any sprinkler system alteration made since last inspection?	❏ Yes	❏ No

What is system type?
- ❏ Wet
- ❏ Dry
- ❏ Preaction
- ❏ Deluge

Is building fully protected with sprinklers?   ❏ Yes   ❏ No
If not, explain: _____

Sprinkler Valves
Do sprinkler valves appear in good working order?	❏ Yes	❏ No	
Is dry pipe valve in heated enclosure?	❏ Yes	❏ No	❏ N/A*
Are spare sprinklers provided?	❏ Yes	❏ No	

Control Valves
Are control valves sealed?	❏ Yes	❏ No	❏ N/A
Are they locked?	❏ Yes	❏ No	❏ N/A
Do they have tamper switches?	❏ Yes	❏ No	❏ N/A

Fire Department Connections
Are fire department connections clear and unobstructed?	❏ Yes	❏ No	❏ N/A
Are protective caps in place?	❏ Yes	❏ No	❏ N/A
Are connections identified?	❏ Yes	❏ No	❏ N/A

Quarterly Inspections and Tests Recorded
Are quarterly inspections and tests recorded?   ❏ Yes   ❏ No

*N/A (not applicable) means there's no such feature in the building.

Copyright © 2002 National Fire Protection Association

Figure 53 A sample sprinkler system inspection form.

For the underground flush test, you will simply indicate that you witnessed the flush and approve it or not. If not, another underground flush test will be needed. For the hydrostatic test, the documentation should show the beginning pressure and the ending pressure, and whether the test is approved or not. The same would hold true for an air test, you should document the beginning air pressure and the ending air pressure, and if the sprinkler system passed. The main drain test requires you to record the static pressure and residual pressure of the riser gauge.

If there are other deficiencies noted in the system, they should be documented and checked again until compliant. Once everything is documented, you should obtain a copy of the above-ground and below-ground testing certificate from the contractor as well as copies of all other system test documents. It is also important that copies of any testing documentation should be given, at a minimum, to the contractor, the building owner, the municipality, and of course, a copy for the occupancy files for historical purposes.

Wrap-Up

Chief Concepts

- Fire suppression systems include automatic sprinkler systems, standpipe systems, and specialized extinguishing systems such as dry chemical systems.
- Prior to any work beginning, the appropriate licensed sprinkler contractor must submit documents for review.
- The importance of a dependable and adequate water supply for fire-suppression systems and fire department operations is critical. There are two basic types of water supply distribution systems: public and private.
- The fire department accesses the public or private water supply via fire hydrants. The two major types of fire hydrants in use today are the dry-barrel hydrant and the wet-barrel hydrant.
- Fire hydrants are located according to local standards and nationally recommended practices. Fire hydrants may be placed a certain distance apart, perhaps every 500′ in residential areas and every 300′ in high-value commercial and industrial areas. Consult your local community codes and standards.
- Fire flow is the flow rate of a water supply, measured at 20 psi (138 kPa) residual pressure, that is available for firefighting.
- The water supply must be sufficient for the sprinkler system plus the additional equipment of the fire department, such as attack hose lines. This is determined by testing a fire flow from a hydrant. A test of a fire hydrant's available water supply is commonly referred to as a hydrant flow test.
- Automatic sprinkler systems are designed to discharge water in sufficient density to control or extinguish a fire in the incipient stage. To accomplish this, the following factors must be considered during the design phase:
  - Occupancy hazards
  - Water supply
- There are four major classifications of automatic sprinkler systems. Each type of system includes a piping system that contains water from a source of supply to the sprinklers in the area under protection. The four major classifications are:
  - Wet-pipe systems
  - Dry-pipe systems
  - Preaction systems
  - Deluge systems
- Sprinkler heads are the working ends of a sprinkler system. In most systems, the heads serve two functions: activate the sprinkler system and apply water to the fire. Sprinkler heads are composed of a body, or frame, which includes the orifice (opening); a release mechanism, which holds a cap in place over the orifice; and a deflector, which directs the water in a spray pattern. Standard sprinkler heads have a ½″ (12.7 mm) orifice, but several other sizes are available for special applications.
- A standpipe system consists of a network of inlets, pipes, and outlets for fire hoses that are built into a structure to provide water for firefighting purposes. It includes one or more inlets supplied by a municipal water supply or fire department hoses using the fire department connection, piping to carry the water closer to the fire, and one or more outlets equipped with valves to which fire hoses can be connected. Standpipe systems are required in many high-rise buildings, and they are found in many other structures as well.
- As the fire inspector, you are not there to conduct the test but rather to observe the test and record the results. A fire sprinkler system must be inspected in stages. First there is the underground flush test. The next test is the hydrostatic test. In a dry-pipe system, there is one additional test, an air test. The last test is the main drain test.

Wrap-Up

■ Hot Terms

Accelerator A device that accelerates the removal of the air from a dry-pipe or preaction sprinkler system.

Alarm valve A valve that signals an alarm when a sprinkler head is activated and prevents nuisance alarms caused by pressure variations.

Automatic sprinkler heads The working ends of a sprinkler system. They serve to activate the system and to apply water to the fire.

Automatic sprinkler system A system of pipes filled with water under pressure that discharges water immediately when a sprinkler head opens.

Butterfly valve A type of indicating valve that moves a piece of metal 90° within the pipe and shows if the water supply is open or closed.

Carbon dioxide extinguishing system A fire suppression system that is designed to protect either a single room or series of rooms by flooding the area with carbon dioxide.

Check valve A valve that allows flow in one direction only. (NFPA 13R)

Chemical-pellet sprinkler head A sprinkler head activated by a chemical pellet that liquefies at a preset temperature.

Class I standpipe A standpipe system designed for use by fire department personnel only. Each outlet should have a valve to control the flow of water and a 2½″ male coupling for fire hose.

Class II standpipe A standpipe system designed for use by occupants of a building only. Each outlet is generally equipped with a length of 1½″ single-jacket hose and a nozzle, which are preconnected to the system.

Class III standpipe A combination system that has features of both Class I and Class II standpipes.

Corrosion-resistant sprinklers Sprinkler heads with special coating or plating such as wax or lead to use in potentially corrosive atmospheres.

Deluge sprinkler system A sprinkler system in which all sprinkler heads are open. When an initiation device, such as a smoke detector or heat detector, is activated, the deluge valve opens and water discharges from all of the open sprinkler heads simultaneously.

Deluge valve A valve assembly designed to release water into a sprinkler system when an external initiation device is activated.

Distributors Relatively small-diameter underground pipes that deliver water to local users within a neighborhood.

Dry chemical extinguishing system An automatic fire extinguishing system that discharges a dry chemical agent.

Dry-barrel hydrant A type of hydrant used in areas subject to freezing weather. The valve that allows water to flow into the hydrant is located underground and the barrel of the hydrant is normally dry.

Dry-pipe sprinkler system A sprinkler system in which the pipes are normally filled with compressed air. When a sprinkler head is activated, it releases the air from the system, which opens a valve so the pipes can fill with water.

Dry-pipe valve The valve assembly on a dry sprinkler system that prevents water from entering the system until the air pressure is released.

Dry sprinkler heads Sprinkler heads that are installed when 40 degrees Fahrenheit cannot be maintained. Dry sprinkler heads are constructed to provide isolation between the head and water supply by use of a cylinder that extends from the head to the threads of the pipe fitting. The threads reside in heated areas or are installed on a dry sprinkler system, and use pendent style dry sprinkler heads.

Early-suppression fast-response (ESFR) sprinkler head A sprinkler head designed to react quickly and suppress a fire in its early stages.

Elevated water storage tower An above-ground water storage tank that is designed to maintain pressure on a water distribution system.

Wrap-Up

Elevation pressure The amount of pressure created by gravity.

Exhauster A device that accelerates the removal of the air from a dry-pipe or preaction sprinkler system.

Fire department connection (FDC) A fire hose connection through which the fire department can pump water into a sprinkler system or standpipe system.

Fire flow The flow rate of a water supply, measured at 20 psi (138 kPa) residual pressure, that is available for firefighting. (NFPA 1141)

Flow pressure The amount of pressure created by moving water.

Flow switch An electrical switch that is activated by water moving through a pipe in a sprinkler system.

Flush sprinkler A sprinkler in which all or part of the body, including the shank thread, is mounted above the lower plane of the ceiling. (NFPA 13)

Frangible-bulb sprinkler head A sprinkler head with a liquid-filled bulb. The sprinkler head activates when the liquid is heated and the glass bulb breaks.

Fusible-link sprinkler head A sprinkler head with an activation mechanism that incorporates two pieces of metal held together by low-melting-point solder. When the solder melts, it releases the link and water begins to flow.

Gravity-feed system A water distribution system that depends on gravity to provide the required pressure. The system storage is usually located at a higher elevation than the end users.

Halon 1301 A liquefied gas-extinguishing agent that puts out a fire by chemically interrupting the combustion reaction between fuel and oxygen. Halon agents leave no residue.

Hydrostatic test A test filling the sprinkler piping with water and pressurizing it, usually to 200 psi (1379 kPa) for two hours, to look for leaks in the pipe work.

Intermediate level sprinklers Sprinklers equipped with integral shields to protect their operating elements from the discharge of sprinklers installed at higher elevations.

Large-drop sprinkler A sprinkler head that generates large drops of water of such size and velocity as to enable effective penetration of a high-velocity fire plume.

Main drain test A test opening the sprinkler system main drain to record the static and residual water pressures. This can indicate if the water supply to the sprinkler system is open or not.

Municipal water system A water distribution system that is designed to deliver potable water to end users for domestic, industrial, and fire protection purposes.

Nozzle sprinkler head Sprinkler heads used in applications requiring special discharge patterns, such as directional spray or fine spray.

Open sprinkler heads A sprinkler that does not have actuators or heat-responsive elements. (NFPA 13)

Ornamental sprinklers Sprinkler that have been painted or plated by the manufacturer.

Outside stem and yoke (OS&Y) valve A sprinkler control valve with a valve stem that moves in and out as the valve is opened or closed.

Pendant sprinkler head A sprinkler head designed to be mounted on the underside of sprinkler piping so that the water stream is directed down.

Pipe schedule system A sprinkler system in which the pipe sizing is selected from a schedule that is determined by the occupancy classification and in which a given number of sprinklers may be supplied from specific sizes of pipe. (NFPA 13)

Pitot gauge A type of gauge that is used to measure the velocity pressure of water that is being discharged from an opening. It is used to determine the flow of water from a hydrant.

Post indicator valve (PIV) A sprinkler control valve with an indicator that reads either open or shut depending on its position.

Preaction sprinkler system A dry sprinkler system that uses a deluge valve instead of a dry-pipe valve and requires activation of a secondary device before the pipes will fill with water.

Primary feeders The largest-diameter pipes in a water distribution system, carrying the greatest amounts of water.

Recessed sprinkler A sprinkler in which all or part of the body, other than the shank thread, is mounted within a recessed housing. (NFPA 13)

Reservoir A water storage facility.

Residential sprinkler system A sprinkler system designed to protect dwelling units.

Residual pressure The pressure that exists in the distribution system, measured at the residual hydrant at the time the flow readings are taken at the flow hydrants.

Riser The vertical supply pipes in a sprinkler system. (NFPA 13)

Wrap-Up

Secondary feeders Smaller-diameter pipes that connect the primary feeders to the distributors.

Shut-off valve Any valve that can be used to shut down water flow to a water user or system.

Sidewall sprinklers A sprinkler that is mounted on a wall and discharges water horizontally into a room.

Sprinkler piping The network of piping in a sprinkler system that delivers water to the sprinkler heads.

Standpipe system A system of pipes and hose outlet valves used to deliver water to various parts of a building for fighting fires.

Static pressure The pressure that exists at a given point under normal distribution system conditions measured at the residual hydrant with no hydrants flowing.

Static water source A water source such as a pond, river, stream, or other body of water that is not under pressure.

Tamper switch A switch on a sprinkler valve that transmits a signal to the fire alarm control panel if the normal position of the valve is changed.

Underground flush test A flushing of the water main supplying the sprinkler system to make certain there is no debris in the supply piping that might clog a sprinkler line.

Upright sprinkler head A sprinkler head designed to be installed on top of the supply piping; it is usually marked SSU ("standard spray upright").

Wall post indicator valve (WPIV) A sprinkler control valve that is mounted on the outside wall of a building. The position of the indicator tells whether the valve is open or shut.

Water-motor gong An audible alarm notification device that is powered by water moving through the sprinkler system.

Water main The generic term for any underground water pipe.

Water supply A source of water.

Wet chemical extinguishing systems An extinguishing system that discharges a proprietary liquid extinguishing agent.

Wet-barrel hydrant A hydrant used in areas that are not susceptible to freezing. The barrel of the hydrant is normally filled with water.

Wet-pipe sprinkler system A sprinkler system in which the pipes are normally filled with water.

Fire Inspector *in Action*

You are inspecting a large industrial complex. The complex owner is new and does not know much about fire equipment. You are pretty much on your own as the person accompanying you is of little help in explaining the fire suppression systems installed in his new complex.

1. The large industrial complex has computer rooms, manufacturing areas, storage areas, large exterior high-voltage electrical transformers filled with PCBs, office space, and a cafeteria. What type of sprinkler should you expect to find in this complex?
   A. Wet-pipe systems
   B. Deluge systems
   C. Special types
   D. All of the above

2. The owner is very proud of the servers in his new computer room. What type of fire suppression system should you expect to find in this room?
   A. Deluge system
   B. Dry-pipe system
   C. Clean agent extinguishing system
   D. Combined dry-pipe and preaction system

3. In the manufacturing area, some corrosive materials are used during the production process. When you look up, what type of sprinkler head should you expect to see?
   A. Upright sprinkler head
   B. Corrosion-resistant sprinkler head
   C. Ornamental sprinklers
   D. Intermediate level sprinkler heads

4. After inspecting the interior of the complex, you are walking the grounds to ensure that the complex has a sufficient water supply. You should find the fire hydrants _____ feet apart.
   A. 300
   B. 500
   C. 200
   D. 450

Portable Fire Extinguishers

CHAPTER 10

NFPA 1031 Standards

Fire Inspector I

4.3.7 Determine the operational readiness of existing portable fire extinguishers, given field observations and test documentation, so that the equipment is in an operational state, maintenance is documented, and deficiencies are identified, documented, and reported in accordance with the policies of the jurisdiction.

(A) Requisite Knowledge. A basic understanding of portable fire extinguishers, including their components and placement, and applicable codes and standards.

(B) Requisite Skills. The ability to observe, make decisions, recognize problems, and read reports.

Fire Inspector II

5.4.3 Review the proposed installation of fire protection systems, given shop drawings and system specifications for a process or operation, so that the system is reviewed for code compliance and installed in accordance with the approved drawings, and deficiencies are identified, documented, and reported in accordance with the applicable codes and standards and the policies of the jurisdiction.

(A) Requisite Knowledge. Proper selection, distribution, location, and testing of portable fire extinguishers; methods used to evaluate the operational readiness of water supply systems used for fire protection; evaluation and testing of automatic sprinkler, water spray, and standpipe systems and fire pumps; evaluation and testing of fixed fire suppression systems; and evaluation and testing of automatic fire detection and alarm systems and devices.

(B) Requisite Skills. The ability to read basic floor plans or shop drawings and identify symbols used by the jurisdiction.

5.4.4 Review the installation of fire protection systems, given an installed system, shop drawings, and system specifications for a process or operation, so that the system is reviewed for code compliance and installed in accordance with the approved drawings, and deficiencies are identified, documented, and reported in accordance with the applicable codes and standards and the policies of the jurisdiction.

(A) Requisite Knowledge. Proper selection, distribution, location, and testing of portable fire extinguishers; methods used to evaluate the operational readiness of water supply systems used for fire protection; evaluation and testing of automatic sprinkler, water spray, and standpipe systems and fire pumps; evaluation and testing of fixed fire suppression systems; and evaluation and testing of automatic fire detection and alarm systems and devices.

(B) Requisite Skills. The ability to read basic floor plans or shop drawings.

Additional NFPA Standards

NFPA 10, *Standard for Portable Fire Extinguishers*

FESHE

There are no FESHE objectives for this chapter.

Knowledge Objectives

After studying this chapter, you will be able to:

1. Define Class A fires.
2. Define Class B fires.
3. Define Class C fires.
4. Define Class D fires.
5. Define Class K fires.
6. Explain the classification and rating system for fire extinguishers.
7. Describe the types of agents used in fire extinguishers.
8. Describe the types of operating systems in fire extinguishers.
9. Describe the basic steps of fire extinguisher operation.
10. Describe how to test the readiness of portable extinguishers in an occupancy.

Skills Objectives

After studying this chapter, you will be able to:

1. Assess the operational readiness of a portable fire extinguisher.

You Are the Fire Inspector

You are inspecting a large home improvement store. Before going inside to inspect the interior, you walk around the entire exterior of the building. In the back of the building, you find large pile of mulch next to a loading dock. You note how close this potential fuel is stored next to the building and begin taking a picture for the inspection report. As you are taking the picture, the owner of the building proudly points out the fire extinguisher that is installed in the loading dock bay.

1. Would it be safe to attack a small mulch fire with a portable fire extinguisher?
2. What type of portable fire extinguisher should be installed?

Introduction

Portable fire extinguishers are required in many types of occupancies. Fire extinguishers are used successfully to put out hundreds of fires every day, preventing millions of dollars in property damage as well as saving lives. Most fire extinguishers are easy to operate and can be used effectively by an individual with only basic training.

Fire extinguishers range in size from models that can be operated with one hand to large, wheeled models that contain several hundred pounds of **extinguishing agent** (material used to stop the combustion process) **Figure 1**. Extinguishing agents include water, water with different additives, dry-chemicals, wet-chemicals, dry-powders, and gaseous agents. Each agent is suitable for specific types of fires.

Fire extinguishers are designed for different purposes and involve different operational methods. As a fire inspector, you must know which is the most appropriate kind of extinguisher to use for different types of fires and which kinds must not be used for certain fires. It is your role to ensure that the fire extinguisher is ready to be used in case of a fire.

Portable Fire Extinguishers

Purposes of Fire Extinguishers

Portable fire extinguishers have one primary use: to extinguish **incipient** fires (those that have not spread beyond the area of origin). They would not typically be thought of as a replacement for a fire suppression system, if one is called for.

Fire extinguishers are placed in many locations so that they will be available for immediate use on small, incipient-stage fires, such as a fire in a wastebasket. A trained individual with a suitable fire extinguisher could easily control this type of fire. As the flames spread beyond the wastebasket to other contents of the room, however, the fire becomes increasingly difficult to control with only a portable fire extinguisher.

Fire extinguishers are also used to control fires where traditional extinguishing methods are not recommended. For example, using water on fires that involve energized electrical equipment increases the risk of electrocution. Applying water to a fire in a computer or electrical control room could cause extensive damage to the electrical equipment. In these cases, it would be better to use a fire extinguisher with the appropriate extinguishing agent. Special extinguishing agents are also required for fires that involve flammable liquids, cooking oils, and combustible metals. The appropriate type of fire extinguisher should be made available in areas containing these hazards.

Classes of Fires

It is essential to match the appropriate type of extinguisher to the type of fire. Fires and fire extinguishers are grouped into classes according to their characteristics. Some extinguishing agents work more efficiently than others on certain types of fires. In some cases, selecting the proper extinguishing agent will mean the difference between extinguishing a fire and being unable to control it.

More importantly, in some cases it is actually dangerous to apply the wrong extinguishing agent to a fire. Using a water extinguisher on an electrical fire, for example, can cause an electrical shock as well as a short-circuit in the equipment. Likewise, a water extinguisher should never be used to fight a grease fire. Burning grease is generally hotter than 212°F (100°C), so the water converts to steam, which expands very rapidly. If the water penetrates the surface of the grease, the steam is produced within the grease. As the steam expands, the hot grease erupts like a volcano and splatters over everything and everyone nearby, potentially resulting in burns or injuries to people and spreading the fire.

Figure 2 Ordinary combustible materials are included in the definition of Class A fires.

Figure 3 Class B fires involve flammable liquids and gases.

Figure 1 Portable fire extinguishers can be large or small. A. A wheeled extinguisher. B. A one-hand fire extinguisher.

Class A Fires

Class A fires involve ordinary combustibles such as wood, paper, cloth, rubber, household rubbish, and some plastics Figure 2 . Natural vegetation, such as grass and trees, is also Class A material. Water is the most commonly used extinguishing agent for Class A fires, although several other agents can be used effectively.

Class B Fires

Class B fires involve flammable or combustible liquids, such as gasoline, oil, grease, tar, lacquer, oil-based paints, and some plastics Figure 3 . Fires involving flammable gases, such as propane or natural gas, are also categorized as Class B fires. Examples of Class B fires include a fire in a pot of molten roofing tar, a fire involving splashed fuel on a hot lawnmower engine, and burning natural gas that is escaping from a gas meter struck by a vehicle. Several different types of extinguishing agents are approved for use in Class B fires.

Class C Fires

Class C fires involve energized electrical equipment, which includes any device that uses, produces, or delivers electrical energy Figure 4 . A Class C fire could involve building wiring and outlets, fuse boxes, circuit breakers, transformers, generators, or electric motors. Power tools, lighting fixtures, household appliances, and electronic devices such as televisions, radios, and computers could be involved in Class C fires as well. The equipment must be plugged in or connected to an electrical source, but not necessarily operating, for the fire to be classified as Class C.

Figure 4 Class C fires involve energized electrical equipment or appliances.

Figure 5 Combustible metals in Class D fires require special extinguishing agents.

Figure 6 Class K fires involve cooking oils and fats.

Electricity does not burn, but electrical energy can generate tremendous heat that could ignite nearby Class A or B materials. As long as the equipment is energized, the incident must be treated as a Class C fire. Agents that will not conduct electricity, such as dry-chemicals or carbon dioxide, must be used on Class C fires.

■ Class D Fires

Class D fires involve combustible metals such as magnesium, titanium, zirconium, sodium, lithium, and potassium. Special extinguishing agents are required to fight combustible metals fires Figure 5 . Normal extinguishing agents can react violently—even explosively—if they come in contact with burning metals. Violent reactions also can occur when water strikes burning combustible metals.

Class D fires are most often encountered in industrial occupancies, such as machine shops and repair shops, as well as in fires involving aircraft and automobiles. Magnesium and titanium—both combustible metals—are used to produce automotive and aircraft parts because they combine high strength with light weight. Sparks from cutting, welding, or grinding operations could ignite a Class D fire, or the metal items could become involved in a fire that originated elsewhere.

■ Class K Fires

Class K fires involve combustible cooking oils and fats Figure 6 . This is a relatively new classification; cooking oil fires were previously classified as Class B combustible liquid fires. The use of high-efficiency modern cooking equipment and the trend toward using vegetable oils instead of animal fats to fry foods in recent years have resulted in higher cooking temperatures. These higher temperatures required the development of a new class of wet-chemical extinguishing agents. Some restaurants continue to use extinguishing agents that were approved for Class B fires, but these agents are not as effective for the new cooking oil fires, as are Class K extinguishers. If a restaurant has a hood and duct extinguishing system, a Class K extinguisher is required in the area of the cooking line. However, the traditional B type extinguisher may still be present for the other possible fires in the kitchen.

Classification of Fire Extinguishers

Portable fire extinguishers are classified and rated based on their characteristics and capabilities. This information is important for selecting the proper extinguisher to fight a particular fire **Table 1**. It is also used to determine which types of fire extinguishers should be placed in a given location so that incipient fires can be quickly controlled.

In the United States, **Underwriters Laboratories, Inc. (UL)** is the organization that developed the standards, classification, and rating system for portable fire extinguishers. The UL system identifies the classes of fires for which each fire extinguisher is both safe and effective.

The classification system for fire extinguishers uses both letters and numbers. The letters indicate the classes of fire for which the extinguisher can be used, and the numbers indicate its effectiveness. Numbers are used to rate an extinguisher's effectiveness only for Class A and Class B fires. Fire extinguishers that are safe and effective for more than one class will be rated with multiple letters. For example, an extinguisher that is safe and effective for Class A fires will be rated with an "A"; one that is safe and effective for Class B fires will be rated with a "B"; and one that is safe and effective for both Class A and Class B fires will be rated with both an "A" and a "B."

The number on the Class A and Class B fire extinguishers indicates the relative effectiveness of the fire extinguisher in the hands of a non-expert user. The higher the number, the greater the extinguishing capability of the extinguisher. To receive a 1-A rating the fire extinguisher must extinguish a certain amount of a wood crib fire. A 2-A rating is able to extinguish twice the fire a 1-A does, a 4-A twice the amount a 2-A does. The effectiveness of Class B extinguishers is based on the approximate area (measured in square feet) of burning fuel they are capable of extinguishing. A 10-B rating indicates that a non-expert user should be able to extinguish a fire in a pan of flammable liquid that is 10 square feet in surface area—that is just over $3' \times 3'$ (91.3 cm × 91.4 cm), not $2' \times 5'$ (61.0 cm × 152.4 cm) or $1' \times 10'$ (30.5 cm × 405.8 cm). An extinguisher rated 40-B should be able to control a flammable liquid pan fire with a surface area of 40 square feet (3.72 m^2). An experienced operator should be able to extinguish 2½ times the extinguisher rating. So an experienced operator should be able to take a 40-B extinguisher and extinguish 100 square feet (9.29 m^2) of fire.

There are no numerical ratings for Class-C fires. Class-C simply means the agent will not conduct electricity. If the fire extinguisher can also be used for Class C fires, it contains an agent proven to be nonconductive to electricity and safe for use on energized electrical equipment. For instance, a fire extinguisher that carries a 2-A:10-B:C rating can be used on Class A, Class B, and Class C fires. It has the extinguishing capabilities of a 2-A extinguisher when applied to Class A fires, has the capabilities of a 10-B extinguisher when applied to Class B fires, and can be used safely on energized electrical equipment.

A Class K extinguisher would not be good to use on live electrical equipment due to the chance of the agent conducting electricity, or on Class D fires for to an explosion possibility. In addition to the Class K rating, many extinguishers may also be rated for Class A fires. Their classification could then be 2A:K.

Standard test fires are used to rate the effectiveness of fire extinguishers. This testing may involve different agents, amounts, application rates, and application methods. Fire extinguishers are rated for their ability to control a specific type of fire as well as for the extinguishing agent's ability to prevent rekindling. Some agents can successfully suppress a fire, but are unable to prevent the material from reigniting. A rating is given only if the extinguisher completely extinguishes the standard test fire and prevents rekindling.

Labeling of Fire Extinguishers

Fire extinguishers that have been tested and approved by an independent laboratory are labeled to clearly designate the classes of fire the unit is capable of extinguishing safely. This traditional lettering system has been used for many years and is still found on many fire extinguishers. More recently, a universal pictograph system, which does not require the user to be familiar with the alphabetic codes for the different classes of fires, has been developed.

Traditional Lettering System

The traditional lettering system uses the following labels **Figure 7**:

- Extinguishers suitable for use on Class A fires are identified by the letter "A" on a solid green triangle. The triangle has a graphic relationship to the letter "A."

Table 1	Types of Fires
Class A	Ordinary combustibles
Class B	Flammable or combustible liquids
Class C	Energized electrical equipment
Class D	Combustible metals
Class K	Combustible cooking media

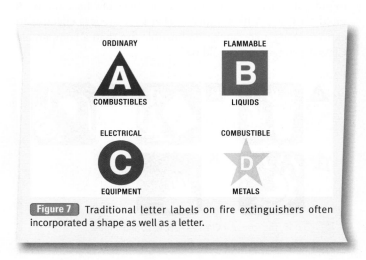

Figure 7 Traditional letter labels on fire extinguishers often incorporated a shape as well as a letter.

- Extinguishers suitable for use on Class B fires are identified by the letter "B" on a solid red square. Again, the shape of the letter mirrors the graphic shape of the box.
- Extinguishers suitable for use on Class C fires are identified by the letter "C" on a solid blue circle, which also incorporates a graphic relationship between the letter "C" and the circle.
- Extinguishers suitable for use on Class D fires are identified by the letter "D" on a solid yellow, five-pointed star.
- Extinguishers suitable for use on Class K (combustible cooking oil) fires are identified by a pictograph showing a fire in a frying pan. Because the Class K designation is new, there is no traditional-system alphabet graphic for it.

Pictograph Labeling System

The pictograph system, such as described for Class K fire extinguishers, uses symbols rather than letters on the labels. This system also clearly indicates whether an extinguisher is inappropriate for use on a particular class of fire. The pictographs are all square icons, each of which is designed to represent a certain class of fire Figure 8 . The icon for Class A fires is a burning trashcan beside a wood fire. The Class B fire extinguisher icon is a flame and a gasoline can; the Class C icon is a flame and an electrical plug and socket. The tanks for Class A, B, and C extinguishers are usually the color red. The icon for Class D extinguishers depicts a metal gear, and the tank is usually colored yellow. Extinguishers rated for fighting Class K fires are labeled with an icon showing a fire in a frying pan and are usually chrome.

Under this pictograph labeling system, the presence of an icon indicates that the extinguisher has been rated for that class of fire. A missing icon indicates that the extinguisher has not been rated for that class of fire. A red slash across an icon indicates that the extinguisher must not be used on that type of fire, because doing so would create additional risk.

An extinguisher rated for Class A fires only would show all three icons, but the icons for Class B and Class C would have a red diagonal line through them. This three-icon array signifies that the extinguisher contains a water-based extinguishing agent, making it unsafe to use on flammable liquid or electrical fires.

Certain extinguishers labeled as appropriate for Class B and Class C fires do not include the Class A icon, but may be used to put out small Class A fires. The fact that they have not been rated for Class A fires indicates that they are less effective in extinguishing a common combustible fire than a comparable Class A extinguisher would be.

Fire Extinguisher Placement

Fire codes and regulations require the installation of fire extinguishers in many areas so that they will be available to fight incipient fires. NFPA 10, *Standard for Portable Fire Extinguishers,* lists the requirements for placing and mounting portable fire extinguishers as well as the appropriate mounting heights.

The regulations for each type of occupancy specify the maximum floor area that can be protected by each extinguisher, the maximum travel distance from the closest extinguisher to a potential fire, and the types of fire extinguishers that should be provided. Two key factors must be considered when determining which type of extinguisher should be placed in each area: the class of fire that is likely to occur and the potential magnitude of an incipient fire.

Extinguishers should be mounted so they are readily visible and easily accessible Figure 9 . Heavy extinguishers should not be mounted high on a wall. If the extinguisher is mounted too high, a smaller person might be unable to lift it off its hook or could be injured in the attempt.

According to NFPA 10, the recommended mounting heights for the placement of fire extinguishers are as follows:

- Fire extinguishers weighing up to 40 lbs (18.14 kg) should be mounted so that the top of the extinguisher is not more than 5′ (1.53 m) above the floor.
- Fire extinguishers weighing more than 40 lbs (18.14 kilograms) should be mounted so that the top of the extinguisher is not more than 3.5′ (1.07 m) above the floor.
- The bottom of an extinguisher should be at least 4″ (10.2 cm) above the floor.

Figure 8   The icons for Classes A, B, C, D, and K fires.

Figure 9   Extinguishers should be mounted in locations with unobstructed access and visibility.

Classifying Area Hazards

Areas are divided into three risk classifications—light, ordinary, and extra hazard—based on the amount and type of combustibles that are present, including building materials, contents, decorations, and furniture. The quantity of combustible materials present is sometimes called a building's <u>fire load</u>. It means the quantity of heat which would be released by the combustion of all the combustible materials. Many objects in a building, such as a large metal cutting press, may weigh a lot, but don't burn, therefore they would not figure into the fire load. The larger the fire load, the larger the potential fire.

The occupancy use category does not necessarily determine the building's hazard classification. The recommended hazard classifications for different types of occupancies are simply guidelines based on typical situations. The hazard classification for each area should be based on the actual amount and type of combustibles that are present.

Light (Low) Hazard

Light (or low) hazard locations are areas where the majority of materials are noncombustible or arranged so that a fire is not likely to spread. Light hazard environments usually contain limited amounts of Class A combustibles, such as wood, paper products, cloth, and similar materials. A light hazard environment might also contain some Class B combustibles (flammable liquids and gases), such as copy machine chemicals or modest quantities of paints and solvents, but all Class B materials must be kept in closed containers and stored safely. Examples of common light hazard environments are most offices, classrooms, churches, assembly halls, and hotel guest rooms Figure 10.

Ordinary (Moderate) Hazard

Ordinary (or moderate) hazard locations contain more Class A and Class B materials than do light hazard locations. Typical examples of ordinary hazard locations include retail stores with on-site storage areas, light manufacturing facilities, auto showrooms, parking garages, research facilities, and workshops or service areas that support light hazard locations, such as hotel laundry rooms or restaurant kitchens Figure 11.

Figure 11 Auto showrooms, hotel laundry rooms, and parking garages are classified as ordinary hazard areas.

Ordinary hazard areas also include warehouses that contain Class I and Class II commodities. Class I commodities include noncombustible products stored on wooden pallets or in corrugated cartons that are shrink-wrapped or wrapped in paper. Class II commodities include noncombustible products stored in wooden crates or multilayered corrugated cartons.

Extra (High) Hazard

Extra (or high) hazard locations contain more Class A combustibles and/or Class B flammables than do ordinary hazard locations. Typical examples of extra hazard areas include woodworking shops; service or repair facilities for cars, aircraft, or boats; and many kitchens and other cooking areas that have deep fryers, flammable liquids, or gases under pressure Figure 12. In addition, areas used for manufacturing processes such as painting, dipping, or coating, and facilities used for storing or handling flammable liquids are classified as extra hazard environments. Warehouses containing products that do not meet the definitions of Class I and Class II commodities are also considered extra hazard locations.

Determining the Most Appropriate Placement of Fire Extinguishers

Several factors must be considered when determining the number and types of fire extinguishers that should be placed in each area of an occupancy. Among these factors are the types of fuels found in the area and the quantities of those materials.

Some areas may need extinguishers with more than one rating or more than one type of fire extinguisher. Environments that include Class A combustibles require an extinguisher rated for Class A fires; those with Class B combustibles require an extinguisher rated for Class B fires; and areas that contain both Class A and Class B combustibles require either an extinguisher that is rated for both types of fires or a separate extinguisher for each class of fire.

Most buildings require extinguishers that are suitable for fighting Class A fires because ordinary combustible materials—such as furniture, partitions, interior finish materials, paper, and packaging products—are so common. Even where other classes

Figure 10 Light hazard areas include offices, churches, and classrooms.

Figure 12 Kitchens, woodworking shops, and auto repair shops are considered possible extra hazard locations.

of products are used or stored, there is still a need to defend the facility from a fire involving common combustibles.

A single multipurpose extinguisher is generally less expensive than two individual fire extinguishers and eliminates the problem of selecting the proper extinguisher for a particular fire. However, it is sometimes more appropriate to install Class A extinguishers in general-use areas and to place extinguishers that are especially effective in fighting Class B or Class C fires near those specific hazards.

Some facilities present a variety of conditions. In these occupancies, each area must be individually evaluated so that extinguisher installation is tailored to the particular circumstances. A restaurant is a good example of this situation. The dining areas contain common combustibles, such as furniture, tablecloths, and paper products, that would require an extinguisher rated for Class A fires. In the restaurant's kitchen, where the risk of fire involves cooking oils, a Class K extinguisher would provide the best defense.

Similarly, within a hospital, extinguishers for Class A fires would be appropriate in hallways, offices, lobbies, and patient rooms. Class B extinguishers should be mounted in laboratories and areas where flammable anesthetics are stored or handled. Electrical rooms should have extinguishers that are approved for use on Class C fires, whereas hospital kitchens would need Class K extinguishers.

Types of Extinguishing Agents

An extinguishing agent is the substance contained in a portable fire extinguisher that puts out a fire. A variety of chemicals, including water, are used in portable fire extinguishers. The best extinguishing agent for a particular hazard depends on several factors, including the types of materials involved and the anticipated size of the fire. Portable fire extinguishers use seven basic types of extinguishing agents:

- Water
- Dry-chemicals
- Carbon dioxide
- Foam
- Wet-chemicals
- Halogenated agents
- Dry-powder

Water

Water is an efficient, plentiful, and inexpensive extinguishing agent. When it is applied to a fire, it quickly converts from liquid into steam, absorbing great quantities of heat in the process. As the heat is removed from the combustion process, the fuel cools below its ignition temperature and the fire stops burning.

Water is an excellent extinguishing agent for Class A fires. Many Class A fuels will absorb liquid water, which further lowers the temperature of the fuel. This also prevents rekindling.

Water is a much less effective extinguishing agent for other classes of fires. Applying water to hot cooking oil, for example, can cause splattering, which can spread the fire and possibly endanger the operator of the fire extinguisher. Many burning flammable liquids will simply float on top of water. Because water conducts electricity, it is dangerous to apply a stream of water to any fire that involves energized electrical equipment. If water is applied to a burning combustible metal, a violent reaction can occur. Because of these limitations, plain water is used only in Class A fire extinguishers.

One notable disadvantage of water is that it freezes at 32°F (0°C). In areas that are subject to below-freezing temperatures, **loaded-stream fire extinguishers** can be used to counteract this limitation **Figure 13**. These extinguishers combine an alkali metal salt with water. The salt lowers the freezing point of water, so the extinguisher can be used in much colder areas.

Wetting agents can also be added to the water in a fire extinguisher. These agents reduce the surface tension of the water, allowing it to penetrate more effectively into many fuels, such as baled cotton or fibrous materials.

Dry Chemicals

Dry-chemical fire extinguishers deliver a stream of very finely ground particles onto a fire. Different chemical compounds are used to produce extinguishers of varying capabilities and characteristics. The dry-chemical extinguishing agents works to interrupt the chemical chain reactions that occur as part of the combustion process.

Figure 13 Loaded-stream fire extinguisher.

Figure 14 Multipurpose dry-chemical extinguishers can be used for Class A, B, and C fires.

Dry-chemical extinguishing agents offer several advantages over water extinguishers:
- They are effective on Class B (flammable liquids and gases) fires.
- They can be used on Class C (energized electrical equipment) fires, because the chemicals are nonconductive.
- They are not subject to freezing.
- They are smaller and less expensive.

The first dry-chemical extinguishers were introduced during the 1950s and were rated only for Class B and C fires. The industry term for these B:C-rated units is "ordinary dry-chemical" extinguishers.

During the 1960s, **multipurpose dry-chemical fire extinguishers** were introduced. These extinguishers are rated for Class A, B, and C fires. The chemicals in these extinguishers form a crust over Class A combustible fuels to prevent rekindling Figure 14 .

Multipurpose dry-chemical extinguishing agents take the form of fine particles and are treated with other chemicals to help maintain an even flow when the extinguisher is being used. Additional additives prevent them from absorbing moisture, which could cause packing or caking and interfere with the extinguisher's discharge.

One disadvantage of dry-chemical extinguishers is that the chemicals—particularly the multipurpose dry-chemicals—are corrosive and can damage electronic equipment, such as computers, telephones, and copy machines. The fine particles are carried by the air and settle like a fine dust inside the equipment. Over a period of months, this residue can corrode metal parts, causing considerable damage. If electronic equipment is exposed to multipurpose dry-chemical extinguishing agents, it should be cleaned professionally within 48 hours after exposure.

Five compounds are used as the primary dry-chemical extinguishing agents:
- Sodium bicarbonate (rated for Class B and C fires only)
- Potassium bicarbonate (rated for Class B and C fires only)
- Urea-based potassium bicarbonate (rated for Class B and C fires only)
- Potassium chloride (rated for Class B and C fires only)
- Ammonium phosphate (rated for Class A, B, and C fires)

Sodium bicarbonate is often used in small household extinguishers. Potassium bicarbonate, potassium chloride, and urea-based potassium bicarbonate all have greater fire-extinguishing capabilities (per unit volume) for Class B fires than does sodium bicarbonate. Potassium chloride is more corrosive than the other dry-chemical extinguishing agents.

Ammonium phosphate is the only dry-chemical extinguishing agent that is rated as suitable for use on Class A fires. Although ordinary dry-chemical extinguishers can also be used against Class A fires, a water dousing is also needed to extinguish any smoldering embers and prevent rekindling.

The selection of which dry-chemical extinguisher to use depends on the compatibility of different agents with one another and with any products that they might contact. Some dry-chemical extinguishing agents cannot be used in combination with particular types of foam.

Carbon Dioxide

Carbon dioxide (CO_2) is a gas that is 1.5 times heavier than air. When carbon dioxide is discharged on a fire, it forms a dense cloud that displaces the air surrounding the fuel. This effect interrupts the combustion process by reducing the amount of oxygen that can reach the fuel. A blanket of carbon dioxide over the surface of a liquid fuel can also disrupt the fuel's ability to vaporize.

In portable **carbon dioxide (CO_2) fire extinguishers**, carbon dioxide is stored under pressure as a colorless and odorless liquid. It is discharged through a hose and expelled on the fire through a horn. When it is released, the carbon dioxide is very cold and forms a visible cloud of "dry ice" when moisture in the air freezes as it comes into contact with the carbon dioxide.

Carbon dioxide is rated for Class B and C fires only. This extinguishing agent does not conduct electricity and has two significant advantages over dry-chemical agents: It is not corrosive and it does not leave any residue.

Carbon dioxide also has several limitations and disadvantages:

- Weight: Carbon dioxide extinguishers are heavier than similarly rated extinguishers that use other extinguishing agents **Figure 15**.
- The ratings on carbon dioxide extinguishers are far less than a typical multi-purpose fire extinguisher.
- Range: Carbon dioxide extinguishers have a short discharge range, which requires the operator to be close to the fire, increasing the risk of personal injury.
- Weather: Carbon dioxide does not perform well at temperatures below 0°F (−18°C) or in windy or drafty conditions, because it dissipates before it reaches the fire.
- Confined spaces: When used in confined areas, carbon dioxide dilutes the oxygen in the air. If it is diluted enough, people in the space can begin to suffocate.
- Suitability: Carbon dioxide extinguishers are not suitable for use on fires involving pressurized fuel or on cooking grease fires.

Foam

Foam fire extinguishers discharge a water-based solution with a measured amount of foam concentrate added. The nozzles on foam extinguishers are designed to introduce air into the discharge stream, thereby producing a foam blanket. Foam extinguishing agents are formulated for use on either Class A or Class B fires.

Class A foam extinguishers for ordinary combustible fires extinguish fires in the same way that water extinguishes fires. This type of extinguisher can be produced by adding Class A foam concentrate to the water in a standard 2.5-gallon (9.5-L),

Figure 15 Carbon dioxide extinguishers are heavy owing to the weight of the container and the quantity of agent needed to extinguish a fire. They also have a large discharge nozzle, making them easily identifiable.

stored-pressure extinguisher. The foam concentrate reduces the surface tension of the water, allowing for its better penetration into the burning materials.

Class B foam extinguishers discharge a foam solution that floats across the surface of a burning liquid and prevents the fuel from vaporizing. This foam blanket forms a barrier between the fuel and the oxygen, extinguishing the flames and preventing reignition. These agents are not suitable for Class B fires that involve pressurized fuels or cooking oils.

The most common Class B additives are **aqueous film-forming foam (AFFF)** and **film-forming fluoroprotein (FFFP) foam** **Figure 16**. Both concentrates produce very effective foams. Which one should be used depends on the product's compatibility with a particular flammable liquid and other extinguishing agents that could be used on the same fire.

Some Class B foam extinguishing agents are approved for use on **polar solvents**—that is, water-soluble flammable liquids such as alcohols, acetone, esters, and ketones. Only extinguishers that are specifically labeled for use with polar solvents should be used if these products are present.

Although they are not specifically intended to extinguish Class A fires, most Class B foams can also be used on

Figure 16 An AFFF extinguisher produces an effective foam for use on Class B fires.

Figure 17 Wet-chemical fire extinguisher.

ordinary combustibles. The reverse is not true, however: Class A foams are not effective on Class B fires. Foam extinguishers are not suitable for use on Class C fires and cannot be stored or used at freezing temperatures.

■ Wet Chemicals

Wet-chemical fire extinguishers are the only type of extinguisher to qualify under the new Class K rating requirements Figure 17 . They use wet-chemical extinguishing agents, which are chemicals applied as water solutions. Before Class K extinguishing agents were developed, most fire-extinguishing systems for kitchens used dry-chemicals. The minimum requirement in a commercial kitchen was a 40-B-rated sodium bicarbonate or potassium bicarbonate extinguisher. These systems required extensive clean-up after their use, which often resulted in serious business interruptions.

All new fixed extinguishing systems in restaurants and commercial kitchens now use wet-chemical extinguishing agents. These agents are specifically formulated for use in commercial kitchens and food-product manufacturing facilities, especially where food is cooked in a deep fryer. The fixed systems discharge the agent directly over the cooking surfaces. There is no numeric rating of their efficiency in portable fire extinguishers. The Class K wet-chemical agents include aqueous solutions of potassium acetate, potassium carbonate, and potassium citrate, either singly or in various combinations. These wet agents convert the fatty acids in cooking oils or fats to a soap or foam, a process known as saponification.

When wet-chemical agents are applied to burning vegetable oils, they create a thick blanket of foam that quickly smothers the fire and prevents it from reigniting while the hot oil cools. The agents are discharged as a fine spray, which reduces the risk of splattering. They are very effective at extinguishing cooking oil fires, and clean-up afterward is much easier, allowing a business to reopen sooner.

Halogenated Agents

Halogenated extinguishing agents are produced from a family of liquified gases, known as halogens, that includes fluorine, bromine, iodine, and chlorine. Hundreds of different formulations can be produced from these elements; these myriad versions have many different properties and potential uses. Although several of these formulations are very effective for extinguishing fires, only a few of them are commonly used as extinguishing agents. The first member of the halon products was Halon 104, more commonly known as carbon tetrachloride.

Halogenated extinguishing agents are commonly called clean agents because they leave no residue and are ideally suited for areas that contain computers or sensitive electronic equipment. Per pound, they are approximately twice as effective at extinguishing fires as is carbon dioxide. The principle behind extinguishing is that clean agents interfere with the combustion process.

Two categories of halogenated extinguishing agents are distinguished: halons and halocarbons. A 1987 international agreement, known as the Montreal Protocol, limited halon production because these agents damage the earth's ozone layer. Halons have since been replaced by a new family of extinguishing agents, halocarbons.

The halogenated agents are stored as liquids and are discharged under relatively high pressure. They release a mist of vapor and liquid droplets that disrupts the molecular chain reactions within the combustion process, thereby extinguishing the fire. These agents dissipate rapidly in windy conditions, as does carbon dioxide, so their effectiveness is limited in outdoor locations. Because halogenated agents also displace oxygen, they should be used with care in confined areas.

Halon agents (bromochlorodifluoromethane) were commonly used many years ago. However, as a result of the Montreal Protocol, the United States banned the production and importation of halon in the mid 1990s. Halon can still be acquired from some companies that recapture the chemical from old systems. Due to the human health concerns, care should be taken around these systems should they be discharged. Currently, four types of halocarbon agents are used in portable extinguishers: hydrochlorofluorocarbon (HCFC), hydrofluorocarbon (HFC), perfluorocarbon (PFC), and fluoroiodocarbon (FIC). Some of the trade names of replacement products used today are Halatron, FM200, Inergen, ECARO-25, and FE 36.

Dry-Powder

<u>Dry-powder extinguishing agents</u> are chemical compounds used to extinguish fires involving combustible metals (Class D fires). These agents are stored in fine granular or powdered form and are applied to smother the fire. They form a solid crust over the burning metal, which both blocks out oxygen (the fuel for the fire) and absorbs heat.

The most commonly used dry-powder extinguishing agent is formulated from finely ground sodium chloride (table salt) plus additives to help it flow freely over a fire. A thermoplastic material mixed with the agent binds the sodium chloride particles into a solid mass when they come into contact with a burning metal.

Another dry-powder agent is produced from a mixture of finely granulated graphite powder and phosphorus-containing compounds. This agent cannot be expelled from fire extinguishers; instead, it is produced in bulk form and applied by hand, using a scoop or a shovel. When applied to a metal fire, the phosphorus compounds release gases that blanket the fire and cut off its supply of oxygen; the graphite absorbs heat from the fire, allowing the metal to cool below its ignition point. Other specialized dry-powder extinguishing agents are available for fighting specific types of metal fires. For details, see NFPA's *Fire Protection Handbook*.

Class D agents must be applied very carefully so that the molten metal does not splatter. No water should come in contact with the burning metal, as even a trace quantity of moisture can cause a violent reaction.

Fire Extinguisher System Readiness

Fire extinguishers must always be in a state of readiness. A standard checklist may help you in assessing fire extinguishers Figure 18. As a fire inspector, you assess the state of readiness by inspecting:

- The travel distances to the fire extinguisher—The travel distance for a Class A fire extinguisher is 75' (22.9 m). For a Class B fire extinguisher, the proper travel distance is determined by the hazard being protected. NFPA 10 will dictate the distance and provides a table stating how many extinguisher are needed for a given floor area.
- The mounting of the fire extinguisher—Most fire extinguishers are mounted with the top not higher that 5' (1.52 m) above the floor. If the fire extinguisher is more than 40 lbs (18.1 kg), then the top must be no higher than 3.5' (1.07 meters) above the floor. These distances make it easy for an operator to pick up the fire extinguisher. In either case, the bottom of the extinguisher must be a minimum of 4" (101.6 mm) off of the floor. This allows for easy cleaning of the floor. The fire extinguisher must be mounted to prevent it from being moved to the back of a shelf or behind a piece of furniture. The fire extinguisher should be mounted near a normal means of egress and by an exit.
- Access to the fire extinguisher—The fire extinguisher should be easily accessible. No item should be blocking the fire extinguisher from view or access.
- The type of fire extinguisher—The proper type of fire extinguisher should be available. It would not be proper to have a water extinguisher in an electric room or a flammable liquids room. To prevent the chance of using the wrong fire extinguisher, most businesses use a multi-purpose extinguisher. These extinguishers are good for the three common classes of fire A, B, and C. An ABC type and a BC type extinguisher look identical, and they only way to tell them apart is to look at the rating or pictographs on the fire extinguisher.

When you inspect fire extinguishers, look for physical damage to the container: ensure that the hose is actually attached, no foreign matter is in the hose, the safety pin is in place, the seal holds the pin in place, and the gauge is showing the proper operating pressure. The location of the fire extinguisher should be considered as well. It should not be placed in an area where it will be subjected to possible damage. This would mean that it should not protrude into a means of egress more than 4" (101.6 mm). A fire extinguisher must have current inspection tags proving that an outside company inspected and tagged each unit. The tag is a visual indicator that the fire extinguisher was functional at the time of the outside company's inspection.

Office of Fire Inspection Portable Fire Extinguisher Checklist

	Code Requirements	Pass/Fail	Required Corrective Action
Travel distance			
Number per floor			
Mounting			
Clear access			
Clear view			
Type			
	Physical Inspection	**Pass/Fail**	**Required Corrective Action**
No visible damage to cylinder			
Hose attached securely			
No obstructions in hose			
Safety pin in place			
Proper gauge pressure			
	Inspection Tag	**Pass/Fail**	**Required Corrective Action**
Current vendor inspection tag			
Tag information matches extinguisher label			

Figure 18 A sample checklist for fire extinguisher assessment.

Documentation and Reporting of Issues

It is common to find deficiencies during the course of a fire inspection. Examples include that fire extinguisher has no pressure, the fire extinguisher was used, or the hose is not connected to the fire extinguisher. These items must be noted in the fire inspection report. In the inspection report, these items are easily addressed by writing that the fire extinguisher must be in proper working order with a new inspection tag. For issues such as fire extinguishers not being mounted, fire extinguishers that cannot be accessed, or fire extinguishers that are missing, write these as violations and re-inspect for code compliance at a later date. A copy of the fire inspection report should be left with the building owner as a reference to correct the violations.

Fire Inspector Tips

Signage should be posted to indicate the location of the fire extinguisher. Many large stores hang the extinguisher on the building columns, and to identify the location to store employees, they paint a red stripe around the top of the column.

Wrap-Up

Chief Concepts

- Portable fire extinguishers have one primary use: to extinguish incipient fires (those that have not spread beyond the area of origin). They would not typically be thought of as a replacement for a fire suppression system, if one is called for.
- It is essential to match the appropriate type of extinguisher to the type of fire. In some cases it is actually dangerous to apply the wrong extinguishing agent to a fire.
- Class A fires involve ordinary combustibles such as wood, paper, cloth, rubber, household rubbish, and some plastics.
- Class B fires involve flammable or combustible liquids, such as gasoline, oil, grease, tar, lacquer, oil-based paints, and some plastics. Fires involving flammable gases, such as propane or natural gas, are also categorized as Class B fires.
- Class C fires involve energized electrical equipment, which includes any device that uses, produces, or delivers electrical energy. A Class C fire could involve building wiring and outlets, fuse boxes, circuit breakers, transformers, generators, or electric motors.
- Class D fires involve combustible metals such as magnesium, titanium, zirconium, sodium, lithium, and potassium. Special extinguishing agents are required to fight combustible metals fires.
- Class K fires involve combustible cooking oils and fats.
- Portable fire extinguishers are classified and rated based on their characteristics and capabilities:
  - Class A—Ordinary combustibles
  - Class B—Flammable or combustible liquids
  - Class C—Energized electrical equipment
  - Class D—Combustible metals
  - Class K—Combustible cooking media
- Fire extinguishers that have been tested and approved by an independent laboratory are labeled to clearly designate the classes of fire the unit is capable of extinguishing safely. This traditional lettering system has been used for many years and is still found on many fire extinguishers. More recently, a universal pictograph system, which does not require the user to be familiar with the alphabetic codes for the different classes of fires, has been developed.
- Fire codes and regulations require the installation of fire extinguishers in many areas so that they will be available to fight incipient fires. NFPA 10, *Standard for Portable Fire Extinguishers,* lists the requirements for placing and mounting portable fire extinguishers as well as the appropriate mounting heights.
- Portable fire extinguishers use seven basic types of extinguishing agents:
  - Water
  - Dry-chemicals
  - Carbon dioxide
  - Foam
  - Wet-chemicals
  - Halogenated agents
  - Dry-powder
- Fire extinguishers must always be in a state of readiness. Fire inspectors assess the state of readiness by inspecting:
  - The travel distances to the fire extinguisher
  - The mounting of the fire extinguisher
  - Access to the fire extinguisher
  - The type of fire extinguisher

Wrap-Up

▪ Hot Terms

Ammonium phosphate An extinguishing agent used in dry-chemical fire extinguishers that can be used on Class A, B, and C fires.

Aqueous film-forming foam (AFFF) A water-based extinguishing agent used on Class B fires that forms a foam layer over the liquid and stops the production of flammable vapors.

Carbon dioxide (CO_2) fire extinguisher A fire extinguisher that uses carbon dioxide gas as the extinguishing agent.

Class A fires Fires involving ordinary combustible materials such as wood, cloth, paper, rubber, and many plastics.

Class B fires Fires involving flammable and combustible liquids, oils, greases, tars, oil-based paints, lacquers, and flammable gases.

Class C fires Fires involving energized electrical equipment where the electrical conductivity of the extinguishing media is of importance.

Class D fires Fires involving combustible metals such as magnesium, titanium, zirconium, sodium, and potassium.

Class K fires Fires involving combustible cooking media such as vegetable oils, animal oils, and fats.

Clean agent A volatile or gaseous fire extinguishing agent that does not leave a residue when it evaporates. Also known as a halogenated agent.

Dry-chemical fire extinguisher An extinguisher that uses a mixture of finely divided solid particles to extinguish fires. The agent is usually sodium bicarbonate-, potassium bicarbonate-, or ammonium phosphate-based, with additives being included to provide resistance to packing and moisture absorption and to promote proper flow characteristics.

Dry-powder extinguishing agent An extinguishing agent used in putting out Class D fires. The common dry-powder extinguishing agents include sodium chloride and graphite-based powders.

Extinguishing agent A material used to stop the combustion process. Extinguishing agents may include liquids, gases, dry-chemical compounds, and dry-powder compounds.

Film-forming fluoroprotein (FFFP) foam A water-based extinguishing agent used on Class B fires that forms a foam layer over the liquid and stops the production of flammable vapors.

Wrap-Up

Fire load The weight of combustibles in a fire area or on a floor in buildings and structures, including either the contents or the building parts, or both.

Halogenated extinguishing agent A liquefied gas extinguishing agent that puts out fires by chemically interrupting the combustion reaction between the fuel and oxygen.

Incipient The initial stage of a fire.

Loaded-stream fire extinguisher A water-based fire extinguisher that uses an alkali metal salt as a freezing-point depressant.

Multipurpose dry-chemical fire extinguisher A fire extinguisher rated to fight Class A, B, and C fires.

Polar solvent A water-soluble flammable liquid such as alcohol, acetone, ester, and ketone.

Saponification The process of converting the fatty acids in cooking oils or fats to soap or foam.

Underwriters Laboratories, Inc. (UL) The U.S. organization that tests and certifies that fire extinguishers (among many other products) meet established standards. The Canadian equivalent is Underwriters Laboratories of Canada (ULC).

Wet-chemical extinguishing agent An extinguishing agent for Class K fires. It commonly consists of solutions of water and potassium acetate, potassium carbonate, potassium citrate, or any combination thereof.

Wet-chemical fire extinguisher A fire extinguisher for use on Class K fires that contains a wet-chemical extinguishing agent.

Fire Inspector *in Action*

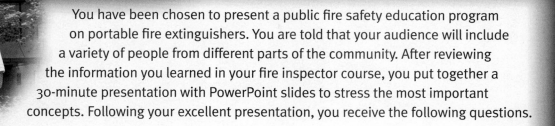

You have been chosen to present a public fire safety education program on portable fire extinguishers. You are told that your audience will include a variety of people from different parts of the community. After reviewing the information you learned in your fire inspector course, you put together a 30-minute presentation with PowerPoint slides to stress the most important concepts. Following your excellent presentation, you receive the following questions.

1. A homeowner who has hardwood floors and carpets asks, "Which type of fire extinguisher would be good to keep near a wood-burning fireplace? What should your answer be?
   A. A halogenated-agent extinguisher
   B. A Class K extinguisher
   C. A 1-B:C-rated extinguisher
   D. A pressurized water extinguisher

2. An older gentleman asks the following question: "I have an unheated woodworking shop behind my house. Which type of extinguisher is best for me?" What should your answer be?
   A. A 2½-gallon pressurized water extinguisher
   B. A 10-B:C-rated extinguisher
   C. A Class K extinguisher
   D. A 2-A:10-B:C-rated extinguisher

3. A woman who manages a computer store asks what would be the best type of extinguishers for her store. What should your answer be?
   A. Class K extinguishers
   B. B:C-rated extinguishers
   C. Pressurized water extinguishers
   D. Halogenated-agent extinguishers

4. The new manager of a fast-food restaurant asks which types of extinguishers he needs. What should your answer be?
   A. Class K extinguishers
   B. Pressurized water extinguishers
   C. A:B:C-rated extinguishers
   D. B:C-rated extinguishers and Class K extinguishers

Electrical and HVAC Hazards

CHAPTER 11

NFPA Standard

Fire Inspector I

4.3.8 Recognize hazardous conditions involving equipment, processes, and operations, given field observations, so that the equipment, processes, or operations are conducted and maintained in accordance with applicable codes and standards and deficiencies are identified, documented, and reported in accordance with the applicable codes and standards and the policies of the jurisdiction. (pp 222–231)

(A) Requisite Knowledge. Practices and techniques of code-compliance inspections, fire behavior, fire prevention practices, ignition sources, safe housekeeping practices, and classification of hazardous materials.

(B) Requisite Skills. The ability to observe, communicate, apply codes and standards, recognize problems, and make decisions.

Fire Inspector II

5.3.6 Evaluate hazardous conditions involving equipment, processes, and operations, given field observations and documentation, so that the equipment, processes, or operations are installed in accordance with applicable codes and standards and deficiencies are identified, documented, and reported in accordance with the policies of the jurisdiction. (pp 222–236)

(A) Requisite Knowledge. Applicable codes and standards, accepted fire protection practices, fire behavior, ignition sources, safe housekeeping practices, and additional reference materials related to protection of hazardous processes and code enforcement. (pp 222–236)

(B) Requisite Skills. The ability to observe, communicate, interpret codes, recognize problems, and make decisions. (pp 222–236)

5.3.12 Verify code compliance of heating, ventilation, air conditioning, and other building service equipment and operations, given field observations, so that the systems and other equipment are maintained in accordance with applicable codes and standards and deficiencies are identified, documented, and reported in accordance with the policies of the jurisdiction. (pp 231–236)

(A) Requisite Knowledge. Types, installation, maintenance, and use of building service equipment; operation of smoke and heat vents; installation of kitchen cooking equipment (including hoods and ducts), laundry chutes, elevators, and escalators; and applicable codes and standards adopted by the jurisdiction. (pp 231–236)

(B) Requisite Skills. The ability to observe, recognize problems, interpret codes and standards, and write reports. (pp 231–236)

Additional NFPA Standards

NFPA 54 *National Fuel Gas Code*
NFPA 70 *National Electrical Code*
NFPA 70B *Recommended Practice for Electrical Equipment Maintenance*
NFPA 77 *Recommended Practice on Static Electricity*
NFPA 85 *Boiler and Combustion Systems Hazards Code*
NFPA 92A *Standard for Smoke-Control Systems Utilizing Barriers and Pressure Differences*
NFPA 92B *Standard for Smoke Management Systems in Malls, Atria, and Large Spaces*
NFPA 211 *Standard for Chimneys, Fireplaces, Vents, and Solid Fuel-burning Appliances*

FESHE Objectives

Fire Plans Review

12. Verify the compliance of a heating, ventilating, and air conditioning (HVAC) system, review sources requiring venting and combustion air, verify the proper location of fire dampers, and evaluate a stairwell pressurization system. (pp 231–236)

Knowledge Objectives

After studying this chapter, you will be able to:

1. Describe the components of an electrical system.
2. Describe the protective practices and equipment of an electrical system.
3. Describe the common hazards of an electrical system.
4. Describe how to identify and seek correction of electrical hazards.
5. Describe the components and operation of heating, ventilation, and air conditioning systems.
6. Describe the potential hazards of heating, ventilation, and air conditioning systems.
7. Describe the impact of heating, ventilation, and air conditioning systems on smoke movement throughout a building.
8. Describe how to identify and seek correction of heating, ventilation, and air conditioning systems.

Skills Objectives

There are no skills objectives for this chapter.

You Are the Fire Inspector

Your assignment is to conduct an inspection in a major industrial building. You feel you are fairly competent in fire protection systems, life safety, and hazardous materials, but you are apprehensive when it comes to electrical installations. You know that the company will be defensive of any criticisms that you will have and violations are sure to be contested. The company has recently seen hard times financially and most likely has neglected electrical maintenance or jury rigged repairs. In short, you will be "walking on eggs on this one" and had better be right.

1. What will the best approach be in preparing for this inspection?
2. Each discrepancy must be accompanied with the appropriate code citation. Where is the best place to start in looking for code citations?
3. After the inspection is complete, you must write the report. How will you be sure that the discrepancies you list are correct?

Introduction

As a fire inspector, you must be aware of potential electrical hazards and the condition of electrical systems that are damaged, defective, or misused. Qualified electricians, electrical engineers, and experienced electrical inspectors have more detailed knowledge of these hazards; however, as a fire inspector, you are more often present in a facility and may be the first to observe an electrical hazard. <u>Electrical inspectors</u> verify compliance with adopted electrical code when buildings are initially constructed, expanded, altered, or renovated. Over the course of time, use or abuse, unpermitted alterations and maintenance, or the lack of adequate maintenance may alter the level of safety that was originally present. You must be aware of the signs and symptoms of electrical system problems.

You may not be qualified to inspect the internal wiring and parts of a facility's electrical system. Your goal should be to ensure that the electrical systems do not provide an ignition source for other materials. Generally, you will verify that enclosures and wiring remain intact, equipment is properly located and maintained, overcurrent devices are installed, and that grounding systems are in place and undamaged. If your inspection raises any questions or concerns, consult with a qualified electrical inspector or electrical engineer.

As a fire inspector, you must also be aware of potential hazards presented by heating, ventilation, and air conditioning (HVAC) systems that may be damaged, inoperative, or misused. While a mechanical engineer will have more detailed knowledge of these hazards, you may be the first to observe problems in the operation of an HVAC system. Without adequate maintenance, service, or repairs, the HVAC system can degrade and improperly operate at the time of a fire, allowing the spread of smoke in a building. A poorly maintained HVAC system may also be the cause of smoke in a building if dirty filters are not changed or fuel-fired equipment is not properly adjusted.

During an inspection, you will observe the general condition of HVAC systems and should have qualified repair and service personnel attend to the equipment if you observe sooting conditions, breached ductwork, missing safety guards on moving equipment, or obvious lack of maintenance. If your inspection raises any concerns or questions as to the proper operation of an HVAC system, consult a qualified mechanical inspector or

Fire Inspector Tips

Recognizing electrical problems is not complicated. If anything seems out of place, it is worthy of investigating. Some warning signs of electrical problems are:

- A slight shock when handling appliances
- Blinking lights or circuits that turn on and off by themselves
- Insulation on wires that are broken or cracked
- A computer monitor, television screen, or light that dims when a major appliance like an air conditioner turns on

Fire Inspector Tips

<u>Never</u> touch an energized wire! Consider <u>all</u> wires to be energized!

mechanical engineer. Fire protection or mechanical engineers can also be a source of information regarding mechanical smoke management systems.

Electrical Systems Overview

Electrical systems are complex arrangements of wires and circuits designed to deliver electrical power to an appliance or electronic device for the comfort and use of occupants in the building. In each electrical system, electricity flows from the main electrical service into the building first through the electric meter, then through the main electrical disconnect for the building, and finally into the circuit breaker box or fuse box. From this point, electricity is distributed to branch circuits and through the building's wiring to supply electricity to outlets **Figure 1**. Protection against drawing too much electrical energy through a circuit, known as overcurrent protection, is provided by circuit breakers or fuses that, when an overcurrent condition is reached, trip or open the circuit, stopping the flow of electrical energy to the outlet.

Some buildings may have multiple electrical sources, such as emergency generators that supply part or all of a building's electrical needs in the event of a failure of the main electrical utility supply. Alternate power sources that augment the normal electrical utility supply may also be found. These power sources may use solar panels, wind generators, or fuel-fired equipment to provide additional power to the building.

The Basics of Electricity

It is important to understand the basic concepts of electricity because you must understand the ratings of extension cords, the minimum requirements for electrical appliances, and proper wiring techniques. In many ways, electricity and the electrical system work much like water and the water distribution system as described in the Chapter, *Fire Flow and Suppression Systems*.

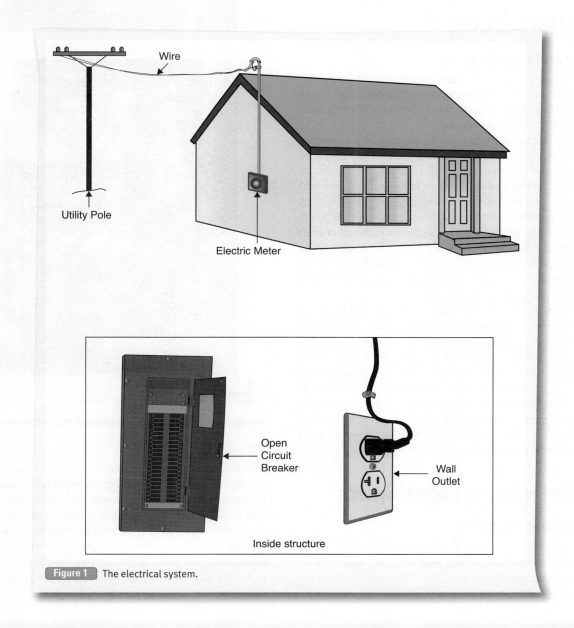

Figure 1 The electrical system.

In order to understand how electricity works, there are four terms which you must understand: volts, current, power, and resistance. The force required to move or conduct electricity is measured in units called volts (V). In many ways, voltage is much like pressure in a water line. Just like in the water distribution system, there is pressure on the electrical system even when the electricity is not flowing. Most houses are wired for 110 V or 220 V. In commercial, industrial, and in the municipal electrical grid system, the voltages are substantially higher. Higher volts (pressure) results in higher electrical flow (current).

The amount of electricity flowing through the system is called current. The volume of electrical flow is measured in ampere (amps or amperage). Amps in the electrical system are much like gallons per minute in the water distribution system. The total electrical power that is available for use is a combination of the volts and the amps, the watt (W). When electricity moves through the electrical system there is resistance, much like there is resistance as water moves through a water pipe. Resistance is measured in Ohms.

Potential Electrical Hazards

Arcing and Overheating

Electrical fires are caused principally by arcing or overheating. Arcing is a high-temperature luminous electric discharge across a gap. Overheating is generally the result of excess current, excessive insulation, or poor connections.

Gaps may be created in the normal operation of equipment, such as in switches or in motors with brushes. They may also be created at loose splices or terminals or where the insulation around the wiring has broken down or has been damaged and the wire is in close proximity to another wire or a grounded metal surface. The arc is commonly referred to a "short" or "short circuit", because the electricity takes a shorter path rather than going through the full circuit. Arcing produces enough heat to ignite nearby combustible materials, such as insulation, and can throw off particles of hot metal that can cause ignition. Arcing can also melt wiring and produce sparks.

Overheating is more subtle, harder to detect, and slower to cause ignition but is equally capable of causing a fire. Conductors and other electrical equipment may generate a dangerous level of heat when they carry excess current. This excess of current may cause wiring to overheat to the point at which the temperature is sufficient to ignite nearby combustible materials, such as wood framing. For example, a lightweight extension cord is run under a rug to power a space heater that uses more electricity than the extension cord is designed to handle. The extension cord overheats and the rug catches on fire.

Protective Practices for Electrical Systems

Overcurrent Protection

Electrical systems have overcurrent protection that opens a circuit if the amount of current will cause an excessive or dangerous temperature. Fuses and circuit breakers are the most commonly used overcurrent devices **Figure 2**.

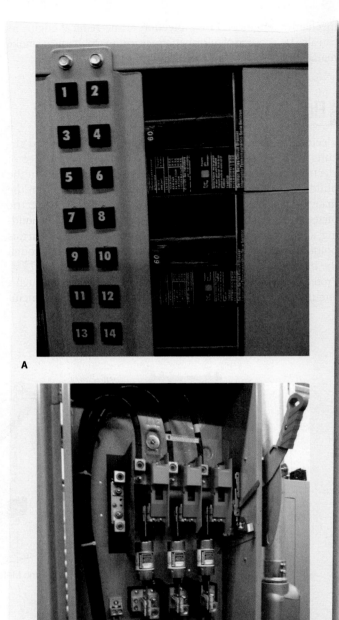

Figure 2 A. Circuit breaker. B. Fuse.

Fuses

A fuse is an overcurrent protective device with a circuit-opening fusible part that is heated and severed by the passage of overcurrent through it. It consists of a thin wire surrounded by a casing. If a wiring short occurs, the wire in the fuse will overheat very quickly and fail, breaking the electrical current before a fire can break out at the point of the arc. Once the fuse wire breaks, the fuse must be replaced. Fuses are rated by their amperage protection. Fuses are located in a central fuse box.

There are two general categories of fuses: plug and cartridge. Plug fuses come in a wide variety of types and styles, which can range from quick-acting to time-delay and from a standard base to a tamper-proof base. Plug fuse boxes have round sockets that look much like a light bulb socket. The fuse looks like a plug that screw into the socket.

The other type of fuse is the cartridge fuse. Like plug fuses, cartridge fuses are available in quick-acting and time-delay types. Some are designed for one-time use where they are thrown away after they trip. The other type of cartridge fuse is the renewable link cartridge fuse where the internal components can be replaced by an electrician and then be reinstalled. The renewable link cartridge fuses can fail to operate properly if not properly serviced, so they should only be serviced by qualified persons, such as experienced electricians or the cartridge fuse manufacturer's technicians.

Circuit Breakers

Another form of overcurrent protection device is the circuit breaker. The circuit breaker operates based upon a variety of principles, but all have the same result: a switch is opened to stop the flow of electricity through the circuit. Because there is a switch rather than a wire that breaks, the circuit breaker can be reset once the fault condition is resolved. Frequently circuit breakers will use more than one type of technology in order to activate the most effectively to the specific fault condition.

Some circuit breakers may have a shunt trip that allows them to be operated from remote locations. For example, a remote circuit breaker may be used to shut down equipment under kitchen exhaust hoods in restaurants or the main disconnect on the incoming utility service feed to a building.

Additional specialized circuit breakers include ground-fault circuit-interrupter (GFCI) protection and arc-fault circuit-interrupter (AFCI) protection. Any of these devices may feel warm under normal loads, but none should be too hot to touch.

Ground-Fault Circuit-Interrupters

Ground-fault circuit-interrupters (GFCI) sense when the current passes to ground through any path other than the proper path. When this occurs, the GFCI trips almost instantly, stopping all current flow in the circuit. Ground-fault circuit-interrupters are extremely important for life protection in wet locations.

GFCI protection is provided by a special circuit breaker located in the panelboard or by a GFCI receptacle installed in the electrical outlet Figure 3. Outlets on the same circuit downstream of the GFCI receptacle may also be protected by the GFCI receptacle if properly installed.

While GFCIs are primarily intended to protect people from shock hazards in outdoor, kitchen, bath or other wet locations, they also see service protecting heat tape used for freeze protection of water piping. Similar devices known as equipment ground-fault protective devices (EGFPD) are used for the same purpose in facilities where pipeline heating or deicing and snowmelting equipment are installed. EGFPD do not provide acceptable shock protection for people and are intended only to protect equipment.

Arc-Fault Circuit Interrupters

Arc-fault circuit-interrupters (AFCI) are much like the GRCI, but they react even more quickly. These devices can detect the presence of arcing faults on circuits even when the current does not rise to a value that will trip the main circuit breakers. They are intended to help prevent fires by disconnecting the individual circuits where great damage may occur. The requirement for these protective devices first appeared in the 2002 edition of NFPA 70, *National Electrical Code*, and applies only to circuits supplying receptacle outlets in bedrooms of dwelling units. Local jurisdictions may require these devices in other locations also.

Grounding

Because electrical systems can hold dangerous levels of electricity that can cause serious and even fatal shocks for anyone who is in very close contact with the electrical system, electrical systems must be properly grounded. Grounding electrical equipment helps to eliminate this shock hazard by providing a path for stray currents and any accumulation of static electricity to flow. It will then be unable to shock a similarly grounded person.

An electrical ground includes a grounding rod imbedded in the earth that is connected to the electrical system. This grounding system stabilizes and provides the path for the voltage potential to reach ground and will also limit voltage increases or spikes due to lightning, line surges, or unintentional

Fire Inspector Tips

Generally, a fuse box will have multiple cartridge fuses installed, and all normally have the same rating. If one fuse has a different rating from the others, this may be an indication of improper fusing or worse, overfusing, and should be investigated further.

Fire Inspector Tips

A visit to the fuse box may reveal a collection of discarded blown fuses or an abundant number of spare fuses. This may indicate that the owners regularly have to replace blown fuses because of overloaded circuits.

Figure 3 A ground-fault circuit interrupter (GFCI).

contact with higher voltage lines. A metallic underground water-piping system must be used as the grounding electrode where it is available and where the buried portion of the pipe is more than 10′ (3 m) long **Figure 4**. If a metal underground water pipe is the only grounding electrode, it must be supplemented by an additional electrode to ensure the integrity of the grounding electrode system. Grounded-steel building frames, concrete-encased electrodes installed in footings, grounding rings or grids, and driven grounding rods are other electrodes that may be used. These may be used either as the main electrode or to supplement the water pipe electrode.

Fire sprinkler system piping is prohibited by NFPA 13, *Standard for the Installation of Sprinkler Systems*, and NFPA 24, *Standard for the Installation of Private Fire Service Mains and Their Appurtenances*, from being used as grounding of electrical systems in buildings; however, this prohibition does not relieve the NFPA 70, *National Electrical Code*, requirement for bonding of metal piping systems. Although sprinkler piping cannot be used for grounding the electrical system, the piping must be connected, or **bonded**, to the system to prevent metallic sprinkler piping from becoming a fire or shock hazard if it is inadvertently energized. Bonding ensures that a pathway is available that will allow enough current to safely flow to operate the fuse or circuit breaker of a circuit that could energize the piping.

There are a wide variety of metal components associated with an electrical system that are not designed to carry any current. This includes components such as the metal boxes, metal sheathing that protects some wiring, and the frames and housings of electrical machinery. Each of these must be grounded. Certain electrical tools and cord-and-plug appliances, such as washers, dryers, air conditioners, pumps and so on, must also be grounded through a third contact in the line plug **Figure 5**.

Generators

Emergency generators are becoming more common in not only industrial and commercial properties, but also residential properties, especially in rural and remote areas **Figure 6**. There are two issues common to most generator installations, first is protection of the fuel source and second is the shock hazard associated with a generator.

Fuels from emergency generators vary from storage tanks of diesel fuel, gasoline or liquefied petroleum gases, to natural gas piped into the generator. Storage tanks must be inspected for spill protection, venting requirements, and fixed-fire protection if the volume of the storage tank exceeds the limits imposed by the local fire code.

Direct connection of the emergency generator to the building's electrical system should only be accomplished through a transfer switch installed in accordance with NFPA 70, *National Electrical Code* **Figure 7**. Transfer switches provide power exclusively from one source or another, without the possibility of backfeed, meaning a reverse of electrical current into

Figure 5 Certain electrical tools and cord-and-plug connected appliances, such as washers, dryers, air conditioners, pumps and so on, must be grounded through a third contact in the line plug.

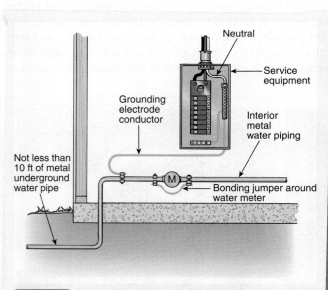

Figure 4 A metallic underground water-piping system must be used as the grounding electrode where it is available and where the buried portion of the pipe is more than 10′ (3 meters) long.

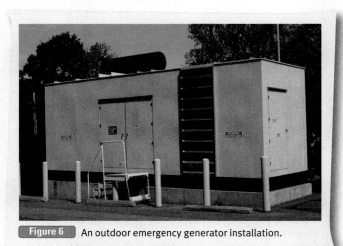

Figure 6 An outdoor emergency generator installation.

Chapter 11 Electrical and HVAC Hazards

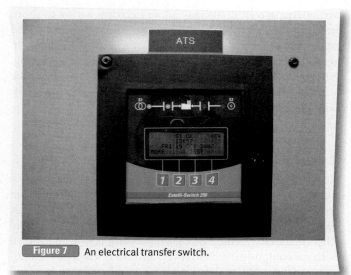

Figure 7 An electrical transfer switch.

Fire Inspector Tips

The presence of generator at a facility must be noted at the main utility disconnect so that responding fire fighters know that another source of energy may automatically start once the main power supply to the building is shut off. A means to shut down any emergency generator must be provided in the same general location as the main utility disconnect.

the general system. This is of particular concern for electrical lineman. Often when there is a power outage, citizens will plug a generator into their houses' electrical systems to provide electricity to their homes. If there is not a transfer switch, the power from the generator will flow through the house wiring and out to the public electrical distribution system. Lineman can be killed while working on lines that should otherwise not be electrified.

Transformers

Transformers are may be found in large industrial facilities that require large amounts of voltage to run mechanical processes. **Dry-type transformers** and **fluid-filled electrical transformers** are used in industrial and commercial occupancies. Dry-type transformers use air as a temperature-maintenance mechanism while fluid-filled transformers typically use a type of oil to remain cool. Dry-type transformers are common in newer commercial facilities with low to moderate current demands Figure 8 . Fluid-filled transformers are more common in industrial facilities and as the main transformer on incoming services where large current loads are present due to their more efficient operation Figure 9 .

In most cases, dry-type transformers do not require a separate room or vault, but they must be separated from combustible materials and provided with adequate ventilation to remove excess heat. Fluid-filled transformers, if not installed outdoors in a fenced yard or substation, usually are required to

Figure 8 A dry-transformer installation.

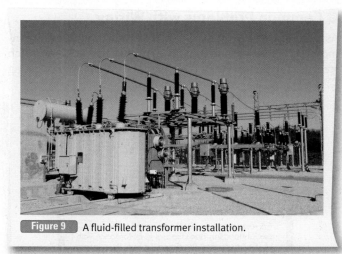

Figure 9 A fluid-filled transformer installation.

be installed in a vault with 3-hour fire-resistive construction and associated opening protection. The vault must also be curbed or of sunken construction to contain the contents of the transformer and any fire protection system water discharge should a spill occur.

Older fluid-filled transformers often used mineral oil as their cooling medium. Newer transformer fluids, classified as less flammable or nonflammable, are now available in addition to mineral oil. When these new fluids are used, the requirements for vaults are reduced or eliminated. Flammable oils may still be found in new installations, so you should inquire as to the type and quantity of material used. <u>Polychlorobiphenol (PCB) fluid</u>, while widely used in the past, is no longer available due to environmental concerns. These transformers are required to be prominently marked with their contents and eventually will be phased out of use Figure 10 .

Under conditions of a full electrical load, transformers operate at elevated temperatures. Many will be too hot to touch for more than a few seconds. All transformers should be provided with adequate ventilation, and the clearance requirements marked on the transformer should be maintained. Materials should not be stored on top of any transformer enclosure.

Outdoor transformers should be located in such a way that leaking fluids will drain away from buildings and prevented from entering environmentally sensitive areas. These transformers should be positioned in such a manner that they will not expose building exits or windows to fire in the event of a transformer failure. Transformers should also be protected from each other to avoid the failure of one transformer from causing the failure of an adjacent one. Protection may involve the construction of a freestanding firewall between transformers, the installation of a water-spray deluge system, or a combination of the two. The size and criticality of service provided by a transformer may dictate additional protection features due to the extended time period to receive a replacement for a damaged or destroyed transformer.

What to Look for During a Fire Inspection

Common Problems with Wiring

As a fire inspector, you need to be on the lookout for the common problems associated with wiring. These problems often stem from either improper installation or damage to existing wiring. These conditions may allow electrical wiring to short, arc, or become overloaded and can become an ignition source, especially in hazardous locations. Be on the lookout for damaged insulation, broken wires, and sloppy repairs to electrical wiring. Wiring must be supported properly along its length and at the point at which it terminates Figure 11 . Wiring should not be exposed to excessive external heat.

Common Problems with Electrical Equipment

Electrical equipment that is improperly used or poorly maintained can be a source of multiple fire hazards. Not only can problems with electrical equipment present ignition sources, but they may also be shock hazards to operating personnel and others in the near vicinity. Be vigilant for dirty, poorly maintained electrical equipment; broken covers and safety features; accumulations of dirt and grease that may prevent adequate cooling; and the improper use of equipment for the task. For example, using a single extension cord to supply power to multiple computers increases risk of current overload.

Cables, Conduits, and Raceways

Obvious hazards are badly deteriorated or improperly supported cables, conduits, or raceways. <u>Conduits</u> are round pipes that hold multiple wires in walls and ceilings. <u>Raceways</u> are narrow channels that hold wires and are threaded through walls and ceilings. Where these items enter main circuit breaker boxes, they should be housed in proper fittings that securely hold them in place. Conduits that are not supported properly may pull apart and expose conductors and insulation to damage Figure 12 .

Cables should be protected from mechanical damage where they pass through walls or floors. They also should be protected from overload, which is not as immediately obvious as mechanical damage. One subjective and preliminary way to determine whether a cable or wiring might be overloaded is to touch an insulated cable or conduit. Depending

Figure 10 A PCB label on a transformer.

Figure 11 Wires supported ongoing into an outlet or junction box on a brick wall.

Figure 12 An example of damaged electrical wiring.

on the load, the cable or conduit may feel warm or even hot. Many common conductors and equipment in normal use may operate at temperatures up to 167°F (75°C), much too hot to touch comfortably. Where overheating is suspected, it should be further investigated by a qualified person, such as an electrical engineer. Your jurisdiction may use infrared scanning equipment to make a more objective assessment of actual and relative temperatures in a system without requiring any physical contact.

Extension Cords

Several unsafe practices involving extension cords may result in fire. One of the most common unsafe practices is the use of extension cords in place of what should be fixed wiring Figure 13 . Extension cords should be used only to connect temporary portable equipment, not as a substitute for permanent wiring. For example, it is perfectly acceptable to use an extension cord to power a drill while working on a project, but it would not be appropriate to use an extension cord to plug in a neon sign in the window of a business.

Extension cords should not be used to supply equipment that will load them beyond their rated capacity.

Because extension cords are not required to have the same current-carrying capacity as the branch-circuit wiring supplying wall receptacles, it is extremely important to ensure that the equipment supplied by the cord has a lower wattage or amperage rating than that of the extension cord. If the extension cord is of lesser diameter than the appliance cord it is supplying, you should carefully examine it to see that it is appropriately rated for the appliance.

It is important that extension cords be well maintained and properly located. Extension cords should never be run where they can be damaged by vehicles, carts, or pedestrians. Coiling can generate a significant amount of heat due to the current flow through the conductors; therefore, extension cords should never be left coiled while in use. For a similar reason, cords should not run under rugs or carpets as any heat generated in the cord cannot be dissipated and may build up sufficiently to ignite the carpet or rug. Extension cords should not be attached to building surfaces, woodwork, pipes, or other equipment, or run through doors; windows; or holes in walls, floors, or ceilings. Damaged extension cords should be replaced and not repaired or spliced.

Outlets and Switches

Outlets and switches are both points in a wiring system with a visible external component. **Outlets**, are points on the wiring system at which current is taken to supply utilization equipment. A **switch** is any set of contacts that interrupts or controls current flow through an electrical circuit. When observing outlets and switches, look for cracked or broken switches and outlets, discolored devices, or covers that indicate overheating. If any of these conditions are observed, their use should be immediately discontinued and they should be promptly repaired or replaced by a qualified electrician.

Electrical System Boxes

Circuit breaker boxes, junction boxes, and electrical cabinets are used to house and protect equipment and connections to contain the sparks, arcs, or hot metal that may be produced by the electrical equipment. All such boxes should be equipped with the proper cover. Boxes and cabinets have pre-punched, circular holes called **knockouts** that can be removed to allow the secure connection of conduit or raceways. Only the knockouts that are necessary to accommodate the conduits or raceways entering the box should be removed; all other openings must be closed. The number of conduits or raceways must not exceed the number for which the box was designed.

Some circuit breaker boxes, especially in industrial settings, may have exposed live parts from which occupants must be protected Figure 14 . This can be accomplished by placing a grounded cage or barrier around these open boards. During a fire inspection, circuit breaker boxes should be checked for any obvious deterioration, dirt, moisture, tracking, or poor maintenance. Also make sure that the surrounding area is kept clear to allow quick and ready access for maintenance, repair, and emergency response. Nothing should be stored in the working space or on top of circuit breaker boxes.

Figure 13 An example of extension cord abuse.

Figure 14 A circuit breaker box.

Lamps and Light Fixtures

Light fixtures are subject to deterioration and poor maintenance. With age, the insulation on fixture wires can dry, crack, and fall away, leaving bare conductors. Sockets may become worn and defective, and the fixture itself may loosen in its mountings. Fixtures should not be mounted directly on ceilings unless specifically listed for such purpose. Since lamps often operate at temperatures high enough to ignite combustible materials, they should be mounted far enough away from materials such as paper or cloth so that their continuous operation does not ignite the combustible materials.

Make sure that aerial lighting is the proper size and type and that a thermal barrier is in place. Discolored light globes can indicate improper lamp size. Most fixtures are marked with the appropriate lamp types and maximum wattage ratings. Recessed light fixtures may have thermal protectors that will turn the fixture off if an incorrect lamp size results in high operating temperatures.

Motors

Motors and other rotating machines should be treated with caution because they can cause physical injuries in addition to the electrical shock hazard. Many motors are automatically controlled and may start without warning; therefore, a motor at rest should be treated as though it were already running.

Motors have many potential fire hazards. Sparks or arcing that results when a motor short circuits can ignite nearby combustibles. Shaft bearings can overheat if they are not properly lubricated. Dust deposits or the accumulation of textile fibers can prevent heat from dissipating from the motor.

The inspection of motors should verify that there are no combustibles in the immediate vicinity of the motor or its controls, that the equipment is cleaned properly and maintained, and that it has the proper overcurrent protection. Motors are designed to operate without overheating under normal conditions, but they also are designed to operate with a significant temperature rise under normal full-load conditions.

A hot casing may indicate a potential problem and should be closely examined by a qualified electrician or the equipment manufacturer's representative.

Hazardous Areas

Electrically hazardous areas are those in which flammable liquids, gases, combustible dusts, or readily ignitable fibers are present in sufficient quantities to represent a fire or explosion hazard. Special electrical equipment is necessary in these areas. The special equipment is intended to keep the electrical system from becoming a source of ignition for the flammable or combustible atmosphere. Electrical equipment used in a hazardous area is specifically listed or identified for use with particular chemicals or hazardous atmospheres. The class, division, and possibly the group of materials or chemicals that creates the hazardous atmosphere will be listed on the label affixed to the electrical equipment. Portable equipment should be similarly listed or identified for use. Complete definitions of the classes and divisions of hazardous locations are covered in Article 500 of NFPA 70 Table 1 .

Static Electricity

Precautions against sparks from static electricity should be taken in locations where flammable vapors, gases, or dusts or easily ignited materials are present. Only qualified persons, such as appropriately trained industrial hygienists, safety engineers, or electrical engineers should be allowed to test for static charges in these locations, since specialized equipment is required and unintended discharges can ignite the hazardous atmosphere or contents. Methods to bring the hazard of static electricity under reasonable control are humidification, bonding, grounding, ionization, conductive floors, or a combination of these methods.

Humidification

Humidity alone is not a completely reliable means of eliminating static charges. To reduce the danger of static electricity, however, the relative humidity meaning the amount of water vapor or moisture held in suspension by gas or air expressed as a percentage of the amount of moisture that would be held in suspension at the same temperature if saturated, should be maintained at a high level. If practical, relative humidity should be as high

Fire Inspector Tips

NFPA 77, *Recommended Practice on Static Electricity*, states that a relative humidity between 30 and 65 percent should be sufficient to provide a surface resistance of most materials that will prevent an accumulation of static electricity. However, humidity alone will not provide adequate protection. The material must absorb the moisture from the air and will only dissipate a charge if there is a conductive path to ground. Some plastics and the surfaces of petroleum liquids are capable of accumulating static charges at 100 percent humidity.

Table 1 Hazardous Electrical Locations

Class I, Division 1	Locations include areas in which ignitable concentrations of flammable gases or vapors exist under normal conditions; areas in which ignitable concentrations of flammable gases or vapors may exist frequently because of repair or maintenance operations or leakage; and areas in which the breakdown or faulty operation of equipment or processes may cause the simultaneous failure of electrical equipment. Electrical equipment used in these locations must be the explosion-proof type or the purged-and-pressurized type enclosures approved for Class I locations.
Class I, Division 2	Locations include areas in which volatile flammable liquids or flammable gases are handled, processed, or used. These flammable substances are normally confined to closed containers or systems and would only escape during accidental rupture, breakdown, or abnormal operation of equipment. Locations also include areas in which failure of ventilation systems could create ignitable and hazardous concentrations of gases or vapors. Areas adjacent to Class I Division 1 locations are also included in this class if ignitable concentrations of gases or vapors could possibly shift from one area to the other.
Class II, Division 1	Locations include areas in which potentially ignitable quantities of combustible dust are, or may be, in suspension in the air continuously, intermittently, or periodically under normal operating conditions. Areas in which mechanical failure or abnormal operation of equipment might result in explosive or ignitable mixtures and provide a source of ignition through simultaneous failure of electrical equipment are also classified as Class II, Division 1, as are areas in which combustible dusts of an electrically conductive nature might be present. Class II, Division 1 locations also include areas in which an accumulation of combustible dust on horizontal surfaces over a 24-hour period exceeds 1/8" (3.1 mm).
Class II, Division 2	Locations include areas in which combustible dust normally is not suspended in the air in quantities sufficient to produce explosive or ignitable mixtures and in which dust accumulations normally are not sufficient to interfere with the normal operation of electrical equipment or other apparatus. They also include areas in which the infrequent malfunction of handling or processing equipment might result in dust suspended in the air that could be ignited by the abnormal operation or failure of electrical equipment or other apparatus. Areas in which an accumulation of combustible dust on horizontal surfaces is 1/8" (3.1 mm) or less, obscuring the surface color of the equipment, are also classified as Class II, Division 2 locations.
Class III, Division 1	Locations include areas in which easily ignitable fibers or materials producing combustible filings are handled or used during the manufacturing process.
Class III, Division 2	Locations include areas in which easily ignitable fibers are stored or handled, except during the manufacturing process.
Class I, Zones 0, 1 and 2	This zone classification system provides an alternate design methodology to the Division 1 and 2 classifications in the Class I locations. This classification system provides for alternate protection methods. The classification of areas must be done by a qualified licensed professional engineer. The area classification should be well documented and the documentation should be readily available. Zones 0 and 1 correspond roughly to Division 1 with Zone 2 corresponding roughly to Division 2.

Source: Table adapted from Article 500 of NFPA 70

as possible, even up to 75 percent, as long as this does not create undue hardship or damage to building construction and insulating materials. Humidification systems must be designed by qualified, licensed mechanical engineers.

Heating, Ventilation, and Air Conditioning Systems Overview

Heating, ventilation, and air conditioning systems, commonly referred to as **HVAC**, vary widely in their types, designs, and safety hazards depending on their use and purpose **Figure 15**. All HVAC systems share two common components: first is the conversion of an energy source into a heating or cooling medium; second is the distribution of the heating or cooling medium throughout the building.

At the heart of an HVAC system are boilers; refrigeration equipment, such as chillers and cooling towers; ground thermal sources; or a combination of these types of equipment will provide the heat or cooling required by the space being conditioned.

Next, a means to circulate the heated or chilled medium is provided to a heat exchange unit. The heat exchange unit will then temper or condition the space through the warming or cooling of air distributed directly by a duct (pipe) system; direct discharge of the air into the space; or by indirect methods such as those found in radiant baseboard systems, radiant flooring, or ceiling panel systems.

■ Energy Conversion

Boilers generate steam, hot water, or hot air by consuming electricity or burning natural gas, coal, or fossil fuels such as kerosene **Figure 16**. Specific combustion safeguards are required by the model building codes depending on the type of fuel used and the energy output of the unit. Refer to NFPA 85, *Boiler and Combustion Systems Hazards Code*, for additional detailed information on boiler operation and safety. A **boiler room** is any room with a boiler of 5 horsepower or greater. Due to the hazardous nature of this operation and the development of elevated pressures within pressure vessels and piping systems, boiler rooms will often be

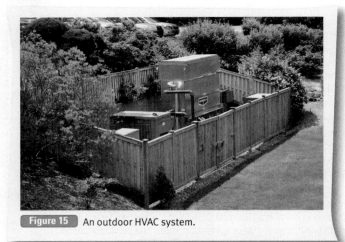

Figure 15 An outdoor HVAC system.

Figure 16 A boiler.

Figure 17 Chiller rooms are detached or separated from adjacent areas by fire-resistive construction.

either detached structures in large complexes or separated from the remainder of the building by fire-resistive construction when the rate of energy used exceeds certain minimum levels in the model building codes.

For the generation of a source of cooling in large commercial and industrial facilities, **chiller plants** using internal refrigeration compressors and an external cooling source, such as cooling towers or large bodies of water acting as a **heat sync**, normally utilize electrical energy. A heat sync is an object that, through conduction, draws heat away from a heat-producing object. Due to the pressures involved in vessels and piping systems and the flammable and potential health hazards of some common refrigerants, chiller rooms are detached or separated from adjacent areas by fire-resistive construction **Figure 17**. Automatic shutdown and mechanical ventilation systems to depressurize and vent leaking refrigerant operated automatically by refrigerant leak detection systems and manually at chiller-room egress doors are required when the quantity of refrigerant or the cooling capacity of the chiller plant exceeds amounts in the model building codes.

With both boiler rooms and chiller rooms, you should be aware of the inherent hazards involving elevated temperatures, pressurized vessels, piping, and the need for continual maintenance and modernization to keep these systems operating efficiently and safely. Large commercial and industrial complexes will often use specialized firms or insurance companies employing trained **boiler inspectors** to routinely inspect the piping, **boiler enclosures**, and combustion safeguards in these installations. Many states and provinces require these inspections on predetermined intervals. As a fire inspector, your role is greater in smaller heating and cooling plants that do not require such specialized inspection services.

Distribution Systems

Once the heated or chilled medium has been generated, it must be distributed to where it is needed. The selection of a particular distribution system by the mechanical engineer designing the system will be dependent on many factors, including fuel source, building type, size and height, climate, heat loads and economics. Either direct or indirect methods may be used or combined to distribute the heating or cooling needed in a building. Circulation of hot water to baseboard units or fan-coil units in individual rooms, or from the boiler room to air handling units serving larger areas of a building, is one common method. Chilled water may also be distributed to fan coil or air handling units for cooling **Figure 18**. Steam distributed to individual radiators or heat exchangers may be found in older installations. **Hydronic heating and cooling systems** circulate hot or chilled water through plastic piping embedded in a gypsum/cement floor layer or imbedded in a concrete floor slab **Figure 19**. The distribution of hot and chilled water and steam in these manners requires only the protected passage of piping systems through fire-resistive construction, such as floors, ceilings, and walls.

Figure 20 A ducted system.

Ducted Systems

Ducted systems are used to distribute conditioned air from air handling units or indirectly heated and cooled air from furnaces, air conditioning coils, and other heat exchangers **Figure 20**. With a ducted air handling system, the air is first filtered, a percentage of outside air is added as required by the model building codes, the air is possibly cleaned **electrostatically** (using electrically charged particles) or humidified, and then finally heated or cooled in an air handling unit. A circulating fan forces the conditioned air into the supply air ductwork. The supply air ductwork provides conditioned air through supply registers in each room or space served by the **air handling unit**. Return air from these spaces is collected by a separate duct, or **plenum system**, and directed back to the air handling unit for conditioning or a portion of the return air may be exhausted from the building. Depending on the size and height of the building and the size of the air handling unit, multiple floors, and large areas of a building may be served by a single air handling unit.

Figure 18 A fan coil unit may distribute chilled water to cool a building.

Figure 19 A hydronic floor heating/cooling system piping is covered with concrete.

Safety Systems

Exhaust Systems

A specialized mechanical system may be used in some industrial, hazardous, or commercial occupancies to remove hazardous fumes, vapors, mists, or particles. These exhaust systems do not normally recirculate the exhausted air. **Dust collection systems** or **chemical treatment systems** may be incorporated in the design of an exhaust system **Figure 21**.

Exhaust systems conveying potentially toxic materials require treatment or neutralization systems that reduce the toxicity to one-half of the level that is **immediately dangerous to life and health (IDLH)** at the point of discharge in the event of a release. The IDLH levels for toxic chemicals are available from the safety data sheets provided by the chemical manufacturer.

Continuous operation of most exhaust systems is required by the model building codes, especially if toxic materials or

Figure 21 A dust collection system.

an accumulation of flammable or combustible gases or dusts may occur. Emergency power backing up normal utility power may also be required by the model building codes for these systems. Dampers, meaning valves or plates for controlling draft or the flow of gases, are not permitted in these systems since the exhausted products are not recirculated in the building. If the exhaust system involves a toxic or hazardous product, the presence of a damper would create a greater hazard by interrupting the exhaust system.

■ Smoke Management Systems

Several methods for managing the movement of smoke in a building may involve HVAC systems in concert with the building's construction. Some design concepts rely on natural ventilation and compartmentalization while others actively use portions of an HVAC system to either contain or remove smoke.

Natural ventilation through the use of gravity vents is most commonly found in some large warehouses. This application of smoke management relies on the buoyancy of the hot products of combustion to open a fusible-link operated vent in a high ceiling. In a few warehouse applications, the gravity vents are replaced by exhaust fans on the roof that automatically activate in the event of a fire. Protection of the fan drive motor, bearings, and electrical supply is required for these fans to operate in high heat conditions.

Older high-rise designs also utilize manually-opened vents or operable or breakable windows so fire fighters can manually provide an avenue for fresh air to enter and smoke to leave an upper story of a building. With this setup, the building's HVAC system usually shuts down the air handling units, if so equipped.

Compartmentation, also referred to as passive smoke management, relies on interior smoke barrier construction in the walls and floor/ceiling assemblies of a building to restrict the free movement of smoke from one smoke compartment to another. In this application, the HVAC system is sent a command from the fire alarm system to shutdown the air handling units and close all supply and return air dampers in and surrounding the affected smoke compartment. The size and location of the smoke compartments in a building is determined by the building's designers. When inspecting a building employing this type of smoke management design, you need to know where smoke barrier walls have been constructed and must verify during subsequent inspections that the integrity of the barrier walls have not been compromised. Initial acceptance testing and annual testing by a qualified person, such as a fire protection engineer, mechanical engineer, or an air-balance technician, is also necessary to ensure that the HVAC control systems, air-handling-unit shutdown features, and smoke/fire dampers operate as designed.

A third smoke management method is intended to contain smoke to its zone of origin by combining compartmentation and using all or a portion of the HVAC system to depressurize the smoke zone. Upon activation, air handling units or dedicated smoke management fans exhaust air from the smoke zone. Smoke/fire dampers close, supply air is shut down, and a negative pressure is developed in the smoke zone where the fire is located. Referred in the model codes as the pressurization method, its name is a misnomer since the system must actively depressurize the zone, otherwise expansion forces from the fire would force smoke through the smoke barriers. With this method, any leakage in the construction of the smoke barriers will cause air to leak into the zone of origin, meaning the portion of the duct system including the smoke zone, instead of having smoke leak out of the zone. Smoke is not actually removed in sufficient quantities to keep the smoke zone habitable. Introducing outside air to flush out the smoke is not recommended: the newly introduced (called make-up air) air may be contaminated, the balance of outside air and exhaust air is subject to weather conditions and seasonal changes, and the reliability of the system is much harder to maintain. Detailed information on this design concept and factors to consider in the design, installation, commissioning, and maintenance of this type of smoke management system can be found in NFPA 92A, *Standard for Smoke-Controlled Systems Utilizing Barriers and Pressure Differences*. Due to its complexity, the design, commissioning, acceptance testing, and annual operational testing of this type of smoke management system should be performed by qualified and experienced engineering, air balance, and technical personnel.

The fourth smoke management system can be found in NFPA 92B, *Standard for Smoke Management Systems in Malls, Atria, and Large Spaces*. This design concept is used in areas with high ceilings and employs HVAC equipment, either dedicated to the smoke management use or part of the normal HVAC system, to remove products of combustion accumulated above a calculated height above the floor. Make-up air, carefully controlled so as to not distort the fire plume, and resulting exhaust calculations, is also an important part of this design methodology. It is the intent of this design concept to maintain a tenable environment for a minimum amount of time necessary to evacuate occupants from the building. Again, due to the complexities involved, only qualified and experienced personnel, such as fire protection engineers and mechanical engineers, should be involved in the engineering and testing of this type of smoke management system.

Stairwell Smoke Management Systems

Aside from the building smoke management systems previously described, stairwell smoke management, independent of the building's smoke management design, is required in high-rise and other special structures. Vestibules, chambers open to the outside air, can be used to keep enclosed stairways free from smoke contamination. Mechanical pressurization of the stairwell, with or without entry vestibules, can also keep smoke from entering the stairwell. The use of vestibules will vary the pressure differential requirement necessary to maintain a clear stairwell. One caution is that the higher the pressure differential developed in the stairwell, the harder it will be to open the exit doors leading into the stairwell. During inspection, you need to ensure that the occupants inside a stairwell can reenter the building quickly should the mechanical pressurization system or building construction features fail to keep smoke from entering the stairwell. Refer to the model building codes for detailed design requirements for smoke protected stair enclosures.

Occasionally pressurization of elevator hoistways may be encountered. This is not a common application and has a number of adverse impacts on such items as elevator safety controls and the potential to force smoke that may enter a hoistway through smoke barrier doors into other floors. Only qualified and experienced engineering personnel should be involved in the design and review of these systems.

In the smoke-management design concepts discussed above, the successful control of a fire by an automatic sprinkler system is critical to the successful management of the products of combustion produced by a fire. An out-of-control fire will produce an out-of-control quantity of smoke that no construction feature or mechanical system can overcome.

Design and testing of smoke management systems requires careful and diligent work by qualified and experienced professional designers. Information and guidance can be found in a number of publications, including the NFPA/SFPE *Fire Protection Engineering Handbook*. Initial testing at the time of construction and recurrent testing at intervals established by either the initial designers or in accordance with NFPA standards should be conducted only by experienced and qualified fire protection engineers, mechanical engineers, and air balance technicians.

Testing

Commissioning, acceptance testing, and related documentation of these systems are especially important as their complexity can be very high. Reliability, likewise, is dependent on the complexity of the design and the maintenance provided during the life of the facility. Annual testing by qualified professional engineers and air balance technicians of these systems is greatly facilitated by good original commissioning documentation.

Potential Hazards

Smoke Distribution through the Ducts

The hazard presented by duct work in a HVAC system is the potential to circulate and distribute contaminated air and smoke throughout a large portion of a building. To protect occupants from this hazard, air handling units circulating 2,000 cubic feet per minute (cfm) or more are required to have a local smoke detector listed for this special application arranged to shutdown the air handling unit in the event smoke is detected. When a building has a fire alarm system, duct detectors, which are designed to detect the presence of smoke, are required to be monitored by the fire alarm system; however, they are categorized as a supervisory signal and not as a fire alarm signal. This is due to a high tendency for false activation in part caused by the dust and particles common in air handling system ductwork.

Additional protection is provided in duct distribution systems by the installation of fire dampers, smoke dampers, or combination smoke/fire dampers at various fire-rated or smoke barriers when penetrated by ducts. Air transfer openings and plenums penetrating these same barriers also require the same protection. The location of required fire and smoke dampers is specified in model building codes.

Activation of fire dampers is accomplished by the use of a fusible link or a thermal sensor in power-operated dampers. These dampers operate only when a high temperature is detected at the individual damper and are intended to stop the spread of fire through the penetrated barrier. Fire dampers are not effective in stopping the spread of smoke. Where the passage of both smoke and fire is not desirable, the codes specify the use of combination smoke/fire dampers. Smoke/fire dampers are tested and listed at three leakage levels, the minimum leakage level required is specified in the model building codes or further restricted by the designer of the system. Operation of smoke/fire dampers includes not only the heat activated element of a fire damper, but also the detection of smoke either internally at the damper or throughout the spaces served by the ductwork system. These dampers will significantly reduce the passage of smoke across a penetrated smoke barrier when properly installed.

> **Fire Inspector Tips**
>
> A major concern in the design of a smoke control system that relies on a differential of pressure between the fire area and surrounding areas is that the operation of the egress doors is affected. If the pressure is too high in surrounding areas and/or the pressure is too low in the fire area the doors will be too difficult to open.

What to Look for During Inspection

In all cases, storage and non-essential materials in boiler and chiller plants should be removed. Combustion air to fuel-fired equipment must be maintained free and unobstructed in order for boilers and furnaces to function efficiently. An adequate supply of combustion air is also necessary to avoid a fuel-rich condition that can lead to inefficient burning, causing sooting or other accumulations of unburned or partially burned materials inside the unit, the **flue** (the passage through which flue gases are conveyed from the combustion chamber to the outer air), or the room. Where obvious tampering, rigging, removal, or poor maintenance of combustion safety equipment is observed, the affected unit should be shutdown until repairs are completed by qualified personnel.

Clearances to combustible materials, whether in the construction of the building or storage nearby, is a significant cause of fires involving HVAC equipment. Furnaces, boilers, chillers, fireplace inserts, flues and exhaust stacks, steam accumulators, and piping and any other piece of equipment operating at an elevated temperature can cause a deterioration of combustible materials over time. This pyrolysis of wood and wood products is visible as a charring and degradation of the surface of the material. Over time, pyrolysis enables the material to ignite at a significantly lower temperature. Installation requirements for listed assemblies will specify clearances from combustible materials and often can be found on the label or in the listing documentation. Where generic materials, such as steam piping, are used, NFPA 211, *Standard for Chimneys, Fireplaces, Vents, and Solid Fuel-burning Appliances*, and NFPA 54, *National Fuel Gas Code*, provides detailed information.

Exhaust stacks, chimneys, and flues for boilers and furnaces also warrant attention from fire inspectors. An exhaust stack is a chimney or ductwork that removes excess heat, fumes or vapors from an area without reuse, to a point of discharge. The general condition and integrity of these items should be visually inspected where accessible. Any indication of damage, such as broken or separated flue piping, damaged masonry, or a leaning exhaust structure, should be further investigated by a qualified mechanical or **structural inspector**.

Of note recently are modern high-efficiency gas furnaces and hot water generators. You may find an increasing number of these units in residential and small to moderate sized commercial properties. A unique aspect of these appliances is the use of plastic piping to supply combustion air and to vent the combustion products. This is possible due to the very low flue gas temperature generated by the high-efficiency unit. Manufacturer's installation and listing documentation will specify the reduced separation, clearance, and exhaust requirements of these units.

Wrap-Up

Chief Concepts

- As a fire inspector, you must be cognizant of potential electrical hazards and the condition of electrical systems that are damaged, defective, or misused.
- Electrical systems are complex arrangements of wires and circuits with the end goal of delivering electrical power to an appliance or electronic device for the comfort and use of occupants in the building.
- Electrical fires are caused principally by arcing or overheating.
- Fuses and circuit breakers are the most commonly used overcurrent devices.
- Obvious hazards are badly deteriorated or improperly supported cables, conduits, or raceways.
- From the simplest HVAC system in a single-family dwelling to a large industrial complex, all HVAC systems share two common components. First is the conversion of an energy source into a heating or cooling medium; second is the distribution of the heating or cooling medium throughout the building.
- A specialized mechanical system may be used in some industrial, hazardous, or commercial occupancies to remove noxious, toxic, or hazardous fumes, vapors, mists, or particles. These exhaust systems do not normally recirculate the exhausted air.

Wrap-Up

■ Hot Terms

Air handling unit A unit installed for the purpose of processing the treatment of air so as to control simultaneously its temperature, humidity, and cleanliness to meet the requirements of the conditioned space it serves.

Amps The measure of the volume of electrical flow.

Arc-fault circuit interrupter (AFCI) A device intended to provide protection from the effects of arc faults by recognizing characteristics unique to arcing and by functioning to de-energize the circuit when an arc fault is detected.

Arcing A high-temperature luminous electric discharge across a gap or through a medium such as charred insulation.

Boiler A closed vessel in which water is heated, steam is generated, steam is superheated, or any combination thereof by the application of heat from combustible fuels in a self-contained or attached furnace. (NFPA 85)

Boiler enclosures The physical boundary for all boiler pressure parts and for the combustion process for a closed vessel in which water is heated, steam is generated, steam is superheated, or any combination thereof by the application of heat from combustible fuels, in a self-contained or attached furnace. (NFPA 85)

Boiler inspector A person who, through formal education and training, is qualified to inspect a boiler and its associated systems against established standards, recommendations, and requirements.

Boiler room Any room with a boiler of 5 horsepower or greater. (NFPA 5000)

Bonding (bonded) The permanent joining of metallic parts to form an electrically conductive path that will ensure electrical continuity and the capacity to conduct any current likely to be imposed. (NFPA 79)

Cartridge fuses A type of overcurrent protective device with a replaceable cartridge containing the fusible part.

Chemical treatment system A system utilizing chemical additives to alter the properties and hazards of a waste stream prior to disposal, reuse, or further chemical processing.

Chiller plants A facility housing refrigeration equipment for the purpose of extracting heat for industrial use, comfort cooling, or other uses.

Circuit breakers A device designed to open and close a circuit by nonautomatic means and to open the circuit automatically on a predetermined overcurrent without damage to itself when properly applied within its rating. (NFPA 70)

Combination smoke fire damper A device that functions as both a fire damper and as a smoke damper. (NFPA 5000)

Compartmentation The subdivision of a building into relatively small areas so that fire or smoke can be confined to the room or section in which it originates. (NFPA 232)

Conduits Round piping where wiring is routed through to provide protection from damage.

Current The flow of electricity, measured in amps.

Dampers A valve or plate for controlling draft or the flow of gases, including air. (NFPA 853)

Dry-type transformers A device that raises or lowers the voltage of alternating current of the original source. (NFPA 70)

Duct detectors A device mounted either inside an HVAC duct or mounted on the outside of the duct with tubing arranged to sample the airflow to respond to the presence of smoke.

Ducted systems A continuous passageway for the transmission of air that, in addition to ducts, includes duct fittings, dampers, fans, and accessory air-management equipment and appliances. (NFPA 853)

Wrap-Up

Dust collection system A pneumatic conveying system that is specifically designed to capture dust and wood particulates at the point of generation, usually from multiple sources, and to convey the particulates to a point of consolidation. (NFPA 664)

Electrical inspector Individuals who verify that electrical systems comply with applicable codes and standards when buildings are initially constructed, altered, or renovated.

Electrostatic Particles or objects electrically charged with either a positive or negative voltage differential.

Equipment ground-fault protective device (EGFPD) A device intended to provide protection of equipment from damage from line-to-ground fault currents by operating to cause a disconnecting means to open all ungrounded conductors of the faulted circuit.

Exhaust stacks Chimney or ductwork that removes excess heat, fumes or vapors from an area without reuse, to a point of discharge.

Fire damper A device installed in an air distribution system, designed to close automatically upon detection of heat to interrupt migratory airflow, and to restrict the passage of flame. (NFPA 221)

Flue The general term for a passage through which flue gases are conveyed from the combustion chamber to the outer air. (NFPA 211)

Fluid-filled electrical transformers A device that raises or lowers the voltage of alternating current of the original source. (NFPA 70)

Fuse An overcurrent protective device with a circuit-opening fusible part that is heated and severed by the passage of overcurrent through it. (NFPA 70)

Gravity vents A component of a type of vent system for the removal of smoke from a fire that utilizes manually or automatically operated heat and smoke vents at roof level and that exhausts smoke from a reservoir bounded by exterior walls, interior walls, or draft curtains to achieve the design rate of smoke mass flow through the vents, and that includes provision for makeup air. (NFPA 204)

Grounding A conducting connection, whether intentional or accidental, between an electrical circuit or equipment and the earth or to some conducting body that serves in place of the earth. (NFPA 70)

Ground-fault circuit-interrupter (GFCI) A device intended for protection of personnel that functions to de-energize a circuit or portion thereof within an established period of time when a fault current-to-ground exceeds some predetermined value that is less than that required to operate the overcurrent protective device of that supply circuit. (NFPA 70)

Heat sync An object that, through conduction, draws heat away from a heat-producing object.

HVAC Stands for heating, ventilation, and air conditioning systems.

Hydronic heating and cooling systems A method of radiant heating or cooling of a space through the circulation of warm or cool water through a system of tubing either imbedded in a floor system or by radiant ceiling panels

Immediately Dangerous to Life or Health (IDLH) Any atmosphere that poses an immediate hazard to life or produces immediate irreversible debilitating effects on health. (NFPA 1670)

Knockouts Pre-punched circular holes in an electrical junction box or other electrical equipment that allows the secure connection of conduit or wiring cables.

Make-up air Air introduced to a space to replace air removed by exhaust systems.

Outlet A point on the wiring system at which current is taken to supply utilization equipment. (NFPA 70)

Polychlorobiphenol (PCB) fluid A class of organic compounds containing one or more chlorine atoms connected to a molecule composed of two linked benzene rings. Trade names associated with PCB's include, but are not limited to, Askarel, Inerteen, and Pyranol/Pyrenol.

Wrap-Up

Plug fuses An overcurrent protective device with a circuit-opening fusible part that is heated and severed by the passage of overcurrent through it. (NFPA 70)

Plenum system An HVAC system that uses a compartment or chamber to which one or more air ducts are connected and that forms part of the air distribution system. (NFPA 90A)

Pressurization method A smoke control method specified in NFPA 92A that employs the development of a pressure differential across a smoke zone boundary to limit the spread of smoke from one smoke zone in a building to another smoke zone.

Raceway An enclosed channel of metal or nonmetallic materials designed expressly for holding wires, cables or busbars, with additional functions as permitted in the electrical code. Raceways include, but are not limited to, rigid metal conduit, rigid nonmetallic conduit, intermediate metal conduit, liquidtight flexible conduit, flexible metallic tubing, flexible metal conduit, electrical nonmetallic tubing, electrical metallic tubing, underfloor raceways, cellular concrete floor raceways, cellular metal floor raceways, surface raceways, wireways, and busways. (NFPA 70)

Relative humidity The amount of water vapor or moisture held in suspension by gas or air and expressed as a percentage of the amount of moisture that would be held in suspension at the same temperature if saturated.

Renewable link cartridge An overcurrent protective device with a circuit-opening fusible part that is heated and severed by the passage of overcurrent through it. (NFPA 70)

Shunt trip A device that remotely causes the manual or automatic opening of an electrical circuit or main panel disconnect.

Smoke damper A device arranged to seal off airflow automatically through a part of an air duct system, to restrict the passage of smoke. A smoke damper is not required to meet all the design functions of a fire damper. (NFPA 221)

Smoke zone The smoke-control zone in which the fire is located. (NFPA 92A)

Stairwell smoke management A design or method that employs architectural design, construction material, mechanical equipment, or a combination thereof with the intent to keep smoke from contaminating a stair shaft with smoke during a fire incident.

Structural inspector A person who, through formal education and training, is qualified to inspect the structural elements of a building against established standards, recommendations, and requirements.

Switch Any set of contacts that interrupts or controls current flow through an electrical circuit.

Vestibule A small room located between two spaces that provides an atmospheric separation for the purposes of controlling airflow or, in a smoke management system, the movement of contaminated air from one space to an adjacent space.

Volt (V) The unit of electrical pressure (or electromotive force) represented by the letter "E"; the difference in potential required to make a current of one ampere flow through the resistance of one ohm. (NFPA 921)

Watt (W) Unit of power or the rate of work represented by a current of one ampere under the potential of one volt.

Zone of origin In the design of a smoke management system, refers to the smoke zone that the fire incident originates.

Fire Inspector *in Action*

Now that you have finished your inspection of a large industrial complex, it is time to write a report detailing the violations that you found and listed in your notes along with photos and sketches. The plant maintenance staff who accompanied you also kept notes and has all of the violation locations noted in detail.

1. If a conductor is carrying more current than it is designed to carry, the result will most likely be that:
    A. The electric bill will be higher than if adequate conductor size were used.
    B. The voltage on the circuit will be increased.
    C. The conductor is subject to excess heating due to the overload.
    D. The resistance on the circuit will increase.

2. When an area needs more illumination, the safest code-compliant way to increase the light level is to:
    A. Increase the size (wattage) of the light bulbs in the existing fixtures.
    B. Add fixtures to the existing circuit
    C. Increase the size of the fixtures on the existing circuit.
    D. Determine that the circuit and the conductors have sufficient capacity to add fixtures to the area.

3. Metal piping systems, including fire sprinkler systems, are required by NFPA 70 to be:
    A. Grounded
    B. Bonded
    C. Grounded and bonded
    D. None of the above

4. Passive and active smoke management systems include:
    A. HVAC
    B. Gravity vents
    C. Compartmentation
    D. All of the above

Ensuring Proper Storage and Handling Practices

CHAPTER 12

NFPA Standard

Fire Inspector I

4.1 General. The Fire Inspector I shall meet the job performance requirements defined in Sections 4.2 through 4.4. In addition, the Fire Inspector I shall meet the requirements of Section 4.2 of NFPA 472. (pp 245–267)

4.3.8 Recognize hazardous conditions involving equipment, processes, and operations, given field observations, so that the equipment, processes, or operations are conducted and maintained in accordance with applicable codes and standards and deficiencies are identified, documented, and reported in accordance with the applicable codes and standards and the policies of the jurisdiction.

(A) Requisite Knowledge. Practices and techniques of code-compliance inspections, fire behavior, fire prevention practices, ignition sources, safe housekeeping practices, and classification of hazardous materials. (pp 245–267)

(B) Requisite Skills. The ability to observe, communicate, apply codes and standards, recognize problems, and make decisions. (pp 245–267)

4.3.12 Verify code compliance for incidental storage, handling, and use of flammable and combustible liquids and gases, given field observations and inspection guidelines from the AHJ, so that applicable codes and standards are addressed and so deficiencies are identified and documented in accordance with the applicable codes, standards, and policies of the jurisdiction. (pp 245–267)

(A) Requisite Knowledge. Classification, properties, labeling, storage, handling, and use of incidental amounts of flammable and combustible liquids and gases. (pp 245–267)

(B) Requisite Skills. The ability to observe, communicate, apply codes and standards, recognize problems, and make decisions. (pp 245–267)

4.3.13 Verify code compliance for incidental storage, handling, and use of hazardous materials, given field observations, so that applicable codes and standards for each hazardous material encountered are addressed and deficiencies are identified, documented, and reported in accordance with the applicable codes and standards and the policies of the jurisdiction. (pp 253–267)

(A) Requisite Knowledge. Classification, properties, labeling, transportation, storage, handling, and use of hazardous materials. (pp 253–267)

(B) Requisite Skills. The ability to observe, communicate, apply codes and standards, recognize problems, and make decisions. (pp 253–267)

Fire Inspector II

5.3.8 Verify code compliance for storage, handling, and use of flammable and combustible liquids and gases, given field observations and inspection guidelines from the authority having jurisdiction, so that deficiencies are identified, documented, and reported in accordance with the applicable codes and standards and the policies of the jurisdiction. (pp 245–267)

(A) Requisite Knowledge. Flammable and combustible liquids properties and hazards, safety data sheet, safe handling practices, applicable codes and standards, fire protection systems and equipment approved for the material, fire behavior, safety procedures, and storage compatibility. (pp 245–267)

(B) Requisite Skills. The ability to identify typical fire hazards associated with processes or operations utilizing flammable and combustible liquids and to observe, communicate, interpret codes, recognize problems, and make decisions. (pp 245–267)

5.3.9 Evaluate code compliance for the storage, handling, and use of hazardous materials, given field observations, so that deficiencies are identified, documented, and reported in accordance with the applicable codes and standards and the policies of the jurisdiction. (pp 253–267)

(A) Requisite Knowledge. Hazardous materials properties and hazards, safety data sheet, safe handling practices, applicable codes and standards, fire protection systems and equipment approved for the material, fire behavior, safety procedures, chemical reactions, and storage compatibility. (pp 253–267)

(B) Requisite Skills. The ability to identify fire hazards associated with processes or operations utilizing hazardous materials and to observe, communicate, interpret codes, recognize problems, and make decisions. (pp 253–267)

Additional NFPA Standards

NFPA 1 *Fire Code*
NFPA 30 *Flammable & Combustible Liquids Code*
NFPA 30A *Code for Motor Fuel Dispensing Facilities and Repair Garages*
NFPA 45 *Standard on Fire Protection for Laboratories Using Chemicals*
NFPA 54 *National Fuel Gas Code*
NFPA 55 *Compressed Gases and Cryogenic Fluids Code*
NFPA 58 *Liquefied Petroleum Gas Code*
NFPA 704 *Standard System for the Identification of the Hazards of Materials for Emergency Response*

FESHE Objectives

There are no FESHE objectives for this chapter.

Knowledge Objectives

After studying this chapter, you will be able to:

1. List the classifications of flammable and combustible liquids.
2. List the classifications of flammable and combustible gases.
3. Describe the classifications of hazardous materials.
4. Describe the labeling system for hazardous materials, including flammable and combustible liquids and gases.
5. Describe how to properly store hazardous materials, including flammable and combustible liquids and gases.
6. Describe how to properly handle hazardous materials, including flammable and combustible liquids and gases.
7. Describe how to ensure that hazardous materials, including flammable and combustible liquids and gases, are stored, handled, and used in accordance with all applicable codes, standards, and policies.
8. Describe the fire protection systems and equipment that are appropriate for use with hazardous materials, including flammable and combustible liquids and gases.

Skills Objectives

There are no skills objectives for this chapter.

You Are the Fire Inspector

A new small manufacturing plant has been built in the industrial park in your community. The company manufactures metal storage cabinets for various companies. The manufacturing process involves cutting and shaping the metal pieces, finish painting and assembly. As the fire inspector you must be prepared to determine:

1. What quantities of flammable and combustible liquids are allowed to be stored on the premises?
2. What safeguards are required for this storage?
3. What quantities of flammable and combustible liquids are allowed to be handled and used in the manufacturing process?

Introduction

As society has become increasingly complex over the last century, more and more special hazards have become commonplace in many occupancies, especially in small quantities. Today it is common to find flammable liquids such as gasoline and flammable gases such as propane in use in many occupancies normally considered to be low hazard, such as mercantile occupancies, small repair shops, and storage sheds **Figure 1**. Also, many industrial facilities utilize flammable and combustible liquids and gases as part of their manufacturing processes, such as paint shops and printing presses. It is important for you, as the fire inspector, to understand the hazards associated with flammable and combustible liquids and flammable gases. It is also important to be able to recognize the presence of other hazardous materials and know where to find the relevant information regarding the hazards of the materials and the required safeguards.

Classification of Flammable and Combustible Liquids

The classification system for flammable and combustible liquids is found in NFPA 30, *Flammable & Combustible Liquids Code*, and is based on the division of flammable liquids into three main categories: Class I liquids, Class II liquids, and Class III liquids **Table 1**.

Table 1 **Main Categories of Flammable and Combustible Liquids**

Class	Flashpoint
Class I liquid	below 100°F (37.8°C)
Class II liquid	100 to 140°F (37.8 to 60°C)
Class III liquid	higher than 140°F (60°C)

Figure 1 It is common to find flammable liquids such as gasoline and flammable gases such as propane in use in many occupancies normally considered to be low hazard.

Class I Liquids

Class I liquids include all flammable liquids, including acetone, alcohols, ethanol, gasoline, and toluene. **Flammable liquids** have flashpoints below 100°F (37.8°C) and vapor pressures not exceeding 40 psi at 100°F (2,068 mm Hg at 37.8°C). Class I liquids are subdivided into:

- <u>Class IA liquids</u>—Include flammable liquids with flashpoints below 73°F (22.8°C) and with boiling points below 100°F (37.8°C). Acetaldehyde, ethyl ether (commonly known as ether), and methyl chloride are included in this classification. Highly volatile due to their low boiling points, these materials often are used as raw materials in chemical manufacturing processes to produce other materials.
- <u>Class IB liquids</u>—Include flammable liquids with flashpoints below 73°F (22.8°C) and with boiling points at or above 100°F (37.8°C). One of the more common classifications of flammable liquids, this range includes such chemicals as acetone, benzene, ethyl alcohol (ethanol), gasoline, heptane, isopropyl alcohol (IPA), methyl ethyl ketone (MEK), and toluene. Chemicals in this classification are commonly used as liquid fuels and as cleaning and degreasing agents in a wide variety of commercial and manufacturing occupancies.
- <u>Class IC liquids</u>—Include flammable liquids with flashpoints at or above 73°F (22.8°C) and below 100°F (37.8°C). Ethyl alcohol in solution with water (20 to 50 percent mixture), nitromethane, and turpentine are examples of a Class IC flammable liquid. Nitromethane is commonly used in motor sports activities while turpentine is a common household cleaner found in hardware stores and homes **Figure 2**.

Class II and Class III Liquids

Class II and Class III liquids include all combustible liquids. <u>Combustible liquids</u> are liquids with flashpoints at or above 100°F (37.8°C). Class II and Class III are subdivided into:

- <u>Class II liquids</u>—Include combustible liquids that have flashpoints at or above 100°F (37.8°C) and below 140°F (60°C). Examples of Class II liquids include mineral spirits and Stoddard Solvent which may be found in parts-cleaning operations at metalworking facilities.
- <u>Class IIIA liquids</u>—Include combustible liquids that have flashpoints at or above 140°F (60°C) and below 200°F (93°C). Heavier combustible liquids in this classification include creosote oil and kerosene and can be found in wood preservative operations or as a fuel source in various types of occupancies.
- <u>Class IIIB liquids</u>—Include combustible liquids that have flashpoints at or above 200°F (93°C). Class IIIB liquids include hydrocarbons and oils such as asphalt, carnuba wax, coal tar pitch, cocoanut oil, ethylene glycol, glycerin, linseed oil, palm oil, peanut oil, propylene glycol and paraffin waxes. These items are generally stable and considered non-hazardous at normal temperatures and pressures and are used in a variety of construction, lubrication, antifreeze or food processes in a wide variety of occupancies **Figure 3**.

Figure 3 Class IIIB liquids like asphalt or coal tar pitch are used in a variety of construction.

Figure 2 Turpentine is a common household cleaner.

Gases

NFPA 55, *Compressed Gases and Cryogenic Fluids Code*, classifies gases as presenting either a physical or health hazard. Those presenting a physical hazard include pyrophoric, oxidizing, flammable, and unstable reactive gases. Those presenting a health hazard include toxic, highly toxic, cryogenic, and corrosive gases.

Classification by Physical Hazard

Gases classified as creating a physical hazard are those that, when subjected to changes in their safe storage environment, can become extremely unstable and burn, explode, or react rapidly and violently with combustible materials. Changes in temperature, pressure, mechanical shock, or exposure to an ignition source can generate a reaction that will cause great physical harm.

Pyrophoric Gases

<u>Pyrophoric gases</u> are flammable gases that spontaneously ignite in air. Examples of phrophoric gases include silane and phosphine. Storage of pyrophoric gases is a special concern

because these gases do not require a source of ignition to burn. Pyrophoric gases are not used widely and are most likely to be found in semiconductor manufacturing facilities.

Oxidizing Gases

Oxidizing gases support combustion. They are generally either oxygen, chlorine, or mixtures of oxygen and other gases, such as oxygen-helium or oxygen-nitrogen mixtures, or certain gaseous oxides, such as nitrous oxide. These mixtures contain considerably more oxygen than is present in the oxygen-nitrogen mixture that comprises air. Oxygen and nitrous oxide are found in hospitals.

Flammable Gases

Every flammable gas has a lower and upper flammable limit. Examples of flammable gases include hydrogen, liquefied-petroleum gas, and methane. Mixtures of a flammable gas and air will ignite only within this range of flammability and when the minimum ignition energy is provided. A flammable gas exists normally at a temperature exceeding its normal boiling point, even when the gas is in the liquid state, as it often is in shipment and storage. A flammable gas exists at a temperature that not only exceeds its flashpoint, but it is usually well above it. In other words, flammable gases can always ignite. The metal-working industry use methane in the form of natural gas to heat-treat objects. Hydrogen and LP-gas can be found in manufacturing plants.

Unstable Reactive Gases

Unstable reactive gases are gases that will undergo violent changes when subjected to shock or changes in temperature or pressure. There are four classes of unstable reactive gases. Class 1 gases are those that are normally stable but will become unstable with increases in temperature and pressure. Class 2 gases will undergo violent changes at higher temperatures and pressures, and Class 3 gases are those that could detonate or explode on their own, but need a strong initiating source or to be heated when confined. Class 4 gases are those that can detonate or explode at normal temperatures and pressures.

■ Classification by Health Hazard

Gases classified as creating a health hazard are those that, if released into the atmosphere and a person is exposed to them, could result in irritation, burns, asphyxiation, and destruction of tissue.

Toxic or Highly Toxic Gases

Certain gases can present a serious life hazard if they are released into the atmosphere. Toxic gases, which are poisonous or irritating when inhaled or contracted, include chlorine, hydrogen sulfide, sulfur dioxide, ammonia, carbon monoxide, and arsine. Toxic gases have a lethal concentration in air between 200 and 2,000 ppm by volume of gas or vapor, or between 2 and 20 mg/L of dust, fume, or mist when tested on rats. Highly toxic gases have a lethal concentration in air 200 ppm by volume of gas or vapor or 2 mg/L or less of dust, fume, or mist when tested on rats. Toxic gases are found in a wide range of industries and are labeled "poison gas" for shipment.

Cryogenic Gases

A cryogenic gas is a liquefied gas that exists in its container at temperatures below −130°F (−90°C). A cryogenic gas cannot be retained indefinitely in a container. Heat from the atmosphere, which can be slowed but not prevented from entering the container, continually raises the container pressure. If the gas is confined, the resulting pressure could greatly exceed any feasible container strength.

Corrosive Gases

A corrosive gas can burn, destroy, or cause irreversible alteration to organic tissue. A corrosive gas can be inhaled or absorbed, with symptoms ranging from irritation to asphyxiation. Corrosive gases can cause damage to any number of materials, but many metals are especially susceptible.

■ Physical States of Gases

The physical properties of flammable and combustible gases and liquids are a primary fire protection concern because they affect the physical behavior of gases while they are inside containers and after any accidental release from containers. Until they are used, gases must be completely confined in containers, including during transportation, transfer, and storage. It is a matter of economic necessity and the ease of usage that gases be packaged in containers that contain as much gas as is practical. These requirements have resulted in transportation and storage of gases in the liquid as well as the gaseous state. Distinguishing between gases in a liquid or gaseous state is important for the application of sound fire prevention and protection practices.

Compressed Gases

A compressed gas is defined in NFPA 55 as a gas, not in solution, that packaged under the charged pressure, is entirely gaseous at a temperature of 68°F (20°C). This means that a compressed gas is one that exists solely in the gaseous state under pressure at all normal atmospheric temperatures inside its container. The pressure depends on the pressure to which the container was originally charged, amount of gas remaining in the container, and the gas temperature. There are no universally defined lower or upper limits to the container pressure. In the United States, the lower limit is customarily considered 25 psi (273 kPa) at normal temperatures of 70 to 100°F (21 to 38°C). The upper limit is restricted only by the economics of container construction and is usually in the range of 1800 to 3600 psi (12,512 to 24,923 kPa).

A container of compressed gas is limited in the weight of gas it can hold. For example, the largest common portable cylinder of compressed oxygen contains only about 20 lb (9 kg) of oxygen Figure 4 .

Liquefied Gases

A liquefied gas is one that, at 68°F (20°C) inside its closed container, exists partly in the liquid state and partly in the gaseous state, and is under pressure as long as any liquid remains in the container. Carbon dioxide is commonly stored as a liquefied gas Figure 5 . The pressure depends on the temperature of the liquid, although the quantity of liquid can affect this under some conditions.

A liquefied gas is much more concentrated than a compressed gas. For example, the compressed oxygen cylinder mentioned previously could hold about 116 lb (53 kg) of liquefied oxygen.

Classification by Usage

Historically, some common gases have been categorized by their principal use. This is reflected in some codes, such as NFPA 54, *National Fuel Gas Code*. This classification scheme is not as precise as classification by physical property, and there is much overlap in uses with the gases.

Fuel Gases

Fuel gases include flammable gases customarily used in appliances to produce light or heat for either comfort or material processing. By far the principal and most widely used fuel gases are natural gas (ethane and methane blends) and liquefied-petroleum gases consisting of blends primarily of butane and propane. Natural gas will usually be provided to a building through a utility distribution system of pipes in a community while LP-gases will be delivered by truck or railcar to storage tanks on the property Figure 6 . Fuel gas usage is very common in almost all occupancies.

Industrial Gases

Industrial gases include the entire range of gases classified by chemical properties customarily used in industrial processes, for welding and cutting, heat treating, chemical processing, refrigeration, and water treatment. Hydrogen, anhydrous ammonia, chlorine, and methane are just a few examples of gases used for industrial purposes.

Medical Gases

By far the most specialized usage classification, medical gases are those gases used for medical purposes, such as anesthesia and respiratory therapy. Oxygen and nitrous oxide are common medical gases and may be found in hospitals Figure 7 .

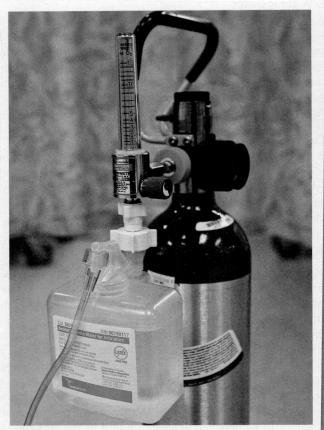

Figure 4 The largest common portable cylinder of compressed oxygen contains only about 20 lb (9 kg) of oxygen.

Figure 5 Carbon dioxide is commonly stored as a liquified gas.

Figure 6 LP-gases will be delivered by truck or rail car to storage tanks on the property.

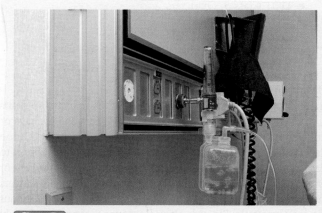

Figure 7 Oxygen and nitrous oxide are common medical gases and may be found in hospitals.

Figure 8 A hazardous material can be found anywhere.

Hazardous Materials

A hazardous material, as defined by the U.S. Department of Transportation (DOT), is a material that poses an unreasonable risk to the health and safety of operating emergency personnel, the public, and/or the environment if it is not properly managed and controlled during handling, storage, manufacture, processing, packaging, use and disposal, or transportation.

One working definition of hazardous materials is "any substance that stores potentially harmful energy or presents a toxicity hazard to people when not contained in its intended container." For our purposes, in this section both definitions are applied throughout. As a fire inspector, you must be able to recognize the presence of hazardous materials, analyze the available information, and ensure that the hazardous materials are stored and handled safely. You must also provide for a method to share the information on the physical or toxic hazards present in a form useable by fire fighters and emergency responders. As the fire inspector, you will often be the first contact between an industrial facility with hazardous materials and your agency. You will also be the most routine visitor to the facility and will be in the best position to verify and update information from past visits.

Hazardous materials can be found anywhere. From hospitals to petrochemical plants, pure chemicals and chemical mixtures are used to create millions of consumer products. Even the small hardware store in your community will have a wide variety of hazardous materials on display, including pesticides, insecticides, flammable paints and thinners, cleaning solvents, and stripping agents. The average grocery store will stock large quantities of chlorine bleach, harsh cleansers, and chemicals to remove rust, unclog drains and sanitize surfaces. More than 80,000 chemicals are registered for use in commerce in the United States, and an estimated 2000 new ones are introduced annually Figure 8 . The bulk of the new chemical substances are industrial chemicals, household cleaners, and lawn care products.

In addition, manufacturing processes sometimes generate hazardous waste. Hazardous waste is the material that remains after a process or manufacturing plant has used some of the material and it is no longer pure or has been chemically altered to an unusable form. Hazardous waste can be just as dangerous as pure chemicals. Hazardous waste can be mixtures of several chemicals, resulting in a hybrid substance. It may be difficult to determine how such a substance will react when it is released or comes into contact with other chemicals.

Labeling

Labels, placards, and other markings on buildings, packages, boxes, and containers often enable fire inspectors to identify a chemical. When used correctly, marking systems indicate the presence of hazardous materials, including flammable liquids and gases and provide clues about the substance. This section provides an introduction to various marking systems being used. It does not cover all the intricacies and requirements for every marking system; it will, however, acquaint you with the most common systems.

Department of Transportation System

The U.S. Department of Transportation (DOT) marking system is characterized by a system of labels and placards. The DOT's Emergency Response Guidebook (ERG) is also part of this system.

Labels and Placards

Placards are diamond-shaped indicators (10 3/4" on each side) that must be placed on all four sides of highway transport vehicles, railroad tank cars, and other forms of transportation carrying hazardous materials Figure 9 . Labels are smaller versions (4" diamond-shaped indicators) of placards and are used on the four sides of individual boxes and smaller packages being transported.

Placards and labels are intended to give you a general idea of the hazardous material inside a particular container. A placard may identify the broad hazard class (flammable, poison, corrosive, etc.) that a tank contains, while the label on a box inside a delivery truck relates only to the potential hazard inside that package Figure 10 .

The ERG can be used during the initial phase of a hazardous materials incident. The book organizes chemicals into nine basic hazard classes, or families, the members of which exhibit similar properties. There also is a "Dangerous" placard, which indicates that more than one hazard class is contained in the same load. The DOT system is a broad-spectrum look at chemical hazards and a valuable resource to fire fighters.

The nine DOT chemical families recognized in the ERG are the following:

- DOT Class 1, Explosives
- DOT Class 2, Gases
- DOT Class 3, Flammable combustible liquids
- DOT Class 4, Flammable solids
- DOT Class 5, Oxidizers
- DOT Class 6, Poisons
- DOT Class 7, Radioactive materials
- DOT Class 8, Corrosives
- DOT Class 9, Other regulated materials (ORM)

Other Considerations

The DOT system does not require that all chemical shipments be marked with placards or labels. In most cases, the package or tank must contain a certain amount of hazardous material before a placard is required. For example, the "1000-pound rule" applies to blasting agents, flammable and nonflammable gases, flammable or combustible liquids, flammable solids, air reactive solids, oxidizers and organic peroxides, poison solids, corrosives, and miscellaneous (Class 9) materials. Placards are required for these materials only when the shipment weighs more than 1000 pounds (45.5 kg).

Some chemicals are so hazardous that shipping any amount of them requires the use of labels or placards. These include Class 1.1, 1.2, or 1.3 explosives, poison gases, water-reactive solids, and high-level radioactive substances. A four-digit identification number, the United Nations (UN) or North American (NA) number, may be required on some placards. This number identifies the specific material being shipped; a list of UN and NA numbers is included in the ERG.

Figure 9 A placard.

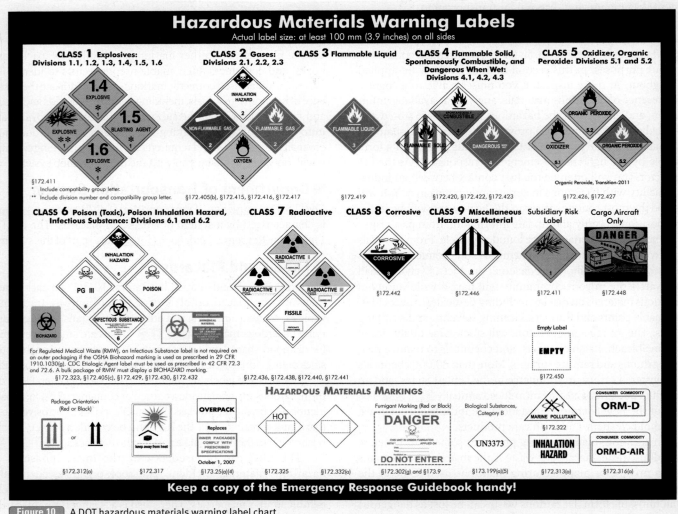

Figure 10 A DOT hazardous materials warning label chart.

Using the ERG

The ERG is divided into four colored sections: yellow, blue, orange, and green.

- **Yellow section**: Chemicals in this section are listed numerically by their four-digit UN or NA identification number. Entry number 1017, for example, identifies chlorine. Use the yellow section when the UN number is known or can be identified. The entries include the name of the chemical and the emergency action guide number.
- **Blue section**: Chemicals in the blue section are listed alphabetically by name. The entry will include the emergency action guide number and the identification number. The same information, organized differently, is in both the blue and yellow sections.
- **Orange section**: This section contains the emergency action guides. Guide numbers are organized by general hazard class and indicate what basic emergency actions should be taken, based on hazard class.
- **Green section**: This section is organized numerically by UN and NA identification number and provides the initial isolation distances for specific materials. Chemicals included in this section are highlighted in the blue or yellow sections. Any materials listed in the green section are always extremely hazardous.

NFPA 704 Marking System

The DOT hazardous materials marking system is used when materials are being transported from one location to another. The National Fire Protection Association (NFPA) NFPA 704 hazard identification system is designed for fixed-facility use. The use of the NFPA 704 marking system is purely voluntary in most jurisdictions, but your state or locally adopted fire and building codes may require its use in some instances. NFPA 704 diamonds are found on the outside of buildings, on doorways to chemical storage areas, and on fixed storage tanks **Figure 11**.

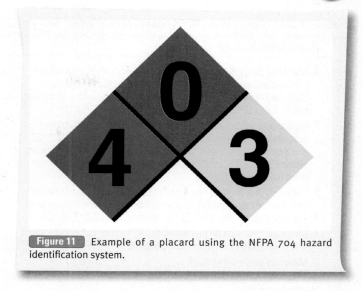

Figure 11 Example of a placard using the NFPA 704 hazard identification system.

The NFPA 704 hazard identification system uses a diamond-shaped symbol of any size, which is itself broken into four smaller diamonds, each representing a particular property or characteristic. The blue diamond at the nine o'clock position indicates the health hazard posed by the material. The top red diamond indicates flammability. The yellow diamond at the three o'clock position indicates reactivity. The bottom white diamond is used for special symbols and handling instructions.

The blue, red, and yellow diamonds will each contain a numerical rating ranging from 0 to 4, with 0 being the least hazardous and 4 being the most hazardous for that type of hazard. The white quadrant will not have a number but may contain special symbols. Among the symbols used are a burning O or the letters OXY (oxidizing capability), a three-bladed fan (radioactivity), and a W with a slash through it (water reactive). **Table 2** provides a description of the numerical ratings in each category.

For more complete information on the NFPA 704 system, consult NFPA 704, *Standard System for the Identification of the Hazards of Materials for Emergency Response*.

Hazardous Materials Information System Marking System

Since 1983, the Hazardous Materials Information System (HMIS) hazard communication program has helped employers comply with the Hazard Communication Standard of the U.S. Occupational Safety and Health Administration (OSHA). The HMIS is similar to the NFPA 704 marking system and uses a numerical hazard rating with similarly colored horizontal columns.

The HMIS is more than just a label; it is a method used by employers to communicate to their personnel the necessary information to work safely around chemicals, and it includes training materials to inform workers of chemical hazards in the workplace. The HMIS is not required by law but is a voluntary system that employers choose to use to comply with OSHA's Hazard Communication Standard. In addition to describing the chemical hazards posed by a particular substance, the HMIS also provides guidance about personal protective equipment employees need to use to protect themselves from workplace hazards. Letters and icons, which are explained in the text, are used to specify the different levels and combinations of protective equipment.

Military Hazardous Materials/WMD Markings

The military has developed its own marking system for hazardous materials and weapons of mass destruction (WMD). The military system has been developed primarily to identify detonation, fire, and special hazards.

In general, hazardous materials within the military marking system are divided into four categories based on the relative detonation and fire hazards. Division 1 materials are considered mass detonation hazards and are identified by a number 1 printed inside an orange octagon. Division 2 materials have explosion-with-fragment hazards and are identified by a number 2 printed inside an orange X. Division 3 materials are mass fire hazards and are identified by a number 3 printed inside an inverted orange triangle. Division 4 materials are moderate fire hazards and are identified by a 4 printed inside an orange diamond.

Chemical hazards in the military system are depicted by colors. Toxic agents, such as sarin or mustard gas, are identified by the color red. Harassing agents, such as tear gas and smoke producers, are identified by yellow. White phosphorous is

Table 2 Hazard Levels in the NFPA Hazard Identification System

Flammability Hazards

4 Materials that will rapidly or completely vaporize at atmospheric pressure and normal ambient temperature, or that are readily dispersed in air and that will burn readily. Liquids with a flashpoint below 73°F (22°C) and a boiling point below 100°F (38°C).

3 Liquids and solids that can be ignited under almost all ambient temperature conditions. Liquids with a flashpoint below 73°F (22°C) and a boiling point above 100°F (38°C) or liquids with a flashpoint above 73°F (22°C) but not exceeding 100°F (38°C) and a boiling point below 100°F (38°C).

2 Materials that must be moderately heated or exposed to relatively high ambient temperatures before ignition can occur. Liquids with flashpoint above 100°F (38°C) but not exceeding 200°F (93°C).

1 Materials that must be preheated before ignition can occur. Liquids that have a flashpoint above 200°F (93°C).

0 Materials that will not burn.

Reactivity Hazards

4 Materials that in themselves are readily capable of detonation or of explosive decomposition or reaction at normal temperatures and pressures.

3 Materials that in themselves are capable of detonation or explosive decomposition or reactions but require a strong initiating source, or that must be heated under confinement before initiation, or that react explosively with water.

2 Materials that readily undergo violent chemical change at elevated temperatures and pressures, or that react violently with water, or that may form explosive mixtures with water.

1 Materials that in themselves are normally stable, but can become unstable at elevated temperatures and pressures.

0 Materials that in themselves are normally stable, even under fire exposure conditions, and are not reactive with water.

Health Hazards

4 Materials that on very short exposure could cause death or major residual injury.

3 Materials that on short exposure could cause serious temporary or residual injury.

2 Materials that on intense or continued, but not chronic, exposure could cause incapacitation or possible residual injury.

1 Materials that on exposure would cause irritation but only minor residual injury.

0 Materials that on exposure under fire conditions would offer no hazard beyond that of ordinary combustible material.

Special Hazards

ACID Acid
ALK   Alkali
COR   Corrosive
OX   Oxidizer
W   Reacts with water

---

identified by white. Specific personal protective gear requirements are identified by pictograms.

Safety Data Sheet (SDS)

A common source of information about a particular chemical is the **safety data sheet (SDS)** specific to that substance **Figure 12**. Essentially, an SDS provides basic information about the chemical makeup of a substance, the potential hazards it presents, appropriate first aid in the event of an exposure, and other pertinent data for safe handling of the material. Generally, an SDS will include:

- Physical and chemical characteristics
- Physical hazards of the material
- Health hazards of the material
- Signs and symptoms of exposure
- Routes of entry
- Permissible exposure limits
- Responsible party contact
- Precautions for safe handling (including hygiene practices, protective measures, and procedures for cleaning up spills or leaks)
- Applicable control measures, including personal protective equipment
- Emergency and first aid procedures
- Appropriate waste disposal

Safety Data Sheets are produced by the chemical product manufacturers and are provided to customers with their orders or upon request. When inspecting a factory or plant, ask the site manager for an SDS for all chemicals on site. All facilities that use or store chemicals are required, by law, to have an SDS on file for each chemical used or stored in the facility.

Shipping papers are required whenever materials are transported from one place to another. They include the names and addresses of the shipper and the receiver, identify the material being shipped, and specify the quantity and weight of each part of the shipment. Shipping papers for road and highway transportation are called **bills of lading** or **freight bills** and are located in the cab of the vehicle. Drivers transporting chemicals are required by law to have a set of shipping papers on their person or within easy reach inside the cab at all times. Shipping papers for rail transport are in the train's consist and are carried by the conductor in the lead locomotive as well as being available through the railroad's dispatch center. The train consist will also indicate the car type and position in the train of the hazardous materials, for instance, the twentieth car being a tank car carrying a certain amount of LP-gas.

A bill of lading may have additional information about a hazardous substance such as its packaging group designation. Packaging group designation is also used by shippers to identify special handling requirements or hazards. Some DOT hazard classes require shippers to assign packaging groups based on the material's flash point and toxicity. A packaging group designation may signal that the material poses a greater hazard than similar

Figure 12 A common source of information about a particular chemical is the SDS specific to that substance.

materials in a hazard class. The three packaging group designations are

- Packaging Group I: high danger.
- Packaging Group II: medium danger.
- Packaging Group III: minor danger.

Where to Find an SDS

SDS are stored in a several different formats depending on the number of chemicals stored within a facility. Facilities with small chemical inventories typically keep the SDS forms in a three-ring binder. In some cases, fire departments require the binder to be stored in a lockbox system located near an entrance. In facilities with a large chemical inventories, the SDS forms may be stored in file cabinets or in multiple binders in a library system format. In some cases, large SDS libraries may be stored electronically on a computer system with a paper backup.

Storage and Handling of Flammable Liquids

Flammable Liquid Tank Storage

Tanks can be installed aboveground, underground, or, under certain conditions, inside buildings Figure 13. Openings and connections to tanks, for venting, gauging, filling, and withdrawing, can present hazards if they are not properly safeguarded. Given substantially constructed, properly installed, and well-maintained tanks, storage of flammable and combustible liquids

A

B

C

Figure 13 Tanks can be installed A. aboveground, B. underground, or C. inside buildings.

should be less dangerous than transferring these liquids because storage does not involve any active steps where human error can lead to spills and where maintaining distance from ignition sources may be less consistently achieveable. The severity of the storage hazard might seem to depend on the quantity stored. As a practical matter, however, the size of the tank or the number of tanks is less important than such factors as the characteristics of the liquid stored, the design of the tank and its foundation and supports, the size and location of vents, and the piping and its connections. Tank installations storing Class I, II, and IIIA flammable and combustible liquids are permitted inside buildings when meeting the requirements of NFPA 30.

Other Storage of Flammable Liquids

Flammable and combustible liquids are packaged, shipped, and stored in bottles, drums, and other containers ranging in size up to 60 gal (225 L). The features of these storage containers are covered in detail later in this chapter. Additionally, liquids are shipped and stored in intermediate bulk containers up to 793 gal (3000 L) and in portable intermodal tanks up to 5500 gal (20,818 L). The features of these transportation containers are covered in detail later in this chapter. Storage requirements for these containers are covered in NFPA 30.

Examples of container types used for the storage of liquids include glass, metal, polyethylene (plastic), and fiberboard. The maximum allowable size for the different types of containers is governed by the class of flammable or combustible liquid to be stored in it **Table 3**.

Container Storage in Buildings

Container storage of flammable and combustible liquids can be found in mercantile and industrial occupancies and in general purpose and flammable liquid warehouses. Any liquid-storage building or room or any portion of a building or room where containers are stored should be designed to protect the containers from exposure fires. This might require the installation of fire walls or partitions or the separation of the containers from other storage arrangements or processes **Figure 14**.

The risk to the occupants of the building involved in the storage, exposure to other buildings, building construction, and the degree of fire protection provided are factors to be considered when evaluating the amount of container storage in buildings. Details and limitations on closed-container storage are given in NFPA 30 and in your state or locally adopted fire and building codes.

Inside Liquid Storage Areas

A room or building used for the storage of flammable and combustible liquids can be considered an inside room, cutoff room, attached building, or liquid warehouse. These differing terms are used to generally describe the scope or quantity involved. An inside room or cutoff room is permitted by NFPA 30 to store only limited quantities while a liquid warehouse is permitted to store unlimited quantities.

Figure 14 Any liquid-storage building or room should be designed to protect the containers from exposure fires. This might require the installation of fire walls or partitions.

Table 3	Maximum Allowable Size—Containers, Intermediate Bulk Containers, and Portable Tanks					
Type		Class IA	Class IB	Class IC	Class II	Class III
Glass		1 pt	1 qt	1.3 gal	1.3 gal	5.3 gal
Metal (other than DOT drums) or approved plastic		1.3 gal	5.3 gal	5.3 gal	5.3 gal	5.3 gal
Safety cans		2.6 gal	5.3 gal	5.3 gal	5.3 gal	5.3 gal
Metal drum (e.g. UN 1A1 / 1A2)		119 gal	119 gal	119 gal	119 gal	119 gal
Approved metal portable tanks and IBCs		793 gal	793 gal	793 gal	793 gal	793 gal
Rigid plastic IBCs (UN 31H1 or 31H2) and composite IBCs (UN31HZ1)		NP	NP	NP	793 gal	793 gal
Polyethylene DOT Specification 34, UN 1H1, or as authorized by DOT exemption		1.3 gal[a]	5.3 gal[a]	5.3 gal[a]	119 gal	119 gal
Fiber drum NMFC or UFC Type 2A; Types 3A, 3B-H, or 3B-L; or Type 4A		NP	NP	NP	119 gal	119 gal

For SI units, 1 pt = 0.473 L ; 1 qt = 0.95 L ; 1 gal = 3.8 L
NP—Not permitted
[a]For Class IB and IC water-miscible liquids, the maximum allowable size of plastic container is 60 gal (227 L), if stored and protected in accordance with Table 4.8(g) of NFPA 30.
Source: NFPA 30: Flammable and Combustible Liquids Code, Table 9.4.3

Flammable Liquid Storage Cabinets

Specially designed storage cabinets are available for storing not more than 120 gal (454 L) of Class I, Class II, and Class IIIA liquids. These cabinets are typically constructed of No. 18 gauge sheet steel consisting of a double wall with 1½″ (38-mm) air space, although wood cabinets constructed in accordance with the requirements of NFPA 30 are permitted. The door should have a three-point latch, with a sill raised to at least 2″ (51 mm) above the bottom of the cabinet to provide a space to capture any small spills or leaks.

The cabinet should be marked in conspicuous lettering: "FLAMMABLE—KEEP FIRE AWAY" **Figure 15**. No more than three such cabinets should be located in a single fire area, except in industrial occupancies where additional cabinets are permitted if separated by 100′ (30.5 m), or where six cabinets are permitted if the building is sprinklered.

■ Handling Methods

One of the safest methods for handling flammable or combustible liquids is to pump the liquid from underground storage tanks, through an adequately designed piping system protected from physical damage, to the dispensing equipment located outdoors or in specially designed inside dispensing rooms. Such rooms should have at least one exterior wall for explosion relief and accessibility for firefighting, interior walls with appropriate fire-resistance ratings, and adequate ventilation and drainage, and be free of sources of ignition.

Where solvents are pumped from storage tanks to the point of use in an industrial building, emergency switches should be located in the dispensing area, at the normal exit door or at other safe locations outside the fire area, and at the pumps to shut down all pumps in case of fire.

Where dispensing by gravity flow, such as in the filling of containers in an industrial operation, a shutoff valve should be installed as closely as practical to the vessel being unloaded. A control valve should be located near the end of the discharge pipe. Additionally, in some filling operations, a heat-actuated valve is desirable to shut off the flow of liquid.

The preferred method of dispensing flammable and combustible liquids from a drum is by use of an approved hand operated pump drawing through the top. An approved self-closing drum faucet can also be used for dispensing from a drum. However, hand-operated pumps are safer than faucets, because the hazard of leakage is reduced.

For handling small quantities of flammable and combustible liquids, safety cans are preferred. Safety cans are substantially constructed to avoid the danger of leakage and are designed to minimize the likelihood of spillage or vapor release and of container rupture under fire conditions. Typical safety cans have pouring outlets with tight-fitting caps or valves normally closed by springs, except when held open by hand, so that contents will not be spilled if a can is tipped over. Safety cans for flammable liquids will also have screens inserted in the pouring outlet to act as flame arresters to avoid ignition of the vapor space inside the safety can. The caps also provide an emergency vent when the cans are exposed to fire **Figure 16**.

Liquids can also be dispensed from the original shipping containers. Open pails or open buckets should never be used for storage. Flammable liquids should always be handled and dispensed in a well-ventilated area free of sources of ignition, and bonding should be provided between the dispensing equipment and the container being filled.

■ Fire Prevention Methods

When flammable and combustible liquids are stored or handled, the liquid is usually exposed to the air at some stage in the operation, except where the storage is confined to sealed containers that are not filled or opened on the premises or where handling is in closed systems and vapor losses are recovered. Even when the storage or handling is in a closed system, there is always the possibility of breaks or leaks, which permit the liquid to escape. Therefore, ventilation is of primary importance to prevent the accumulation of flammable vapors. It is also good practice to eliminate sources of ignition in places where low flashpoint flammable liquids are stored, handled, or used, even though no vapor may ordinarily be present.

Whenever possible in manufacturing processes involving flammable or combustible liquids, equipment, such as compressors, stills, towers, and pumps, should be located in the open. This will lessen the fire potential created by the escape and accumulation of flammable vapors. Gasoline and almost all other flammable liquids produce heavier-than-air vapors, which tend to settle on the floor or in pits or depressions. Such vapors may flow along the floor or ground for long distances, be ignited at some remote point, and flash back. The removal method of such vapors at the floor level (including pits) is usually the proper form of ventilation. Convection currents of heated air or normal vapor diffusion may carry even heavy vapors upward, and, in such

Figure 15 A metal flammable liquid storage cabinet should be marked in conspicuous lettering: "FLAMMABLE—KEEP FIRE AWAY"

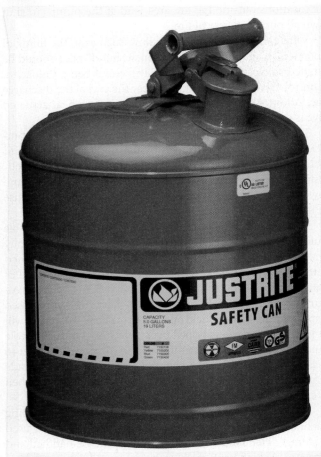

Figure 16 Safety cans for flammable liquids will also have screens inserted in the pouring outlet to act as flame arresters to avoid ignition of the vapor space inside the safety can.

the containers are closed and excessive pressures can develop when they are exposed to rather nominal heat sources and fire (or by overfilling, in the case of liquefied gas containers), overpressure protection is usually needed.

Gas Containers

In North America, there are two types of gas containers: cylinders and tanks. The specific features of these types of storage and transportation containers are covered in detail later in this chapter. Originally, the distinction between cylinders and tanks was based on size. Cylinders, being the smaller, were considered portable; tanks, being the larger, were essentially used in stationary service **Figure 17**. There was also a distinction that reflected the pressures in the containers. Cylinders were thought of as being used for high pressures and tanks for low or moderate

A

B

Figure 17 Originally, the distinction between cylinders and tanks was based on size. **A.** Cylinders, being the smaller, were considered portable; **B.** tanks, being the larger, were essentially used in stationary service.

instances, ceiling ventilation may also be desirable. Ventilation to eliminate flammable vapors may be either natural or artificial.

Although natural ventilation, which depends on temperature and wind conditions, has the advantage of not being dependent on manual starting or on power supply, it is not so easily controlled as is mechanical ventilation. Mechanical ventilation should be used wherever there are extensive indoor operations involving flammable and combustible liquids. In rooms or buildings where possible explosions of flammable vapors may occur, it is recommended that relief through explosion venting be provided for at least Class IA liquids and unstable liquids.

Storage of Gases

A gas must be stored in a container that is gastight for the range of temperature and pressure conditions present at the storage location and the conditions that will be present in the transportation environment. The storage container should contain the source of energy needed to remove the gas from storage and move it to the point of use. This energy is the pressure of the gas in the container. Gas containers are closed pressure vessels containing considerable energy per unit of volume, and they require careful design, manufacture, and maintenance. Furthermore, because pressures. Over the years, these distinctions have lessened so that, today, the only real distinction lies in the regulations or codes under which the container is built.

Practically all gases must be transported from the manufacturer to the user, making the safety of the container in transportation a matter of primary concern. As a result, criteria for many gas containers reflect transportation safety conditions. Because it could be hazardous as well as uneconomical to require gases to be transferred from a shipping container to a separate container for other use, every effort has been made to utilize the same container whenever feasible. This is generally the procedure used for the smaller containers.

Storage Safety Considerations

Fire protection safeguards for gas storage reflect the hazards of the container/gas combination and the hazards of the gas when it escapes from the container.

Container/Gas Hazard Safeguards

The major container/gas hazard is the hazard of a **boiling liquid expanding vapor explosion (BLEVE)**. The BLEVE hazard is restricted to containers of liquefied gases and the major cause of such BLEVEs in storage is fire exposure. BLEVEs resulting from corrosion of a container are far less frequent and impact-caused BLEVEs even less frequent for containers in storage.

A basic BLEVE safeguard is to reduce the chances of fire exposure to the container. This safeguard is also applicable to containers of nonliquefied gases (compressed gases) because, although by definition they are not subject to a BLEVE, they can still fail violently from fire exposure.

To reduce the chance of fire exposure, the quantity of combustibles in the vicinity of gas containers must be limited. This applies whether the storage is indoors or outdoors. Except where small quantities of gas are involved (e.g., one or two cylinders of compressed gas), it is highly desirable that storage rooms or docks be of noncombustible or limited combustible construction. If the building that houses the storage room presents a substantial fire load, the storage room walls should have a suitable fire-resistance rating.

Containers of flammable gases should not be stored with nonflammable oxidizing gases. Oxidizing gases, such as oxygen and nitrous oxide, increase the speed of burning of all flammable gases, resulting in higher flame temperatures. Explosions of flammable gas–oxidizing gas mixtures are also possible.

NFPA 55, *Compressed Gases and Cryogenic Fluids Code*, requires separation of flammable gases and nonflammable, oxidizing gases by means of a 20′ (6100 mm) distance or by a wall with a ½-hr fire-resistance rating.

The next piece in BLEVE prevention is the use of container overpressure limiting devices **Figure 18**. These devices are vital to controlling the BLEVE or compressed gas container failure hazard. Even though these devices cannot by themselves always prevent container failure, they do extend the time until vessel failure in all cases and can prevent failure under many fire exposure conditions. It is essential that the device not be blocked closed by corrosion, paint deposits, and other obstructions, not be damaged mechanically, and not be disabled, plugged or

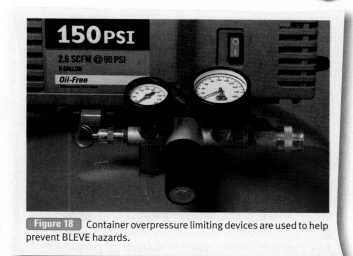

Figure 18 Container overpressure limiting devices are used to help prevent BLEVE hazards.

removed. Portable containers should be checked for this every time they are taken into the facility and whenever they are connected to consuming equipment or are filled.

Care in Handling

Also reflecting the container/gas hazard, it is important that the containers not be subjected to physical abuse. Although quite sturdy as a result of their design as pressure vessels, any dent or gouge in the container can reduce safety factors and, at the least, shorten failure times from fire exposure or lead to impact failures upon subsequent movement. If the valve is sheared off on some smaller portable compressed gas containers, the nozzle reaction from escaping gas can be sufficient to propel the cylinder violently. Where the container is designed for a valve-protecting cap or collar, these always should be in place during storage or movement.

Safeguards for Escaping Gas

The major escaping gas hazard is combustion of flammable gas and is, in turn, manifest as either a fire or a combustion explosion. Fire can also lead to explosive container failures.

Inspection for Leakage

All storage containers should be inspected periodically for leakage from container appurtenances or from the container itself **Figure 19**. Portable containers should not be placed into storage if they are leaking. The senses of sight, sound, and smell are invaluable leak detectors. Although most gases are invisible, some do have color (e.g., chlorine). Liquefied gases escaping as liquids can lead to the formation of a visible cloud of condensed water vapor. Because gases are stored under pressure, a leak can be accompanied by a hissing sound. Although many gases are odorless, some do possess strong odors (e.g., chlorine and anhydrous ammonia).

Natural Gas and LP-Gas

Natural gas and LP-gas usually have an odorant added to them. Natural gas usually has ethyl mercaptan, an odorant derived from a chemical found in the glands of a skunk, while LP-gas has a pungent chemical derived from the oil found in almonds added so that the presence of an otherwise odorless gas can be detected by occupants before a dangerous level is reached.

Figure 19 All storage containers should be inspected periodically for leakage from container appurtenances or from the container itself.

Due to the toxicity of many industrial gases such as chlorine or the ability to displace oxygen such as natural gas, LP-gas, nitrogen, or carbon dioxide, your senses of sight, smell, and sound should not be relied upon to find leaks. Many gas detection instruments are available as either built-in monitoring systems or as portable gas analyzers. Leak detection solutions are also available that, when applied to small leaks, show bubbles.

Ventilation of Spaces

Indoor storage areas should be ventilated, regardless of the chemical hazard of the gas. In areas used solely for storage (no filling of containers), the amount of ventilation need not be great. Storage areas can be ventilated in different ways. Enclosed storage areas that have 25 percent of the available wall area open are considered the same as outdoor storage. Indoor storage areas are required by NFPA 55 to have natural or mechanical ventilation that provides a minimum of 1 cfm/ft2 (0.3 m^3/min m^2) of floor area.

Controlling Ignition Sources

Ignition sources should be controlled in flammable gas storage areas. The vapor density of the gas, in part, determines the extent of the area in which ignition sources should be eliminated or controlled. Many gases are heavier than air at all times. Others will be temporarily heavier than air when they are released in liquid form and vaporize. In general, flammable gas storage areas with no container filling are classified as Division 2 locations for purposes of installing electrical equipment because of the ventilation provided and the nominal leakage potential.

Electrical equipment installed in storage areas used for flammable gases must be selected based on the properties of the gases to be stored. Gases are assigned to a group by testing that determines the gap through which burning gas will propagate. It determines the design, construction, and testing of electrical equipment enclosures for use in classified areas. Most flammable gases are Group C or Group D materials, which allow the largest gaps. Acetylene is a Group A material and hydrogen is a Group B material. Because of the limited availability of electrical equipment for use with Groups A and B materials, electrical equipment for use where Group A or B materials are used might have to be purged and pressurized. For further information on this subject refer to NFPA 70, *National Electrical Code®*, or NFPA 496, *Standard for Purged and Pressurized Enclosures for Electrical Equipment*.

Hazardous Materials Storage and Handling

Safe storage and handling of hazardous materials, including flammable and combustible liquids and gases, requires knowledge of all of the hazardous properties of the material, which can be obtained from the manufacturer's SDS. Safe storage also depends on the quantity, size, and nature of the containers and their storage arrangement. Principles of good storage include segregating the material from other materials in storage, from processing and handling operations, and from incompatible materials; protecting containers from physical damage; using a hazard identification system; and providing fire protection based specifically on the nature of the hazard.

■ Containers

A <u>container</u> is any vessel or receptacle that holds material, including storage vessels, pipelines, and packaging. Often, container type, size, and material provide important clues about the nature of the substance inside. However, do not rely solely on the type of container, however, when making a determination about hazardous materials.

Red phosphorus from a drug lab, for example, might be found in an unmarked plastic jug. Acetone or other solvents may be stored in 55-gallon steel drums with two capped openings on the top. Sulfuric acid, at 97 percent concentration, is typically found in a polyethylene drum, but that drum could be black, red, white, or blue. The same sulfuric acid might also be found in a 1-gallon amber glass container. Hydrofluoric acid, on the other hand, is incompatible with silica (glass) and would be stored in a plastic container.

■ Container Type

Hazardous materials can be found in many different types of containers, ranging from 1-gallon glass containers to 5000-gallon steel storage tanks. Steel or polyethylene plastic drums, bags, high-pressure gas cylinders, railroad tank cars, plastic buckets, above-ground and underground storage tanks, truck tankers, and pipelines are all types of containers used with hazardous materials.

Some very recognizable chemical containers, such as 55-gallon drums and compressed gas cylinders, can be found in almost every commercial building. Materials in cardboard drums are usually in solid form. Stainless steel containers hold particularly dangerous chemicals, and cryogenic liquids are kept in thermos-like **Dewar containers** designed to maintain the appropriate temperature.

Containers may be large or small. Smaller containers may be stored and shipped in larger containers. One way to distinguish containers is to divide them into two separate categories: bulk storage containers and nonbulk storage vessels.

Container Volume

<u>Bulk storage containers</u>, or large-volume containers, are defined by their internal capacity based on the following measures:
- Liquids: more than 119 gallons (451 L)
- Solids: more than 882 pounds (401 kg)
- Gases: more than 882 pounds (401 kg)

Bulk storage containers include fixed tanks, large transportation tankers, totes, and intermodal tanks. In general, bulk storage containers are found in occupancies that rely on and need to store large quantities of a particular chemical. Most manufacturing facilities have at least one bulk storage container. Often, these bulk storage containers are surrounded by a supplementary containment system to help control an accidental release. <u>Secondary containment</u> is an engineered method to control spilled or released product if the main containment vessel fails. A 5000-gallon (18.9 kL) vertical storage tank, for example, may be surrounded by a series of short walls that form a catch basin around the tank. The basin typically can hold the entire volume of the tank and accommodate water from hose lines or sprinkler systems in the event of fire. Many storage vessels, including 55-gallon (208 L) drums, may have secondary containment systems.

Large-volume horizontal tanks are also common. When stored above ground, these tanks are referred to as aboveground storage tanks (ASTs); if they are placed underground, they are known as underground storage tanks, (USTs). These tanks can hold a few hundred gallons to several thousand gallons of product and are usually made of aluminum, steel, or plastic. USTs and ASTs can be pressurized or nonpressurized. Nonpressurized horizontal tanks are usually made of steel or aluminum. Because it is difficult to relieve internal pressure in these tanks, they can be dangerous when exposed to fire and are required to have normal venting to allow air to enter or exit as the tank is normally filled or product is withdrawn and also have emergency venting capabilities when exposed to the high heat of a fire. Typically, they hold flammable or combustible materials such as gasoline, oil, or diesel fuel.

Pressurized horizontal tanks have rounded ends and large vents or pressure-relief stacks. The most common aboveground pressurized tanks are liquid propane and liquid ammonia tanks, which can hold a few hundred gallons to several thousand gallons of product. These tanks also contain a small vapor space; 10 to 15 percent of total capacity can be vapor. Refer to the discussion of liquefied gases in the chapter on the properties and effects of hazardous materials for more information on these materials.

Another common bulk storage vessel is the <u>tote</u>. Totes are portable plastic tanks surrounded by a stainless steel web that adds both structural stability and protection or are constructed of aluminum or stainless steel. They can hold a few hundred gallons of product and may contain any type of chemical, including flammable liquids, corrosives, food-grade liquids, and oxidizers.

Shipping and storing totes can be hazardous. They often are stacked atop one another and are moved with a forklift. A mishap with the loading or moving process can compromise the container. Because totes have no secondary containment, any leak will create a large spill of either a liquid or powder. Additionally, the steel webbing around plastic totes makes leaks difficult to patch. Totes are shipped from the product producer to the user and returned for refilling and reuse. As such, they can be damaged due to the normal wear and tear of shipping and use. Inspection of their condition prior to reuse is necessary.

<u>Intermodal tanks</u> are both shipping and storage vehicles **Table 4**. They hold between 5000 and 6000 gallons (18.9 to 22.7 kL) of product and can be either pressurized or nonpressurized. In most cases, an intermodal tank is shipped to a facility, where it is stored and used, and then it is returned to the shipper for refilling. Intermodal tanks can be shipped by any mode of transportation: air, sea, or land. These horizontal round tanks are surrounded by, or are part of, a boxlike steel framework for shipping. There are three basic types of intermodal (IM or IMO) tanks:
- IM-101 containers have a 6000-gallon (22.7 kL) capacity, with internal working pressures between 25 pounds per square inch (psi) (172 kPa) and 100 psi (690 kPa). These containers typically carry mild corrosives, food-grade products, or flammable liquids.
- IM-102 containers have a 6000-gallon (22.7 kL) capacity, with internal working pressures between 14 psi (96 kPa) and 30 psi (207 kPa). They primarily carry flammable liquids and corrosives.
- IMO Type 5 containers are high-pressure vessels with internal pressures of several hundred psi that carry liquefied gases like propane and butane.

Table 4 Common Bulk Storage Vessels, Locations, and Contents

Tank Shape	Common Locations	Hazardous Materials Commonly Stored
Underground tanks	Residential, commercial	Fuel oil and combustible liquids
Covered floating roof tanks	Bulk terminals and storage	Hightly volatile flammable liquids
Cone roof tanks	Bulk terminal and storage	Combustible liquids
Open floating roof tanks	Bulk terminal and storage	Flammable and combustible liquids
Dome roof tanks	Bulk terminal and storage	Combustible liquids
High-pressure horizontal tanks	Industrial storage and terminal	Flammable gases, chlorine, ammonia
High-pressure spherical tanks	Industrial storage and terminal	Liquid propane gas, liquid nitrogen gas
Cryogenic liquid storage tanks	Industrial and hospital storage	Oxygen, liquid nitrogen gas

Nonbulk Storage Vessels

Nonbulk storage vessels are all other types of containers. Nonbulk storage vessels can hold a few ounces to many gallons and include drums, bags, carboys, compressed gas cylinders, cryogenic containers, and more. Nonbulk storage vessels hold commonly used commercial and industrial chemicals such as solvents, industrial cleaners, and compounds. This section describes the most common nonbulk storage vessels.

Drums

Drums are easily recognizable, barrel-like containers. They are used to store a wide variety of substances, including food-grade materials, corrosives, flammable liquids, and grease. Drums may be constructed of low-carbon steel, polyethylene, cardboard, stainless steel, nickel, or other hybrid materials Figure 20. Generally, the nature of the material dictates the construction of the storage drum. Steel utility drums, for example, hold flammable liquids, cleaning fluids, oil, and other noncorrosive chemicals. Polyethylene drums are used for corrosives such as acids, bases, oxidizers, and other materials that cannot be stored in steel containers. Cardboard drums hold solid materials such as soap flakes, sodium hydroxide pellets, and food-grade materials. Stainless steel or other heavy-duty drums generally hold materials too aggressive for either plain steel or polyethylene.

Closed-head drums have a permanently attached lid with one or more small openings called bungs. Typically, these openings are threaded holes sealed by caps that can be removed only by using a special wrench called a bung wrench. Closed-head drums usually have one 2-inch bung and one 3/4-inch bung. The larger hole is used to pump product from the drum, while the smaller bung functions as a vent.

An open-head drum has a removable lid fastened to the drum with a ring. The ring is tightened with a clasp or a threaded nut-and-bolt assembly. The lid of an open-head drum may or may not have bung-type openings.

Bags

Bags are commonly used to store solids and powders such as cement powder, sand, pesticides, soda ash, and slaked lime. Storage bags may be constructed of plastic, paper, cloth (burlap) or plastic-lined paper. Bags come in different sizes and weights, depending on their contents.

Pesticide bags must be labeled with specific information Figure 21. Fire fighters can learn a great deal from the label, including

- Name of the product.
- Statement of ingredients.
- Total amount of product in the container.
- Manufacturer's name and address.
- U.S. Environmental Protection Agency (EPA) registration number, which provides proof that the product was registered with the EPA.
- The EPA establishment number, which shows where the product was manufactured.
- Signal words to indicate the relative toxicity of the material:
  - Danger: Poison: Highly toxic by all routes of entry.
  - Danger: Severe eye damage or skin irritation.
  - Warning: Moderately toxic.
  - Caution: Minor toxicity and minor eye damage or skin irritation.
- Practical first aid treatment description.
- Directions for use.
- Agricultural use requirements.
- Precautionary statements such as mixing directions or potential environmental hazards.
- Storage and disposal information.
- Classification statement on who may use the product.

In addition, every pesticide label must have the statement, "Keep out of reach of children."

Carboys

Some corrosives and other types of chemicals are transported and stored in vessels called carboys Figure 22. A carboy is a glass, plastic, or steel container that holds 5 to 15 gallons of product. Glass carboys often have a protective wood or fiberglass box to help prevent breakage.

Nitric, sulfuric, and other strong acids are transported and stored in thick glass carboys protected by a wooden or Styrofoam

Figure 20 Drums may be constructed of cardboard, polyethylene, or stainless steel.

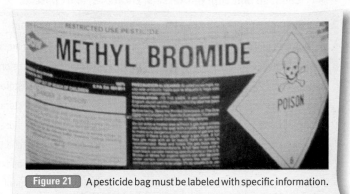

Figure 21 A pesticide bag must be labeled with specific information.

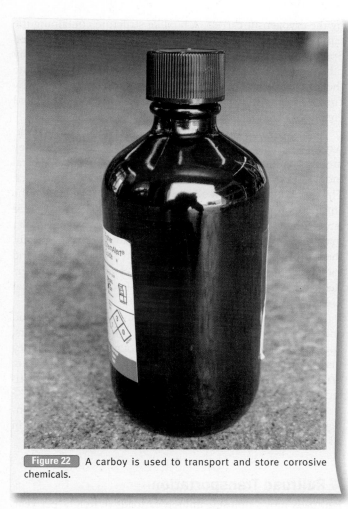

Figure 22 A carboy is used to transport and store corrosive chemicals.

Figure 23 A small cryogenic Dewar.

crate to shield the glass container from damage during normal shipping.

Cylinders

As discussed previously, **cylinders** are used to store combustible and flammable liquids and gases. Uninsulated compressed gas cylinders are used to store gases such as nitrogen, argon, helium, and oxygen. They have a range of sizes and pressure readings. A typical oxygen cylinder used for medical purposes, for example, has a gas pressure reading of about 2000 psi. Very large compressed gas cylinders found at a fixed facility may have pressure readings of 5000 psi or greater.

The high pressures exerted by these cylinders are potentially dangerous. If the cylinder is punctured or the valve assembly fails, the rapid release of compressed gas will turn the cylinder into a deadly missile. Additionally, if the cylinder is heated rapidly, it could explode with tremendous force, spewing product and metal fragments for long distances. Compressed gas cylinders do have pressure-relief valves, but those valves may not be sufficient to relieve the pressure created during a fast-growing fire. The result would be a catastrophic explosion.

The low-pressure Dewar is another common cylinder type. Dewars are thermos-like vessels designed to hold cryogenic liquids (cryogens), gaseous substances that have been chilled until they liquefy Figure 23 . Typical cryogens are oxygen, helium, hydrogen, argon, and nitrogen. Under normal conditions, each of these substances is a gas. A complex process turns them into liquids that can be stored in Dewar containers. Nitrogen, for example, becomes a liquid at −320°F (−160°C) and must be kept that cold to remain a liquid.

Cryogens pose a substantial threat if the Dewar fails to maintain the low temperature. Cryogens have large expansion ratios, even larger than the expansion ratio of propane. Cryogenic helium, for example, has an expansion ratio of approximately 750 to 1. If one volume of liquid helium is warmed to room temperature and vaporized in a totally enclosed container, it can generate a pressure of more than 14,500 psi (99,973 kPa). Therefore, cryogenic containers usually have two pressure-relief devices: a pressure-relief valve and a frangible (easily broken) disk.

■ Hazardous Material Storage Lockers

Hazardous material storage lockers have become very popular in recent years, as they meet the environmental concerns dictated by the special handling of hazardous chemicals, some of which may also be flammable and combustible liquids. These lockers are movable, modular, prefabricated storage buildings that provide a safe and cost-effective means of storing hazardous materials. The lockers are equipped with spill containment features and can be provided with electricity, mechanical ventilation, and may be pre-piped for fire suppression systems. Generally, the lockers are located outside, but NFPA 30 permits the lockers to be utilized as inside liquid storage rooms, as long as the lockers are constructed with the fire resistive ratings and applicable requirements listed in NFPA 30. NFPA 30 limits the gross floor area of the lockers to 1500 sq ft (140 m^2) Figure 24 .

Figure 24 A hazardous materials storage locker.

Transporting Hazardous Materials

Although rail, air, and sea transport is used to deliver chemicals and other hazardous materials to their destinations, the largest volume is transported over land, by highway transportation vehicles. Even when another method of transport is used, vehicles often carry the shipments from the station, airport, or dock to the factory or plant.

One of the most common chemical tankers is the flammable liquid tanker, also known as the MC-306 tanker Figure 25A. It typically carries gasoline or other flammable and combustible materials. The oval-shaped tank is pulled by a diesel tractor and can carry between 6000 and 10,000 gallons (22.7 to 37.9 kL). The MC-306 is nonpressurized, usually made of aluminum, and offloaded through valves at the bottom of the tank. It is a common highway sight and a reliable way to transport chemicals.

A similar vehicle is the MC-307 chemical hauler, which has a round or horseshoe-shaped tank and typically carries 6000 to 7000 gallons (22.7 to 26.5 kL) Figure 25B. The MC-307 is used to transport flammable liquids, mild corrosives, and poisons. Tanks that transport corrosives may have a rubber lining.

The MC-312 corrosives tanker is used for concentrated sulfuric and nitric acids and other corrosive substances Figure 25C. Due to the much heavier weight per volume of concentrated acids versus other liquids transported by tankers, tankers used for this purpose have smaller diameters than either the MC-306 or the MC-307 and may be, but not always, characterized by several reinforcing rings around the tank. These rings provide structural stability during transportation and in the event of a rollover. The inside of an MC-312 tanker operates at approximately 75 psi and holds approximately 6000 gallons (22.7 kL).

The MC-331 pressure cargo tanker carries materials like ammonia, propane, and butane Figure 25D. The tank has rounded ends, typical of a pressurized vessel, and is commonly constructed of steel with a single tank compartment. The MC-331 operates at approximately 300 psi, and could be a significant explosion hazard if it accidentally rolls over or is threatened by fire.

The MC-338 cryogenic tanker operates much like the Dewar containers described earlier and carries many of the same substances Figure 25E. This low-pressure tanker relies on tank insulation to maintain the low temperatures required by the cryogens it carries. A boxlike structure containing the tank control valves is typically attached to the rear of the tanker. Special training is required to operate valves on this and any other tanker. An untrained individual who attempts to operate the valves may disrupt the normal operation of the tank, thereby compromising its ability to keep the liquefied gas cold and creating a potential explosion hazard.

Tube trailers carry compressed gases such as hydrogen, oxygen, helium, and methane Figure 25F. Essentially, they are high-volume transportation vehicles comprising several individual cylinders banded together and affixed to a trailer. The individual cylinders on the tube trailer are much like the smaller compressed gas cylinders previously discussed. These large-volume cylinders operate at 3000 to 5000 psi; one trailer may carry several different gases in individual tubes. Typically, there is a valve control box toward the rear of the trailer, and each individual cylinder has its own relief valve. These trailers frequently can be seen at construction sites or at facilities that use great quantities of these materials.

Dry bulk cargo tanks also are commonly seen on the road and carry dry bulk goods such as powders, pellets, fertilizers, and grain Figure 25G. These tanks are not pressurized, but they may use pressure to offload product. Dry bulk cargo tanks are generally V-shaped and have rounded sides that funnel the contents to the bottom-mounted valves.

Intermodal tanks can be laid out in a wide variety of configurations similar to those mentioned above and are found in all modes of transportation Figure 25H.

Railroad Transportation

Railroads move almost 2 million carloads of freight per year, with relatively few hazardous materials incidents. Railway tank cars carry volumes up to 30,000 gallons (113.7 kL) and have the potential to create large leaks or vapor clouds. Hazardous materials incidents involving railroad transportation, although relatively rare, can be extremely dangerous.

Fortunately, there are only three basic railcar configurations that fire inspectors should recognize: nonpressurized, pressurized, and special use. Each has a distinct profile that can be recognized from a long distance. Additionally, railcars are usually labeled on both sides with the volume and maximum working pressure inside the tank. Dedicated haulers often have the chemical name clearly visible.

Nonpressurized (general service) railcars typically carry general industrial chemicals, consumer products such as corn syrup, flammable and combustible liquids, and mild corrosives. Nonpressurized railcars are easily identified by looking at the top of the car: nonpressurized railcars will have visible valves and piping without a dome cover Figure 26.

Pressurized railcars will have an enclosed dome on the top of the railcar. These cars transport materials such as propane, ammonia, ethylene oxide, and chlorine. Pressurized cars have internal working pressures ranging from 100 to 500 psi and are equipped with relief valves, similar to those on bulk storage tankers. Unfortunately, the high volumes carried in these cars can generate long-duration, high-pressure leaks that may be impossible to stop Figure 27.

Figure 25 **A.** An MC-306 flammable liquid tanker. **B.** An MC-307 chemical hauler. **C.** An MC-312 corrosives tanker. **D.** An MC-331 pressure-cargo tanker. **E.** An MC-338 cryogenic tanker. **F.** A tube trailer. **G.** A dry bulk cargo tank. **H.** Intermodal tanks.

Figure 26 Nonpressurized rail tank car.

Figure 28 Special-use railcars can carry hazardous materials.

Figure 27 Pressurized rail tank car.

Figure 29 Information about the pipe's contents and owner is also often found.

Special-use railcars include boxcars, flat cars, cryogenic and corrosive tank cars, and high-pressure compressed gas tube cars. In each case, the hazard will be unique to the particular railcar and its contents. Do not assume that only the chemical tank cars pose a threat; until you know what is in a particular car, assume a hazardous situation may exist in any rail car found at any facility **Figure 28**.

Regardless of the mode or container type used to deliver hazardous materials to a property, as the fire inspector, you must be familiar with each method. The unloading or transloading of materials upon delivery presents a potential for chemical release, potential fire or explosion, or health hazards. The unloading area must be thoroughly evaluated to determine if there may be a hazardous atmosphere requiring hazardous location electrical equipment, if suppression of a fire or the cooling of the transporting container is necessary to avoid a breach, or if the tanker, trailer, rail car, or other vehicle is being used in place of a permanent storage tank or warehouse facility.

Pipelines

Of the various methods used to transport hazardous materials, the high-volume <u>pipeline</u> is rarely involved in emergencies. Fire inspectors visiting pipeline facilities, such as pump stations or compressor stations along the pipeline, terminal facilities where product enters or leaves the pipeline system, need to be aware of the operations and hazards presented by these facilities.

In many areas, large-diameter pipelines transport natural gas, crude oil, refined petroleum products such as gasoline, diesel fuel, or jet fuel and other products from delivery terminals to distribution facilities. In regions with large chemical and petrochemical plants, pipelines may also carry a variety of intermediate chemicals used in the production of finished products such as plastic resins and refined petroleum products. Pipelines are often buried underground, but they may be above ground in remote areas. The <u>pipeline right-of-way</u> is an area, patch, or roadway that extends a certain number of feet on either side of the pipe itself. This area is maintained by the company that owns the pipeline. The company is also responsible for placing warning signs at regular intervals along the length of the pipeline.

Pipeline warning signs include a warning symbol, the pipeline owner's name, and an emergency contact phone number **Figure 29**. Pipeline emergencies are complex events that require specially trained responders. You can assist emergency response personnel in your jurisdiction by documenting and forwarding emergency contact, facility site plans, emergency shutdown, and facility access information to the fire department. Prelocated caches of response equipment, such as foam concentrate, spare SCBA air bottles, and fire hose can be established in cooperation with the pipeline owner at key locations in their facilities. You can be a vital link in communications between the pipeline owner and the fire fighters responding to an incident.

Information about the pipe's contents and owner is also often found at regular intervals along a pipeline's right-of-way and at the vent pipes where the pipeline passes under roadways, railroads, or water courses. These inverted J-shaped tubes provide natural venting of the pipe sleeve placed around the pipeline as it passes under roads, rails or streams. **Vent pipes** are clearly marked and stand approximately 3 feet above the ground or higher in areas with high snowfall accumulations Figure 30 .

Fire Protection Systems

Fire and explosion prevention measures are based on one or more of the following techniques or principles:

- Exclusion of sources of ignition
- Exclusion of air (oxygen)
- Storage of liquids in closed containers or systems
- Ventilation to prevent the accumulation of vapor within the flammable range
- Use of an atmosphere of inert gas instead of air

Extinguishing methods for flammable and combustible liquid fires involve shutting off the fuel supply, excluding air by various means, cooling the liquid to stop evaporation, or a combination thereof.

Although many flammable and combustible liquids can be classed as normal or stable liquids, others introduce the problem of instability or reactivity. The storage, handling, and use of unstable (reactive) flammable or combustible liquids require special attention. It may be necessary to increase the distances to property lines from storage tanks and between adjacent tanks or to provide extra fire protection. For example, it would be poor practice to locate heat-reactive and water-reactive flammable or combustible liquids tanks adjacent to each other. In the event of a ground fire, water applied to the heat-reactive tank for protection might penetrate the tank of water-reactive liquid and cause a violent reaction.

Fire Protection Systems and Equipment for Flammable Liquids

The design of fire protection systems for flammable liquids is dependant on how the liquid is being used or stored. Either cooling, suppression of vapors, or depriving the fire of oxygen may be used to extinguish most flammable liquid fires. Automatic sprinkler protection may be effective in areas where small to moderate quantities of flammable liquids are in use by cooling and diluting the flammable liquid to the point that the fire is extinguished.

Local application carbon dioxide systems may extinguish a fire in a localized manufacturing or process area by reducing the availability of oxygen to the fire, such as in a paint spray booth or on a printing press. In larger use applications or in storage warehouses, additives to the sprinkler system, such as foam concentrate, may be employed to provide a barrier over the surface of a potential spill fire. High- and low-expansion foam systems may also be employed to control a fire in a very large flammable liquid pool fire to suppress vapors thus avoiding their ignition and some cooling effect from the water component of the foam discharge, such as that presented by an aircraft repair hangar.

Fire Protection Systems and Equipment for Flammable Gases

The protection of occupancies using flammable gases relies on one or a combination of factors:

- Cooling—Water spray systems may used alone or in combination with other detection and controls to provide a cooling spray or stream of water allowing flammable gas fires to consume their fuel supply without causing damage to surrounding steel support structures, tanks or other process equipment, motors, or fire barriers Figure 31 .
- Removal of the fuel source—In process piping systems handling flammable gases, automatic or manual shutoff of the gas supply will remove the fuel from the fire. Fire detection systems that sense heat, products of combustion or radiated emissions, such as infrared or ultraviolet light, may be used to monitor the process equipment and arranged to shutoff the supply of flammable gases upon detection of a fire.
- Removal of oxygen from the fire—Total flooding carbon dioxide systems may be used in normally unoccupied closed spaces to reduce the oxygen concentration available to a flammable gas fire Figure 32 .

Combustion Explosion Safeguards

Basic combustion explosion prevention safeguards are designed to limit flammable gas–air mixture accumulation in a structure. Fundamental safeguards are the use of rugged

Figure 30 Vent pipes must be clearly marked and stand approximately 3 feet above the ground or higher in areas with high snowfall accumulations.

Figure 31 Water spray systems may used alone or in combination with other detection and controls to provide a cooling spray or stream of water allowing flammable gas fires to consume their fuel supply without causing damage to surrounding steel support structures, tanks or other process equipment, motors, or fire barriers.

Figure 32 Total flooding carbon dioxide systems may be used in normally unoccupied closed spaces to reduce the oxygen concentration available to a flammable gas fire.

Ignition source control is also fundamental to combustion explosion prevention. This safeguard is limited mainly to industrial operations where the conduct of employees can be controlled. It is impossible to prevent smoking, use of candles, and so on in residential occupancies.

In addition to combustion explosion prevention safeguards, the severity of the explosion can be reduced by special structural design whereby some elements of the structure are designed to dislocate at lower pressures, and other elements are designed to stay in place at the lower pressures that result. This is known as "deflagration venting" and is covered in some detail in NFPA 68, *Standard on Explosion Protection by Deflagration Venting*. Such a building design is not practical for most ordinary buildings, and this practice is applied to industrial structures.

Fire Protection Systems for Hazardous Materials

Fire protection systems for hazardous materials will vary greatly depending on the chemical composition of the hazard. Differing hazardous materials may have very different reactions to fire extinguishing agents. Highly reactive metal hydrides will react violently with water, for example.

Containment of hazardous materials will be a concern in the design of any fire extinguishing system. The runoff caused by the leak from a tank, container, or piping system may be aggravated and multiplied many times by the application of large hose streams to a fire. Impounding contaminated runoff from a facility using or warehousing hazardous materials should be considered in the planning stage of any project.

Foam-water automatic sprinkler systems and high- and low-expansion foam systems, if they are chemically compatible with the hazardous material can provide not only effective fire suppression capability, but also can suppress potentially hazardous or toxic vapors.

Gaseous agents, both local application and total flooding systems, may be employed to protect hazardous materials. A wide range of gaseous agents are available, including carbon dioxide, halocarbon clean agents and inert clean agents. Each agent will have its advantages and its drawbacks. NFPA 2001, *Standard on Fire Clean Agent Extinguishing Systems*, should be consulted for design requirements.

Due to the complexities involved in the design of a fire protection system for hazardous materials, a fire protection engineer should be consulted to review the hazards present and the performance of any proposed design.

Ensuring Code Requirements

Fire prevention codes generally allow nominal amounts of flammable and combustible liquids and gases and other hazardous materials to be stored and used in "non high hazard" occupancies for normal maintenance or daily operations. These materials may also be on display in a wholesale or retail sales setting. Codes limit the amount of hazardous materials, including flammable and combustible liquids and gases, that are allowed to be stored, used, handled, dispensed or displayed to specific maximum allowable quantities. The codes

containers and equipment that minimize the chance of leakage, the minimal quantity of gas released by emergency flow control devices, and the use of limiting orifices to minimize the quantity released.

Burners are often equipped with flame failure devices that shut off the flow of gas if the flame is extinguished for any reason.

also prescribe the safeguards required in the immediate area within which the hazardous materials are to be stored or used. A "control area" concept is used to allow quantities of materials presenting a physical or health hazard to be separated by either fire-resistive construction or larger distances between one maximum allowable quantity grouping and another. Quantities in excess of these amounts within the same control area require increasing levels of protection built into the facility, including a higher fire-resistive construction classification, the provision of automatic sprinkler protection, the increased fire-resistive separation of other occupancies, reduced exit distances, detection and notification systems, secondary containment and runoff provisions among other protection requirements. Exceeding the maximum allowable quantity in the codes also may require the issuance of a specific permit by the Authority Having Jurisdiction.

NFPA 1, *Fire Code* contains a series of tables indicating the maximum allowable quantities of hazardous materials presenting a physical or health hazard allowed within a control area for a specific occupancy classification.

NFPA also publishes the Building and Fire Code Classifications of Hazardous Materials. The *Fire Code* requires that the degree of hazard of the hazardous material control area be marked using the NFPA 704, *Standard System for the Identification of the Hazards of Materials for Emergency Response*, that Safety Data Sheets (SDS) be available for each hazardous material on premises and that possible ignition sources be controlled, such as no open flames and no smoking in the control area.

Mercantile, hazardous materials storage centers, and industrial occupancies in general are allowed to have greater amounts of flammable and combustible liquids and gases and hazardous materials but in turn are required to provide a higher level of fire protection systems such as fire walls, fire doors, ventilation systems, flammable liquid storage rooms, explosion proof electrical equipment and automatic extinguishing systems.

As the complexity of the hazards of these occupancies increases, especially for industrial, hazardous materials, and storage facilities, you may need the help of a fire protection engineer to determine the extent of the hazards and the protection required.

Wrap-Up

Chief Concepts

- The classification system for flammable and combustible liquids is found in NFPA 30, *Flammable & Combustible Liquids Code*, and is based on the division of flammable liquids into three main categories: Class I liquids, Class II liquids, and Class III liquids.
- Class I liquids include all flammable liquids, including acetone, alcohols, ethanol, gasoline, and toluene.
- Class II and Class III liquids include all combustible liquids.
- Gases are classified by five categories:
  - Toxic
  - Pyrophoric
  - Oxidizing
  - Flammable
  - Nonflammable
- Toxic gases, which are poisonous or irritating when inhaled or contacted, include chlorine, hydrogen sulfide, sulfur dioxide, ammonia, carbon monoxide, and arsine.
- Pyrophoric gases are flammable gases that spontaneously ignite in air. Examples of phrophoric gases include silane and phosphine
- Oxidizing gases support combustion. They are generally either oxygen, chlorine, or mixtures of oxygen and other gases, such as oxygen-helium or oxygen-nitrogen mixtures, or certain gaseous oxides, such as nitrous oxide.
- Every flammable gas has a lower and upper flammable limit. Examples of flammable gases include hydrogen, liquefied-petroleum gas, and methane.
- The physical properties of flammable and combustible gases and liquids are a primary fire protection concern because they affect the physical behavior of gases while they are inside containers and after any accidental release from containers. Until they are used, gases must be completely confined in containers, including during transportation, transfer, and storage.
- As a fire inspector, you must be able to recognize the presence of hazardous materials, analyze the available information, and ensure that the hazardous materials are stored and handled safely. You must also provide for a method to share the information on the physical or toxic hazards present in a form useable by fire fighters and emergency responders.
- Labels, placards, and other markings on buildings, packages, boxes, and containers often enable fire inspectors to identify a chemical. When used correctly, marking systems indicate the presence of hazardous materials, including flammable liquids and gases and provide clues about the substance.
- Tanks can be installed aboveground, underground, or, under certain conditions, inside buildings. Openings and connections to tanks, for venting, gauging, filling, and withdrawing, can present hazards if they are not properly safeguarded.
- A gas must be stored in a container that is gastight for the range of temperature and pressure conditions present at the storage location and the conditions that will be present in the transportation environment.
- Safe storage and handling of hazardous materials, including flammable and combustible liquids and gases, requires knowledge of all of the hazardous properties of the material, which can be obtained from the manufacturer's SDS.

Wrap-Up

Hot Terms

Bills of lading Shipping papers for roads and highways.

Boiling liquid expanding vapor explosion (BLEVE) An explosion that occurs when a tank containing a volatile liquid is heated.

Bulk storage containers Large-volume containers that have an internal volume greater than 119 gallons (451 L) for liquids and greater than 882 pounds (401 kg) for solids and a capacity of greater than 882 pounds (401 kg) for gases.

Class IA liquids Those liquids that have flashpoints below 73°F (22.8°C) and boiling points below 100°F (37.8°C).

Class IB liquids Those liquids that have flashpoints below 73°F (22.8°C) and boiling points at or above 100°F (37.8°C).

Class IC liquids Those liquids that have flashpoints at or above 73°F (22.8°C) and boiling points but below 100°F (37.8°C).

Class II liquids Those liquids that have flashpoints at or above 100°F (37.8°C) and below 140°F (60°C).

Class IIIA liquids Those liquids that have flashpoints at or above 140°F (60°C) and below 200°F (93.4°C).

Class IIIB liquids Those liquids that have flashpoints at or above 200°F (93.4°C).

Combustible liquids Any liquid that has a closed-cup flash point at or above 100°F (37.8°C). {NFPA 306}

Compressed gas Any material or mixture having, when in its container, an absolute pressure exceeding 40 psia (an absolute pressure of 276 kPa) at 70°F (21.1°C) or, regardless of the pressure at 70°F (21.1°C), having an absolute pressure exceeding 104 psia (an absolute pressure of 717 kPa) at 130°F (54.4°C). (NFPA 58)

Container Any vessel or receptacle that holds material, including storage vessels, pipelines, and packaging.

Corrosive gas A gas that can cause burns as well as destroy or cause irreversible harm to organic tissue and metals.

Cryogenic gas A refrigerated liquid gas having a boiling point below −130°F (−90°C) at atmospheric pressure. (NFPA 1992)

Cylinders A portable compressed-gas container.

Department of Transportation (DOT) marking system A unique system of labels and placards that, in combination with the North American Emergency Response Guide, offers guidance for first responders operating at a hazardous materials incident.

Dewar containers Containers designed to preserve the temperature of the cold liquid held inside.

Drums Barrel-like containers built to DOT Specification 5P (1A1).

Dry bulk cargo tanks Tanks designed to carry dry bulk goods such as powders, pellets, fertilizers, or grain; they are generally V-shaped with rounded sides that funnel toward the bottom.

Emergency Response Guidebook (ERG) The guidebook developed by the Department of Transportation to provide guidance for first responders operating at a hazardous materials incident in coordination with DOT's labels and placards marking system.

Flammable gas Any substance that exists in the gaseous state at normal atmospheric temperature and pressure and is capable of being ignited and burned when mixed with the proper proportions of air, oxygen, or other oxidizers. (NFPA 99)

Freight bills Shipping papers for roads and highways.

Fuel gases Any gas used as a fuel source, including natural gas, manufactured gas, sludge gas, liquefied petroleum gas–air mixtures, liquefied petroleum gas in the vapor phase, and mixtures of these gases. See NFPA 54, National Fuel Gas Code. (NFPA 97)

Flammable liquids Flammable liquids shall be or shall include any liquids having a flash point below 100°F (37.8°C) and having a vapor pressure not exceeding 40 psi (276 kPa) (absolute) at 100°F (37.8°C). Flammable liquids shall be subdivided as follows: (a) Class I liquids shall include those having flash points below 100°F (37.8°C) and shall be subdivided as follows: 1. Class IA liquids shall include those having flash points below 73°F (22.8°C) and having a boiling point below 100°F (37.8°C). 2. Class IB liquids shall include those having flash points below 73°F (22.8°C) and having a boiling point above 100°F (37.8°C). 3. Class IC liquids shall include those having flash points at or above 73°F (22.8°C) and below 100°F (37.8°C). Combustible liquids shall be or shall include any liquids having a flash point at or above 100°F (37.8°C). They shall be subdivided as follows: (a) Class II liquids shall include those having flash points at or above 100°F (37.8°C) and below 140°F (60°C). (b) Class IIIA liq-

Wrap-Up

uids shall include those having flash points at or above 140°F (60°C) and below 200°F (93.3°C). (NFPA 11)

Hazardous material Any materials or substances that pose an unreasonable risk of damage or injury to persons, property, or the environment if not properly controlled during handling, storage, manufacture, processing, packaging, use and disposal, or transportation.

Hazardous Materials Information System (HMIS) A color-coded marking system by which employers give their personnel the necessary information to work safely around chemicals.

Industrial gases The entire range of gases classified by chemical properties customarily used in industrial processes, for welding and cutting, heat treating, chemical processing, refrigeration, and water treatment.

Intermodal tanks Bulk containers that can be shipped by all modes of transportation—air, sea, or land.

Liquefied gas A gas, other than in solution, that in a packaging under the charged pressure exists both as a liquid and a gas at a temperature of 20°C (68°F). (NFPA 30)

MC-306 flammable liquid tanker Commonly known as a gasoline tanker, this tanker typically carries gasoline or other flammable and combustible materials.

MC-307 chemical hauler A tanker with a rounded or horseshoe-shaped tank.

MC-312 corrosives tanker A tanker that will often carry aggressive acids like concentrated sulfuric and nitric acid, having reinforcing rings along the side of the tank.

MC-331 pressure cargo tanker A tank commonly constructed of steel with rounded ends and a single open compartment inside; there are no baffles or other separations inside the tank.

MC-338 cryogenic tanker A low-pressure tanker designed to maintain the low temperature required by the cryogens it carries.

Medical gases A patient medical gas or medical support gas. (NFPA 99)

NFPA 704 hazard identification system A hazardous materials marking system designed for fixed-facility use.

Nonbulk storage vessels Containers other than bulk storage containers.

Oxidizing gases Gases that support combustion; generally either oxygen, chlorine, or mixtures of oxygen and other gases, such as oxygen-helium or oxygen-nitrogen mixtures, or certain gaseous oxides, such as nitrous oxide.

Pipeline A length of pipe, including pumps, valves, flanges, control devices, strainers, and similar equipment, for conveying fluids and gases.

Pipeline right-of-way An area, patch, or roadway that extends a certain number of feet on either side of the pipe itself and that may contain warning and informational signs about hazardous materials carried in the pipeline.

Pyrophoric gases Flammable gases that spontaneously ignite in air; examples of phrophoric gases include silane and phosphine.

Safety Data Sheet (SDS) A form, provided by manufacturers and compounders (blenders) of chemicals, containing information about chemical composition, physical and chemical properties, health and safety hazards, emergency response, and waste disposal of the material.

Secondary containment Any device or structure that prevents environmental contamination when the primary container or its appurtenances fail. Examples of secondary containment are dikes, curbing, and double-walled tanks.

Shipping papers A shipping order, bill of lading, manifest, or other shipping document serving a similar purpose and usually including the names and addresses of both the shipper and the receiver as well as a list of shipped materials with quantity and weight.

Signal words Information on a pesticide label that indicates the relative toxicity of the material.

Tote Portable tanks, usually holding a few hundred gallons of product, characterized by a unique style of construction.

Tube trailers High-volume transportation devices made up of several individual compressed gas cylinders banded together and affixed to a trailer.

Unstable reactive gas A gas that can undergo violent changes when subjected to shock or changes in temperature or pressure.

Vent pipes Inverted J-shaped tubes that allow for pressure relief or natural venting of the pipeline for maintenance and repairs.

Fire Inspector *in Action*

As a fire inspector, you are preparing to inspect a number of small manufacturing plants located in an industrial park. Knowing that most of these facilities will have one or more Special Hazards in the form of flammable or combustible liquids, gases or other hazardous materials you prep for your inspections by reviewing your Special Hazards knowledge.

1. The most commonly used means to determine the relative hazard of flammable and combustible liquids is:
   A. boiling point
   B. autoignition temperature
   C. flashpoint
   D. vapor pressure

2. The flashpoint of a liquid is the _____ temperature at which the _____ pressure of the liquid will produce an ignitable mixture and resultant flame
   A. lowest, total
   B. highest, vapor
   C. average, nominal
   D. lowest, vapor

3. Liquids with flashpoints above 100°F (37.8°C) are referred to as:
   A. combustible liquids
   B. flammable liquids
   C. class 1A liquids
   D. class 1B liquids

4. Fire protection systems for flammable gases rely on:
   A. cooling
   B. removal of the fuel source
   C. removal of oxygen from the fire
   D. all of the above

Safe Housekeeping Practices

CHAPTER 13

NFPA 1031 Standard

Fire Inspector I

4.3.8 Recognize hazardous conditions involving equipment, processes, and operations, given field observations, so that the equipment, processes, or operations are conducted and maintained in accordance with applicable codes and standards and deficiencies are identified, documented, and reported in accordance with the applicable codes and standards and the policies of the jurisdiction. (pp 274–283)

(A) Requisite Knowledge. Practices and techniques of code compliance inspections, fire behavior, fire prevention practices, ignition sources, safe housekeeping practices, and classification of hazardous materials. (pp 274–283)

(B) Requisite Skills. The ability to observe, communicate, apply codes and standards, recognize problems, and make decisions. (pp 274–283)

Fire Inspector II

5.3.6 Evaluate hazardous conditions involving equipment, processes, and operations, given field observations and documentation, so that the equipment, processes, or operations are installed in accordance with applicable codes and standards and deficiencies are identified, documented, and reported in accordance with the policies of the jurisdiction. (pp 274–283)

(A) Requisite Knowledge. Applicable codes and standards, accepted fire protection practices, fire behavior, ignition sources, safe housekeeping practices, and additional reference materials related to protection of hazardous processes and code enforcement. (pp 274–283)

(B) Requisite Skills. The ability to observe, communicate, interpret codes, recognize problems, and make decisions. (pp 274–283)

Additional NFPA Standards

- **NFPA 1** *Fire Code*
- **NFPA 13** *Standard for the Installation of Sprinkler Systems*
- **NFPA 30** *Flammable and Combustible Liquids Code*
- **NFPA 31** *Standard for the Installation of Oil-Burning Equipment*
- **NFPA 33** *Standard for Spray Application Using Flammable or Combustible Materials*
- **NFPA 54** *ANSI Z223.1-2015 National Gas Fuel Code*
- **NFPA 80** *Standard for Fire Doors and Other Opening Protectives*
- **NFPA 101** *Life Safety Code®*
- **NFPA 115** *Standard for Laser Fire Protection*
- **NFPA 211** *Standard for Chimneys, Fireplaces, Vents, and Solid Fuel-Burning Appliances*
- **NFPA 654** *Standard for the Prevention of Fire and Dust Explosions from the Manufacturing, Processing, and Handling of Combustible Particulate Solids*
- **NFPA 921** *Guide for Fire and Explosion Investigations*
- **NFPA 1143** *Standard for Wildland Fire Management*
- **NFPA 1144** *Standard for Reducing Structure Ignition Hazards from Wildland Fire*
- **NFPA 1405** *Guide for Land-Based Fire Departments That Respond to Marine Vessel Fires*

FESHE Objectives

There are no FESHE objectives for this chapter.

Knowledge Objectives

After studying this chapter, you will be able to:

1. Describe the importance of good housekeeping practices in fire prevention.
2. List the three requirements of good housekeeping.
3. Identify and evaluate the common fire and life safety hazards related to housekeeping practices outside of buildings:
   - Obstructions to fire protection equipment
   - Fire exposure threats
   - Spontaneous ignition threats
4. Identify and evaluate the common fire and life safety hazards related to housekeeping practices inside of buildings:
   - Spontaneous ignition threats
   - Dust and lint accumulation
   - Mechanical equipment
   - Combustible materials storage
   - Flammable and combustible liquids
   - Painting, coating, and finishing operations
   - Floor cleaning and treatments
   - Fire-protection equipment obstructions
   - Commercial cooking operations
   - Compressed gas cylinders
5. Describe the methods for reducing the risk presented by unsafe housekeeping practices.

Skills Objectives

There are no skills objectives for this chapter.

You Are the Fire Inspector

As a new fire inspector, you are tasked by your department to go out to a factory in your community to conduct a routine inspection. This business is a major employer in town. Upon arrival, you notice a large pile of wooden pallets being stored outside the building, adjacent to numerous doors, windows, and similar building openings. Inside the building you notice that many exit aisles and doors are obstructed or blocked with the storage of raw materials and finished products. The facility's floors have lots of debris—such as cardboard and sawdust—from their process. In addition you see evidence of discarded smoking materials throughout the building.

1. How can you impress upon plant management the need for better control of their housekeeping practices?
2. How much time will it take to correct these items?
3. What strategies should they implement right away and what strategies are worthy of additional time to implement?

Introduction

Good housekeeping practices are an effective fire prevention measure which can best be described as plain common sense. An extensive background in fire safety or fire protection is not needed in order to recognize poor housekeeping practices that potentially pose a fire safety risk. Poor housekeeping practices should serve as a caution to the fire inspector; places that have poor housekeeping practices often have other fire safety deficiencies **Figure 1**.

Safe housekeeping practices—both indoor and outdoor—accomplish four major fire and life safety objectives:

1. Eliminate unwanted fuels, helping to control fire growth and making extinguishment easier.
2. Remove obstructions or impediments to egress.
3. Control sources of ignition.
4. Improve safety for firefighting and emergency response personnel.

Certain aspects of housekeeping are common to almost all types of occupancies while other aspects are unique to certain occupancies. For example, manufacturing occupancies may generate large quantities of combustible dusts that require frequent cleaning. Other occupancies may generate large amounts of trash, waste, or other by-products that need to be removed. In almost all occupancies, the storage of materials can cause obstructions to egress.

As a fire inspector, you must be able to recognize housekeeping problems and take actions to eliminate them. You can use these opportunities to educate the property owners on the importance of reducing these types of hazards. By taking the time to educate property owners, you will find that the incidence of housekeeping problems will decrease over time.

Housekeeping Overview

A business that keeps its operation neat and clean will have a substantially reduced risk of fire and may benefit in related areas such as lower worker injury rates and improved employee morale. Three basic requirements of good housekeeping are:

Figure 1 Poor housekeeping practices should serve as a caution; places that have poor housekeeping practices often have other fire safety deficiencies.

- Equipment arrangement and layout—This includes properly cleaning, servicing, and placing devices that generate combustible dusts or waste materials. Sometimes the layout or location of equipment can pose additional problems, such as equipment that is producing dust located close to an ignition source or an air intake.
- Material storage and handling—This includes trash and waste disposal or recycling operations that are found in buildings or on the property outside of a building.
- Operational neatness, cleanliness, and orderliness—This includes emptying trash and waste on frequent intervals to avoid accumulation. The frequency of trash removal may vary substantially depending on the type, form, and amount of trash or waste generated. Some occupancies, such as large retail stores, have very complex and involved trash and waste handling operations to remove, compact, and bale materials for recycling or disposal.

Exterior Issues

You should always inspect the exterior of the property in addition to the inside. In some cases, need for exterior housekeeping may not be apparent to the property owner because these areas are not viewed as often as the inside of the building. Poor housekeeping practices outside of the building can result in obstructions to the site or building, obstructions to fire protection equipment, fire exposure threats, wildfire concerns and an unattractive nuisance easily ignited by vandals or juveniles.

Blocked, Obstructed, or Impaired Access

Exterior housekeeping issues can obstruct access for responding vehicles and fire fighters. These access obstructions can delay or hinder firefighting operations. Examples of access obstructions include roads, driveways, fire lanes, or similar accesses being blocked by trash, debris, or dumpsters. Trees or bushes can also become overgrown and block exit doors.

<u>Premise identification</u>, the posting of an address, can aid emergency responders in finding the building. Sometimes the identification numbers may be missing or may not be easily visible from the road. In other cases, the address may have changed. Larger complexes may have separate streets or roads and incorporate their own numbering system. It is critical that fire, law-enforcement, and emergency-medical personnel be able to locate the building quickly in an emergency. For this reason the NFPA 1, *Fire Code* requires that legible address numbers be placed on all buildings. These address numbers should be visible from the road.

In some areas of the United States, snow accumulation can pose additional fire problems. Snow and ice must be removed to make fire hydrants and fire protection equipment accessible. For larger buildings, it may be necessary to plow snow from fire lanes and fire apparatus access roads. One of the most common problems, however, is the failure to keep exits and outside egress paths shoveled or cleared of snow **Figure 2**.

Obstructions to Fire Protection Equipment

Housekeeping issues can obstruct fire department access to critical firefighting equipment, such as fire hydrants and fire department connections. This can include the presence of trees, bushes, landscaping, or snow that can block fire protection equipment. In some situations, signs can be added to assist fire fighters in locating these devices. These signs can identify the location of critical fire protection devices that are otherwise blocked or obstructed from normal view.

In some cases, fire department connections may be obstructed by the addition of a fence **Figure 3**. You should

Figure 2 Snow cannot be allowed to accumulate and block an exit.

> **Fire Inspector Tips**
>
> Proper housekeeping procedures are the responsibility of the property owner. As the fire inspector, you must be able to communicate the need for safe housekeeping practices to the property owner to implement.

> **Fire Inspector Tips**
>
> Different types of occupancies or businesses may have different housekeeping problems depending on the nature and type of processes taking place. Some retail and office occupancies may generate large amounts of combustible trash and recycling materials. Certain industrial and factory occupancies may produce combustible dusts. Other occupancies may have oily rags that need proper disposal.

> **Safety Tips**
>
> Obstructed exits also mean obstructed fire department access into and through a building.

ensure that action is taken to ensure that fire fighters have access to firefighting equipment by moving or removing the fence or relocating the equipment. In some rare instances where site security is a higher-than-normal concern, you may also consider instructing the property owner to install a gate in the fence with a key located in a secure fire department keybox.

Protection of Flammable Liquid and Gas Equipment

When gasoline dispensers, utility gas meters and piping, liquid propane (LP-Gas) tanks, or flammable-liquid tanks are located in areas where they could be struck by vehicles, they should be protected against impact. The lack of impact protection is fairly common with natural gas meters and piping Figure 4. Protection is commonly accomplished by installing posts or bollards to minimize the possibility of these items being damaged. Other options include installing other barriers—such as guard rails or "jersey barriers"—or installing a high curb to protect these items.

Fire Exposure Threats

Another exterior risk is the accumulation of combustible materials that can pose a fire exposure threat to nearby buildings or equipment. This may involve excessive amounts of trash or debris or may be a by-product of normal business operations. Combustible waste materials from industrial and manufacturing operations are commonly stored on-site before being hauled away. One common example for manufacturing and storage occupancies is the accumulation of idle wood pallets. NFPA 1 has separation requirements for idle pallets Figure 5. The requirements vary based on the number of pallets and the construction of the building. The minimum separation distance for 50 or more pallets is 30′ from buildings; this distance is increased to as much as 50′ if there are relatively large quantities of pallets (over 200). Fencing these outdoor storage areas in order to limit access of unwanted persons is recommended. Ignition sources, such as smoking debris, cutting or welding operations, and hot-work tools such as flame torches, should be kept at a safe distance of 25′. NFPA 1 requires 35′ of distance between certain types of hot-work operations and combustible materials. A fire involving these pallets could quickly spread into the facility through the structure's doors and windows.

Another common fire safety problem is the storage of waste-rubber tires outside a building. This is a dangerous practice because this material burns very intensely and can quickly cause severe damage to the building or its contents. NFPA 1 contains limits on the size and number of tires allowed to be stored, in addition to distances that the tires should be kept from buildings.

Combustible waste, rubbish, and debris should never be stored or allowed to accumulate next to flammable or combustible liquid or flammable gas storage tanks or containers. Combustible storage in the proximity of flammable liquid gas tanks or containers can easily be ignited and can cause the failure of the containers and a disastrous fire Figure 6. Tall grass and weeds should never be allowed to grow near flammable or combustible liquid storage or liquefied propane (LP Gas) installations. A fire involving the grass or weeds would have a disastrous outcome if it reached the tanks containing the combustible product. The use of herbicides should be carefully considered because many

Figure 3 An obstructed fire department connection is an example of poor outdoor housekeeping.

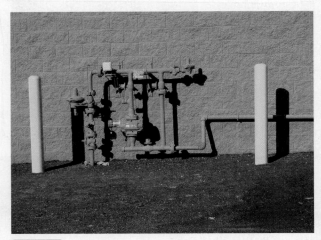

Figure 4 If natural-gas meters are not protected against vehicle impact, then they and could be easily struck and damaged.

Figure 5 An excessive pile of wooden pallets could allow fire spread into the building.

Figure 6 Combustible debris stored next to an LP-Gas tank poses a dangerous hazard for the occupancy.

of these materials contain chlorate compounds that are oxidizers and can contribute to rapid-fire conditions, especially during dry periods. The use of herbicides is generally limited to pre-emergent grass and weed killers, while any large stand of tall grass or weeds is removed mechanically.

NFPA 1 requires that dumpsters and similar waste receptacles should be located at least 10′ from combustible buildings and should not be placed under roof eaves or overhangs. If located under eaves or next to combustible buildings or building openings, a dumpster fire can quickly become a building fire. Dumpster fires are relatively common due to discarded ignition materials present in dumpsters, as well as easy access for arsonists.

Tall grass and weeds should never be allowed to grow near flammable or combustible liquid storage or liquefied propane (LP Gas) installations. A fire involving the grass or weeds would have a disastrous outcome if it reached the tanks containing the combustible product. The use of herbicides should be carefully considered because many of these materials contain chlorate compounds that are oxidizers and can contribute to rapid-fire conditions, especially during dry periods. The use of herbicides is generally limited to pre-emergent grass and weed killers while any large stand of tall grass or weeds is removed mechanically.

■ Wildland Interface

Buildings and facilities constructed near large wildland areas without an intervening firebreak are of growing concern in some areas of the country. These areas are sometimes called a **wildland/urban interface**. Homes, businesses, and apartments have been lost as a result of fast-moving wildland or forest fires. In some jurisdictions, specially developed fire prevention standards or wildland urban interface codes have been adopted to protect against this type of fire spread.

The zone between a structure and an area of native vegetation is known as the wildland-urban interface. Facilities constructed in or adjacent to wildland areas without sufficient **defensible space** or area cleared of combustibles, are at an additional risk of fire. This risk is a two-directional consideration. First, sufficient defensible space is necessary to stop a wildland fire of trees, brush, grasses, or weeds to spread to outdoor storage or the building itself. Second, a major fire involving a structure or outdoor storage cannot be allowed to spread to the wildland area where a much larger fire may endanger a community.

Breaking the fuel ladder between buildings and outdoor storage and the wildland areas is the essential intent of a defensible space. The **fuel ladder** is a continuous progression of fuels that allows fire to move from brush to limbs to tree crowns or structures. Depending on a number of factors, including **slope**, meaning the upward or downward slant of the land; **aspect**, the compass direction toward which a slope faces; and environmental factors such as weather and winds conditions expected during fire events, the defensible space may need to vary from as little as 30′ (9 m) to as much as 200′ (61 m). It is not necessary to remove all vegetation in this space: modifying existing vegetation and fire-safe landscaping treatments appropriate for the climate can achieve a practical solution Figure 7 . Where conditions cannot provide the necessary defensible space, relocating or removing outdoor storage or protecting the structure through improved construction materials, especially roofing, can provide the necessary protection. NFPA 1143, *Standard for Wildland Fire Management*, provides more detailed information on this subject.

Aside from the wildland/urban interface issues, overgrown vegetation, such as weeds, tall grass, brush, shrubs and trees, can cause other concerns for a fire inspector. These materials can obstruct views or impair access to fire hydrants and fire department connections Figure 8 . Trees and bushes can become so overgrown that they block exit doors and pathways.

■ Interior Issues

Housekeeping issues inside a building can increase the possibilities of ignition by introducing a heat source, producing larger or more rapidly developing fires due to the additional fuel loads, or can hamper occupant egress or firefighting access by blocking exits and aisles. You should check for general cleanliness in the building. The level of cleanliness is relative to the operation or type of occupancy. Some businesses, such as wood shops, repair garages, and agricultural mills, generate dusts and will be messier than a typical school or office occupancy. A business that keeps its operation neat and clean will have less risk of a fire.

Figure 7 Ladder fuels in the wildland/urban interface area need to be broken.

Figure 8 Fire department connections blocked by overgrown vegetation are a hazard for fire fighters.

■ Oily Waste, Towels, or Rags

Spontaneous ignition can occur with oily waste, towels, or rags. Spontaneous ignition is the combustion of a material by an internal chemical or biological reaction that has produced sufficient heat to ignite the material. Some of the more common materials subject to spontaneous ignition are linseed oil, Tung oil, charcoal, and certain vegetable oils—most notably peanut oil, which is often used in cooking operations. Oily waste and rags are commonly found in restaurants, for cleaning cooking equipment; vehicle repair garages; industrial occupancies; paint-spraying operations; and building maintenance areas.

Oily waste, towels, and rags should be stored in metal containers with tight-fitting covers. Commercially made containers are available for this purpose. These are metal containers that have a self-closing lid that inhibits fire development by restricting oxygen.

■ Dust and Lint Accumulation

Accumulations of combustible dust can be a major concern in certain types of manufacturing, storage, and industrial operations. A combustible dust is a finely divided solid material that presents a fire or explosion hazard when dispersed and ignited in air. Some dusts pose an extreme explosion risk. Examples of businesses that have a dust explosion risk include agricultural operations where crops are milled or crushed; food product manufacturing involving flour, starches, and sugars; woodworking operations generating sawdust; and coal-handling operations.

It is imperative that these operations minimize dust accumulations by removing dust from equipment and structural components. This should be done by vacuuming or suction; every effort should be made to avoid putting these products in suspension in the air. In most cases, the dust accumulation can be removed using vacuum-cleaning equipment with an explosion-proof motor. Specially made dust collection systems are also manufactured for this purpose. These systems can also incorporate automatic fire detection and suppression systems.

In many cases the dust collection storage vessel, often referred to as the hopper or cyclone, is located outside of the building. You should view the ductwork and hopper to make sure there are no leaks where dust can escape and to minimize exposure to ignition sources. One of the common problems in dust collection systems is the introduction of a spark that causes an explosion. The spark can be from cutting, welding, or grinding operations or from an electrical source.

Processes that generate lint, especially clothes-drying operations, need to be cleaned frequently to minimize build-up of this easily ignitable material. This can especially be a problem with commercial laundry operations that handle large volumes and have dryers heated to higher temperatures to dry clothes more quickly. You should look for evidence of lint accumulation inside the laundry equipment, exhaust ductwork, and in the room itself.

Timber, woodworking, textile or agricultural grain-processing facilities can generate large amounts of combustible dusts or fibers that, when airborne, can explosively ignite or rapidly spread a fire faster than a conventional sprinkler system can control. Dust and fibers can accumulate on walls, ceilings, motors, heating equipment, and structural members; under tables; and inside ducts and conveying equipment. Removing these dusts or fibers can be a dangerous process because an explosive fire can occur if not done correctly. In most cases dust, lint, and fibers should be removed by way of a vacuum system employing dust-collection equipment and safe electrical hardware Figure 9. In a few cases, damp cloths are used to remove dust or fibers, and then are properly disposed in metal containers if spontaneous combustion is a possibility. Compressed air or blowers should never be used to clean dusty areas as this only suspends the material in the air, making it easy to ignite.

■ Mechanical Equipment

Another interior housekeeping concern is excessive lubrication on motors, engines, compressors, and similar equipment that can attract dirt and dust, resulting in overheating of the equipment. Excessive accumulations of oil, grease, and similar

Figure 9 A woodworking shop must have a dust collection system to help prevent fires.

lubrication should be removed from the equipment. If you see oil or grease on motors or compressors, you should require that the property owner remove these accumulations.

Combustible Materials and Storage

Many buildings lack adequate storage provisions, leading people to store materials unsafely. Sometimes, carts or powered equipment used to transport materials within a building can be left in undesirable locations. One of the more common problems seen by fire inspectors is storage of combustibles and carts in the egress system, causing obstructions and impediments to prompt exiting in an emergency **Figure 10**.

Another egress concern is combustible storage under stairways. Should these materials be ignited, the fire can damage the stairs rendering them unusable. In other situations the smoke generated by these burning materials would contaminate the stairs and other egress components, impeding the ability of occupants to use these egress paths. The area under stairs should not be used for storage unless it is separated from the stairs by fire-resistant rated construction.

Storage must also be separated from potential ignition sources, such as boilers, furnaces, water heaters, kilns, heat-producing appliances, and space heaters. In residential settings, such as a nursing home, it is common to find boxes, newspapers, and clothes kept too close to a water heater or furnace. A minimum of 18″ is specified in NFPA 54, *National Gas Fuel Code*, and NFPA 211, *Standard for Chimneys, Fireplaces, Vents, and Solid Fuel-Burning Appliances*, for gas and electric heaters and 36″ (914 mm) for high-heat producing appliances, such as boilers, incinerators, and solid-fuel burning appliances.

High-Piled Combustible Storage

Many storage and retail occupancies, sometimes referred to as big-box stores, utilize high-piled storage arrangements to maximize space. **High-piled storage** is often defined as solid-piled, palletized, rack storage, bin box, or shelf storage in excess of 12′ (366 cm) in height. In some cases the storage arrangement poses risks to firefighting personnel by creating very narrow aisles and unstable piles. This arrangement could hamper firefighting operations, and the piles could collapse on fire fighters, especially when boxes became saturated with water or weakened by fire damage.

NFPA 1 contains specific minimum aisle dimensions and maximum pile height and sizes. The minimum aisle dimensions and storage pile dimensions are based on a number of factors, including the commodity being stored and level of fire protection provided. Aisles between racks of storage or solid piles on pallets should be a minimum of 4′ according to NFPA 13, *Standard for the Installation of Sprinkler Systems*. This distance is needed for safe forklift operations inside the building. In some rack storage conditions, NFPA 13 requires aisles to be a minimum of 8′ (244 cm).

Another concern that occurs in high-piled storage operations involves safety concerns for products that expand with the absorption of water. Where large quantities of paper products, especially rolled paper, are stored, it is necessary to keep an aisle between the paper storage and the sidewall of the building. During firefighting operations, either from automatic fire sprinklers or firefighting hoselines, these products can soak up water, expand, and push against the outside walls of the building, causing severe structural damage or collapse. For this reason, NFPA 13 requires a 24″ (610 mm) aisle between exterior walls of the building and materials that will absorb water and expand.

Trash or Recycling Issues

Emptying trash and waste at frequent intervals to prevent accumulation is one example of an effective housekeeping practice. The frequency of trash removal may vary substantially depending on the amount of trash or waste generated. Some occupancies, such as large retail stores or large office buildings, have waste-handling operations to remove, compact, and/or bale materials generated in their everyday operation.

Many states and jurisdictions encourage or require businesses and employers to participate in environmental recycling

> **Fire Inspector Tips**
>
> In rack storage configurations, the aisles are typically better defined by the presence of the rack themselves: the racks form the aisles. In solid-piled storage, you may have to work closely with the property owner to define aisles and designated storage areas.

Figure 10 An exit door that is blocked is a disaster waiting to happen.

programs. Recycling programs in many areas encourage or require participation for an increasing number of recylable materials, such as plastic, and waste considered hazardous to normal waste disposal streams, such as foam products. Although materials involved in recycling programs are not considered trash or waste in the classic sense, they represent the same types of housekeeping issues—they are often combustible and are frequently stored in undesirable arrangements or locations. The number and size of containers for recycling programs can also be much greater. Individual containers for segregation of recycling materials may be located in each work area, with larger collection points in key areas.

■ Packing and Shipping Materials

Cardboard, paper, styrofoam, expanded plastics, excelsior, straw, and similar materials used for packaging and shipping. Loose packing material, known as **dunnage**, is used to pack, support, and brace products within shipping containers, inside rail cars, on flatbed trucks, and inside a ship's holds. They are considered clean waste and may often be reused or recycled.

All packaging and shipping materials are combustible and represent a significant fire risk. These materials are relatively easy to ignite and have relatively high rates of heat release; some of the plastic materials have rates of heat release similar to flammable liquids. Very large quantities should be protected and kept in separate fire areas.

If the business being inspected uses lots of packaging or shipping materials, these materials should be kept in special fire-rated storage rooms or vaults until they are being used. You should note the condition of shipping and receiving rooms. Pay particular attention to large quantities of accumulated waste near packaging or unloading operations. If this is discovered, a regularly scheduled program for cleaning and waste removal should be instituted immediately. NFPA 1 requires that combustible rubbish not stored in fire-rated rooms or vaults must be removed from the building by the property owner daily. Once removed from the building, these materials should be stored in accordance with requirements for outdoor storage in dumpster or trash areas at a safe distance from the building.

■ Flammable/Combustible Liquids

Flammable and combustible liquids are commonly found when conducting inspections. The model fire codes typically allow small quantities to be used inside buildings; the amounts will vary with the type of occupancy involved. When not in use, most model fire codes require that flammable and combustible liquids, and other types of hazardous materials, be stored in approved containers or cabinets. A flammable-liquid storage cabinet is used for the storage of flammable and combustible liquids; it protects liquids and their vapors from ignition sources.

Wherever flammable or combustible liquids are handled or used, there is a risk of spills occurring. Businesses that handle flammable or combustible liquids should have an emergency management plan for handling spills and leaks. Both NFPA 1 and federal environmental laws require some sort of hazardous materials management plan. One means of complying with spill control requirements for relatively small quantities of flammable liquids—under 5 gallons (19 L) in maximum anticipated spill quantity—is to have an adequate supply of absorptive materials and tools, which can be used to control, contain, or clean up spills. A common type of absorptive material is a granular product similar to cat litter; it is readily available from automotive parts stores, hardware stores, and safety supply stores.

If larger spills are anticipated, spill control is usually accomplished using some sort of containment system or diking to prevent the flammable or combustible liquids from spreading into the sewer or ground. You should ensure that spill-control measures exist where flammable or combustible liquids are being used.

Disposal of many flammable and combustible liquids has been made more difficult and expensive due to state and federal environmental concerns and laws. Draining into sewers or dumping on the ground are not acceptable disposal methods for combustible materials. Flammable and combustible liquid spills should be cleaned up on site and the materials kept in a separate container for disposal by a hazardous waste disposal contractor. Environmental laws require that they be disposed of by specialized hazardous waste disposal companies.

In certain types of occupancies, such as automotive repair garages and service stations, used motor oil may be stored in tanks on the property. This used motor oil is sometimes referred to as waste oil. It may merely be stored in tanks or drums until it can properly be disposed of or it may be connected to fuel-burning equipment for heating parts of the building. NFPA 1 and NFPA 31 contain specific requirements for waste-oil burners. They can only be installed in industrial occupancies and must be specifically listed to burn waste oil. The tanks must also be listed. Waste oil used in these systems should never contain gasoline because they are designed to burn combustible liquids, such as oil, and not flammable liquids, such as gasoline.

The use of flammable cleaning solvents is becoming fairly rare because of development of nonflammable solvents that have no flash points, are very stable, and have limited toxicity problems. Flammable liquids are still used for some cleaning purposes: examples are alcohols and paint thinners. Flammable liquids used for cleaning should be stored in safety cans with tight-fitting lids that are used only for dispensing small quantities. Flammable liquids should not be stored in open pails, buckets, dip tanks, or containers that may be degraded by the liquid.

As part of a routine fire inspection, you should review or discuss the flammable liquid disposal plans or procedures that the property uses. The plans or procedures often involve contracting with a hazardous waste disposal company.

Paintings, Coatings, Finishes, and Lubricants

Many operations in manufacturing, factory, industrial, vehicle repair garages, and similar businesses use paintings, coatings, or finishes that are sources of combustible residue. A **spray booth** is a power-ventilated enclosure around a spraying operation or process that limits the escape of the material being sprayed, and directs these materials to an exhaust system.

Spray booths, exhaust ducts, fans, and motors need to be cleaned on a regular basis to prevent the dangerous accumulation of these residues. Filters in spray finish operations, such as spray booths and spray rooms, need to be installed and replaced regularly to minimize residue accumulation. These filters trap the residues to prevent them from accumulating in the ductwork and on the fans. Filters with large quantities of residue accumulation that no longer stay in their proper position or that are physically damaged or torn need to be replaced.

Because of the high number of fires associated with spray finishing operations, the area or spray booth is required to have an automatic fire suppression system. You need to note how sprinklers are protected against residue accumulation in spray finish booths and ducts. One method is to cover the sprinkler with cellophane plastic or paper bag that is changed on a regular basis to avoid excessive accumulation or residue.

Floor Cleaning and Treatment

Floor cleaning, treatment, or refinishing can be a fire hazard if flammable solvents or finishes are used. In addition, the removal and refinishing of floor surfaces, especially wood floors, can generate combustible dusts and residues. This risk has been reduced recently because the finishing industry has developed newer products that are not classified as flammable liquids.

If a floor cleaning or refinishing operation is conducted, there should be adequate ventilation, and only materials having a flash point above the highest room temperature should be used. In addition, nonsparking equipment should be used, and there should be no open flames in the area. The containers for the solvents or cleaning compounds should be labeled as to their flammability or combustibility. If no labels are present, Safety Data Sheets (SDS) should be reviewed.

Some floor treatments and dressings contain oils or compounds that can spontaneously ignite. Oily mops, towels, or rags used to apply these treatments or dressings should be stored in the same manner as oily rags, in metal containers with tight-fitting lids.

■ Obstructions to Fire Protection Equipment

Improperly stored materials can obstruct access to fire protection equipment, such as fire extinguishers, control valves, and fire alarm pull stations. Improper storage can also impair the proper operation of passive fire protection equipment, such as fire doors.

In sprinkler-protected buildings, NFPA 1 and NFPA 13 require that storage be kept at least 18″ (457 mm) below sprinklers. This distance allows sprinklers to develop their characteristic "umbrella" spray pattern to effectively extinguish a fire Figure 11 . Storage too close can hamper or impair ceiling-mounted sprinklers.

■ Kitchen Cooking Hoods, Exhaust Ducts, and Equipment

Most commercial kitchens are equipped with a cooking **hood and exhaust system**. The hood and exhaust system are installed above a cooking appliance to direct and capture grease-laden vapors and exhaust gases. Grease accumulation on cooking hoods, on the hoods' grease filters, or inside the exhaust duct represents a serious fire safety risk. This accumulated grease can

Figure 11 In sprinkler-protected buildings, NFPA 1 and NFPA 13 require that storage be kept at least 18″ (457 mm) below sprinklers.

be ignited from sparks or heat from the cooking operation or from a small fire on the cooking surface.

The kitchen hood and filters should be inspected and cleaned on a very regular basis. Visual inspections will reveal how much grease has accumulated in the ductwork. The amount of grease accumulated will determine how often cleaning will need to occur; it is recommended that the property owner conduct inspections at least weekly. It may have to be inspected and cleaned daily depending on the amount of food cooked and the method of cooking.

The entire hood, grease-removal appliances, exhaust ducts, fans, and related equipment need to undergo a thorough cleaning on a regular basis. This can be a very messy and difficult job, especially in the ductwork itself. Commercial firms that specialize in this type of cleaning should be utilized. You may ask the property owner to produce records showing how often the hood and ductwork is being cleaned. The amount of food cooked and the method of cooking will determine the necessary cleaning frequency. If any grease is observed dripping from the kitchen hood, filters, exhaust duct, or from the exterior of the building, an immediate cleaning is needed Figure 12 . It is also important that filters be in place and clean to ensure proper operation.

■ Compressed Gas Cylinders

Many industrial facilities, automotive repair garages, and other similar shops need to use **compressed gas cylinders** as part of their operations. One of the most common uses of compressed gas cylinders is an oxygen-acetylene cutting torch for cutting and welding of metals. Compressed gas cylinders are portable pressure vessels of 100 lb (45 kg) water capacity or less, designed to contain a gas or liquid at gauge pressures over 40 psi (276 kPa). There are restrictions on the amounts of certain types of compressed gases, such as flammable gases, permitted in an area. Compressed gas cylinders should also be secured in a manner to prevent them from being knocked over Figure 13 . If the cylinder tips over and the neck of the cylinder ruptures, the cylinder can be propelled like a rocket and do considerable damage.

Figure 12 Excessive accumulations of grease require immediate cleaning.

states and communities have enacted smoking regulations that limit smoking in public buildings and workplaces. These factors have led to a decrease in the emphasis that you need to place on the control of smoking. It has, however, also led to an increase in people smoking in areas where smoking represents a fire hazard. You should be aware of hidden ash cans or cigarette butts in areas where smoking is prohibited. Outdoor smoking areas should also be surveyed to ensure adequate separation from combustible materials and vegetation, especially in wildland-interface hazard areas.

There are situations where smoking should definitely be prohibited and "No Smoking" signs should be prominently displayed. Smoking should always be forbidden near flammable liquids, both indoors and outdoors; near flammable gases, such as LP Gas and acetylene; and in areas where there are large quantities of combustibles, such as retail and mercantile occupancies. Smoking should also be prohibited in areas where dust accumulations are present, such as woodworking plants, and in areas where there are combustible decorations.

In areas where smoking is allowed, approved smoking receptacles should be provided. Receptacles should not allow the cigarette to fall outside the container and should self-extinguish any combustible materials added to the container. These smoking receptacles can also be filled with sand to assist in extinguishing the smoking materials. Specifically-designed receptacles are available from commercial sources such as hardware stores and safety supply companies **Figure 14**.

Figure 13 A compressed gas cylinder should be secured against being tipped or knocked over.

Figure 14 A cigarette receptacle.

Control of Smoking

By controlling smoking, a common ignition source can be eliminated. Smoking introduces a flame (lighter or matches) and a smoldering heat source (the cigarette or cigar). Cigarette smoking is generally on the decline in the United States. In addition, many

Fire Inspector Tips

No Smoking Areas

There are definite situations where smoking is prohibited and "No Smoking" signs should be posted prominently. Smoking should always be prohibited indoors or outdoors in close proximity to the following:

1. The storage or use of any flammable or combustible gases, such as acetylene, LP-gas, CNG, or hydrogen
2. The storage, dispensing, or use of any flammable or combustible liquid
3. Locations where organic dust is generated or present, such as grain mills and woodworking shops
4. Locations such as warehouses or stockrooms, where large amounts of combustible materials, including dunnage and other packing materials, are stored
5. Diplayed or stored combustible decorations, permanent or seasonal

Public safety and community education programs can also incorporate smoking safety messages for those who choose to smoke. Education on unsafe smoking practices, such as smoking in bed or sneaking a smoke in a hazardous location; support for safer smoking materials; and availability of smoking cessation programs can be communicated to the community as part of an all-risk approach to fire and life safety.

Correction of Housekeeping Issues

Correcting identified housekeeping issues is often a minor expense for the business or property owner. The correction, however, often involves staff time so some owners are sometimes reluctant or slow to make the corrections if it takes staff away from duties that generate income for the business. In the case of repeated, chronic, or on-going housekeeping issues, the long-term solution may involve a change in the internal operation, such as daily removal of all trash, or a remodeling or renovation of the building. The remodeling or renovation may create or add storage space or relocate equipment or operations so that the condition, such as a blocked exit door, no longer exists.

The strategies for correcting housekeeping problems should involve the following considerations:

- Reducing or eliminating sources of ignition:
  - Combustibles too close to ignition sources—A safe separation distance should be maintained between heat sources and combustible materials. The distances are spelled out in NFPA 1. As a rule of thumb, 18″ (457 mm) of separation is standard from typical heat producing appliances, such as water heaters, furnaces, and stoves. A minimum 36″ (914 mm) of separation is required from high-heat producing appliances, such as ceiling-mounted heaters and solid fuel-burning appliances.
  - Spontaneous ignition of materials—Materials subject to spontaneous heating should be stored in metal containers with tight-fitting lids. This includes oil-soaked rags, especially those used with tung oil, linseed oil, corn oil, peanut oil, soybean oil, or fish oil. The NFPA's Fire Protection Handbook has an extensive list of materials that are subject to spontaneous ignition.
  - Heat due to friction or overheating—Cleaning and proper maintenance can often prevent these fires from happening. Cleaning away oil, grease, and dirt accumulations allows the equipment to dissipate heat and run at lower temperatures. Lubrication of bearings or moving parts reduces friction and overheating.
- Controlling the fuel load—The less there is to burn, the smaller the fire will be. You should work with the property owner to reduce the amount of waste and combustibles stored in the building. If indoor storage is absolutely necessary, NFPA 1 requires that storage be neat and orderly and that it not obstruct access to exiting. Outdoor storage is usually preferable to inside storage but should still be kept away from building openings and overhangs to prevent an exterior fire from spreading into the building.
- Providing access to the site or building for emergency responders—Premise and building address are critical so emergency responders can find the building. Emergency vehicle access to the site, including parking areas, is critical for firefighting purposes.
- Providing access to fire protection systems and equipment—Fire fighters need quick access to fire hydrants, fire sprinkler connections, fire extinguishers, standpipe connections, fire sprinkler controls, and fire alarm control panels. Storage should not be blocking access to these critical fire safety features. Fire inspectors will often find this type of equipment blocked or obstructed. Delays getting to these devices can allow the fire to grow or, in some cases, can cause additional water damage if sprinklers cannot be turned off after a fire has been extinguished.
- Providing and maintaining adequate egress—Life safety should always be your number one priority. Anything that blocks, obstructs, or impedes an occupant's ability to get out of the building quickly in a fire or other emergency needs to be corrected quickly. Storage obstructing egress paths or blocking exit doors are common housekeeping problems.

Wrap-Up

Chief Concepts

- Good housekeeping practices are an effective fire-prevention measure that can best be described as plain common sense.
- Three basic requirements of good housekeeping are:
  - Equipment arrangement and layout
  - Material storage and handling
  - Operational neatness, cleanliness, and orderliness
- Poor housekeeping practices outside of the building can result in obstructions to the site or building, obstructions to fire protection equipment, fire exposure threats, wildfire concerns, and an unattractive nuisance easily ignited by arsonists.
- Housekeeping issues inside a building can increase the possibilities of ignition by introducing a heat source, produce larger or more rapidly developing fires due to the additional fuel loads, or hamper occupant egress or firefighting access by blocking exits and aisles. A business that keeps its operation neat and clean will have less risk of a fire.
- The level of cleanliness is relative to the operation or type of occupancy. Some businesses, such as wood shops, repair garages, and agricultural mills, generate dust and will be messier than a typical school or office occupancy.
- Spontaneous ignition can occur with oily waste, towels, or rags.
- Accumulations of combustible dust can be a major concern in certain types of manufacturing, storage, and industrial operations.
- Timber, woodworking, textile, or agricultural grain-processing facilities can generate large amounts of combustible dusts or fibers that, when airborne, can explosively ignite or rapidly spread a fire faster than a conventional sprinkler system can control.

Wrap-Up

- Another interior housekeeping concern is excessive lubrication on motors, engines, compressors, and similar equipment that can attract dirt and dust and result in overheating of the equipment.
- Storage must also be separated from potential ignition sources, such as boilers, furnaces, water heaters, kilns, heat-producing appliances, and space heaters. In residential settings, such as nursing homes, it is common to find boxes, newspapers, and clothes kept too close to a water heater and furnace.
- The strategies for correcting housekeeping problems should involve the following considerations:
  - Reducing or eliminating sources of ignition
  - Controlling the fuel load
  - Providing access to the site or building for emergency responders
  - Providing access to fire protection systems and equipment
  - Providing and maintaining adequate egress

Wrap-Up

■ Hot Terms

Aspect Compass direction toward which a slope faces. (NFPA 1144)

Combustible dust Any finely divided solid material that is 16.5 mil (420 µ) or smaller in diameter (material passing a U.S. No. 40 Standard Sieve) and presents a fire or explosion hazard when dispersed and ignited in air. (NFPA 654)

Compressed gas cylinders Portable pressure vessels of 100 lb (45.3 kg) water capacity or less designed to contain a gas or liquid at gauge pressures over 40 psi (276 mPa).

Defensible space An area as defined by the AHJ [typically a width of 30′ or (9.14 m) more] between an improved property and a potential wildland fire where combustible materials and vegetation have been removed or modified to reduce the potential for fire on improved property spreading to wildland fuels or to provide a safe working area for fire fighters protecting life and improved property from wildland fire. (NFPA 1144)

Dunnage Loose packing material (usually wood) protecting a ship's cargo from damage or movement during transport. (NFPA 1405)

Fuel ladder A continuous progression of fuels that allows fire to move from brush to limbs to tree crowns or structures.

High-piled storage Solid-piled, palletized, rack storage, bin box, and shelf storage in excess of 12′ (3.7 m) in height. (NFPA 13)

Hood and exhaust system Devices installed above a cooking appliance to direct and capture grease-laden vapors and exhaust gases.

Premise identification Posting of an address for emergency responders.

Slope Upward or downward incline or slant, usually calculated as a percentage. (NFPA 1144)

Spontaneous ignition Initiation of combustion of a material by an internal chemical or biological reaction that has produced sufficient heat to ignite the material. (NFPA 921)

Spray booth A power-ventilated enclosure for a spray application operation or process that confines and limits the escape of the material being sprayed, including vapors, mists, dusts, and residues that are produced by the spraying operation and conducts or directs these materials to an exhaust system. (NFPA 33)

Wildland/urban interface Any area where wildland fuels threaten to ignite combustible homes and structures. (NFPA 1143)

Fire Inspector *in Action*

You have arrived at a multiple-tenant commercial property to conduct a regularly scheduled annual inspection. The name above the door of the first business is "Fred's Grille," the business located in the center of the building is "Tom's Computers," and the last business is "Mike's Sporting Goods."

1. During your inspection of the first business, "Fred's Grille," the following item is of significant fire and life safety interest:
   A. The condition and functionality of the door hardware on all egress doors from the dining area and the kitchen.
   B. Evidence of grease accumulations around cooking equipment and kitchen exhaust systems.
   C. Accumulations of empty food-packaging containers and wrappers adjacent to cooking equipment in the kitchen.
   D. All of the above.

2. The least valuable information an inspection of the roof of the building will provide the inspector is:
   A. A view of surrounding terrain and distant topographic features.
   B. Evidence of any grease accumulations at the discharge of the kitchen exhaust fan above "Fred's Grille."
   C. The general condition of any lightening arrestor system installed on the building.
   D. An indication of the general level of care and preventative maintenance provided to the building by its owner.

3. Upon entering "Tom's Computers," you notice a large inventory of older personal computer equipment on display. The store has repeatedly marked-down the products hoping for a quick sale. Which of the following concerns you as a fire inspector:
   A. The business owner may be willing to donate a portion of his inventory to the local school system.
   B. The owner of the business is not concerned with carrying current, state of the art computer equipment in his inventory.
   C. This is an indication of a struggling business with a large, outdated and expensive inventory and should be watched closely for suspicious fire activities.
   D. The inventory has little insurable value and is not worth protecting.

4. "Mike's Sporting Good's" has posted a special annual sale in its windows and is opening early to accommodate a large anticipated crowd of shoppers. Which of the following is not a fire and life safety concern:
   A. Prompt removal of shipping cartons and boxes as the special sale goods are unpacked and displayed for sale.
   B. Maintenance of aisles for customers and employees to safely move around the store and access exits.
   C. Enforcement of parking time limits to move traffic in and out of the parking lot.
   D. Checking for the use of damaged electrical to supply special display equipment and fixtures brought in for this sale.

Writing Reports and Keeping Records

CHAPTER 14

NFPA Objectives

Fire Inspector I

4.2 Administration. This duty involves the preparation of correspondence and inspection reports, handling of complaints, and maintenance of records, as well as participation in legal proceedings and maintenance of an open dialogue with the plan examiner and emergency response personnel, according to the following job performance requirements. (pp 290–302)

4.2.1 Prepare inspection reports, given agency policy and procedures, and observations from an assigned field inspection, so that the report is clear and concise and reflects the findings of the inspection in accordance with the applicable codes and standards and the policies of the jurisdiction. (pp 290–302)

(A) Requisite Knowledge. Applicable codes and standards adopted by the jurisdiction and policies of the jurisdiction. (pp 290–302)

(B) Requisite Skills. The ability to conduct a field inspection, apply codes and standards, and communicate orally and in writing. (pp 290–302)

4.2.5 Identify the applicable code or standard, given a fire protection, fire prevention, or life safety issue, so that the applicable document, edition, and section are referenced. (p 299)

(A) Requisite Knowledge. Applicable codes and standards adopted by the jurisdiction. (p 299)

(B) Requisite Skills. The ability to apply codes and standards.

4.2.6 Participate in legal proceedings, given the findings of a field inspection or a complaint and consultation with legal counsel, so that all information is presented and the inspector's demeanor is professional. (p 302)

(A) Requisite Knowledge. The legal requirements pertaining to evidence rules in the legal system and types of legal proceedings. (p 302)

(B) Requisite Skills. The ability to maintain a professional courtroom demeanor, communicate, listen, and differentiate facts from opinions. (p 302)

Fire Inspector II

5.2 Administration. This duty involves conducting research, interpreting codes, implementing policy, testifying at legal proceedings, and creating forms and job aids, according to the following job performance requirements. (pp 290–302)

FESHE

Principles of Code Enforcement

5. Describe the importance of thorough documentation. (pp 290–302)

Knowledge Objectives

After studying this chapter, you will be able to:

1. Describe the common elements of effective written documentation.
2. Discuss the barriers that affect written communication.
3. Describe how field notes, sketches, diagrams, and photographs are used to complete a fire inspection report.
4. Describe the comment elements in all fire inspections reports.
5. Describe the proper documentations methods to use when resolving complaints.
6. Describe how to properly reference a code or standard.
7. Describe how to maintain records in accordance with applicable laws and department practices.
8. Describe how to comply with Freedom of Information Act requests.
9. Describe how to present evidence during a legal proceeding.

Skills Objectives

There are no skills objectives for this chapter.

You Are the Fire Inspector

As a new fire inspector, you are asked to handle your first complaint, a report from a concerned parent about locked and blocked exits at a public school during an after-hours band competition. When you arrive, you find several students milling about the hallways and common spaces, waiting their turn to compete. In an effort to prevent students from roaming the hallways throughout the schools, the maintenance staff has secured various roll-down gates, stairway doors, and exterior exits. The principal tells you that he does not have the staff necessary to patrol every hallway and corridor and that he was told to by the school superintendent to lock certain doors.

1. How will you document your findings and observations?
2. How can you be sure that your report accurately reflects the findings of your inspection?

Introduction

As a fire inspector, you must be able to document what you observed during the inspection and what needs to be corrected, as well as communicate to the owner or the owner's representative the importance and necessity of correcting any deficiencies found during the inspection. You must also be able to state the code sections that apply and express what the code requires, both verbally and in writing. Finally, you must provide a compliance date based on the severity of the deficiency.

It is important that you document the findings and observations of each fire inspection clearly, concisely, and accurately. Every fire inspection should have a written record of the fire inspection. Written records, from standard inspection forms to detailed inspection reports, providing documentation about what existed, what was observed, and what happened, are the basis for corrective actions and verification. Inspection records may also be used in appeals hearings, legal proceedings, fire inspector training sessions, budget preparation, and code modification meetings.

Since all written communications are official and could be made available to the public, state law requires you to keep them on file for a certain time before they can be discarded. Some items must be kept for three years, until obsolete, or superseded. Others are classified as permanent records and must be kept forever. Since different rules apply to the actual retention periods for different types of records, it is important to know the laws in your state. Fire inspection records should be kept as long as the building exists because fire inspection records are generally subject to open records requests.

Written Documentation

Written records are a vital part of every fire inspection activity. Your written records, from standard inspection forms to detailed inspection reports, are often the only documentation of existing conditions, code violations, and corrective actions that need to be performed. Common uses of written records include:
- Plans review and pre-construction design meetings
- Fire inspections, both initial and follow-up
- Appeal hearings

Fire Inspector Tips

Information is valuable only if the meaning is communicated effectively. There are many barriers to effective communication. You can use this simple acronym, **AT WAR**, as a reminder to add value to all communications, both verbal and written.

- **Accurate** – The information must be factual and accurately reflect the situation. Examples include the results of plans review or fire inspection, the applicable referenced codes or standards and the required corrective actions.
- **Timely** – The information must be well-timed. Delays may result in lost opportunities or cost money; hurried reports may be incomplete, inaccurate, or otherwise flawed.
- **Well-presented** – The information must be clear and concise to make sure its meaning is not lost. Use pictures and diagrams where they will help clarify or emphasize a point or observation.
- **Accessible** – All interested stakeholders must be able to have the information. Ask the questions, "Who knows what I need to know?" "What do I know that others need to know?" and "Who else needs to know what I know?" to ensure that no one is left out.
- **Relevant** – The information must be appropriate, in both content and style. It must have meaning for, and be understood by, the intended audience.

- Legal proceedings
- Other uses, such as fire investigations, staff scheduling and job assignments, training and education, trend analysis, media reports, development of policies and procedures, and budget resource justification

Most written records result from an initial inspection, reinspection, or a complaint. You should prepare a written record for every inspection, even the ones where you are denied entry. Written records document the conditions that existed at the time of your inspection and your observations **Figure 1**. It is important to record the following:

- Deficiencies and code violations
- Modifications to systems and equipment
- Maintenance and house-keeping of equipment, processes, or operations
- Hazardous materials storage, handling, and use
- Other hazardous conditions
- Emergency plans, drills, and exercises

Simply stated, a written record provides evidence of any fire and life safety hazards that you identified and the building's degree of code compliance. A written record should clearly and accurately document observed deficiencies and fire hazards, relevant code sections, any required corrective actions, compliance date, and the status of those corrections (in progress and completed).

■ Barriers that Affect Communication

Written records eliminate one common barrier to communication—trying to remember exactly what was said and by whom. Written records are superior to verbal communications for documentation purposes; however, there are several things that can get in the way of effective communication, even with written records.

Written documentation can be incomplete or inaccurate if observations are not recorded properly, problems overlooked, or the wrong code provision is cited. Errors may lead to uncorrected problems or improper corrections. One of the reported factors of The Station nightclub fire in Rhode Island was the lack of documentation about the use of combustible acoustic materials around the stage area. Another was the allegation that the pyrotechnics used had been approved by an authority, but there was no documentation of this in the files.

Any suggestion of personal bias or opinion could affect the perception of the reader and influence how they react. Refrain from inflammatory, offensive, or potentially harassing remarks and from slang. For example:

Poor word choice	Preferred word choice
"I told them a thousand times."	"Repeated reminders"
"It was a hole-in-the-wall man cave."	"Small tenant space"
"It was a dumb thing to do."	"Did not meet code requirements"

Likewise, improper grammar, word usage, or spelling could reflect poorly on you and your agency. If your writing looks unprofessional, it may raise questions about your competency. On the other hand, if your writing is too formal or uses too many "big" words, the reader may not understand the terms used and may tune out your message.

Anything that interferes with the clarity of the message creates a communication barrier. Examples include the use of long sentences or thoughts that do not lead to clear conclusions. For example, write "The fire extinguishers in Room 146 must be located no more than 75 feet apart …" instead of, "The fire extinguishers, light blue in color and last serviced by ABC Fire Protection four years ago, are located in all processing rooms, including Room 146, and as described in NFPA 10 (current edition as adopted by the jurisdiction on 1 JAN 2010) have been found to be mounted in places that appear to be inconsistent with distance separations for code requirements for existing buildings or new construction—refer to Reference Photo No. 146-01 as documentation of this deficiency needing correction …"

The use of technical jargon or acronyms may be confusing to the reader. This is most often a concern in detailed, formal fire inspection reports. Remember who your audience is, and write for your audience. For example, use "expensive" instead of "requires significant capital expenditure" or "fire protection system" instead of "FPS." Avoid the use of jargon and a style that is too structured or formal for the audience. For example, use "lighted for night use" instead of "adequately illuminated for nocturnal operations."

One way to overcome communication barriers is to practice writing the way you speak. This style will come most naturally and if you are an effective speaker, your words will have a ring of truth and sincerity. Be professional, be accurate and concise, but be personable.

■ Field Notes

Good field notes are the basis for thorough and accurate records. Field notes are the original, on-scene description of conditions that existed at the time of the fire inspection. They document what you observed and serve as reminders to perform additional research and documentation. Good field notes are complete, concise, neat, well written, and free from grammatical and spelling errors. Well-written field notes can make the difference in legal actions. For example, a set of good field notes can be used to re-trace or re-create your actions, support your testimony, and act as evidence.

Most field notes are handwritten, but the growing use of handheld computers provides another way to record field notes and attach digital photographs, maps, and other records **Figure 2**. Some fire inspectors use portable recorders to dictate and voice record their observations. If you voice record your field notes, speak clearly and slowly, be aware of background noises, and keep your comments free from technical jargon, inappropriate or offensive comments, and discussions that do not support the facts **Figure 3**.

Fire Inspector Tips

Your written records could be the difference between a successful court case and an unsuccessful one. It is vital that you write clear, accurate, and consistent information on every form of written documentation, from standard inspection forms to detailed fire reports.

Figure 1 A sample written record.

Chapter 14 Writing Reports and Keeping Records

Figure 2 Digital cameras can document conditions in the field.

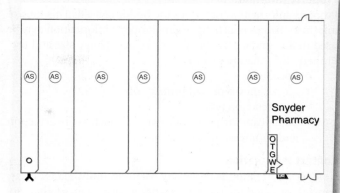

Figure 4 Use field sketches as rough drafts to provide graphic representations of room layout and dimensions, egress locations, identified fire hazards, fire protection system features, and similar information.

Figure 3 Some fire inspectors use portable recorders to dictate and voice record their observations.

Fire Inspector Tips

An effective photograph tells a story, accurately and representatively. Two elements of a good photo are composition and lighting. Chapter 15 of NFPA 921, *Guide for Fire and Explosion Investigations*, contains valuable guidelines about how to use photographs to document your observations, and how to improve the photographs you take.

when you need proof that a specific condition existed at the time of the inspection. Photographs are often used as evidence in legal proceedings, or to document before and after conditions. As with diagrams, number all photographs so they can be easily referenced and accessed.

Documenting the Fire Inspection

Fire Inspection Report

The fire inspection report is the most common of all written documentation. The fire inspection report can take several forms, from checklists to detailed reports. Regardless of the format of the fire inspection report, all reports should contain these common elements:

- Be professional and well-written.
- Be focused and concise
- Present facts, free from bias, opinion, or criticism.
- Use correct grammar, word usage, spelling, punctuation, and style. Computer-generated reports and templates can ensure consistent appearance, format, and quality.
- Whenever possible, avoid passive sentences. Use "The property owner shall correct the following" instead of, "The following shall be corrected."

All type of fire inspection reports should contain specific information. The order of the information is not important, but it should be consistent from report to report.

Sketches, Diagrams, and Photographs

Sketches, diagrams, and photographs provide visual documentation of observed conditions. Use field sketches as rough drafts to provide graphic representations of room layout and dimensions, egress locations, identified fire hazards, fire protection system features, and similar information Figure 4. Do not worry about drawing your sketch to scale, but try to keep things in proportion, and write dimensions on the sketch where appropriate. Be sure to record the location of all hazards and noted deficiencies.

Diagrams are the final inspection drawings. Whenever possible, diagrams should be drawn to scale, or include dimensioning measurements and information, and use standard mapping symbols such as those found in NFPA 170, *Standard for Fire Safety and Emergency Symbols*. Be sure to include a legend that illustrates each symbol and its use. Number all drawings so they can be easily referenced and accessed in a written record.

Photographs are effective reminders of what you actually saw during your inspection, and can often convey unsafe conditions better than words can. Photographs are especially helpful

Property Information

Property information describes the building itself. This type of information does not change often. Property information may be used to document and changes to the occupancy size and use. Property information includes:

- Location
- Occupancy type and primary use
- Construction type
- Significant fire protection features, such as fire detection and suppression systems

Contact Information

Contact information is especially useful in multi-tenant occupancies, or when the property owner is located in another city or area of the country and is represented locally by a management company, agent, or other representative. You can use this information in an emergency, to schedule appointments, or to track changes in ownership. Contact information includes the following information for each property owner and tenant:

- Company name and address
- Primary point of contact information, including telephone number, fax number, website, and e-mail address

Inspection-Specific Information

Each fire inspection report should completely and accurately document the findings of the fire inspection. This information includes your name, the name of the property representative, along with the date and time of your inspection. The report should describe the purpose of the inspection and all deficiencies found, including applicable code references. You should record any corrective actions taken or required, and the time allowed for correction to be made. Finally, include any recommendations not required by code at the end of the report and clearly identify them as recommendations to the building owner.

Follow-up Information

The final part of the report should include follow-up information. Typical follow-up information for the building owner includes reinspection dates, means of requesting additional information or assistance, and a reminder about the importance of completing required corrections. Many reports include a closing thank-you statement with the follow-up information.

■ Reinspection Reports

During reinspections, you do not need to inspect the entire occupancy again. Likewise, you do not need to write a completely new fire inspection report. You should inspect and document only the problem areas noted during your initial fire inspection. Be sure to praise corrective actions and note any remaining deficiencies in your reinspection report.

Additional reinspections may be required, usually when many deficiencies need correcting, when corrections are complex, or when there are multiple correction timelines. If additional reinspections are required, include the planned reinspection date and your contact information.

Always end your reinspection report on a positive note. Thank the property owner or representative for completing any

> **Fire Inspector Tips**
>
> Remember, reinspection is not a completely new inspection; however, if you identify additional problems, include them in the reinspection report as items separate from the original issues.

corrective actions and provide a reminder to complete remaining corrections.

■ Final Notice

A <u>final notice</u> should usually be reserved for times that the owner/occupant makes little or no effort to correct problems and deficiencies; however, you should consider using a final notice immediately in exigent, dangerous, or potentially life-threatening situations. Examples of circumstances where a final notice is appropriate include locked or blocked exits, exposed electrical connections, and structural damage caused by fire. The final notice should include consequences of failure to comply with requirements, including legal actions `Figure 5`.

Choosing the Right Type of Documentation

It is important to choose the right type of written documentation. You should use the type that is appropriate to the situation, for example, a courtesy reminder of an upcoming annual inspection for a licensed facility, a detailed report of a complex inspection with many observed deficiencies, or a final notice. Some types, such as a courtesy letter, should be friendly and written as if they were spoken. Other types, such as a final notice, require a more formal approach and may require a specific format or presentation to be used.

■ Letters

Letters are commonly used for final notices and reminders, simple inspection reports, or cover letters for more detailed inspection reports. Each letter you write is a reflection of your agency, so be professional and courteous. It is also important that you write as if you were speaking directly with the person. Develop a personal style that reflects the way you speak, with a focus on the reader. This style will keep your letter personable, yet professional.

Common elements of a successful letter start with a clean, consistent format. Use standard paper size and fonts. A-size, 8.5" by 11" paper is preferred. Twelve-point type is preferred for legibility; do not use a font size less than 11 point for letters. Every letter should start with salutations and greetings and end with a closing and signature. Make generous use of white space, and start a new paragraph for every subject. The letter should be one page if possible, but should not be longer than two pages. If more space is needed, consider using a checklist or detailed report attached to a cover letter.

A letter and its appearance are a reflection of your agency. Use a consistent format for all agency letters. Spell out the month

Township of Upper Moreland
Office of the Fire Marshal
117 Park Avenue • Willow Grove, Montgomery County, Pennsylvania 19090
Phone: 215-659-3100 • Fax: 215-659-1364 • Email: uppermoreland.org

FIRE CODE INSPECTION REPORT

OCCUPANCY: _____ CONTROL: _____

ADDRESS: _____ OWNER/MGR/PRINCIPAL: _____

ADDRESS: _____

PHONE: _____

ICC USE: _____ INSPECTOR: _____ DATE: _____ DAY: _____ TIME: _____

✔ VIOLATION SEE REMARKS ↓
- ❏ PERMIT INSPECTION
- ❏ BLOCK INSPECTION
- ❏ RESIDENTIAL INSPECTION
- ❏ TANK INSTALL/REMOVAL
- ❏ FOLLOW-UP

✔ VIOLATION SEE REMARKS ↓
- ❏ TEST _____
- ❏ COMPLAINT _____
- ❏ OCCUPANCY _____
- ❏ MAR _____
- ❏ OTHER _____

1. Fire Extinguishers due to be inspected.
2. Fire Extinguishers not provided/installed properly.
3. Keep EXIT signs illuminated and visible.
4. Provide additional EXIT signs (high and low).
5. Maintain adequate aisle width.
6. Keep fire exits unlocked and free of obstructions.
7. Repair/install panic hardware.
8. Removed obstructions from fire towers or escapes.
9. Maintain egres slighting.
10. Provide/maintain emergency lighting.
11. Maintain 24" clearance at ceiling with combustibles.
12. Repair voids in the ceiling: holes, tiles, etc.
13. Remove combustible storage from boiler room.
14. Maintain proper housekeeping.
15. Remove grease from ranges, hoods, duct, fans, etc.
16. Keep electrical appliances clear of combustibles.
17. Repair improper wiring, fuses, grounding, etc.
18. Provide proper storage of compressed gas cylinders.
19. Oxidizers stored separately from flammable liquids, corrosive liquids, combustible materials.
20. Provide proper containers/storage of flammable liquids.
21. Bond wires shall be used when dispensing Class I or II liquids from metal to metal containers.
22. Discontinue smoking and post proper signs.
23. NFPA signs posted at entrance to buildings where hazardous materials are used or stored.
24. Provide/maintain fire zone signs.
25. Install temporary fire zone signs.
26. Discontinue Improper/Illegal burning.
27. Dumpster/combustibles too close to building.
28. Post address on building/sign.
29. Street address posted on rear of strip occupancy.
30. Occupancy limit sign(s) missing/Improperly posted.
31. Failure to obtain required permit.
32. Required test or drill logs missing/incomplete.
33. Required emergency plan missing oroutofdate.
34. Fire alarm system annual test report.
35. Fire sprinkler system annual test report.
36. Sprinklers obstructed.
37. Other suppression system test report. (List in remarks).
38. Knox Box Key Check.
39. Other (List in remarks).

REMARKS: _____
_____
_____
_____

❏ Haz Mati Non-Permit/Req'd. Site   ❏ Sprinkler   ❏ Standpipe   ❏ Cooking System   ❏ AFA   ❏ UST/AGT

IN THE INTEREST OF FIRE SAFETY AND TO COMPLY WITH THE UPPER MORELAND TOWNSHIP FIRE CODE, THE ABOVE VIOLATIONS MUST BE CORRECTED IMMEDIATELY. FAILURE TO COMPLY WILL RESULT IN PENALTIES AS SET FORTH IN THE FIRE CODE OF UPPER MORELAND TOWNSHIP. ARE INSPECTION WILL BE ON OR ABOUT 30 DAYS. RECEIPT OF NOTICE ACKNOWLEDGED:

SIGNATURE: _____ INSPECTOR: _____ BADGE: _____

PRINT NAME: _____ FIRE MARSHAL: _____ DATE: _____

THIS REPORT IS BASED UPON OBSERVATIONS AT THE TIME OF SURVEY WHICH MAY NOT DISCOVER ALL HAZARDS.

Figure 5 The final notice should include consequences of failure to comply with requirements, including legal actions.

in every date. Use formal salutations such as Mr., Ms., and Dr. where appropriate. Use a subject or reference line to help your reader understand what the letter is about. Each letter should include these basic design elements:

Header	Agency letterhead or return address Date Inside address Subject/Reference:
Salutation	Dear Mr., Ms., or other recognized and appropriate form
Body	Start with an introduction, followed by the main topic(s) of the letter and ending with a closing paragraph or statement. Start a new paragraph for every major point.
Closing/Signature	Begin with "Sincerely" or "Regards," and end with your signature, printed name, and title.
Enclosures	List all documents or other information included with your letter.
Copies	List the name of everyone who received a copy of this letter, followed by their company or agency name.

Email

A documentation tool that is growing in popularity is e-mail. Most often, e-mails are used to respond to a request for information or code interpretation. E-mail can be a convenient, useful, and inexpensive way to communicate with property owners, design professionals, and others. Not quite as immediate as a telephone call, but more timely and less expensive than a detailed report delivered by post or courier, e-mails are commonly used by fire inspectors; however, you should remember that e-mails are considered legal documents, work products, or official records subject to the Freedom of Information Act and similar state Public Information Act laws. Your email correspondence should be professional and follow the same standards as your correspondence via letters. During legal proceedings, e-mails can be subpoenaed. Thus all e-mails that contain information that is relevant to your work should be saved according to your agency's records retention policies, just like any other written document.

Checklists

Common uses for checklists include simple fire inspection forms that list common and/or critical deficiencies and preliminary inspections of multiple locations with similar characteristics, such as schools or roadside vendors. One advantage of checklists is that they can be used during a fire inspection as a reminder of common issues. However, it is easy to forget or overlook less common problems, so you must be careful that you inspect "to code" rather than "to the checklist."

You can use a checklist as part of your post-inspection interview, and leave a copy of the checklist with the owner Figure 6. In many cases, a well-designed checklist is the only inspection record needed. In some cases, you may want to include diagrams or photographs to illustrate specific points of your inspection. You can also follow-up the checklist with a formal written report if needed.

A successful checklist starts with a clean, consistent format with easy-to-read groupings. Use standard paper size and fonts. A-size, 8.5" by 11" paper is preferred. Twelve-point type is preferred for legibility; but smaller fonts, down to 8 point, are acceptable. Be sure to check the font type and size for legibility. Margins and white space are not as critical for checklists. Margins often are as close to the edge of the page as a printer or photocopier will allow, usually 0.5". However, remember to leave adequate margin for binding or hole punching to ensure that critical information is not lost.

Checklists are commercially available, but you may want to develop your own checklists. Developing your own checklists and forms allows you to customize them to meet your needs based on the types of occupancies and conditions that are common in your community.

Detailed Reports

Detailed reports should be used to document serious deficiencies, information that will not fit easily into a letter or checklist format, numerous violations, or corrective actions required. You should also consider a detailed report when multiple occupancies or multiple tenants are part of the same inspection process. Detailed reports often include field notes, sketches, diagrams, and photographs.

Reports may be handwritten. A handwritten report may appear less threatening or informal, and lends an air of immediateness to an inspection. As an added advantage, you can leave a copy with the owner after the inspection; however, a handwritten report may appear poorly written, sloppy, or ignorable.

A typed report can be more time consuming to prepare but appears more official and authoritative. Most formal reports are accompanied by a cover letter and support documentation, such as diagrams, photographs, equations, and detailed analyses.

An effective report should use a clean, consistent format with easy-to-read groupings. Use standard paper size and fonts. A-size 8.5" by 11" paper is preferred. Twelve-point type is preferred for legibility; but smaller fonts, down to 8 point, may be appropriate is some cases. Check the font type and size for legibility. Long or detailed reports often use tabs and appendices to improve readability, provide additional reference information, or to provide a common place for tables, figures, charts, and photographs.

Recording Complaints

A complaint occurs when a member of the public indicates that there is, in their opinion, a safety issue at a building. You then need to investigate the complaint. If you determine that there are no code issues during your investigation, then the complaint is unfounded. If you note an issue that must be corrected, you will need to document the issue properly.

Depending on the seriousness of the issue, the violation must be corrected immediately or within a specific, articulated time frame. If an exit door is locked, the time frame may be immediate. If there are empty fire extinguishers, you might allow one or two days for a company to come and refill them, or for

Inspection Checklist
Inspection Procedures

PREINSPECTION CHECKLIST

Equipment: _____
_____

General
- ❏ Identification (photo ID)
- ❏ Business work hours

Clothing
- ❏ Coveralls
- ❏ Overshoes
- ❏ Boots

Personal Protective Equipment (PPE)
- ❏ Hard hat
- ❏ Safety shoes
- ❏ Safety glasses
- ❏ Gloves
- ❏ Ear protection
- ❏ Respiratory protection

Tools
- ❏ Flashlight
- ❏ Tape measure(s)
- ❏ Pad (graph paper) and pen or pencil
- ❏ Magnifying glass

Test gauges
- ❏ Combustible gas detector
- ❏ Pressure gauges
- ❏ Pitot tube or flow meter

Plans and Reports
- ❏ Previous reports
- ❏ Violation notices
- ❏ Previous surveys
- ❏ Applicable codes and standards

Notes: _____
_____
_____

SITE INSPECTION

Property Name: _____

Address: _____
_____

Occupancy Classification
- ❏ Assembly
- ❏ Educational
- ❏ Day care
- ❏ Health care
- ❏ Ambulatory health care
- ❏ Detention and correctional
- ❏ One- and two-family dwelling
- ❏ Lodging and rooming
- ❏ Hotel/Motel/Dormitory
- ❏ Apartment
- ❏ Residential board and care
- ❏ Mercantile
- ❏ Business
- ❏ Industrial
- ❏ Storage
- ❏ Mixed

Copyright © 2002 National Fire Protection Association (Page 1 of 2)

Figure 6 A sample checklist.

(Continues)

Hazard of Contents
- ❏ Light (low)
- ❏ Mixed
- ❏ Ordinary (moderate)
- ❏ Special hazards
- ❏ Extra (high)

Exterior Survey
- ❏ Housekeeping and maintenance

Building construction type
- ❏ Type I (fire resistive)
- ❏ Type IV (heavy timber)
- ❏ Type II (noncombustible)
- ❏ Type V (wood frame)
- ❏ Type III (ordinary)
- ❏ Mixed

Construction problems
Building height _____ feet _____ stories
- ❏ Potential exposures
- ❏ Outdoor storage
- ❏ Hydrants

Fire department connection
- ❏ Vehicle access
- ❏ Drainage (flammable liquid and contaminated runoff)
- ❏ Fire lanes marked
- ❏ Is it obstructed?
- ❏ Is it identified?

Building Facilities
- ❏ HVAC systems
- ❏ Gas distribution systems
- ❏ Conveyor systems
- ❏ Electrical systems
- ❏ Refuse handling systems
- ❏ Elevators

Fire Detection and Alarm Systems
See Form A-8.

Fire Suppresion Systems
See Form A-10.

Closing Interview
- ❏ Imminent fire safety hazards
- ❏ Housekeeping issues
- ❏ Maintenance issues
- ❏ Overall evaluation

Items to be researched:
- ❏ _____
- ❏ _____
- ❏ _____

Report
- ❏ Draft
- ❏ Review
- ❏ Final

Notes: _____

Copyright © 2002 National Fire Protection Association (Page 2 of 2)

Figure 6 A sample checklist.

(Continued)

> **Fire Inspector Tips**
>
> Complaint inspections do not take the place of any scheduled inspections. The complaint requires only looking for one specific item. The routine inspection is a full top-to-bottom look at the building.

the owner to purchase new ones at a store. If the violation is not dealt with correctly, reinspections must occur until the problem is corrected.

Regardless of whether a specific complaint form is used or not, certain information should be gathered. There are the basics such as the date of the complaint and the location of the business. When possible, get the complainant's name and contact information so you can follow up with the complainant if needed. Even if the complainant is anonymous, the complaint must still be investigated. Document what the complainant states the problem is. You should date when your complaint inspection is conducted and document the results of your inspection. Note any course of action the business needs to take and the time frame allowed to correct the violation.

When conducting the reinspection, the date of the reinspection should be noted, as well as what was found. If the violation is corrected, the complaint form is now completed, and it can be filed.

In most cases a standard inspection form is sufficient in noting a complaint. In the case where problems are life threatening or there are a great number of issues, then a letter listing all of the specific issues may be a better method to convey the results of the complaint inspection. For example, you conduct a complaint inspection at an apartment complex. Since all of the buildings are under one management company, you write a letter listing each building and the violations associated with each building. A letter also gives the opportunity to begin the paper trail of documentation. A letter can be sent by certified mail, with return receipt requested.

Code References

The proper way to document a violation is to indicate both the violation and what must be done to correct it. While it may be correct to state that the rear door does not meet code requirements, it is more helpful to indicate that deadbolts must be removed or the door cannot be blocked. To give the owner an opportunity to thoroughly investigate the violation, a code reference should be given. A typical reference might look like "This violates NFPA 13 section 10.3.2.1." In the cases when violations are noted and fixed immediately during the inspection, the code reference would not be needed; however, the violation should be documented, and you should indicate that the issue was resolved during inspection.

Record-Keeping Practices

Every fire inspection should include written documentation about the inspection. The documentation should, at minimum, note existing conditions and the correction of, or failure to correct, identified violations. Each piece of documentation should become part of the permanent record about a particular building or occupancy **Figure 7**. The inspection files are an important tool to improve fire safety and, if needed, can be used in legal proceedings.

Records Retention Requirements

All inspection files are subject to records retention laws and must be kept for a specific length of time, depending on their content. Most model codes include records retention requirements, usually three years. If a building has been involved in an arson fire, the legal requirements may change, depending on the statute of limitations and court orders to expunge certain information from arson investigation files.

Figure 7 Each piece of documentation should become part of the permanent record about a particular building or occupancy.

> **Fire Inspector Tips**
>
> In the codes, you will notice the use of the following words, "shall, may, and is authorized to". "Shall" is mandatory, and is used when an action is required; no latitude or discretion is allowed. By contrast, the use of "may" allows discretion, and you are not necessarily required to perform the action. Likewise, use of the term "is authorized to" indicates that you have the power to do something but are not obligated to do so.

> **Fire Inspector Tips**
>
> All inspection files, with few exceptions, are official government documents. This includes forms, checklists, detailed inspection reports, field notes, e-mails, sketches, photographs, and related materials.

Common classifications of public documents and retention schedules include:

- Archival or permanent — Forever
- Essential — Short range to archival
- Official public records — Short range to archival
- Office files and memoranda — Short range to archival

Since different rules apply to actual retention periods for different types of records, it is important to know the laws in your state. On example of a public document retention schedule is shown in **Table 1**.

Since inspection files form a permanent record of a building and its occupancies that often outlasts any statute of limitations, inspection files are often kept long after legal requirements expire. In fact, it's a good idea to maintain building files for the life of the building. Most agencies maintain active files for some period, and then archive older files to storage. The decision to place files in an archive instead of an active file location is more of an agency decision than a legal requirement.

Freedom of Information Requests

With very few exceptions, inspection reports and related documents are subject to open records requests under the **Freedom of Information Act (FOIA)**. The Freedom of Information Act was signed into law on July 4, 1966 by President Lyndon B. Johnson. This law applies only to federal government documents. Since that time, many states have passed similar legislation to govern the release of documents at the state and local level, including records of counties, parishes, cities, and other political subdivisions such as school districts, emergency services districts, and other special-purpose districts. Freedom of Information (FOI) laws are strict, and penalties for non-compliance can be severe, so it's best to

Table 1 Sample Public Records Retention Schedule

Item No.	Series Title and Description	OPR or OFM	Primary (Source Document)	Secondary (All Copies)	Disposition Authority Number	Remarks
12.2	ALARM INSPECTION LOG	OFM	3 years	Destroy when obsolete or superseded	GS53-04-01	
12.4	ALARM SYSTEMS TEST AND MAINTENANCE RECORD	OFM	3 years	Destroy when obsolete or superseded	GS53-04-03	
12.6	BUILDING INSPECTION REPORTS	OFM	6 years	Destroy when obsolete or superseded	GS53-04-05	May include records pertaining to the decommissioning of underground fuel tanks
12.7	BURNING PERMITS	OPR	3 years	GS53-04-06	Destroy when obsolete or superseded	A record of temporary permits authorizing the recipient to burn on a specific site for a specific period
12.12	FIRE ALARM SYSTEMS DRAWINGS	OFM	Life of facility	Destroy when obsolete or superseded	GS53-04-11	
12.13	FIRE CODE VIOLATION NOTIFICATION	OPR	6 years	Destroy when obsolete or superseded	GS53-04-21	Official notice of violation and statement of required corrective action
12.14	FIRE HYDRANT AND WATER MAIN INSPECTION LOG	OFM	3 years	Destroy when obsolete or superseded	GS53-04-12	
12.15	FIRE HYDRANT AND WATER MAIN INSPECTION REPORTS	OFM	3 years	Destroy when obsolete or superseded	GS53-04-13	
12.23	INSPECTOR'S TEST OF FIRE FIGHTING EQUIPMENT	OFM	Life of equipment	Destroy when obsolete or superseded	GS53-04-18	

OFM: Public records which have been designated as "Office Files and Memoranda" for the purposes of RCW 40.14.010.
OPR: Public records which have been designated as "Official Public Records" for the purposes of RCW 40.14.010.

adopt a policy and ensure that all employees understand how to follow through on any FOI requests. While the exact provisions of each state law differ, most include provisions about:
- How the public can initiate an FOI request
- How long a government agency has to respond to an FOI request
- If the agency can charge a fee to process the request
- If the agency can ask why the information is wanted
- The penalties for failure to comply with an FOI request

When a FOI is received, it is a legal request and all appropriate laws must be followed. Most agencies will have a FOI officer. This person will be the one to receive the request, route it to the appropriate person, receive the requested documents, review those documents, and return the results to the requester, all the while making sure that laws are being followed.

Most documents are public records and susceptible to disclosure. Some of the exceptions to disclosure would be company trade secrets, internal-agency memos, active criminal investigations, medical records, and most personnel information. Because there are certain pieces of information that cannot be released, it is important that the FOI officer redacts information as needed. Your role as the fire inspector is to simply supply the information asked for in the FOI request. You should advise the FOI officer if you suspect any sensitive information is present in the documents. For example, on your inspection form you may have listed the emergency contact information for a business owner. This information should be pointed out to the FOI officer so that the FOI officer can determine whether or not to release it.

The FOI request must be filled in a specific amount of time. If the request is not filled, it is considered an automatic denial. If the FOI officer denies the request for any reason, the reason for denial must be given, and the requestor has the right to appeal the denial. You may need to relay the information given to you by the FOI officer to the requester.

■ Organizational Practices

File arrangement is subject to agency preference, but most agencies arrange and index their files by address. This is because the occupancy type, owner, or tenant may change many times during a building's lifetime, but the address rarely changes. This method ensures that all information about a particular building remains with the file.

Fire Inspector Tips

Fire inspection information is often subject to Community Right-to-Know legislation. At the national level the Emergency Planning and Community Right-to-Know Act of 1986 (EPCRA, commonly known as "SARA Title III") was signed into law on October 17, 1986 by President Ronald Reagan. This act amended the Comprehensive Environmental Response, Compensation, and Liability Act of 1980 (CERCLA, commonly known as "Superfund"). One of EPCRA's purposes is to provide the public with information about potential chemical hazards in their community. Any occupancy that stores or uses more than the threshold amounts of chemicals or other hazardous substances is required to report that information periodically to state and local officials.

Cross-references are important, too. Businesses may move locations, but keep the same owners and operators. Such occupancy changes may require application of current code standards if the new location is a conversion from another occupancy type. Often violations, especially for poor housekeeping and operating practices, move with the business and require correction.

You should use a consistent file format. Two common file formats are chronological and functional. The chronological format arranges file contents by date, usually with the most recent activity at the front of the file **Figure 8**. This file format is easy to build; simply place the new information on top of the last entry, and return the file to its storage location. The difficulty is finding specific or related activities if you do not know the date of the activity.

A functional file format groups information into section by activity type. Within each section, you can sort activities chronologically, with the most recent activities at the front of each section **Figure 9**. This file format takes more effort to establish and maintain but makes it easier to find information about specific functional activities. One functional file format uses the groupings listed below:

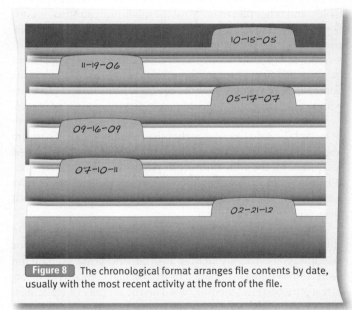

Figure 8 The chronological format arranges file contents by date, usually with the most recent activity at the front of the file.

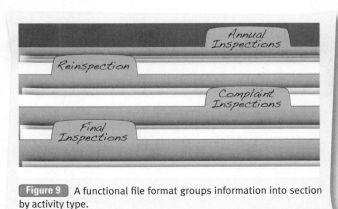

Figure 9 A functional file format groups information into section by activity type.

- Applicable inspection code or standard (sometimes listed on the file folder cover)
- Applications for inspections, permits, or certificates
- Results of plans review
- Permits, licenses, and certificates of compliance and/or occupancy
- Documentation of hazardous operations and hazardous materials inventories
- Fire detection, notification, and suppression systems
- Approved modifications, exceptions, and variances
- Complaints
- Inspection reports and written correspondence
- Formal violation notices
- Fire investigations
- Legal actions—civil and/or criminal
- Field notes and other supporting documents

Reporting Systems

Electronic reporting systems are becoming more popular. The primary advantages to electronic reporting systems are the ability to research code issues in the field, reduce errors, and provide on-scene printing. With practice, you can attach various photos, tables, and other files to improve the quality of your final report. Some electronic reporting systems allow you to enter data while performing the inspection. Others require you to transfer handwritten or checklist information into a database. Reports can be generated using predetermined or custom-designed formats. Some software systems provide only basic formatting and report-editing functions; others provide sophisticated reporting features.

Each feature comes with a cost; as a general rule, the more powerful the software, the higher the purchase price. There are also learning curves and hardware purchase costs associated with making the transition from paper to electronic reports. It is important for your agency to establish policies about who has access to the reports and ensure adequate back-ups, in both electronic and paper format.

Another key feature of electronic reporting systems is the ability to perform complex sorting and grouping functions. For example, you can easily identify all occupancies that contain fire protection systems maintained by a particular company or prioritize a reinspection schedule based on the severity of violations.

Fire Inspector Tips
Do not take legal proceedings personally.

Legal Proceedings

A code violation begins with a notation on a fire inspection form and may end in a courtroom. If you have poor documentation, presenting concrete material in court may be difficult; however, with good, detailed documentation, testimony will be supported by what you wrote and photographed during the fire inspection.

Whenever you conduct a fire inspection, all documentation and correspondence is subject to use in a legal proceeding, so document thoroughly. You never know when findings from a routine inspection may end up in court. If a fire or injury occurs in the building, all of its records, including past inspection reports, will be pulled. Everyone from fire officials, building officials, government officials, and attorneys will be examining the records to see what was noted on previous inspections. They may be looking for a trend of recurring violations or a violation that was never documented.

Presenting Evidence

If you have to go to court to present evidence, you should review all a building's records until you are an expert on that building, from type of occupancy to any violations listed during the its lifetime. Review the codes to make certain that you completely understand exactly what the code requires. For example, say you are testifying about a fire that occurred after an owner converted a lumber warehouse to a tire warehouse without notifying the city. Due to inadequate sprinkler coverage, the fire spread to neighboring buildings and destroyed an entire storage complex. Before the court case, you would examine the records to see if the owner had a history of violations, and you would review the codes to know sprinkler system requirements for the occupancy.

When in court, act and look professional. Wear your uniform or professional attire. The witness stand is not the stage to be clever or witty; simply answer the questions you are asked. For example, if you are asked, "Do you know what day it is?" your answer should be "Yes," not "It's Monday, July 10." If the attorney wants to know the date, the attorney will ask, "What is the date?"

When giving testimony, only state information you know is fact. If you cannot remember a fact, simply state that you are uncertain or that you need to refresh your memory and would like to look at your paperwork. If you do not know the answer to a question, say that you do not know. Keep your answers short. There is no need to expand on a yes or no answer, unless the answer needs clarification. Refrain from giving your opinion—you are in court as a fire official to present the facts.

Wrap-Up

Chief Concepts

- As a fire inspector, you must be able to document what you observed during the inspection and what needs to be corrected, as well as communicate to the owner or the owner's representative the importance and necessity of correcting any deficiencies found during the inspection.
- Your written records, from standard inspection forms to detailed inspection reports, are often the only documentation of existing conditions, code violations, and corrective actions that need to be performed.
- Written records eliminate one common barrier to communication—trying to remember exactly what was said and by whom. Written records are superior to verbal communications for documentation purposes
- Good field notes are the basis for thorough and accurate records. Field notes are the original, on-scene description of conditions that existed at the time of the fire inspection.
- Sketches, diagrams, and photographs provide visual documentation of observed conditions. Use field sketches as rough drafts to provide graphic representations of room layout and dimensions, egress locations, identified fire hazards, fire protection system features, and similar information. Diagrams are the final inspection drawings. Photographs are visual evidence of what was seen during the inspection.
- All fire inspection reports should include property information, contact information, inspection-specific information, and follow-up information.
- Reinspection reports often just need to document if the violations found during the initial inspection were properly addressed.
- A complaint occurs when a member of the public indicates that there is, in their opinion, a safety issue at a building. You then need to investigate the complaint. If there are no code issues during your investigation, then the complaint is unfounded but still must be documented.
- To give the owner an opportunity to thoroughly investigate the violation, a code reference should be given.

Wrap-Up

- All inspection files are subject to records retention laws and must be kept for a specific length of time, depending on their content. Most model codes include records retention requirements, usually three years.
- File arrangement is typically by agency preference, but most agencies arrange and index their files by address. This is because the occupancy type, owner, or tenant may change many times during a building's lifetime, but the address rarely changes.
- A code violation may begin with a notation on a fire inspection form and may end in a courtroom. If you have poor documentation, presenting material in court may be difficult, as you must rely almost completely on your memory.

■ Hot Terms

<u>Final notice</u> Written correspondence used when violations are not corrected to notify the owner that legal action may be taken to ensure code compliance.

<u>Freedom of Information Act (FOIA)</u> Signed into law on July 4, 1966 by President Lyndon B. Johnson, allows public access to government records.

Fire Inspector *in Action*

It is your first day as a fire inspector. You have been assigned a department vehicle and issued the personal equipment necessary to perform routine fire inspections. The inventory list of personal equipment includes the following:
- Camera, digital
- Checklists for common occupancy locations
- Clipboard, with ruled and graph paper, pencils, and ruler
- Measuring tape, 100-foot
- Hard hat, leather gloves, latex gloves, and safety glasses
- Portable radio and cellular telephone
- Local codes and standards

1. Later that day, you are assigned to complete an initial fire inspection of a new daycare center. Which of the following pieces of equipment will you most likely leave in your car during this visit?
   A. Checklist listing the common violations in daycare centers
   B. Digital camera
   C. Clipboard, with ruled and graph paper, pencils, and ruler
   D. Local codes

2. While you inspect the day care center, you notice a burned-out light bulb in an exit sign above an emergency exit door. There are no other violations. You should:
   A. Document the deficiency, the required corrected action, and the follow-up inspection date, all in writing.
   B. Ignore it; it's not necessary to correct if only one problem exists.
   C. Close the day care and arrest the operator for a serious code violation.
   D. Verbally explain the problem and verbally ask the operator to correct it before the next annual visit.

3. On your annual visit a year later, you discover a new addition with two rooms and a kitchen, but no commercial vent or suppression system. None of these modifications is shown on the original site plans in the permanent records. You should:
   A. Issue a Fire Marshal's Order to supply architect-certified drawings for the additions.
   B. Use field notes, sketches, and photographs to document the additions as part of your inspection process.
   C. Use a simple cover letter to document the inspection and required actions.
   D. Both A and B.

4. You check your e-mails and find a request from the local television station's investigative reporter. He says he has received a call from a parent of a child at the day care that has complained about "severe fire hazards" at the day care. He asks for an on-camera interview and for you to send him a copy of "any and all records of any and all inspections." He includes his contact information, including cell phone number. You should:
   A. Ignore the request because the e-mail request is not on station letterhead.
   B. Call the reporter and ask why he wants the information.
   C. Follow your department's open records request policies and procedures.
   D. Agree to the on-camera interview, but do not provide any written reports.

Life Safety and the Fire Inspector

CHAPTER 15

NFPA 1031 Standard

There are no NFPA 1031 Standards for this chapter.

FESHE Objectives

Fire and Life Safety Education

1. Differentiate between Public Education, Public Information, and Public Relations/Marketing. (pp 308–309)
2. Demonstrate the need for establishing fire and life safety education as a value within the fire service culture. (pp 309–311)
3. Identify stakeholders; develop partnerships and coalitions to work on fire and life safety education activities. (pp 311–312)
4. Identify and use local, regional, and national sources of data for fire and injury prevention programs. (p 312)
5. Identify budget needs for program delivery and the process for requesting funds. (pp 312–313)
6. Select, design, implement, and evaluate fire and life safety education programs that address specific community risk issues. (p 312)
7. Develop an accountability system to measure program delivery. (p 313)

Knowledge Objectives

After studying this chapter, you will be able to:

1. Discuss the role of the fire inspector in promoting fire and life safety education.
2. Discuss the techniques that the fire inspector may use create a fire and life safety program.
3. Discuss the techniques that the fire inspector may use to encourage partnerships and cooperation in stakeholders.
4. Describe how to determine the community's fire and life safety risks.
5. Describe how to create a fire and life safety program to meet the needs of the local community.
6. Describe how to measure the effectiveness of a fire and life safety program.

Skills Objectives

There are no skills objectives for this chapter.

You Are the Fire Inspector

Your community has experienced three serious fires in commercial buildings over the past six months. The Fire Chief has asked that you to prepare a report on the recent commercial fires in your community. In the report, you are to gather the relevant factors that contributed to the fire and then develop and make recommendations for appropriate intervention strategies. You walk out of his office, wondering where to begin. You wonder if you have the necessary skills to meet the task that lies ahead.

1. What data should you collect?
2. How are you going to change the behavior of an entire community?

Introduction

One of the roles of fire inspector is being a fire and life safety educator. **Fire and life safety educators** are personnel that teach fire safety messages to target audiences **Figure 1**. This is one of the most valuable members of the fire service, yet one of the least understood. Traditionally, these educators have implemented public education programs based upon gut feeling and tradition, with little understanding of whether the programs are truly effective. To create a successful fire and life safety education program, you must understand how to assess risks and their root causes, develop relationships to support your life and safety education programs, and how to measure your programs' effectiveness in preventing fires. The fire and life safety educator is a professional who is committed to saving lives, reducing injuries, and saving property.

Figure 1 Fire and life safety educators are personnel who teach fire safety messages to target audiences.

Role of the Fire Inspector in Fire and Life Safety Education

Ask virtually anyone in the fire service to summarize their core mission and they will respond, "To save lives and property." Prevention is an effective and efficient way to achieve this mission. Fire and life safety education is one of the core tenets of effective fire prevention.

As a fire inspector, you must have the ability to develop programs that will address the risks of fire. Your programs should:
- Target behaviors that contribute to the likelihood of fires occurring
- Work to reduce injuries and damage if a fire does occur

Public Education, Public Information, and Public Relations

Public education is the presentation of a safety message with the sole goal of reinforcing good behaviors or changing undesirable behaviors in order to make the community safer. **Public relations** are activities focused on promoting the organization by creating a public perception that is positive and that creates public support for the organization. Often the term marketing is used as a synonym for public relations. It is easy for the line between public education and public relations to blur because providing high quality fire and life safety education will build a positive image of the entire fire department, including any

Fire Inspector Tips

Today, fire and life safety education expands beyond just fire prevention. In order to meet the mission of saving lives and property, you may present programs on the importance of car seats, carbon monoxide detectors, or bicycle helmets.

independent fire inspection agencies. Providing high quality public relations will not provide effective public education. For example, having a crew of fire fighters attend a neighborhood block party is public relations—there is little expectation that the attendees will change their behavior based upon the visit **Figure 2**. When a fire inspector teaches school children how to create an escape plan, this is public education that will benefit public relations **Figure 3**.

Public information is another term that is sometimes interchanged with public education and public relations. **Public information** is information that is disseminated to the public. Fire departments do this in a variety of ways such as news releases and interviews. Public information may include information about incidents that have occurred, events that will be taking place, or other matters of public concern. For example, a news release giving the details of the cause and origin of a fire is considered public information. Proactive public information will generally result in a positive public image of the fire department and will also provide a safety message. For example, when

Figure 4 Proactive public information will result in a positive public image of the fire department and will provide a safety message.

Figure 2 Public relations focus on promoting the organization by creating a public perception that is positive and that creates public support for the organization.

the incident commander tells the press that the occupants were able to escape because of a working smoke alarm, the public information event reinforces the use of smoke alarms **Figure 4**.

Creating a Fire and Life Safety Education Program

When creating a fire and life safety education program, you should follow a series of steps to ensure that your resources are used wisely. Too often, there is a tendency to adopt another community's program without fully knowing if it is a wise choice for your community.

■ Indentifying the Problem

Often anecdotal evidence is used to determine the hazards in the community. **Anecdotal evidence** is evidence that may or may not be true and is used to generalize when there is insufficient data. For example, the fire department notices that they have had several fires that were caused by children playing with matches and lighters. The fire inspector chooses to focus public education on this issue, visiting daycare centers to deliver a message about the dangers of playing with matches and lighters to the

Figure 3 Public education is the presentation of a safety message with the sole goal of reinforcing good behaviors or changing undesirable behaviors in order to make the community safer.

children; however, no information is given to the parents about this community problem. This shows that the inspector does not understand the underlying causes and means of prevention that need participation of multiple parties. This method frequently fails to understand and address the underlying causes go undetected altogether.

Collecting Data

A better method is to collect data to help identify problems in the community. Anecdotal evidence might guide some of the data that is collected, but the data should help identify the real problem. In addition, the data allows you to identify the root causes of the problems. Problems are simply the symptoms, so failure to identify the root causes of the problem may result in the problem not being resolved. For example, in the community with a rash of fires caused by matches and lighters, the root cause of the problem may actually be the parents and caregivers leaving the children unattended where matches and lighters are accessible. The public education message would be more effective by addressing parents and caregivers in addition to the children. The parents and caregivers should be reminded not to leave children unattended and store matches and lighters in a safe location.

There are numerous sources of data available. The best choice of data is local data. For example, being able to look at the leading cause of residential fires in your community would be more relevant to your community than national statistics; however, sometimes communities lack local data or there are so few incidents that it is difficult to determine whether a single incident is statistically valid.

Truly useful data provides more than just the basics: time, date, location, type of event, and who responded. This basic information lacks the details to help you to determine common fire events and the underlying factors that cause the events to occur. This is one of the reasons why the federal government encourages all fire departments to use the <u>National Fire Incident Reporting System (NFIRS)</u>. This national database collects data on the many underlying factors of a fire. This data is essential to making good public education decisions at the local level.

There are many state sources of data, including the State Fire Marshal's office. The State Fire Marshal amasses the data from the local fire departments as part of NFIRS. In many states, they also provide fire investigations for many local jurisdictions and for particularly serious fires. Non-fire data may be available from the state's Health Department, Homeland Security Department, Family Services Department, and Transportation Department. Likewise, there are many state organizations which collect statewide data, such poison control centers.

There is a vast amount of information at the national level, starting with the United States Fire Administration. Their information is available to the public through print and online sources. The United States Fire Administration has a vast amount of resources focused on fire and life safety prevention and has a section dedicated to statistics that you can use to determine local risks. Likewise, the Federal Emergency Management Agency's (FEMA) Learning Resource Center (LRC) provides a wealth of information. Its national library is located at the National Fire Academy in Emmitsburg, Maryland and is available online.

The United States Department of Health and Human Service collects data and provides prevention information for many of the medical incidents to which the fire service responds. Likewise, the United States Department of Transportation provides data and educational information about a wide variety of transportation issues.

One of the most important data sets available is the United States Census Data. Every ten years, the Census Bureau collects census information. This information covers a wide variety of data and ties it to very small geographic areas. For example, community data can reveal income levels, as well as certain habits that are occurring within a small section of a community, such as smoking. This information can help the fire inspector to determine if a cigarette safety program is needed in the community. The Census Bureau will also provide support in helping interested parties sorting through, assessing, and applying the collected data.

Risk: Frequency versus Severity

Once the data is collected, you will quickly see that some events occur more often (<u>frequency</u>) than others and that some of the problems are more significant when they do occur (<u>severity</u>) than others. In order to maximize available resources, create a matrix plotting the problems based upon frequency and severity Figure 5 . The most common method is to plot the frequency of a specific risk on a horizontal axis. It is recommended to

Fire Inspector Tips

Another resource is the National Fire Protection Association (NFPA). One specific area of focus of the NFPA is the evaluation of fire and life safety data, the production of technical reports about incidents and their causes, as well as the development of prevention programs.

Fire Inspector Tips

Making the community safe is more than just preventing fires. To succeed in the mission of saving lives and property, you must also explore the data on accidents and injuries. This data is often available from local hospitals, safety councils, emergency managers, and other local agencies/organizations that track local incidents themselves. In addition, many local libraries have a wealth of information and will be very helpful in assisting with your search for common accidents and injuries in the community.

Fire Inspector Tips

Generally, local fire data is maintained in the local fire department's records systems. Data will generally be collected in an electronic spreadsheet, but some small departments may only have paper records that you may have to sort through to collect the data.

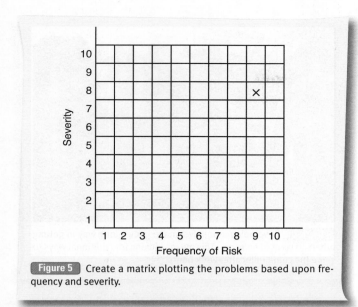

Figure 5 Create a matrix plotting the problems based upon frequency and severity.

Fire Inspector Tips

A dilemma can occur with problems that are low frequency, but high severity (upper left corner of the graph) and those that are high frequency, but low severity (lower right corner of the graph). Generally, focus pulls toward those that are higher frequency but lower severity, but this may vary.

use a rating scale, such as 1 to 10, to give the relative degree of frequency. Then, each problem is plotted by its severity on the vertical axis, again using a 1 to 10 scale. The goal of creating the matrix is to focus the public education message on those issues with the greatest risk to the community.

Once the problems have been plotted, you will find some problems arise more frequently and have a high degree of severity. These types of problems will be in the upper right corner of the graph. Those with low frequency and low severity will be in the lower left corner of the graph. This allows you to quickly determine the highest priority of problems to address. For example, a community in Kansas might have tornados as a high frequency, high severity event while categorizing an earthquake as a low frequency, high severity event because they are not near a fault line. This might highly contrast with a community along the California coast, where tornados would be very low frequency while earthquakes would have a much higher frequency.

Developing the Fire and Life Safety Education Program

For the fire and life safety education program to be successful, it must be well thought out with goals established. Some of the questions to consider include:
- What do you want to achieve?
- Do you want to educate every student in a certain age group?
- Will you provide refresher education and training as this group matures?
- Are you going to go to the students or have them come to you?
- Are you going to use videos or distribute gifts with a fire safety message?
- Who will present the educational programs?
- Who will train the educators?
- How will the program be financed?
- How long will the program run?
- When will it start?

These questions can only be answered once the goals and objectives have been established.

Support

The best made plans will fail if there are inadequate resources to carry them out. You can create a great fire and life safety education program; however, if there are not adequate funds to purchase the required materials, or there are not enough people to present the program to the community, all of your work will be of little value. Therefore, your first step is to determine the level of commitment from the fire department to support your program. You will need to obtain a general idea of the resources and funds available for your program.

This will require the input of the person who controls the fire prevention budget. This might be the Fire Marshal, an Assistant Chief, or even the Fire Chief, depending upon the individual organization. There are many competing demands for resources. While the need for a first class fire and life safety education program may be recognized, the available resources may not be available due to other higher priorities.

The conversation with the responsible individual might help guide you on how to develop the fire and life safety education program. For example, if you are tied up with other commitments, but fire fighters are available to present the program, you might delegate the presentation responsibilities to them. If fire fighters are not available, you might delegate a stakeholder group to deliver the program. Likewise, if you have the time but lack the financial resources, you might focus on obtaining grants, donations, or partnerships to provide the funding.

Identifying Stakeholders

You can use stakeholder groups to develop support and resources, to solve an identified problem. A **stakeholder** is an individual or group that is impacted by an issue. Returning to the example of the fires started by unattended matches and lighters, City Council members are stakeholders because the public looks to them to resolve community problems. The hospital's burn unit is a stakeholder because they have to treat the children who are burned as a result of a fire. The insurance companies are stakeholders because they pay the claims for the property damage as well as the heath care. Neighborhood groups are stakeholders because high fire incidents in the neighborhood reduce property values. As you can see, there are many stakeholders for one problem.

You should develop a list of all of the stakeholders for each issue you wish to address. Once the list is developed, it should be prioritized by those stakeholders who are impacted the greatest by the problem, or those with the most interest in resolving the problem. For example, a parenting coalition in your community may be focused on resolving the issue of children playing with matches and lighters and may be eager to work with you on your fire and life safety education program.

Developing Relationships

Once the most important stakeholders are identified, you must work to educate them about the problem and then encourage them to assist you in taking action. This is easier if there is already a strong relationship between the fire department and the stakeholder organization. Strong relationships take time to develop. Simply approaching a stakeholder "cold turkey" may work, but often it will not. Relationships are about people: two or more people are attracted together because they like each other on a personal basis. So how do you get people to like you? It boils down as simply as being nice and helping others. If you are nice and genuinely want to help others, others in turn, want to help you Figure 6 .

One way you can demonstrate being nice and helping others is through public relations. Developing a positive image goes a long way in getting others to want to help you. This means being professional at all times and looking for additional ways to serve the community after work hours Figure 7 .

Developing Goals and Objectives

Once the root cause of the problem is identified, the educational focus is identified, and the fire department and stakeholders are on board, you need to determine the goals and objectives of the fire and life safety education program. Goals are broad in nature and reflect the overall direction of the program, while the objectives are specific and measurable and provide the means for evaluating the success of the program. In order to foster "buy in" or "ownership" in the new program, a wide variety of personnel should be included in setting the goals and objectives. Take the time to meet a varied selection of fire personnel from fire fighters to the Fire Marshal.

Figure 6 Building strong relationships with stakeholders is as simple as being nice and helping others.

Figure 7 Developing a positive image goes a long way in getting others to want to help you. This means looking for additional ways to serve the community after work hours.

Content Development

Once the goals and objectives have been established and you know what the available resources are, the content can be developed. Many professionally created programs are available at no charge through the United States Fire Administration. There are also many commercially developed programs and materials available. If there are adequate financial resources available, and a commercial program addresses your problem, this may be the best choice. A commercial program should be able to give you metrics to prove that it is effective in presenting its message.

If a commercial product is not available or you lack the resources to purchase it, you can create your own fire and life safety education program or look to other fire departments that have faced the same issue. FEMA's LRC has many research papers about the success of various fire and life safety education programs in fire departments across the country. Often fire departments are more than happy to share their successes.

Developing your own fire and life safety program can be very time consuming and takes a significant understanding of educational techniques to ensure that it is created properly. You might consider soliciting the help of local teachers, or colleges and universities for technical expertise on developing curriculum and creating an educational program.

Budget

Once the fire and life safety education program is fully developed, the actual budget must be developed. This requires a careful study of all the facets of the program. You must consider both the direct and indirect costs. **Direct costs** include expenses that must be paid but would otherwise not have occurred without the program such as printing, supplies, and fuel. **Indirect costs** are those costs that the fire department would have incurred whether the program existed or not. An example would be the value of fire personnel time spent presenting the program. These costs could be significant in a career fire department.

The indirect costs should include the cost of benefits as well. While these funds will be expended whether the fire personnel are providing public education or not, knowing these costs allows for a valid cost/benefit analysis to be conducted. Any overtime costs should be included as a direct cost since the cost is in addition to the fire department's normal expenditures.

Once the expenses are determined, the revenues should also be considered. All of the revenue may come from a public education line item in the fire department's normal operating budget. If this is not the case, other sources of funds should be identified such as grants or donations. Once the revenues and expenses are identified, the budget should balance and provide a roadmap for the program. The budget should ensure that the program stays within its projections and is accountable for its funds.

Presenting the Material

Once the fire and life safety education program is purchased or developed, it's time to implement it. You may exclusively present fire and life safety education program. More often these programs are delegated to firefighting personnel to present. Fire officers may not understand their role or the effort that is necessary to present a comprehensive safety program to the public. Therefore, you must create a reason for them to want to participate. This can be done in a variety of ways. One way is to show them the goals of the program and how it will make the community a safer place to live. You can also tap into their own self-interest by showing how it will make firefighting less dangerous for them. You may also try to make it fun, such as developing a program that will allow any fire personnel with the acting bug to present skits demonstrating proper fire behavior Figure 8 .

Effectiveness Metrics

Virtually every fire department provides some form of public education; however, the effectiveness of these programs is difficult to prove without effectiveness metrics. Effectiveness measures how well a program is meeting its goals, and metrics are the data proving the claim. Because there are limited resources, the fire department must quantify the return they are getting on their public education investment. The question

Figure 8 Fire and life safety education programs can be entertaining and fun for the presenters.

always comes around to, "How do you measure something that didn't happen?" but there are ways to quantify results.

To measure the effectiveness of a fire and life safety program, first, measure the behaviors driving the problem before the fire and life safety public education program begins. Then examine local data to see if there is a change in behavior. The results should be compared to the goals and objectives of the program. If it isn't successful, it should be modified or abandoned in favor of other programs that can show their benefits. For example, after a series of fire deaths, you might assess the percentage of homes without a working smoke alarm either due to non-existence or a dead or missing battery. You would then present your fire and life safety educational program to the community. After the presentation, you would reassess the percentage of homes without a working smoke alarm. Any positive change should be attributable to your program. More smoke alarms *should* translate into fewer fire deaths.

Another example is evaluating children on how they react if awoken to a smoke alarm going off in their home. Once the initial assessment is made by talking with school children, the fire and live safety education program is presented in the classroom. After a period of time, the children are reassessed to see how they react to smoke alarms.

Wrap-Up

■ Chief Concepts

- One of the roles of fire inspector is being a fire and life safety educator. The fire and life safety educator is a professional who is committed to saving lives, reducing injuries, and saving property.
- Fire and life safety education is one of the core tenets of effective fire prevention.
- Fire and life safety education programs should target behaviors that contribute to the risk of fires and reduce injuries and damage if a fire occurs.
- Public education is used to change attitudes or behaviors. Public relations are activities to improve the image of an organization. Public information is information disseminated to the public.
- Effective public education can provide positive public relation benefits. Public information can create a positive image and reinforce public education messages.
- Creating a fire and life safety program begins with identifying the problem. To help identify the greatest fire and injury hazards in the community, you should collect data to help identify the problem and its root causes.
- Once the problem is identified, determine the goals and objectives of the fire and life safety educational program.
- Developing relationships with stakeholders is essential in obtaining the resources to create a fire and life safety education program.
- Developing a successful relationship begins long before there is a request for assistance. In order to build strong relationships within the community, simply be professional, friendly, and helpful.
- Fire and life safety programs may be locally developed, modeled after programs by other fire departments, or commercially developed.
- To measure the effectiveness of a fire and life safety program, first, measure the behaviors driving the problem before the fire and life safety public education program begins. Then examine local data to see if there is a change in behavior. The results should be compared to the goals and objectives of the program.

■ Hot Terms

<u>Anecdotal evidence</u> Evidence that may or may not be true, used to generalize when there is insufficient data to base it upon.

<u>Direct costs</u> Expenses that must be paid but would otherwise not have occurred without the program such as printing, supplies, or fuel.

<u>Fire and life safety educators</u> Personnel that teach fire safety messages to target audiences.

<u>Frequency</u> How often a particular type of incident occurs.

<u>Indirect costs</u> Costs which the fire department would have incurred whether the fire and life safety education program existed or not, such as labor and apparatus costs.

<u>National Fire Incident Reporting System (NFIRS)</u> A national database which collects data about fire incidents including the collection of many of the underlying factors that caused the fire.

<u>Public education</u> Teaching a safety message with the goal of reinforcing good behaviors, or changing undesirable behaviors, in order to make the community safer.

<u>Public information</u> Information that is disseminated to the public via the fire department.

<u>Public relations</u> Activities that are focused on building a positive image of any organization.

<u>Stakeholder</u> An individual or group that is impacted by an issue.

<u>Severity</u> The amount of death, injury, or damage that is the result of an incident.

Fire Inspector *in Action*

The Fire Chief approaches you with his concern about the fire department's fire and life safety education program. He says that he thinks the fire department needs to participate in the upcoming community block parties because the dispatch records indicate that fires in the area are on the rise. He asks if you think this is the best use of the department's resources.

1. Would the block party be a good method to reduce the number of house fires?
   A. Yes, block parties put you in contact with many children, which is a high risk group.
   B. Yes, block parties put you in contact with many children and adults, so you can focus on changing behaviors.
   C. No, block parties are primarily a public relations event where minimal public education occurs.
   D. No, block parties tend to draw business people, so the audience would be wrong if you want to reduce house fires.

2. Do you feel you have enough information to determine an appropriate fire and life safety education program?
   A. Yes, the Fire Chief has enough anecdotal evidence to suggest an appropriate program.
   B. No, the dispatch records do not indicate the root cause of the fires, so an effective program cannot be developed.
   C. Yes, the dispatch records provide data which indicates house fires are on the rise.
   D. No, the dispatch records are not sufficient to indicate there really is a problem.

3. The first step in the development of a fire and life safety program is:
   A. Identify the problem.
   B. Collect data to determine the root cause.
   C. Create a budget.
   D. Develop the goals and objectives.

4. Which of the following will provide local data about the demographics of the community, which will assist in focusing your public education message?
   A. National Fire Incident Reporting System (NFIRS)
   B. US Census
   C. National Fire Protection Association (NFPA)
   D. US Department of Health and Human Services

APPENDIX A: Changing Codes and Standards

NFPA 1031 Standard

5.2.4 Recommend modifications to the adopted codes and standards of the jurisdiction, given a fire safety issue, so that the proposed modifications address the problem, need, or deficiency.

(A) Requisite Knowledge. State statutes or local ordinances establishing or empowering the agency to adopt, enforce, and revise codes and standards; the legal instruments establishing or adopting codes and standards; and the development and adoption process for fire and life safety legislation or regulations.

(B) Requisite Skills. The ability to recognize problems, collect and develop potential solutions, and identify cost/risk benefits.

Development of NFPA Documents

NFPA is the publisher of almost 300 codes, standards, recommended practices, and guides that are related to fire and life safety. All NFPA documents are voluntary documents, which means that they do not have the power of law unless an AHJ adopts them. A number of NFPA's codes and standards have been adopted into law in various jurisdictions around the world, including the following:

- NFPA 1, *Fire Code*™
- NFPA 54, *National Fuel Gas Code*
- NFPA 70, *National Electrical Code*®
- NFPA 101, *Life Safety Code*®

One of the distinctive features of all NFPA documents is that they are developed through a consensus process, which brings together technical committee volunteers representing varied viewpoints and interests to achieve consensus on the document content. There are more than 200 of these technical committees, and they are governed by a 13-member Standards Council, which is also composed of volunteers.

It is the technical committees—not NFPA staff—who are responsible for the document content. In contrast to a common misconception, the staff liaisons do not write the material that is contained in the document. Their role is to coordinate the development process, and the NFPA's function is to serve as a mechanism for publishing the material that is written or approved by the technical committee volunteers.

Technical Committees

More than 6000 volunteers serve on NFPA's technical committees, with selection of these volunteers based on their background and expertise. The committees are balanced to ensure that no single interest has an overriding representation that may unfairly influence the outcome of the development process.

NFPA uses the following membership categories to fill and balance a committee:

- Manufacturer
- User
- Installer/maintainer
- Labor representative
- Enforcing authority
- Insurance representative
- Special expert
- Consumer
- Applied research/testing laboratory

Technical committees review and respond to all proposed changes to existing documents. Consensus occurs when a majority of a committee accepts a proposed change to the document it oversees. Acceptance by a committee—and its subsequent recommendation for change—requires at least a two-thirds majority vote by written ballot. Ultimately, committee recommendations are voted on at the NFPA's annual meeting, at which the entire association membership has the opportunity to approve or reject the committee recommendations for change to the documents.

Oversight of Technical Committees

The work of the committees is coordinated by NFPA staff liaisons. NFPA staff members appointed to serve as staff liaisons to the technical committees are responsible for ensuring that the committees follow the rules governing committee activities and for coordinating their meetings. The staff liaisons also record the meeting actions and coordinate the publication of committee reports as well as the final document.

All of the committee meetings and the NFPA annual meeting are open to the public, and anyone, whether an NFPA member or not, can submit proposals to change a document or make comments on proposals the committee accepts. The only step in the process that is limited to NFPA members is the voting that takes place at the annual meeting.

Revision Cycles

All NFPA documents are revised on a staggered basis in two annual revision cycles—a fall revision cycle and an annual revision cycle that centers on the annual meeting in June. Typically, individual documents are updated on regular cycles of 3 to 5 years, with anywhere from 20 to 45 documents reporting in a given revision cycle.

Most revision cycles have five distinct steps:
- Step 1: Call for Proposals
- Step 2: Report on Proposals
- Step 3: Report on Comments
- Step 4: Technical Committee Report Session
- Step 5: Standards Council Issuance

Documents in the fall revision cycle for which an intent to make an amending motion has not been filed are issued on a "consent agenda" in January and are not voted on by the membership at the annual meeting.

Call for Proposals

The first step for any NFPA document that is entering its revision cycle occurs when a call for proposals is issued. An indication that a document is entering its revision cycle, along with the dates for accepting proposed changes, appears in NFPA publications, on the NFPA Web site (www.nfpa.org), and in various professional and governmental publications that serve parties interested in the subject matter. Anyone, whether a member of NFPA or not, can submit a proposal for the technical committee to consider during a 20-week window. The form for submitting proposals for change includes a section in which the proposer must provide substantiation for the change. The proposal forms can be found at the back of NFPA documents or on the NFPA Web site.

Tools and techniques are constantly being updated and modified. Consequently, newer methods that are not covered in the current edition might be used at a fire scene. As new principles and technologies are evaluated by the NFPA process, they may or may not be included in the regular cycle revisions of the document. For this reason, it is important that fire investigation professionals be diligent in making proposals and comments to the document during its revision cycles.

Report on Proposals

After the proposal closing date, the committee meets to act on the proposals, to develop its own proposals, and to prepare its report. Through a series of meetings, the committee determines what material should be added, deleted, or modified and then determines the language that should be used. A document titled Report on Proposals (ROP) is generated as a result of these meetings. The ROP is published and is available, free of charge, to anyone who requests one. In addition, the ROP can be obtained from the NFPA Web site.

The ROP is circulated for 10 weeks to allow the public to review all of the proposals for change and the related committee actions. The ROP establishes a window of approximately 60 days during which the public can submit comments on proposed changes to the document as published in the ROP. The committee then votes on its actions relative to each proposal by letter ballot. A two-thirds affirmative vote by the committee is required before an accepted proposal can move forward to the next step.

If the committee revises or rejects a public proposal (either in whole or in part), the committee must provide a reason for doing so.

Report on Comments

After the closing date for public comments, the committee reconvenes to discuss the comments and determine what course of action should be taken in response to each. The committee again votes by letter ballot. The comments and the balloted committee actions are published in a document called the Report on Comments (ROC). This report is also publicly available for a 7-week period. The ROC establishes a closing date for receipt of a Notice of Intent to Make a Motion (NITMAM). Documents in the fall revision cycle that do not receive a notice are issued in January based on a consent agenda. Documents that do receive such a notice are held over to the Annual revision cycle for membership vote at the June meeting.

Technical Committee Report Session

Following these steps, the ROP and ROC are then submitted for consideration and vote at the NFPA's annual meeting in June. During this meeting, any registered attendee can comment on the document, whether or not that individual is an NFPA member; however, only NFPA members can vote on whether the ROP or ROC actions should be accepted or sent back to the committee for further action.

Standards Council Issuance

Appeals to membership action taken at the association meeting may be submitted to the Standards Council within 20 days following the meeting. The Standards Council meets approximately 5 weeks after the association meeting to consider the appeals. The council then directs appropriate actions to be taken according to the appeals it accepts and officially issues the documents. The documents become effective 21 days from the date of issuance. Anyone who still has a grievance about a change to the document has an additional 3 weeks to appeal to the NFPA Board of Directors.

Corrections to Documents

Between revisions of the document, errata and Temporary Interim Amendments (TIAs) may be issued. For example, if a publishing error is found in some part of the document, the NFPA can issue an erratum correcting the error. In the next revision of the document, the error is corrected.

If there are concept changes that must occur between revisions, a TIA is placed before the committee to address the issue. The committee must conduct a letter ballot on proposed TIAs, and if there is two-thirds agreement, the Standards Council issues the TIA. The TIA then becomes a proposal during the next cycle of the document.

Formal Interpretation of Codes and Standards

A Formal Interpretation (FI) is a mechanism for providing an explanation of the meaning or intent of any specific provision that is included in an issued NFPA code or standard. FIs are processed through the technical committee that is responsible for the document and must be clearly worded to solicit a "Yes" or "No" answer from the committee.

Confirmation from the technical committee for an FI is achieved through a letter ballot. If a three-fourths majority is not achieved, the FI fails, and the item is placed on the committee's next meeting agenda. FIs of NFPA guides, though not unheard of, are unusual, since NFPA guides do not contain mandatory provisions.

APPENDIX B

NFPA 1031 Correlation Guide

Chapter 4: Fire Inspector

Objectives	Corresponding Chapter	Corresponding Pages
4.1	1, 12	4–9, 245–267
4.2	14	290–302
4.2.1	14	290–302
4.2.2	1	8
4.2.3	6	108
4.2.4	5	95
4.2.5	14	299
4.2.6	14	302
4.3	5	79–93
4.3.1	3, 5	42–52, 81
4.3.2	3, 7	42–52, 119–122
4.3.3	7	124–140
4.3.4	2, 5	14–36, 86–93
4.3.5	9	184–196
4.3.6	8	150–163
4.3.7	10	204–215
4.3.8	11, 12, 13	222–231, 245–267, 274–283
4.3.9	9	171, 184–196
4.3.10	7	140–143
4.3.11	5, 7	87, 124–140
4.3.12	12	245–267
4.3.13	12	253–267
4.3.14	4, 5	58–72, 88–92
4.3.15	4, 5	58–72, 79–95
4.3.16	9	176–179
4.4	6	108

Chapter 5: Fire Inspector II

Objectives	Corresponding Chapter	Corresponding Pages
5.1	1	4–9
5.2	14	290–302
5.2.1	1	8
5.2.2	6	108–113
5.2.3	5	94–95
5.2.4	App. A	316–318
5.2.5	5	95
5.3	5	79–93
5.3.1	3, 7	52, 119–123
5.3.2	3, 5	52, 81
5.3.3	6	108–112
5.3.4	9	171, 184–196
5.3.5	7	124–140
5.3.6	11, 13	222–236, 274–283
5.3.7	7	140–143
5.3.8	12	245–267
5.3.9	12	253–267
5.3.10	4, 5	58–72, 79–95
5.3.11	6, 9	108–112, 171, 184–196
5.3.12	5, 11	91, 231–236
5.4	6	112–113
5.4.1	6	108–113
5.4.2	7	119–123
5.4.3	6, 9, 10	110–111, 171, 176–179, 194–196, 214–215
5.4.4	6, 9, 10	108–112, 171, 184–196, 214–215
5.4.5	7	124–140
5.4.6	6	108–112

NFPA 1031, *Standard for Professional Qualifications for Fire Inspector and Plan Examiner*, 2014 Edition, ProBoard Matrix

INSTRUCTIONS: In the column titled 'Cognitive/Written Test' place the number of questions from the Test Bank that are used to evaluate the applicable JPR, RK, RS, or objective. In the column titled 'Manipulative/Skill Station' identify the skill sheets that are used to evaluate the applicable JPR, RS, or objective. When the Portfolio or Project method is used to evaluate a particular JPR, RK, RS, or objective, identify the applicable section in the appropriate column and provide the procedures to be used as outlined in the NBFSPQ Operational Procedures, COA-5. Evaluation methods that are not cognitive, manipulative, portfolio, or project based should be identified in the 'Other' column.

NFPA 1031 - Fire Inspector I - 2014 Edition

Section	Objective / JPR, RK, RS — Abbreviated Text	Cognitive Written Test	Manipulative Skills Station	Portfolio	Projects	Other
1.3.1	behavior of fire, fire prevention principles, written and oral communications, public relations, and basic mathematics					Chapter 3 (pp 42–52), Chapter 4 (pp 58–72), Chapter 9 (pp 176–179), Chapter 14 (pp 290–302), Chapter 15 (pp 308–313)
4.1	requirements of Sections 4-2 of NFPA 472					Chapter 1 (pp 4–9), Chapter 13 (pp 274–283)
4.2.1	Prepare inspection reports					Chapter 5 (pp 83–85), Chapter 13 (pp 290–302)
4.2.1(A)	RK: Applicable codes and standards					Chapter 13 (pp 290–302)
4.2.1(B)	RS: interpret codes and standards					Chapter 13 (pp 290–302)
4.2.2	Recognize the need for a permit					Chapter 1 (p 8)
4.2.2(A)	RK: Permit policies					Chapter 1 (p 8)
4.2.2(B)	RS: communicate verbally and in writing					Chapter 1 (p 8)
4.2.3	Recognize the need for plan review					Chapter 6 (p 108)
4.2.3(A)	RK: Plan review policies					Chapter 6 (p 108)
4.2.3(B)	RS: communicate verbally and in writing					Chapter 6 (p 108)
4.2.4	Investigate common complaints					Chapter 5 (p 95)
4.2.4(A)	RK: Applicable codes and standards					Chapter 5 (p 95)
4.2.4(B)	RS: interpret codes and standards					Chapter 5 (p 95)
4.2.5	applicable code or standard					Chapter 14 (p 299)
4.2.5(A)	RK: Applicable codes and standards					Chapter 14 (p 299)

(Continues)

APPENDIX C

NFPA 1031 - Fire Inspector I - 2014 Edition *(Continued)*

Section	Objective / JPR, RK, RS — Abbreviated Text	Cognitive Written Test	Manipulative Skills Station	Portfolio	Projects	Other
4.2.5(B)	RS: interpret codes					Chapter 14 (p 299)
4.2.6	participate in legal proceedings					Chapter 14 (p 302)
4.2.6(A)	RK: legal requirements pertaining to evidence rules					Chapter 14 (p 302)
4.2.6(B)	RS: courtroom demeanor, communicate, listen, and differentiate facts from opinions					Chapter 14 (p 302)
4.3.1	occupancy classification of a single-use occupancy					Chapter 3 (pp 42–52), Chapter 5 (p 81)
4.3.1(A)	RK: occupancy classification types					Chapter 3 (pp 42–52), Chapter 5 (p 81)
4.3.1(B)	RS: make observations and correct decisions					Chapter 3 (pp 42–52), Chapter 5 (p 81)
4.3.2	Compute the allowable occupant load of a single-use occupancy or portion thereof					Chapter 3 (pp 42–52), Chapter 7 (pp 119–122)
4.3.2(A)	RK: Occupancy classification types; applicable codes, regulations, and standards					Chapter 3 (pp 42–52), Chapter 7 (pp 119–122)
4.3.2(B)	RS: calculate accurate occupant loads					Chapter 3 (pp 42–52), Chapter 7 (pp 119–122)
4.3.3	Inspect means of egress elements					Chapter 7 (pp 124–140)
4.3.3(A)	RK: Applicable codes and standards					Chapter 7 (pp 124–140)
4.3.3(B)	RS: recognize problems					Chapter 7 (pp 124–140)
4.3.4	Verify the type of construction for an addition or remodeling project					Chapter 2 (pp 14–18), Chapter 5 (pp 86–93)
4.3.4(A)	RK: Applicable codes and standards					Chapter 2 (pp 18–36), Chapter 5 (pp 86–93)
4.3.4(B)	RS: read plans, make decisions, and interpret codes					Chapter 5 (pp 86–93)
4.3.5	Determine the operational readiness of existing fixed fire suppression systems					Chapter 9 (pp 194–196)

(Continues)

NFPA 1031 - Fire Inspector I - 2014 Edition (Continued)

Section	Objective / JPR, RK, RS — Abbreviated Text	Cognitive Written Test	Manipulative Skills Station	Portfolio	Projects	Other
4.3.5(A)	RK: components and operation of fixed fire suppression systems					Chapter 9 (pp 194–196)
4.3.5(B)	RS: observe, make decisions, recognize problems, and read reports					Chapter 9 (pp 194–196)
4.3.6	Determine the operational readiness of existing fire detection and alarm systems					Chapter 8 (pp 150–163)
4.3.6(A)	RK: components and operation of fire detection and alarm systems					Chapter 8 (pp 150–163)
4.3.6(B)	RS: observe, make decisions, recognize problems, and read reports					Chapter 8 (pp 150–163)
4.3.7	Determine the operational readiness of existing portable fire extinguishers					Chapter 10 (pp 214–215)
4.3.7(A)	RK: portable fire extinguishers, components, and placement					Chapter 10 (pp 204–215)
4.3.7(B)	RS: observe, make decisions, recognize problems, and read reports					Chapter 10 (pp 214–215)
4.3.8	Recognize hazardous conditions involving equipment, processes, and operations					Chapter 11 (pp 222–231), Chapter 13 (pp 274–283)
4.3.8(A)	RK: code compliance inspections					Chapter 12 (pp 245–267), Chapter 13 (pp 274–283)
4.3.8(B)	RS: observe, communicate, interpret codes, recognize problems, and make decisions					Chapter 12 (pp 245–267), Chapter 13 (pp 274–283)
4.3.9	Compare an approved plan to an installed fire protection system					Chapter 9 (pp 171, 184–196)
4.3.9(A)	RK: Fire protection symbols and terminology					Chapter 9 (pp 171, 184–196)
4.3.9(B)	RS: plans for fire protection systems					Chapter 9 (pp 171, 184–196)
4.3.10	emergency planning and preparedness					Chapter 7 (pp 140–143)
4.3.10(A)	RK: emergency evacuation drills					Chapter 7 (pp 140–143)

(Continues)

NFPA 1031 - Fire Inspector I - 2014 Edition *(Continued)*

Section	Objective / JPR, RK, RS — Abbreviated Text	Cognitive Written Test	Manipulative Skills Station	Portfolio	Projects	Other
4.3.10(B)	RS: emergency evacuation requirements					Chapter 7 (pp 140–143)
4.3.11	Inspect emergency access for a site					Chapter 5 (pp 87), Chapter 7 (pp 124–140)
4.3.11(A)	RK: codes and standards					Chapter 5 (pp 87), Chapter 7 (pp 124–140)
4.3.11(B)	RS: emergency access					Chapter 5 (pp 87), Chapter 7 (pp 124–140)
4.3.12	code compliance for incidental storage					Chapter 12 (pp 245–267)
4.3.12(A)	RK: flammable and combustible liquids and gases					Chapter 12 (pp 245–267)
4.3.12(B)	RS: observe, communicate, interpret codes, recognize problems, and make decisions					Chapter 12 (pp 245–267)
4.3.13	Verify code compliance for incidental storage, handling, and use of hazardous materials					Chapter 12 (pp 253–267)
4.3.13(A)	RK: hazardous materials.					Chapter 12 (pp 253–267)
4.3.13(B)	RS: observe, communicate, interpret codes, recognize problems, and make decisions					Chapter 12 (pp 253–267)
4.3.14	Recognize a hazardous fire growth potential in a building or space					Chapter 5 (pp 88–92)
4.3.14(A)	RK: fire behavior					Chapter 4 (pp 58–72)
4.3.14(B)	RS: recognize hazardous conditions, and make decisions					Chapter 5 (pp 88–92)
4.3.15	determine code compliance					Chapter 5 (pp 79–95)
4.3.15(A)	RK: fire behavior, flame spread and smoke development					Chapter 4 (pp 58–72), Chapter 5 (pp 79–95)
4.3.15(B)	RS: observe, communicate, interpret codes, recognize problems, and make decisions					Chapter 5 (pp 79–95)
4.3.16	verify fire flows					Chapter 9 (pp 176–179)
4.3.16(A)	RK: types of water distribution systems					Chapter 9 (pp 177–179)
4.3.16(B)	RS: use pitot tube and gauges					Chapter 9 (pp 176–179)

INSTRUCTIONS: In the column titled 'Cognitive/Written Test' place the number of questions from the Test Bank that are used to evaluate the applicable JPR, RK, RS, or objective. In the column titled 'Manipulative/Skill Station' identify the skill sheets that are used to evaluate the applicable JPR, RS, or objective. When the Portfolio or Project method is used to evaluate a particular JPR, RK, RS, or objective, identify the applicable section in the appropriate column and provide the procedures to be used as outlined in the NBFSPQ Operational Procedures, COA-5. Evaluation methods that are not cognitive, manipulative, portfolio, or project based should be identified in the 'Other' column.

NFPA 1031 - Fire Inspector II - 2014 Edition

Section	Abbreviated Text	Cognitive Written Test	Manipulative Skills Station	Portfolio	Projects	Other
5.1	Fire Inspector I prerequisite					Chapter 1 (pp 4–9)
5.2.1	Process a permit application,					Chapter 1 (p 8)
5.2.1(A)	RK: Permit application process					Chapter 1 (p 8)
5.2.1(B)	RS: application of the requisite knowledge.					Chapter 1 (p 8)
5.2.2	Process a plan review application					Chapter 6 (pp 108–113)
5.2.2(A)	RK: Plan review application process					Chapter 6 (pp 108–113)
5.2.2(B)	RS: communicate verbally and in writing					Chapter 6 (pp 108–113)
5.2.3	Investigate complex complaints					Chapter 5 (pp 94–95)
5.2.3(A)	RK: codes and standards					Chapter 5 (pp 94–95)
5.2.3(B)	RS: interpret codes and standards					Chapter 5 (pp 94–95)
5.2.4	Recommend modifications to codes and standards of the jurisdiction					Appendix A (pp 316–318)
5.2.4(A)	RK: adopt, enforce, and revise codes and standards					Appendix A (pp 316–318)
5.2.4(B)	RS: recognize problems, communicate, and identify cost/risk benefits					Appendix A (pp 316–318)
5.2.5	Recommend policies and procedures for the delivery of inspection services					Chapter 5 (pp 95)
5.2.5(A)	RK: code enforcement					Chapter 5 (pp 95)
5.2.5(B)	RS: construction methods and materials					Chapter 5 (pp 95)
5.3.1	Compute the occupant load of a multiuse building					Chapter 3 (p 52), Chapter 7 (pp 119–123)

(Continues)

NFPA 1031 - Fire Inspector II - 2014 Edition *(Continued)*

Section	Abbreviated Text	Cognitive Written Test	Manipulative Skills Station	Portfolio	Projects	Other
5.3.1(A)	RK: occupant loads					Chapter 3 (p 52), Chapter 7 (pp 119–123)
5.3.1(B)	RS: occupant loads					Chapter 3 (p 52), Chapter 7 (pp 119–123)
5.3.2	occupancy classification					Chapter 3 (p 52), Chapter 5 (p 81)
5.3.2(A)	RK: Occupancy classification types					Chapter 3 (p 52), Chapter 5 (p 81)
5.3.2(B)	RS: code requirements					Chapter 5 (p 81)
5.3.3	determine building area, height, occupancy class and construction type					Chapter 6 (pp 108–112)
5.3.3(A)	RK: building construction					Chapter 6 (pp 108–112)
5.3.3(B)	RS: identify characteristics of each type of construction					Chapter 6 (pp 108–112)
5.3.4	fire protection systems and equipment					Chapter 9 (pp 184–196)
5.3.4(A)	RK: codes and standards for fire protection systems					Chapter 9 (pp 184–196)
5.3.4(B)	RS: use codes and standards, and read reports, plans, and specifications					Chapter 9 (pp 184–196)
5.3.5	Analyze the egress elements of a building or portion of a building					Chapter 7 (pp 124–140)
5.3.5(A)	RK: Acceptable means of egress devices					Chapter 7 (pp 124–140)
5.3.5(B)	RS: egress requirements					Chapter 7 (pp 124–140)
5.3.6	Evaluate hazardous conditions involving equipment, processes, and operations					Chapter 11 (pp 222–236), Chapter 13 (pp 274–283)
5.3.6(A)	accepted fire protection practices					Chapter 11 (pp 222–236), Chapter 13 (pp 274–283)
5.3.6(B)	observe, communicate, interpret codes, recognize problems, and make decisions					Chapter 11 (pp 222–236), Chapter 13 (pp 274–283)
5.3.7	Evaluate emergency planning and preparedness procedures					Chapter 7 (pp 140–143)
5.3.7(A)	RK: Occupancy requirements					Chapter 7 (pp 140–143)

(Continues)

NFPA 1031 - Fire Inspector II - 2014 Edition (Continued)

Objective / JPR, RK, RS		Cognitive Written Test	Manipulative Skills Station	Portfolio	Projects	Other
Section	Abbreviated Text					
5.3.7(B)	RS: compare submitted plans and procedures with applicable codes and standards					Chapter 7 (pp 140–143)
5.3.8	storage, handling, and use of flammable and combustible liquids and gases					Chapter 12 (pp 245–267)
5.3.8(A)	RK: Flammable and combustible liquids					Chapter 12 (pp 245–267)
5.3.8(B)	RS: processes or operations utilizing flammable and combustible liquids					Chapter 12 (pp 245–267)
5.3.9	Evaluate code compliance for the storage, handling, and use of hazardous materials					Chapter 12 (pp 253–267)
5.3.9(A)	RK: Hazardous materials properties and hazards					Chapter 12 (pp 253–267)
5.3.9(B)	RS: processes or operations utilizing hazardous materials					Chapter 12 (pp 253–267)
5.3.10	Determine fire growth potential in a building or space					Chapter 5 (pp 79–95)
5.3.10(A)	RK: decorations, decorative materials, furnishings, and safe housekeeping practices					Chapter 5 (pp 79–95)
5.3.10(B)	RS: recognize hazardous conditions, and make decisions					Chapter 5 (pp 79–95)
5.3.11	Verify compliance with construction documents					Chapter 6 (pp 108–112), Chapter 9 (pp 171, 184–196)
5.3.11(A)	RK: Applicable codes and standards for installation and testing of fire protection systems					Chapter 6 (pp 108–112), Chapter 9 (pp 171)
5.3.11(B)	RS: ability to witness and document tests					Chapter 6 (pp 108–112), Chapter 9 (pp 194–196)
5.3.12	code compliance of heating, ventilation, air conditioning					Chapter 5 (p 91), Chapter 11 (pp 231–236), Chapter 12 (pp 245–267)
5.3.12(A)	RK: Types, installation, maintenance, and use of building service equipment					Chapter 5 (p 91), Chapter 11 (pp 231–236), Chapter 12 (pp 245–267)

(Continues)

NFPA 1031 - Fire Inspector II - 2014 Edition (Continued)

Section	Objective / JPR, RK, RS Abbreviated Text	Cognitive Written Test	Manipulative Skills Station	Portfolio	Projects	Other
5.3.12(B)	RS: ability to observe, recognize problems					Chapter 5 (p 91), Chapter 11 (pp 231–236), Chapter 12 (pp 245–267)
5.4.1	Classify the occupancy type					Chapter 6 (pp 112–113)
5.4.1(A)	RK: Occupancy classification types					Chapter 6 (pp 112–113)
5.4.1(B)	RS: read plans					Chapter 6 (pp 112–113)
5.4.2	Compute the occupant load					Chapter 7 (pp 119–123)
5.4.2(A)	RK: occupant loads for an occupancy and building use					Chapter 7 (pp 119–123)
5.4.2(B)	RS: calculate accurate occupant loads					Chapter 7 (pp 119–123)
5.4.3	review proposed installation of a fire protection system					Chapter 6 (pp 110–111), Chapter 9 (pp 171, 194–196)
5.4.3(A)	RK: selection, distribution, location, and testing of portable fire extinguishers					Chapter 6 (pp 110–111), Chapter 9 (pp 176–179, 194–196)
5.4.3(B)	RS: read basic floor plan or shop drawings and identify symbols					Chapter 6 (pp 110–111), Chapter 9 (p 171)
5.4.4	Review the installation of fire protection systems					Chapter 6 (pp 110–111), Chapter 9 (pp 171, 184–196)
5.4.4(A)	RK: Proper selection, distribution, location, and testing of portable fire extinguishers					Chapter 6 (pp 110–111), Chapter 9 (pp 171–196)
5.4.4(B)	RS: basic floor plans or shop drawings.					Chapter 6 (pp 110–111), Chapter 9 (p 171)
5.4.5	Verify that means of egress elements are provided					Chapter 7 (pp 124–140)
5.4.5(A)	RK: Applicable codes and standards					Chapter 7 (pp 124–140)
5.4.5(B)	RS: ability to read plans and research codes and standards.					Chapter 7 (pp 124–140)
5.4.6	Verify the construction type of a building or portion thereof					Chapter 6 (pp 108–112)
5.4.6(A)	RK: Building construction					Chapter 6 (pp 108–112)
5.4.6(B)	RS: identify characteristics of each type of building construction.					Chapter 6 (pp 108–112)

APPENDIX D

Fire and Emergency Service Higher Education (FESHE) Correlation Guide

FESHE Objective	Chapter	Pages
Principles of Code Enforcement	1	pp 4 – 8
1. Explain the code enforcement system and the fire inspector's role in that system.	1	pp 6 – 7
2. Describe the development and adoption processes for codes and standards.	1	p 7
3. Describe the difference between prescriptive and performance-based codes.	1	p 8
4. Describe the legal authority and limitations relevant to fire code inspections.	1	p 9
6. Recognize ethical practices for the code enforcement officer.	1	p 9
7. Explain the application, the interrelationship of codes, standards, recommended practices, and guides.	1	pp 6 – 7
10. Describe the political, business, and other interests that influence the code enforcement process.	1	p 7
11. Identify the professional development process for code enforcement practitioners.	1	p 9

Principles of Code Enforcement

FESHE Objective	Chapter	Pages
1. Describe the differences in how codes apply to new and existing structures.	5	p 92
2. Identify appropriate codes and their relationship to other requirements for the built environment.	5	pp 79 – 95

Fire Plans Review

FESHE Objective	Chapter	Pages
1. Describe at least three reasons for performing plan checks, the objectives of a proposed plans review program, the impact of such a program, and how the program will enhance current fire prevention programs.	6	p 103
2. Develop a graphic illustration of a model plans review system, identifying at least four components involved in the system including the use of plans review checklists.	6	pp 108 – 112
3. List three methods to monitor and evaluate the effectiveness of code requirements according to applicable standards.	6	pp 108 – 112
4. Determine fire department access, verify appropriate water supply, and review general building parameters.	6	pp 108 – 112
5. Determine occupancy classification and construction type; calculate occupant load, height, and area of a building.	6	pp 108 – 112
6. Determine the appropriateness of the three components of a building's egress system (exit access, exit, and exit discharge), verify building compartmentation and the proper enclosure of vertical openings.	6	pp 108 – 112
7. Identify special hazards, verify interior finish and establish the proper locations for pre-engineered fire extinguishing systems.	6	pp 108 – 112
8. Verify the compliance of a heating, ventilating, and air conditioning (HVAC) system; review sources requiring venting and combustion air; verify the proper location of fire dampers; and evaluate a stairwell pressurization system.	6	p 110
9. Verify the proper illumination for exit access and the arrangement of exit lighting; perform a life safety evaluation of the egress arrangement of a building.	6	pp 108 – 112

FESHE Objective	Chapter	Pages
10. Verify the design of a fire alarm and detection system, and an offsite supervisory system for compliance with applicable standards.	6	pp 108 – 112
Fire Plans Review		
1. Determine occupancy classification, construction type: calculate occupant load and, the height and area of a building.	7	pp 119 – 123
2. Determine the appropriateness of the three components of a building's egress system (exit access, exit, and exit discharge), verify building compartmentation and the proper enclosure of vertical openings.	7	pp 124 – 140
13. Verify the proper illumination for exit access, the arrangement of exit lighting and perform a life safety evaluation of the egress arrangement of a building.	7	pp 124 – 140
Fire Plans Review		
1. Verify the design of a fire alarm and detection system and an offsite supervisory system for compliance with applicable standards.	8	pp 150 – 163
Fire Plans Review		
8. Determine fire department access, verify appropriate water supply, and review general building parameters.	9	pp 176 – 179
11. Identify special hazards, verify interior finish and establish the proper locations for pre-engineered fire extinguishing systems.	9	pp 184 – 196
Fire Plans Review		
12. Verify the compliance of a heating, ventilating, and air conditioning (HVAC) system, review sources requiring venting and combustion air, verify the proper location of fire dampers, and evaluate a stairwell pressurization system.	11	pp 231 – 236

FESHE Objective	Chapter	Pages
Principles of Code Enforcement		
5. Describe the importance of thorough documentation.	14	pp 290 – 302
Fire and Life Safety Education		
1. Differentiate between Public Education, Public Information, and Public Relations/Marketing.	15	pp 308 – 309
2. Demonstrate the need for establishing fire and life safety education as a value within the fire service culture.	15	pp 309 – 311
3. Identify stakeholders; develop partnerships and coalitions to work on fire and life safety education activities.	15	pp 311 – 312
4. Identify and use local, regional, and national sources of data for fire and injury prevention programs.	15	pp 312
5. Identify budget needs for program delivery and the process for requesting funds.	15	pp 312 – 313
6. Select, design, implement, and evaluate fire and life safety education programs that address specific community risk issues.	15	pp 312
7. Develop an accountability system to measure program delivery.	15	pp 313

Glossary

Accelerator A device that accelerates the removal of the air from a dry-pipe or preaction sprinkler system.

Air handling unit A unit installed for the purpose of processing the treatment of air so as to control simultaneously its temperature, humidity, and cleanliness to meet the requirements of the conditioned space it serves.

Air sampling detector A system that captures a sample of air from a room or enclosed space and passes it through a smoke detection or gas analysis device.

Alarm initiation device An automatic or manually operated device in a fire alarm system that, when activated, causes the system to indicate an alarm condition.

Alarm matrix A chart showing what will happen with the fire alarm system when an initiating device is activated.

Alarm notification appliance An audible and/or visual device in a fire alarm system that makes occupants or other persons aware of an alarm condition.

Alarm valve A valve that signals an alarm when a sprinkler head is activated and prevents nuisance alarms caused by pressure variations.

Alternative clause This clause allows for the code provisions to be altered and an alternative offered that would not reduce the level of safety within the building.

Ambulatory health care occupancy A building or portion thereof used to provide services or treatment simultaneously to four or more patients that, on an outpatient basis. (NFPA 101, *Life Safety Code*)

Ammonium phosphate An extinguishing agent used in dry-chemical fire extinguishers that can be used on Class A, B, and C fires.

Amps The measure of the volume of electrical flow.

Anecdotal evidence Evidence that may or may not be true, used to generalize when there is insufficient data to base it upon.

Annealed The process of forming standard glass.

Annual inspections Inspections performed as part of the regular inspection cycle.

Apartment building is a building or portion thereof containing three or more dwelling units with independent cooking and bathroom facilities. (NFPA 5000)

Aqueous film-forming foam (AFFF) A water-based extinguishing agent used on Class B fires that forms a foam layer over the liquid and stops the production of flammable vapors.

Arc-fault circuit interrupter (AFCI) A device intended to provide protection from the effects of arc faults by recognizing characteristics unique to arcing and by functioning to de-energize the circuit when an arc fault is detected.

Architectural plan A drawing showing floor plans, elevation drawings, and features of a proposed building's layout and construction.

Arcing A high-temperature luminous electric discharge across a gap or through a medium such as charred insulation.

Area of refuge An area that is either (1) a story in a building where the building is protected throughout by an approved, supervised automatic-sprinkler system and has not less than two accessible rooms or spaces separated from each other by smoke-resisting partitions; or (2) a space located in a path of travel leading to a public way that is protected from the effects of fire, either by means of separation from other spaces in the same building or by virtue of location, thereby permitting a delay in egress travel from any level. (NFPA 101)

As built diagrams A set of drawings provided by a contractor showing how a system was actually installed, which may be different from the approved plans.

Aspect Compass direction toward which a slope faces. (NFPA 1144)

Assembly occupancies Buildings (1) used for a gathering of 50 or more persons for deliberation, worship, entertainment, eating, drinking, amusement, awaiting transportation, or similar uses; or (2) used as a special amusement building regardless of occupant load. (NFPA 101, *Life Safety Code*)

Automatic sprinkler heads The working ends of a sprinkler system. They serve to activate the system and to apply water to the fire.

Automatic sprinkler system A system of pipes filled with water under pressure that discharges water immediately when a sprinkler head opens.

Auxiliary system A fire alarm system that sounds an alarm in the building and transmits a signal to the fire department via a public alarm box system.

Awning windows Windows that have one large or two medium-size panels, which are operated by a hand crank from the corner of the window.

Backdraft The sudden explosive ignition of fire gases when oxygen is introduced into a superheated space previously deprived of oxygen.

Balloon-frame construction An older type of wood frame construction in which the wall studs extend vertically from the basement of a structure to the roof without any fire stops.

Beam detector A smoke detection device that projects a narrow beam of light across a large open area from a sending unit to a receiving unit. When the beam is interrupted by smoke, the receiver detects a reduction in light transmission and activates the fire alarm.

Bills of lading Shipping papers for roads and highways.

Bimetallic strip A device with components made from two distinct metals that respond differently to heat. When heated, the metals will bend or change shape.

Board of Appeals A group of persons appointed by the governing body of the jurisdiction adopting the code for the purpose of hearing and adjudicating differences of opinion between the authority having jurisdiction and the citizenry in the interpretation, application, and enforcement of the code. (NFPA 1)

Boiler A closed vessel in which water is heated, steam is generated, steam is superheated, or any combination thereof by the application of heat from combustible fuels in a self-contained or attached furnace. (NFPA 85)

Boiler enclosures The physical boundary for all boiler pressure parts and for the combustion process for a closed vessel in which water is heated, steam is generated, steam is superheated, or any combination thereof by the application of heat from combustible fuels, in a self-contained or attached furnace. (NFPA 85)

Boiler inspector A person who, through formal education and training, is qualified to inspect a boiler and its associated systems against established standards, recommendations, and requirements.

Boiler room Any room with a boiler of 5 horsepower or greater. (NFPA 5000)

Boiling liquid expanding vapor explosion (BLEVE) An explosion that occurs when a tank containing a volatile liquid is heated.

Bonding (bonded) The permanent joining of metallic parts to form an electrically conductive path that will ensure electrical continuity and the capacity to conduct any current likely to be imposed. (NFPA 79)

Bowstring trusses Trusses that are curved on the top and straight on the bottom.

Bulk storage containers Large-volume containers that have an internal volume greater than 119 gallons (451 L) for liquids and greater than 882 pounds (401 kg) for solids and a capacity of greater than 882 pounds (401 kg) for gases.

Business license or change of occupancy inspections Inspections that occur when the building department is notified of a new business requesting permission to open

Business occupancy An occupancy used for the transaction of business other than mercantile. (NFPA 101, *Life Safety Code*)

Butterfly valve A type of indicating valve that moves a piece of metal 90° within the pipe and shows if the water supply is open or closed.

Carbon dioxide (CO_2) fire extinguisher A fire extinguisher that uses carbon dioxide gas as the extinguishing agent.

Carbon dioxide extinguishing system A fire suppression system that is designed to protect either a single room or series of rooms by flooding the area with carbon dioxide.

Cartridge fuses A type of overcurrent protective device with a replaceable cartridge containing the fusible part.

Casement windows Windows in a steel or wood frame that open away from the building via a crank mechanism.

Central station An off-premises facility that monitors alarm systems and is responsible for notifying the fire department of an alarm. These facilities may be geographically located some distance from the protected building(s).

Check valve A valve that allows flow in one direction only. (NFPA 13R)

Chemical energy Energy that is created or released by the combination or decomposition of chemical compounds.

Chemical treatment system A system utilizing chemical additives to alter the properties and hazards of a waste stream prior to disposal, reuse, or further chemical processing.

Chemical-pellet sprinkler head A sprinkler head activated by a chemical pellet that liquefies at a preset temperature.

Chiller plants A facility housing refrigeration equipment for the purpose of extracting heat for industrial use, comfort cooling, or other uses.

Circuit breakers A device designed to open and close a circuit by nonautomatic means and to open the circuit automatically on a predetermined overcurrent without damage to itself when properly applied within its rating. (NFPA 70)

Class A fires Fires involving ordinary combustible materials, such as wood, cloth, paper, rubber, and many plastics.

Class B fires Fires involving flammable and combustible liquids, oils, greases, tars, oil-based paints, lacquers, and flammable gases.

Class C fires Fires that involve energized electrical equipment, where the electrical conductivity of the extinguishing media is of importance.

Class D fires Fires involving combustible metals such as magnesium, titanium, zirconium, sodium, and potassium.

Class I standpipe A standpipe system designed for use by fire department personnel only. Each outlet should have a valve to control the flow of water and a 2½″ male coupling for fire hose.

Class IA liquids Those liquids that have flashpoints below 73°F (22.8°C) and boiling points below 100°F (37.8°C).

Class IB liquids Those liquids that have flashpoints below 73°F (22.8°C) and boiling points at or above 100°F (37.8°C).

Class IC liquids Those liquids that have flashpoints at or above 73°F (22.8°C) and boiling points but below 100°F (37.8°C).

Class II liquids Those liquids that have flashpoints at or above 100°F (37.8°C) and below 140°F (60°C).

Class II standpipe A standpipe system designed for use by occupants of a building only. Each outlet is generally equipped with a length of 1½″ single-jacket hose and a nozzle, which are preconnected to the system.

Class III standpipe A combination system that has features of both Class I and Class II standpipes.

Class IIIA liquids Those liquids that have flashpoints at or above 140°F (60°C) and below 200°F (93.4°C).

Class IIIB liquids Those liquids that have flashpoints at or above 200°F (93.4°C).

Class K fires Fires involving combustible cooking media, such as vegetable oils, animal oils, and fats.

Clean agent A volatile or gaseous fire extinguishing agent that does not leave a residue when it evaporates. Also known as a halogenated agent.

Code A standard that is an extensive compilation of provisions covering broad subject matter or that is suitable for adoption into law independently of other codes and standards.

Code analysis A summary of the features of fire protection and building characteristics in a plan set.

Coded system A fire alarm system design that divides a building or facility into zones and has audible notification devices that can be used to identify the area where an alarm originated.

Combination smoke fire damper A device that functions as both a fire damper and as a smoke damper. (NFPA 5000)

Combustibility The property describing whether a material will burn and how quickly it will burn.

Combustible dust Any finely divided solid material that is 16.5 mil (420 µ) or smaller in diameter (material passing a U.S. No. 40 Standard Sieve) and presents a fire or explosion hazard when dispersed and ignited in air. (NFPA 654)

Combustible liquids Any liquid that has a closed-cup flash point at or above 100°F (37.8°C). (NFPA 306)

Commissioning The time period of plant testing and operation between initial operation and commercial operation.

Common path of travel The portion of exit access that must be traversed before two separate and distinct paths of travel to two exits are available. (NFPA 101, *Life Safety Code*) (NFPA 101)

Compartmentation The subdivision of a building into relatively small areas so that fire or smoke can be confined to the room or section in which it originates. (NFPA 232)

Complaint form Form that lists in detail any complaint that is lodged with the fire inspection agency and is investigated.

Complaint inspections Inspections that occur when someone registers a concern of a possible code violation.

Compressed gas Any material or mixture having, when in its container, an absolute pressure exceeding 40 psia (an absolute pressure of 276 kPa) at 70°F (21.1°C) or, regardless of the pressure at 70°F (21.1°C), having an absolute pressure exceeding 104 psia (an absolute pressure of 717 kPa) at 130°F (54.4°C). (NFPA 58)

Compressed gas cylinders Portable pressure vessels of 100 lb (45.3 kg) water capacity or less designed to contain a gas or liquid at gauge pressures over 40 psi (276 mPa).

Conditional approval Grants partial approval for work being done, often issued for the final certificate of occupancy when violations are minor and do not pose a hazard.

Conduction Heat transfer to another body or within a body by direct contact.

Conduits Round piping where wiring is routed through to provide protection from damage.

Construction or final inspections Inspections that are conducted as a building is being constructed, including sprinkler systems, fire alarm systems, and fire pumps

Container Any vessel or receptacle that holds material, including storage vessels, pipelines, and packaging.

Continuous beam A beam supported at three or more points. Structurally advantageous because if the span between two supports is overloaded, the rest of the beam assists in carrying the load.

Convection Heat transfer by circulation within a medium such as a gas or a liquid.

Corrosion-resistant sprinklers Sprinkler heads with special coating or plating such as wax or lead to use in potentially corrosive atmospheres.

Corrosive gas A gas that can cause burns as well as destroy or cause irreversible harm to organic tissue and metals.

Cryogenic gas A refrigerated liquid gas having a boiling point below −130°F (−90°C) at atmospheric pressure. (NFPA 1992)

Current The flow of electricity, measured in amps.

Curtain walls Nonbearing walls that are used to separate the inside and outside of the building, but that are not part of the support structure for the building.

Curved roofs Roofs that have a curved shape.

Cylinders A portable compressed-gas container.

Dampers A valve or plate for controlling draft or the flow of gases, including air. (NFPA 853)

Day-care occupancy An occupancy in which four or more clients receive care, maintenance, and supervision, by other than their relatives or legal guardians, for less than 24 hours per day. (NFPA 101, *Life Safety Code*)

Dead end corridor A passageway from which there is only one means of egress. (NFPA 301)

Dead load The weight of a building. It consists of the weight of all materials of construction incorporated into a building, including but not limited to walls, floors, roofs, ceilings, stairways, built-in partitions, finishes, cladding, and other similarly incorporated architectural and structural items, as well as fixed service equipment, including the weight of cranes.

Decay phase The phase of fire development in which the fire has consumed either the available fuel or oxygen and is starting to die down.

Defensible space An area as defined by the AHJ [typically a width of 30′ or (9.14 m) more] between an improved property and a potential wildland fire where combustible materials and vegetation have been removed or modified to reduce the potential for fire on improved property spreading to wildland fuels or to provide a safe working area for fire fighters protecting life and improved property from wildland fire. (NFPA 1144)

Deluge head A sprinkler head that has no release mechanism; the orifice is always open.

Deluge sprinkler system A sprinkler system in which all sprinkler heads are open. When an initiation device, such as a smoke detector or heat detector, is activated, the deluge valve opens and water discharges from all of the open sprinkler heads simultaneously.

Deluge valve A valve assembly designed to release water into a sprinkler system when an external initiation device is activated.

Department of Transportation (DOT) marking system A unique system of labels and placards that, in combination with the Emergency Response Guide, offers guidance for first responders operating at a hazardous materials incident.

Detail view A view on a drawing of a specific element of construction or building feature in a larger scale to provide further clarity.

Detention and correctional occupancy An occupancy used to one or more persons under varied degrees of restraint or security where such occupants are mostly incapable of self-preservation because of security measures not under the occupant's control. (NFPA 101, *Life Safety Code*)

Dewar containers Containers designed to preserve the temperature of the cold liquid held inside.

Direct costs Expenses that must be paid but would otherwise not have occurred without the program such as printing, supplies, or fuel.

Distributors Relatively small-diameter underground pipes that deliver water to local users within a neighborhood.

Dormitory A building or space in a building in which group sleeping accommodations are provided for more than 16 persons who are not members of the same family in one room, or a series of closely associated rooms, under joint occupancy and single management, with or without meals, but without individual cooking facilities. (NFPA 101, *Life Safety Code*)

Double-action pull-station A manual fire alarm activation device that requires two steps to activate the alarm. The person must push in a flap, lift a cover, or break a piece of glass before activating the alarm.

Double-hung windows Windows that have two movable sashes that can go up and down.

Double-pane glass A window design that traps air or inert gas between two pieces of glass to help insulate a house.

Drums Barrel-like containers built to DOT Specification 5P (1A1).

Dry bulk cargo tanks Tanks designed to carry dry bulk goods such as powders, pellets, fertilizers, or grain; they are generally V-shaped with rounded sides that funnel toward the bottom.

Dry chemical extinguishing system An automatic fire extinguishing system that discharges a dry chemical agent.

Dry sprinkler heads Sprinkler heads that are installed when 40 degrees Fahrenheit cannot be maintained. Dry sprinkler heads are constructed to provide isolation between the head and water supply by use of a cylinder that extends from the head to the threads of the pipe fitting. The threads reside in heated areas or are installed on a dry sprinkler system, and use pendent style dry sprinkler heads.

Dry-barrel hydrant A type of hydrant used in areas subject to freezing weather. The valve that allows water to flow into the hydrant is located underground and the barrel of the hydrant is normally dry.

Dry-chemical fire extinguisher An extinguisher that uses a mixture of finely divided solid particles to extinguish fires. The agent is usually sodium bicarbonate-, potassium bicarbonate-, or ammonium phosphate-based, with additives being included to provide resistance to packing and moisture absorption and to promote proper flow characteristics.

Dry-pipe sprinkler system A sprinkler system in which the pipes are normally filled with compressed air. When a sprinkler head is activated, it releases the air from the system, which opens a valve so the pipes can fill with water.

Dry-pipe valve The valve assembly on a dry sprinkler system that prevents water from entering the system until the air pressure is released.

Dry-powder extinguishing agent An extinguishing agent used in putting out Class D fires. The common dry-powder extinguishing agents include sodium chloride and graphite-based powders.

Dry-type transformers A device that raises or lowers the voltage of alternating current of the original source. (NFPA 70)

Duct detector A smoke detection device mounted either inside an HVAC duct or mounted on the outside of the duct with tubing arranged to sample the airflow to respond to the presence of smoke.

Ducted systems A continuous passageway for the transmission of air that, in addition to ducts, includes duct fittings, dampers, fans, and accessory air-management equipment and appliances. (NFPA 853)

Dunnage Loose packing material (usually wood) protecting a ship's cargo from damage or movement during transport. (NFPA 1405)

Dust collection system A pneumatic conveying system that is specifically designed to capture dust and wood particulates at the point of generation, usually from multiple sources, and to convey the particulates to a point of consolidation. (NFPA 664)

Early-suppression fast-response (ESFR) sprinkler head A sprinkler head designed to react quickly and suppress a fire in its early stages.

Educational occupancies Buildings used for educational purposes through the twelfth grade by six or more persons for 4 or more hours per day or more than 12 hours a week. (NFPA 101, *Life Safety Code*)

Electrical energy Heat that is produced by electricity.

Electrical inspector Individuals who verify that electrical systems comply with applicable codes and standards when buildings are initially constructed, altered, or renovated.

Electrical plans Design documents in a plan set showing the power layout and lighting plan of a proposed building.

Electrostatic Particles or objects electrically charged with either a positive or negative voltage differential.

Elevated water storage tower An above-ground water storage tank that is designed to maintain pressure on a water distribution system.

Elevation pressure The amount of pressure created by gravity.

Elevation view A view in a drawing showing the exterior of the building.

Emergency plan A document that outlines procedures for occupants to deal with all types of building-related emergency situations.

***Emergency Response Guidebook* (ERG)** The guidebook developed by the Department of Transportation to provide guidance for first responders operating at a hazardous materials incident in coordination with DOT's labels and placards marking system.

Enabling legislation Legislation in which local jurisdiction adopt a specific set of codes.

Endothermic Reactions that absorb heat or require heat to be added.

Equipment ground-fault protective device (EGFPD) A device intended to provide protection of equipment from damage from line-to-ground fault currents by operating to cause a disconnecting means to open all ungrounded conductors of the faulted circuit.

Equivalencies The use of systems, methods, or devices of equivalent or superior quality, strength, fire resistance, effectiveness, durability, to those prescribed by a code or standard.

Evacuation plan A document that is part of an emergency plan and outlines procedures to vacate building occupants in a safe, orderly, and efficient manner.

Exhaust stacks Chimney or ductwork that removes excess heat, fumes or vapors from an area without reuse, to a point of discharge.

Exhauster A device that accelerates the removal of the air from a dry-pipe or preaction sprinkler system.

Exigent circumstance An immediate life safety issue which requires that immediate actions be taken.

Exit access That portion of a means of egress that leads to an exit. (NFPA 101, *Life Safety Code*)

Exit discharge That portion of a means of egress between the termination of an exit and a public way. (NFPA 101, *Life Safety Code*)(NFPA 101)

Exit That portion of a means of egress that is separated from all other spaces of a building or structure by construction or equipment as required to provide a protected way of travel to the exit discharge. (NFPA 101, *Life Safety Code*) (NFPA 101)

Exothermic Reactions that result in the release of energy in the form of heat.

Extinguishing agent A material used to stop the combustion process. Extinguishing agents may include liquids, gases, dry-chemical compounds, and dry-powder compounds.

Film-forming fluoroprotein (FFFP) foam A water-based extinguishing agent used on Class B fires that forms a foam layer over the liquid and stops the production of flammable vapors.

Final notice Written correspondence used when violations are not corrected to notify the owner that legal action may be taken to ensure code compliance.

Final or construction inspection form A form used when inspecting specialized systems such as fire alarm, sprinkler, hood, and duct suppression systems, as well as for other types of construction phase inspections, such as a ceiling inspection or a final inspection prior to issuance of a certificate of occupancy.

Fire A rapid, persistent chemical reaction that releases both heat and light.

Fire alarm control panel The component in a fire alarm system that controls the functions of the entire system.

Fire and life safety education specialist A member of the fire department who deals with the public on education, fire safety, and juvenile fire safety programs.

Fire and life safety educators Personnel that teach fire safety messages to target audiences.

Fire damper A device installed in an air distribution system, designed to close automatically upon detection of heat to interrupt migratory airflow, and to restrict the passage of flame. (NFPA 221)

Fire department connection (FDC) A fire hose connection through which the fire department can pump water into a sprinkler system or standpipe system.

Fire enclosure A fire-rated assembly used to enclose a vertical opening such as a stairwell, elevator shaft, and chase for building utilities.

Fire flow The flow rate of a water supply, measured at 20 psi (138 kPa) residual pressure, that is available for firefighting. (NFPA 1141)

Fire inspection A visual inspection of a building and its property to determine if the building complies with all pertinent statutes and regulations of the jurisdiction.

Fire inspector I An individual at the first level of progression who has met the job performance requirements specified in this standard for Level I. The fire inspector I conducts basic fire inspections and applies codes and standards. (NFPA 1031)

Fire inspector II An individual at the first level of progression who has met the job performance requirements specified in this standard for Level II. The Fire Inspector II conducts most types of inspections and interprets applicable codes and standards. (NFPA 1031)

Fire investigator An individual who has demonstrated the skills and knowledge necessary to conduct, coordinate, and complete an investigation. (NFPA 1033)

Fire load The weight of combustibles in a fire area or on a floor in buildings and structures, including either the contents or the building parts, or both.

Fire marshal A member of the fire department who inspects businesses and enforces laws that deal with public safety and fire codes.

Fire partition An interior wall extending from the floor to the underside of the floor above.

Fire protection engineer A member of the fire department who works with building owners to ensure that their fire suppression and detection systems will meet code and function as needed.

Fire tetrahedron A geometric shape used to depict the four components required for a fire to occur: fuel, oxygen, heat, and chemical chain reactions.

Fire triangle A geometric shape used to depict the three components of which a fire is composed: fuel, oxygen, and heat.

Fire wall A wall with a fire-resistive rating and structural stability that separates buildings or subdivides a building to prevent the spread of fire.

Fire window A window or glass block assembly with a fire-resistive rating.

Fixed-temperature heat detector A sensing device that responds when its operating element is heated to a predetermined temperature.

Flame detector A sensing device that detects the radiant energy emitted by a flame.

Flame point (fire point) The lowest temperature at which a substance releases enough vapors to ignite and sustain combustion.

Flameover (rollover) A condition in which unburned products of combustion from a fire have accumulated in the ceiling layer of gas to a sufficient concentration (i.e., at or above the lower flammable limit) such that they ignite momentarily.

Flammability limits (explosive limits) The upper and lower concentration limits (at a specified temperature and pressure) of a flammable gas or vapor in air that can be ignited, expressed as a percentage of the fuel by volume.

Flammable gas Any substance that exists in the gaseous state at normal atmospheric temperature and pressure and is capable of being ignited and burned when mixed with the proper proportions of air, oxygen, or other oxidizers. (NFPA 99)

Flammable liquids Flammable liquids shall be or shall include any liquids having a flash point below 100°F (37.8°C) and having a vapor pressure not exceeding 40 psi (276 kPa) (absolute) at 100°F (37.8°C). Flammable liquids shall be subdivided as follows: (a) Class I liquids shall include those having flash points below 100°F (37.8°C) and shall be subdivided as follows: 1. Class IA liquids shall include those having flash points below 73°F (22.8°C) and having a boiling point below 100°F (37.8°C). 2. Class IB liquids shall include those having flash points below 73°F (22.8°C) and having a boiling point above 100°F (37.8°C). 3. Class IC liquids shall include those having flash points at or above

73°F (22.8°C) and below 100°F (37.8°C). Combustible liquids shall be or shall include any liquids having a flash point at or above 100°F (37.8°C). They shall be subdivided as follows: (a) Class II liquids shall include those having flash points at or above 100°F (37.8°C) and below 140°F (60°C). (b) Class IIIA liquids shall include those having flash points at or above 140°F (60°C) and below 200°F (93.3°C). (NFPA 11)

Flash point The minimum temperature at which a liquid or a solid releases sufficient vapor to form an ignitable mixture with the air.

Flashover A condition in which all combustibles in a room or confined space have been heated to the point at which they release vapors that will support combustion, causing all combustibles to ignite simultaneously.

Flat roofs Horizontal roofs often found on commercial or industrial occupancies.

Flow pressure The amount of pressure created by moving water.

Flow switch An electrical switch that is activated by water moving through a pipe in a sprinkler system.

Flue The general term for a passage through which flue gases are conveyed from the combustion chamber to the outer air. (NFPA 211)

Fluid-filled electrical transformers A device that raises or lowers the voltage of alternating current of the original source. (NFPA 70)

Flush sprinkler A sprinkler in which all or part of the body, including the shank thread, is mounted above the lower plane of the ceiling. (NFPA 13)

Frangible-bulb sprinkler head A sprinkler head with a liquid-filled bulb. The sprinkler head activates when the liquid is heated and the glass bulb breaks.

Freedom of Information Act (FOIA) Signed into law on July 4, 1966 by President Lyndon B. Johnson, allows public access to government records.

Freight bills Shipping papers for roads and highways.

Frequency How often a particular type of incident occurs.

Fuel All combustible materials. The actual material that is being consumed by a fire, allowing the fire to take place.

Fuel gases Any gas used as a fuel source, including natural gas, manufactured gas, sludge gas, liquefied petroleum gas–air mixtures, liquefied petroleum gas in the vapor phase, and mixtures of these gases. See NFPA 54, National Fuel Gas Code. (NFPA 97)

Fuel ladder A continuous progression of fuels that allows fire to move from brush to limbs to tree crowns or structures.

Fully developed phase The phase of fire development in which the fire is free-burning and consuming much of the fuel.

Fuse An overcurrent protective device with a circuit-opening fusible part that is heated and severed by the passage of overcurrent through it. (NFPA 70)

Fusible-link sprinkler head A sprinkler head with an activation mechanism that incorporates two pieces of metal held together by low-melting-point solder. When the solder melts, it releases the link and water begins to flow.

Gas detector A device that detects and/or measures the concentration of dangerous gases.

Gas One of the three phases of matter. A substance that will expand indefinitely and assume the shape of the container that holds it.

Girder A beam that supports other beams.

Glass blocks Thick pieces of glass that are similar to bricks or tiles.

Glazed Transparent glass.

Gravity vents A component of a type of vent system for the removal of smoke from a fire that utilizes manually or automatically operated heat and smoke vents at roof level and that exhausts smoke from a reservoir bounded by exterior walls, interior walls, or draft curtains to achieve the design rate of smoke mass flow through the vents, and that includes provision for makeup air. (NFPA 204)

Gravity-feed system A water distribution system that depends on gravity to provide the required pressure. The system storage is usually located at a higher elevation than the end users.

Ground-fault circuit-interrupter (GFCI) A device intended for protection of personnel that functions to de-energize a circuit or portion thereof within an established period of time when a fault current-to-ground exceeds some predetermined value that is less than that required to operate the overcurrent protective device of that supply circuit. (NFPA 70)

Grounding A conducting connection, whether intentional or accidental, between an electrical circuit or equipment and the earth or to some conducting body that serves in place of the earth. (NFPA 70)

Growth phase The phase of fire development in which the fire is spreading beyond the point of origin and beginning to involve other fuels in the immediate area.

Gypsum A naturally occurring material composed of calcium sulfate and water molecules.

Gypsum board The generic name for a family of sheet products consisting of a noncombustible core primarily of gypsum with paper surfacing.

Halogenated extinguishing agent A liquefied gas extinguishing agent that puts out fires by chemically interrupting the combustion reaction between the fuel and oxygen.

Halon 1301 A liquefied gas-extinguishing agent that puts out a fire by chemically interrupting the combustion reaction between fuel and oxygen. Halon agents leave no residue.

Hardware The parts of a door or window that enable it to be locked or opened.

Hazardous material Any materials or substances that pose an unreasonable risk of damage or injury to persons, property, or the environment if not properly controlled during handling, storage, manufacture, processing, packaging, use and disposal, or transportation.

Hazardous Materials Information System (HMIS) A color-coded marking system by which employers give their personnel the necessary information to work safely around chemicals.

Health care occupancy An occupancy used for purposes of medical or other treatment or care of four or more persons where such occupants are mostly incapable of self-preservation due to age, physical or mental disability, or because of security measures not under the occupant's control. (NFPA 101, *Life Safety Code*)

Heat detector A fire alarm device that detects abnormally high temperature, an abnormally high rate-of-rise in temperature, or both.

Heat sync An object that, through conduction, draws heat away from a heat-producing object.

High hazard contents Contents that are likely to burn with extreme rapidity or from which explosions are likely (NFPA 520)

High-piled storage Solid-piled, palletized, rack storage, bin box, and shelf storage in excess of 12′ (3.7 m) in height. (NFPA 13)

Hollow-core A door made of panels that are honeycombed inside, creating an inexpensive and lightweight design.

Hood and exhaust system Devices installed above a cooking appliance to direct and capture grease-laden vapors and exhaust gases.

Horizontal exit An exit between adjacent areas on the same deck that passes through an A-60 Class boundary that is contiguous from side shell to side shell or to other A-60 Class boundaries. (NFPA 301)

Horizontal-sliding windows Windows that slide open horizontally.

Hotel A building or group of buildings under the same management in which there are sleeping accommodations for more than 16 people and is primarily used by transients for lodging with or without meals. (NFPA 101, *Life Safety Code*)

HVAC Stands for heating, ventilation, and air conditioning systems.

Hydronic heating and cooling systems A method of radiant heating or cooling of a space through the circulation of warm or cool water through a system of tubing either imbedded in a floor system or by radiant ceiling panels.

Hydrostatic test A test filling the sprinkler piping with water and pressurizing it, usually to 200 psi (1379 kPa) for two hours, to look for leaks in the pipe work.

Hypoxia A state of inadequate oxygenation of the blood and tissue.

Ignition phase The phase of fire development in which the fire is limited to the immediate point of origin.

Ignition temperature The minimum temperature at which a fuel, when heated, will ignite in air and continue to burn.

Immediately Dangerous to Life or Health (IDLH) Any atmosphere that poses an immediate hazard to life or produces immediate irreversible debilitating effects on health. (NFPA 1670)

Incipient The initial stage of a fire.

Indirect costs Costs which the fire department would have incurred whether the fire and life safety education program existed or not, such as labor and apparatus costs.

Industrial gases The entire range of gases classified by chemical properties customarily used in industrial processes, for welding and cutting, heat treating, chemical processing, refrigeration, and water treatment.

Industrial occupancy An occupancy in which products are manufactured or in which processing, assembling, mixing, packaging, finishing, decorating, or repair operations are conducted. (NFPA 101, *Life Safety Code*)

Interior finish Any coating or veneer applied as a finish to a bulkhead, structural insulation, or overhead, including the visible finish, all intermediate materials, and all application materials and adhesives.

Intermediate level sprinklers Sprinklers equipped with integral shields to protect their operating elements from the discharge of sprinklers installed at higher elevations.

Intermodal tanks Bulk containers that can be shipped by all modes of transportation—air, sea, or land.

Ionization smoke detector A device containing a small amount of radioactive material that ionizes the air between two charged electrodes to sense the presence of smoke particles.

Jalousie windows Windows made of small slats of tempered glass, which overlap each other when the window is closed. Often found in trailers and mobile homes, jalousie windows are held together by a metal frame and operated by a small hand wheel or crank found in the corner of the window.

Jamb The part of a doorway that secures the door to the studs in a building.

Joist A beam.

Knockouts Pre-punched circular holes in an electrical junction box or other electrical equipment that allows the secure connection of conduit or wiring cables.

Laminated glass Glass manufactured with a thin vinyl core covered by glass on each side of the core.

Laminated wood Pieces of wood that are glued together.

Large-drop sprinkler A sprinkler head that generates large drops of water of such size and velocity as to enable effective penetration of a high-velocity fire plume.

Line detector Wire or tubing that can be strung along the ceiling of large open areas to detect an increase in heat.

Liquefied gas A gas, other than in solution, that in a packaging under the charged pressure exists both as a liquid and a gas at a temperature of 20°C (68°F). (NFPA 30)

Liquid One of the three phases of matter. A nongaseous substance that is composed of molecules that move and flow freely and that assumes the shape of the container that holds it.

Live load The weight of the building contents.

Load-bearing wall A wall that is designed to provide structural support for a building.

Loaded-stream fire extinguisher A water-based fire extinguisher that uses an alkali metal salt as a freezing-point depressant.

Lodging or rooming house Building or portion thereof that does not qualify as a one- or two-family dwelling, that provides sleeping accommodations for a total of 16 or fewer people on a transient or permanent basis, without personal care services, with or without meals, but without separate cooking facilities for individual occupants. (NFPA 101, *Life Safety Code*)

Low hazard contents Contents of such low combustibility that no self-propagating fire therein can occur (NFPA 520)

Lower explosive limit (LEL) The minimum amount of gaseous fuel that must be present in the air mixture for the mixture to be flammable or explosive.

Main drain test A test opening the sprinkler system main drain to record the static and residual water pressures. This can indicate if the water supply to the sprinkler system is open or not.

Make-up air Air introduced to a space to replace air removed by exhaust systems.

Manual pull-station A device with a switch that either opens or closes a circuit, activating the fire alarm.

Masonry A built-up unit of construction or combination of materials such as brick, clay tiles, or stone set in mortar.

Master-coded alarm An alarm system in which audible notification devices can be used for multiple purposes, not just for the fire alarm.

Matter Made up of atoms and molecules.

MC-306 flammable liquid tanker Commonly known as a gasoline tanker, this tanker typically carries gasoline or other flammable and combustible materials.

MC-307 chemical hauler A tanker with a rounded or horseshoe-shaped tank.

MC-312 corrosives tanker A tanker that will often carry aggressive acids like concentrated sulfuric and nitric acid, having reinforcing rings along the side of the tank.

MC-331 pressure cargo tanker A tank commonly constructed of steel with rounded ends and a single open compartment inside; there are no baffles or other separations inside the tank.

MC-338 cryogenic tanker A low-pressure tanker designed to maintain the low temperature required by the cryogens it carries.

Means of egress A continuous and unobstructed way of exit travel from any point in a building or structure to a public way, consisting of three separate and distinct parts: (a) the exit access, (b) the exit, and (c) the exit discharge. A means of egress comprises the vertical and horizontal travel and includes intervening room spaces, doorways, hallways, corridors, passageways, balconies, ramps, stairs, enclosures, lobbies, escalators, horizontal exits, courts, and yards. (NFPA 101)

Mechanical energy Heat that is created by friction.

Mechanical plans Drawings in a plan set showing the proposed plumbing, HVAC, or other mechanical systems for a building.

Medical gases A patient medical gas or medical support gas. (NFPA 99)

Mercantile occupancy An occupancy used for the display and sale of merchandise. (NFPA 101, *Life Safety Code*)

Mixed occupancy A multiple occupancy where the occupancies are intermingled. (NFPA 101, *Life Safety Code*)

Multiple occupancy A building or structure in which two or more classes of occupancy exist. (NFPA 101, *Life Safety Code*)

Multipurpose dry-chemical fire extinguisher A fire extinguisher rated to fight Class A, B, and C fires.

Municipal water system A water distribution system that is designed to deliver potable water to end users for domestic, industrial, and fire protection purposes.

National Fire Incident Reporting System (NFIRS) A national database which collects data about fire incidents including the collection of many of the underlying factors that caused the fire.

NFPA 704 hazard identification system A hazardous materials marking system designed for fixed-facility use.

Nonbearing wall A wall that is designed to support only the weight of the wall itself.

Nonbulk storage vessels Containers other than bulk storage containers.

Noncoded alarm An alarm system that provides no information at the alarm control panel indicating where the activated alarm is located.

Nonflammable gases Those gases that will not burn in any concentration of air or oxygen.

Nozzle sprinkler head Sprinkler heads used in applications requiring special discharge patterns, such as directional spray or fine spray.

Obscuration rate A measure of the percentage of light transmission that is blocked between a sender and a receiver unit.

Occupancy The intended use of a building.

Occupant load The number of people who might occupy a given area.

One- or two-family dwelling A building that contains no more than two dwelling units with independent cooking and bathroom facilities. (NFPA 5000)

Open sprinkler heads A sprinkler that does not have actuators or heat-responsive elements. (NFPA 13)

Ordinary hazard contents Those contents likely to burn with moderate rapidity and give off a considerable volume of smoke.

Ornamental sprinklers Sprinkler that have been painted or plated by the manufacturer.

Outlet A point on the wiring system at which current is taken to supply utilization equipment. (NFPA 70)

Outside stem and yoke (OS&Y) valve A sprinkler control valve with a valve stem that moves in and out as the valve is opened or closed.

Oxidizing gases Gases that support combustion; generally either oxygen, chlorine, or mixtures of oxygen and other gases, such as oxygen-helium or oxygen-nitrogen mixtures, or certain gaseous oxides, such as nitrous oxide.

Panic hardware A door-latching assembly incorporating a device that releases the latch upon the application of a force in the direction of egress travel. (NFPA 101B)

Parallel chord trusses Trusses in which the top and bottom chords are parallel.

Parapet wall Walls on a flat roof that extend above the roofline.

Partition wall A non-load-bearing wall that subdivides spaces within any story of a building or room.

Party wall A wall constructed on the line between two properties.

Pendant sprinkler head A sprinkler head designed to be mounted on the underside of sprinkler piping so that the water stream is directed down.

Performance-based code Outlines the requirement that a design has to meet, but does not state that a particular method or material must be used to meet the requirement.

Performance-based design A design process whose fire safety solutions are designed to achieve a specified goal for a specified use or application. (NFPA 914)

Permit A document issued by the authority having jurisdiction for the purpose of authorizing performance of a specified activity. (NFPA 1)

Photoelectric smoke detector A device to detect visible products of combustion using a light source and a photosensitive sensor.

Pipe schedule system A sprinkler system in which the pipe sizing is selected from a schedule that is determined by the occupancy classification and in which a given number of sprinklers may be supplied from specific sizes of pipe. (NFPA 13)

Pipeline A length of pipe, including pumps, valves, flanges, control devices, strainers, and similar equipment, for conveying fluids and gases.

Pipeline right-of-way An area, patch, or roadway that extends a certain number of feet on either side of the pipe itself and that may contain warning and informational signs about hazardous materials carried in the pipeline.

Pitched chord truss Type of truss typically used to support a sloping roof.

Pitched roof A roof with sloping or inclined surfaces.

Pitot gauge A type of gauge that is used to measure the velocity pressure of water that is being discharged from an opening. It is used to determine the flow of water from a hydrant.

Plan set Created by design professionals, plan sets include a series of drawings detailing how a proposed building will be built. Also known as plans, blueprints, construction documents, or shop drawings.

Plan view A view on a drawing where a horizontal slice is made in the building or area and everything above or below the slice is shown.

Plate glass A type of glass that has additional strength so it can be formed in larger sheets, but will still shatter upon impact.

Platform-frame construction Construction technique for building the frame of the structure one floor at a time. Each floor has a top and bottom plate that acts as a firestop.

Plenum system An HVAC system that uses a compartment or chamber to which one or more air ducts are connected and that forms part of the air distribution system. (NFPA 90A)

Plug fuses An overcurrent protective device with a circuit-opening fusible part that is heated and severed by the passage of overcurrent through it. (NFPA 70)

Plume The column of hot gases, flames, and smoke that rises above a fire. Also called a convection column, thermal updraft, or thermal column.

Polar solvent A water-soluble flammable liquid such as alcohol, acetone, ester, and ketone.

Polychlorobiphenol (PCB) fluid A class of organic compounds containing one or more chlorine atoms connected to a molecule composed of two linked benzene rings. Trade names associated with PCB's include, but are not limited to, Askarel, Inerteen, and Pyranol/Pyrenol.

Post indicator valve (PIV) A sprinkler control valve with an indicator that reads either open or shut depending on its position.

Preaction sprinkler system A dry sprinkler system that uses a deluge valve instead of a dry-pipe valve and requires activation of a secondary device before the pipes will fill with water.

Premise identification Posting of an address for emergency responders.

Prescriptive code Defines the specifics of a material of construction or action to be taken; such as the type of electrical wiring to use, based on the anticipated usage or requirement to conduct evacuation drills in a structure.

Pressurization method A smoke control method specified in NFPA 92A that employs the development of a pressure differential across a smoke zone boundary to limit the spread of smoke from one smoke zone in a building to another smoke zone.

Primary feeders The largest-diameter pipes in a water distribution system, carrying the greatest amounts of water.

Projected windows Windows that project inward or outward on an upper hinge; also called factory windows. They are usually found in older warehouses or commercial buildings.

Proprietary supervising system A fire alarm system that transmits a signal to a monitoring location owned and operated by the facility's owner.

Protected premises fire alarm system A fire alarm system that sounds an alarm only in the building where it was activated. No signal is sent out of the building.

Public education Teaching a safety message with the goal of reinforcing good behaviors, or changing undesirable behaviors, in order to make the community safer.

Public information Information that is disseminated to the public via the fire department.

Public relations Activities that are focused on building a positive image of any organization.

Pyrolysis The destructive distillation of organic compounds in an oxygen-free environment that converts the organic matter into gases, liquids, and char.

Pyrophoric gases Flammable gases that spontaneously ignite in air; examples of phrophoric gases include silane and phosphine.

Raceway An enclosed channel of metal or nonmetallic materials designed expressly for holding wires, cables or busbars, with additional functions as permitted in the electrical code. Raceways include, but are not limited to, rigid metal conduit, rigid nonmetallic conduit, intermediate metal conduit, liquidtight flexible conduit, flexible metallic tubing, flexible metal conduit, electrical nonmetallic tubing, electrical metallic tubing, underfloor raceways, cellular concrete floor raceways, cellular metal floor raceways, surface raceways, wireways, and busways. (NFPA 70)

Radiation The combined process of emission, transmission, and absorption of energy traveling by electromagnetic wave propagation between a region of higher temperature and a region of lower temperature.

Rafters Joists that are mounted in an inclined position to support a roof.

Rate-of-rise heat detector A fire detection device that responds when the temperature rises at a rate that exceeds a predetermined value.

Recessed sprinkler A sprinkler in which all or part of the body, other than the shank thread, is mounted within a recessed housing. (NFPA 13)

Reinspection An inspection performed to determine if code violations have been corrected.

Relative humidity The amount of water vapor or moisture held in suspension by gas or air and expressed as a percentage of the amount of moisture that would be held in suspension at the same temperature if saturated.

Remote annunciator A secondary fire alarm control panel in a different location than the main alarm panel; it is usually located near the front door of a building.

Remote supervising station system A fire alarm system that sounds an alarm in the building and transmits a signal to the fire department or an off-premises monitoring location.

Renewable link cartridge An overcurrent protective device with a circuit-opening fusible part that is heated and severed by the passage of overcurrent through it. (NFPA 70)

Reservoir A water storage facility.

Residential board and care occupancy A building or portion thereof that is used for lodging and boarding of four or more residents, not related by blood or marriage to the owners or operators, for the purpose of providing personal care services. (NFPA 101, *Life Safety Code*)

Residential sprinkler system A sprinkler system designed to protect dwelling units.

Residual pressure The pressure that exists in the distribution system, measured at the residual hydrant at the time the flow readings are taken at the flow hydrants.

Riser The vertical supply pipes in a sprinkler system. (NFPA 13)

Safety Data Sheet (SDS) A form, provided by manufacturers and compounders (blenders) of chemicals, containing information about chemical composition, physical and chemical properties, health and safety hazards, emergency response, and waste disposal of the material.

Saponification The process of converting the fatty acids in cooking oils or fats to soap or foam.

Secondary containment Any device or structure that prevents environmental contamination when the primary container or its appurtenances fail. Examples of secondary containment are dikes, curbing, and double-walled tanks.

Secondary feeders Smaller-diameter pipes that connect the primary feeders to the distributors.

Sectional view On a drawing in a plan set, a vertical slice of a building showing the internal view of the building.

Self inspections Inspection performed by the building owner or occupant.

Separated occupancy A multiple occupancy where the occupancies are separated by fire resistance-rated assemblies. (NFPA 101, *Life Safety Code*)

Severity The amount of death, injury, or damage that is the result of an incident.

Shipping papers A shipping order, bill of lading, manifest, or other shipping document serving a similar purpose and usually including the names and addresses of both the shipper and the receiver as well as a list of shipped materials with quantity and weight.

Shunt trip A device that remotely causes the manual or automatic opening of an electrical circuit or main panel disconnect.

Shut-off valve Any valve that can be used to shut down water flow to a water user or system.

Sidewall sprinklers A sprinkler that is mounted on a wall and discharges water horizontally into a room.

Signal words Information on a pesticide label that indicates the relative toxicity of the material.

Simple beam Supported at two points neat its ends. In simple beam construction, the load is delivered to the two reaction points and the rest of the structure renders no assistance in an overload.

Single-action pull-station A manual fire alarm activation device that takes a single step—such as moving a lever, toggle, or handle—to activate the alarm.

Single-station smoke alarm A single device usually found in homes that detects visible and invisible products of combustion and sounds an alarm.

Site plan A drawing showing the building and surrounding area, including items such as roads, driveways, and hydrants.

Slope Upward or downward incline or slant, usually calculated as a percentage. (NFPA 1144)

Smoke An airborne particulate product of incomplete combustion that is suspended in gases, vapors, or solid or liquid aerosols.

Smoke damper A device arranged to seal off airflow automatically through a part of an air duct system, to restrict the passage of smoke. A smoke damper is not required to meet all the design functions of a fire damper. (NFPA 221)

Smoke detector A device that detects smoke and sends a signal to a fire alarm control panel.

Smoke zone The smoke-control zone in which the fire is located. (NFPA 92A)

Smokeproof enclosures A stair enclosure designed to limit the movement of products of combustion produced by a fire. (NFPA 101, *Life Safety Code*)

Solid One of the three phases of matter. A substance that has three dimensions and is firm in substance.

Solid-core A door design that consists of wood filler pieces inside the door. This construction creates a stronger door that may be fire rated.

Spalling Chipping or pitting of concrete or masonry surfaces.

Specifications book A collection of all of the information about a project that may be provided to the fire inspector in addition to the plan set during a plan review.

Spontaneous ignition Initiation of combustion of a material by an internal chemical or biological reaction that has produced sufficient heat to ignite the material. (NFPA 921)

Spot detector A single heat-detector device; these devices are often spaced throughout an area.

Spray booth A power-ventilated enclosure for a spray application operation or process that confines and limits the escape of the material being sprayed, including vapors, mists, dusts, and residues that are produced by the spraying operation and conducts or directs these materials to an exhaust system. (NFPA 33)

Sprinkler piping The network of piping in a sprinkler system that delivers water to the sprinkler heads.

Stairwell smoke management A design or method that employs architectural design, construction material, mechanical equipment, or a combination thereof with the intent to keep smoke from contaminating a stair shaft with smoke during a fire incident.

Stakeholder An individual or group that is impacted by an issue.

Standard A document, the main text of which contains only mandatory provisions using the word "shall" to indicate requirements and that is in a form generally suitable for mandatory reference by another standard or code or for adoption into law. Nonmandatory provisions shall be located in an appendix or annex, footnote, or fineprint note and are not to be considered a part of the requirements of a standard.

Standpipe system A system of pipes and hose outlet valves used to deliver water to various parts of a building for fighting fires.

Static pressure The pressure that exists at a given point under normal distribution system conditions measured at the residual hydrant with no hydrants flowing.

Static water source A water source such as a pond, river, stream, or other body of water that is not under pressure.

Stop work order A form used when contractors do not have the clearance for performing the work, or when work must be corrected prior to performing additional work.

Storage occupancy An occupancy used primarily for the storage or sheltering of goods, merchandise, products, vehicles, or animals. (NFPA 101, *Life Safety Code*)

Structural inspector A person who, through formal education and training, is qualified to inspect the structural elements of a building against established standards, recommendations, and requirements.

Structural plan A drawing showing the proposed building's load-bearing components.

Supervised Electronically monitoring the alarm system wiring for an open circuit.

Switch Any set of contacts that interrupts or controls current flow through an electrical circuit.

T tapping Improper wiring of an initiating device so that it is not supervised.

Tamper switch A switch on a sprinkler valve that transmits a signal to the fire alarm control panel if the normal position of the valve is changed.

Tempered glass Glass that is much stronger and harder to break than ordinary glass.

Temporal-3 pattern A standard fire alarm audible signal for alerting occupants of a building.

Thermal column A cylindrical area above a fire in which heated air and gases rise and travel upward.

Thermal conductivity A property that describes how quickly a material will conduct heat.

Thermal layering The stratification (heat layers) that occurs in a room as a result of a fire.

Thermal radiation How heat transfers to other objects.

Thermoplastic material A plastic material capable of being repeatedly softened by heating and hardened by cooling and, that in the softened state, can be repeatedly shaped by molding or forming.

Thermoset material A plastic material that, after having been cured by heat or other means, is substantially infusible and cannot be softened and formed.

Tote Portable tanks, usually holding a few hundred gallons of product, characterized by a unique style of construction.

Truss A collection of lightweight structural components joined in a triangular configuration that can be used to support either floors or roofs.

Tube trailers High-volume transportation devices made up of several individual compressed gas cylinders banded together and affixed to a trailer.

Type I construction (fire resistive) Buildings with structural members made of noncombustible materials that have a specified fire resistance.

Type II construction (noncombustible) Buildings with structural members made of noncombustible materials without fire resistance.

Type III construction (ordinary) Buildings with the exterior walls made of noncombustible or limited-combustible materials, but interior floors and walls made of combustible materials.

Type IV construction (heavy timber) Buildings constructed with noncombustible or limited-combustible exterior walls, and interior walls and floors made of large-dimension combustible materials.

Type V construction (wood frame) Buildings with exterior walls, interior walls, floors, and roof structures made of wood.

Underground flush test A flushing of the water main supplying the sprinkler system to make certain there is no debris in the supply piping that might clog a sprinkler line.

Underwriters Laboratories, Inc. (UL) The U.S. organization that tests and certifies that fire extinguishers (among many other products) meet established standards. The Canadian equivalent is Underwriters Laboratories of Canada (ULC).

Unstable reactive gas A gas that can undergo violent changes when subjected to shock or changes in temperature or pressure.

Upper explosive limit (UEL) The maximum amount of gaseous fuel that can be present in the air mixture for the mixture to be flammable or explosive.

Upright sprinkler head A sprinkler head designed to be installed on top of the supply piping; it is usually marked SSU ("standard spray upright").

Vapor density The weight of an airborne concentration (vapor or gas) as compared to an equal volume of dry air.

Variance A waiver allowing a condition that does not meet a recognized code or standard to continue to exist legally.

Vent pipes Inverted J-shaped tubes that allow for pressure relief or natural venting of the pipeline for maintenance and repairs.

Vestibule A small room located between two spaces that provides an atmospheric separation for the purposes of controlling airflow or, in a smoke management system, the movement of contaminated air from one space to an adjacent space.

Volatility The ability of a substance to produce combustible vapors.

Volt (V) The unit of electrical pressure (or electromotive force) represented by the letter "E"; the difference in potential required to make a current of one ampere flow through the resistance of one ohm. (NFPA 921)

Wall post indicator valve (WPIV) A sprinkler control valve that is mounted on the outside wall of a building. The position of the indicator tells whether the valve is open or shut.

Water main The generic term for any underground water pipe.

Water supply A source of water.

Water-motor gong An audible alarm notification device that is powered by water moving through the sprinkler system.

Watt (W) Unit of power or the rate of work represented by a current of one ampere under the potential of one volt.

Wet-barrel hydrant A hydrant used in areas that are not susceptible to freezing. The barrel of the hydrant is normally filled with water.

Wet chemical extinguishing agent An extinguishing agent for Class K fires. It commonly consists of solutions of water and potassium acetate, potassium carbonate, potassium citrate, or any combination thereof.

Wet chemical extinguishing systems An extinguishing system that discharges a proprietary liquid extinguishing agent.

Wet chemical fire extinguisher A fire extinguisher for use on Class K fires that contains a wet chemical extinguishing agent.

Wet-pipe sprinkler system A sprinkler system in which the pipes are normally filled with water.

Wildland/urban interface Any area where wildland fuels threaten to ignite combustible homes and structures. (NFPA 1143)

Wired glass Glass made by molding tempered glass around a special wire mesh.

Wood panels Thin sheets of wood glued together.

Wood trusses Assemblies of small pieces of wood or wood and metal.

Wooden beams Load-bearing members assembled from individual wood components.

Zone of origin In the design of a smoke management system, refers to the smoke zone that the fire incident originates.

Zoned coded alarm A fire alarm system that indicates which zone was activated both on the alarm control panel and through a coded audio signal.

Zoned noncoded alarm A fire alarm system that indicates the activated zone on the alarm control panel.

Zoned system A fire alarm system design that divides a building or facility into zones so that the area where an alarm originated can be identified.

Index

A

Aboveground storage tanks (ASTs), 259
AHJ. *See* Authority Having Jurisdiction
Air-duct insulation, 71
Air handling unit, 233, 234
Air sampling detector, 158
Alarm
   initiation devices, 151, 155, 157
      automatic, 155
      manual, 155
   matrix, 151
   notification
      appliances, 159
      device, 151, 159, 164
   valve, 181
Alternative clause, 113
Ambulatory health care occupancy, 46–47
American Society for Testing and Materials (ASTM), 72
Ammonium phosphate, 211
Ampere/amps/amperage, 224
Anecdotal evidence, 309–310
Annealed glass, 32
Annual inspections, 79
Apartment building, 45
Aqueous film-forming foam (AFFF), 212
Arc-fault circuit-interrupters (AFCI), 225
Architectural plans, 105–106, 109
Arcing, 224
Area hazards, 209
Areas of refuge, 138
Aspect, 277
Assembly occupancies, 50–51
ASTM. *See* American Society for Testing and Materials
Authority Having Jurisdiction (AHJ), 8, 120, 267
Automatic initiation devices
   air sampling detectors, 158
   flame detectors, 158
   gas detectors, 158
   heat detectors, 157
      fixed-temperature, 157
      line, 157–158
      rate-of-rise, 157
   smoke detectors, 156–157
Automatic sprinkler heads, 188–189
Automatic sprinkler systems, 159, 178–179
   occupancy hazards, 179–180
Auxiliary systems, 161
Awning windows, 34

B

Backdraft, 68
Bags, 260
Balloon-frame construction, 21
Beam detector, 156
Bills of lading, 252
Bimetallic strip, 157
BLEVE. *See* Boiling-liquid, expanding vapor explosion
Board of Appeals, 112
Boiler enclosures, 232
Boiler inspectors, 232

Boiler room, 231
Boilers, 231, 232
Boiling-liquid, expanding vapor explosion (BLEVE), 70, 257
Bonding of metal piping systems, 226
Bowstring trusses, 26
Building
   components
      beams, 29
      columns, 28–29
      doors, 29–31
      fire doors, 34–35
      fire windows, 34–35
      floor coverings, 35–36
      floors and ceilings, 22–23
      foundations, 22
      girders, 29
      interior finish, 35–36
      joists, 29
      rafters, 29
      roof, 23–26
      walls, 26–28
      windows, 31–34
   construction elements, 72
   container storage in, 254–255
   features, 90–91
   hazard, 70
Bulk storage containers, 259
Business license, 80
Business occupancy, 48
Butterfly valve, 181

C

Cables, 228–229
Carbon dioxide, 212, 248
Carbon dioxide extinguishing systems, 188
Carbon dioxide (CO_2) fire extinguishers, 212
Carboys, 260–261
Cardboard drums, 260
Cartridge fuses, 225
Casement windows, 34, 35
Ceiling, 71
   assemblies, 23
Central stations, 161–162
Chemical energy, 59
Chemical hazards in military system, 251
Chemical-pellet sprinkler heads, 192
Chemical suppression system plans, 111
Chemical treatment systems, 233
Chiller plants, 232
Circuit breakers, 224, 225, 229, 230
Class I flammable liquids, 245–246
Class IA liquids, 246
Class IB liquids, 246
Class IC liquids, 246
Class II liquids, 246
Class III liquids, 246
Class IIIA liquids, 246
Class IIIB liquids, 246
Class I Standpipes, 193
Class II Standpipes, 193–194
Class A fires, 63
Class B fires, 63

Class C fires, 63
Class D fires, 63
Class K fires, 63, 64
Clean agents, 213
   extinguishing systems, 187–188
Closed-head drums, 260
Code analysis, 104–105
Coded system, 160
Code references, 299
Codes, 6–7
Code violations, 94
   interior inspection, conducting, 88
Combination smoke/fire dampers, 235
Combustibility, 14
Combustible dust, 278
Combustible gases, physical properties of, 247
Combustible liquids, 245–246, 280–281
Combustion
   air, 236
   explosion safeguards, 265–266
   products of, 60–61
Commissioning, 113
Common path of travel, 133
Compartmentation, 234
Complaint form, 83
Complaint inspections, 79
Compressed gas, 247
   cylinders, 261, 281
Concrete floors, 15, 22
Conditional approval, 80
Conduction, 61
Conduits, 228–229
Construction, 79–80
   considerations, 92
   inspection forms, 83
   materials, types of, 14–18
   types of, 18–22
Container
   hazard safeguards, 257
   type, 258
   volume, 259
Continuous beam, 29
Convection, 61
Convergence, 128
Cooking equipment, 91
Corrosion-resistant sprinklers, 189
Corrosive gas, 247
Cryogenic gas, 247
Cryogens, 261
Current, 224
Curtain walls, 28
Curved roofs, 23–24
Cylinders, 256, 261

D

Dampers, 234
Day-care occupancy, 47
Dead end corridor, 129
Dead load, 22
Decay phase, 65, 67
Decorations, 71
Defensible space, 277
Deflagration venting, 266

Deluge sprinkler systems, 185–186
Deluge valve, 181
Department of Transportation (DOT) marking system, 249
 chemical families recognized in ERG, 250
Detail view, 107–108
Detention and correctional occupancy, 51–52
Dewar containers, 258, 261
Direct costs, 312
Distribution systems, 233
Documentation
 checklists, 296
 detailed reports, 296
 email, 296
 letters, 294–296
Door construction, 29–30
Dormitory, 44–45
DOT marking system. *See* Department of Transportation marking system
Double-action pull-station, 155, 156
Double-hung windows, 33
Double-pane glass, 32
Drums, 260
Dry bulk cargo tanks, 262, 263
Dry chemical extinguishing systems, 187
Dry-powder extinguishing agents, 214
Dry-type transformers, 227
Duct detectors, 158, 235
Ducted systems, 233
Dunnage, 280
Dust accumulation, 278
Dust collection systems, 233, 234

E

Early-suppression fast-response (ESFR) sprinkler heads, 190, 191
Educational occupancies, 47–48
Egress, 105, 124–126
 capacity, 128, 134
Electrical cords, 89
Electrical energy, 59
Electrical equipment, problem with, 228
Electrical inspectors, 222
Electrical insulation, 71
Electrical plans, 106, 110
Electrical system, 223–224
 boxes, 229
 protective practices for, 224–226
Electrical transfer switch, 227
Electricity, basics of, 223–224
Electrostatic, 233
Elevation pressure, 176
Elevation view, 107
Elevators, 92, 138
Emergency lighting, 139
Emergency plans, 140–143
Emergency Response Guidebook (ERG), 249
 DOT chemical families recognized in, 250
 using, 251
Enabling legislation, 81
Endothermic heat, 59

Energy
 conservation of, 59–60
 conversion, 231–232
 types of, 59
Equipment ground-fault protective devices (EGFPD), 225
Equivalencies, 113
ERG. *See Emergency Response Guidebook*
Escalators, 92, 138
Ethyl alcohol, 246
Evacuation drills, 143
Evacuation leaders, 142–143
Evacuation plans, 140–143
Exhaust stacks, 236
Exhaust systems, 233–234
Exigent circumstance, 86
Exit, 125
 access, 124–125
 discharge, 125–126
 lighting, 139–140
 signs, 139–140
 travel distance, measurement, 129–133
Exothermic heat, 59
Explosive limits, 69
Extension cords, 229
Exterior inspection, 87
Extinguishing agents
 carbon dioxide, 212
 dry chemicals, 210–212
 dry-powder, 214
 foam, 212–213
 water, 210
 wet chemicals, 213
Extinguishment, methods of, 62
Extra/high hazard area, 209

F

FDC. *See* Fire department connection
Federal Emergency Management Agency (FEMA), 310
Field notes, 291
Film-forming fluoroprotein (FFFP) foam, 212
Final notice, 294
Fire
 characteristics of
  gas-fuel, 69–70
  liquid-fuel, 68–69
  solid-fuel, 64–68
 chemistry of, 58–64
 Class A, 205
 Class B, 205
 Class C, 205, 206
 Class D, 206
 classes of, 63–64
 Class K, 206
 conditions needed for, 60
 and life safety education specialist, 5
Fire alarm control panel, 151
Fire and Life Safety Education Program
 budget, 312–313
 creating, 309–311

 developing, 311–313
 role of, 308–309
Fire and life safety educators, 308
Fire codes, 152
Fire dampers, 235
Fire department
 content development, 312
 developing
  goals and objectives, 312
  relationships, 312
 effectiveness metrics, 313
 fire prevention roles of, 5
 job of, 4, 5
 notification, local alarm systems, 160–161
Fire department connection (FDC), 180, 194
Fire detection system, 88
 components of, 151
 inspection of, 150
 maintenance, 162
 plans, 111
 and suppression systems, 89
 testing, 150, 162–163
Fire enclosure, 27
Fire exit hardware, 135
Fire exposure threats, 276–277
Fire extinguishers, 89
 classification of, 207
 labeling of, 207
 placement of, 208
 system readiness, 214
Fire flow rates
 calculations of, 176
 flow and pressure, 176
Fire growth/spread, 61–62, 70
 building construction elements, 72
 building/contents hazard, 70
 furnishings, 72
 hidden building elements, 70–71
 interior finish, 71
Fire hydrant locations, 176
Fire inspection, 4, 5
 construction considerations, 92
 exterior, 87
 and fire service, 4–5
 interior, 88–92
 legal authority for, 8
 permit, 8
 preplan sketch, 92
 presentation, 86–87
 remodeling considerations, 92–93
 report, 293–294
 three E's of fire prevention, 4
Fire inspector
 career development, 9
 ethics and, 9
 legal proceedings, 8
Fire inspector I, roles and responsibilities for, 5–6
Fire inspector II
 range of authority for, 6
 roles and responsibilities for, 6
Fire investigator, 5

Fire marshal, 5
Fire partition, 27
Fire point, 69
Fire prevention
  codes, 266–267
  methods, 255–256
Fire prevention program, 103
Fire protection engineer, 5, 113
Fire protection equipment, 275–276, 281
Fire protection features, 88–89
Fire protection systems, 105
  for flammable gases, 265–266
  for flammable liquids, 265
  for hazardous materials, 266
  plans, 110–111
Fire resistance, 18
Fire resistive construction, 18–19
Fire-resistive floors, 22
Fire-retardant–treated wood, 17
Fire suppression system, 89
  alarm initiation by, 158–159
  testing
    air, 195
    hood and duct system, 195
    hydrostatic, 195
    main drain, 195
    underground flush, 195–196
Fire tetrahedron, 60
Fire triangle, 60
Fire wall, 15
Fire windows, 34–35
Fixed-temperature heat detector, 157
Flame detector, 158
Flameover, 68
Flame point, 69
Flame spread, 70
  and smoke development ratings, 72
Flammability limits, 69
Flammable gas, 247
  physical properties of, 247
Flammable liquids, 276, 280–281
  classification of, 245–246
  handling of, 255–256
  physical properties of, 247
  storage
    cabinet, 255
    container, 254–255
    tank, 253–254
Flashlight, 82
Flashover, 67
Flash point, 69
Flat roofs, 24–25
Floors, 71
Flow pressure, 176
Flue gases, 236
Fluid-filled electrical transformers, 227
Flush sprinkler head, 189, 190
Foam, 212–213
Foamed-plastic insulation, 71
FOIA. *See* Freedom of Information Act
Formal Interpretation (FI) of codes and standards, 318

Freedom of Information Act (FOIA), 300–301
Freight bills. *See* Bills of lading
Frequency, 310, 311
Fuel, 58–59
Fuel gases, 248
Fuel ladder, 277
Fully developed phase, 65, 67
Fuses, 224–225

G
Gas, 58, 59
  care in handling, 257
  categories, 246–247
  classification by usage, 248
  compressed, 247
  containers, 256–257
  detector, 158
  equipment, 276
  hazard safeguards, 257
  inspection for leakage, 257–258
  storage, 256–258
    safety considerations, 257
Generators, 226–227
Girder, 29
Glass blocks, 16
Glass construction, 31–33
Glass doors, 30
Glazed, 31
Gravity vents, 234
Ground-fault circuit-interrupters (GFCI), 225
Grounding, 225–226
Growth phase, 65, 66–67
Gypsum, 16
Gypsum board, 16

H
Halogenated extinguishing agents, 213
Halon 1301, 188
Hardware, 29
Hazardous areas, electrically, 230, 231
Hazardous material, 249
  container, 258–259
  military, 251–252
  nonbulk storage vessels, 260–261
  railroad transportation, 262–264
  storage and handling of, 258
  storage lockers, 261
  transporting, 262
  warning labels, 250
Hazardous Materials Information System (HMIS), 251
Hazardous waste, 249
Hazard recognition, 89
Health care occupancy, 46
Heat detectors, 157
Heating, ventilation, and air conditioning (HVAC) systems, 23, 91, 222
  components of, 231–233
  potential hazards of, 235–236
  safety systems, 233–235
  smoke distribution through ducts in, 235
Heat sync, 232

Heavy timber construction, 20
Hidden building elements, 70–71
High hazard contents, 89
High-piled storage, 279
High-rise buildings, special structures and, 52
Hollow-core doors, 30
Hood and exhaust system, 281
Horizontal exits, 135–136
Horizontal-sliding windows, 37
Hotel, 44–45
Hotel and Motel Safety Act (1990), 7
Housekeeping, 274–275, 283
Humidification, 230
HVAC. *See* Heating, ventilation, and air conditioning systems
Hydrant flow test, 178
Hydrant testing procedure, 177–178
Hydrogen sulfide, 247
Hydronic heating and cooling systems, 233
Hypoxia, 61

I
Ignition phase, 64, 66
Ignition sources, controlling, 258
Ignition temperature, 59
IM-101 containers, 259
IM-102 containers, 259
Immediately dangerous to life and health (IDLH) levels, 233
IMO type 5 containers, 259
IM tanks. *See* Intermodal tanks
Indirect costs, 312
Indoor gas storage areas, 258
Industrial gases, 248
Industrial occupancy, 48
Inspections
  documentation, 93–94
  interior, conducting, 88–92
  process, 95
  types of, 79–80
Interior finish, 35–36, 71
Intermediate level sprinklers, 189, 190
Intermodal (IM) tanks, 259, 262, 263
International Association of Fire Chiefs (IAFC), 153
International Building Code, 42
Inward-opening doors, 30
Ionization smoke detector, 153, 154

J
Jalousie windows, 33, 34
Jamb, 29
Joist, 29

K
Knockouts, 229

L
Labeling for hazardous materials, 249–253
Laminated glass, 33
Laminated wood, 17
Lamps, 230
Laundry/garbage chutes, 91

Learning Resource Center (LRC), 310, 312
Ledge doors, 30
Legal proceedings, presenting evidence, 302
LEL. *See* Lower explosive limit
Light energy, 59
Light fixtures, 230
Light/low hazard area, 209
Line detectors, 157–158
Lint accumulation, 278
Liquefied gas, 247–248
Liquid, 58, 59, 68, 69
   combustible, 245–246
Live load, 22
Load-bearing wall, 26, 27
Loaded-stream fire extinguishers, 210
Local alarm system, 160
Lodging/rooming house, 43–44
Lower explosive limit (LEL), 69
Low hazard contents, 90
LP-gases, 248, 257–258
LRC. *See* Learning Resource Center

M

Make-up air, 234, 235
Manual pull-station, 155
Marking system
   HMIS, 251
   military, 251–252
   NFPA 704, 251
Masonry, 14–15
Master-coded alarm, 160
Matter, 58, 59
Maximum travel distance, 124
MC-307 chemical hauler, 262, 263
MC-312 corrosives tanker, 262, 263
MC-338 cryogenic tanker, 262, 263
MC-306 flammable liquid tanker, 262, 263
MC-331 pressure cargo tanker, 262, 263
Means of egress, 124–126
   elements and arrangements
      area of refuge, 138
      doors, 134–135
      exit passageways, 137
      exit stairs, 136
      fire escape stairs, 137
      horizontal exits, 135–136
      panic hardware, 135
      ramps, 136–137
      ropes and ladders, 138
      smokeproof enclosure, 136
      windows, 138–139
   evaluation of, 126
      common path of travel, 133
      corridor capacity, 128
      egress capacity, 128, 134
      minimum width, 128–129
      remoteness, exit, 129
      travel distance, exits, 129–130
      maintenance of, 140
      measurement of, 128
      number of, 129
Measuring device, 82
Mechanical energy, 59
Mechanical equipment, 278–279
Mechanical plans, 110
Medical gases, 248
Mercantile occupancy, 49
Metal doors, 30
Metallic underground water-piping system, 226
Metals, 16
Military hazardous materials, 251–252
Mill construction, 20
Mixed occupancy, 52
Model code organizations, 7
Motors, 230
Multiple occupancy, 52
Multipurpose dry chemical fire extinguishers, 210

N

National Fire Incident Reporting System (NFIRS), 310
National Fire Protection Association (NFPA), 7, 251
   documents, development, 316–318
Natural gas, 248, 257–258
NFIRS. *See* National Fire Incident Reporting System
NFPA. *See* National Fire Protection Association
NFPA/SFPE *Fire Protection Engineering Handbook,* 235
NFPA 1, *Fire Code,* 267, 275, 316
NFPA 10, *Standard for Portable Fire Extinguishers,* 208
NFPA 13, *Standard for the Installation of Sprinkler Systems,* 90, 111, 113, 226, 279
NFPA 24, *Standard for the Installation of Private Fire Service Mains and Their Appurtenances,* 226
NFPA 30, *Flammable & Combustible Liquids Code,* 245, 254, 255, 261
NFPA 54, *National Fuel Gas Code,* 236, 248, 279, 316
NFPA 55, *Standard for the Storage, Use, and Handling of Compressed and Liquefied Gases in Portable Cylinders,* 246, 247, 257
NFPA 68, *Guide for Venting of Deflagrations,* 266
NFPA 70, *National Electrical Code,* 139, 225, 226, 230, 258, 316
NFPA 72, *National Fire Alarm Code,* 110, 111
NFPA 77, *Recommended Practice on Static Electricity,* 230
NFPA 80, *Standard for Fire Doors and Fire Windows,* 34, 35
NFPA 85, *Boiler and Combustion Systems Hazards Code,* 231
NFPA 92A, *Standard for Smoke-control Systems,* 234
NFPA 92B, *Standard for Smoke Management Systems in Malls, Atria, and Large Areas,* 234
NFPA 101, *Life Safety Code,* 4, 42–44, 46–52, 89, 104, 119, 316
NFPA 170, *Fire Safety and Emergency Symbols,* 104
NFPA 211, *Standard for Chimneys, Fireplaces, Vents, and Solid Fuel-Burning Appliances,* 236, 279
NFPA 220, *Standard on Types of Building Construction,* 18, 22, 81, 104
NFPA 251, *Standard Methods of Tests of Fire Resistance of Building Construction and Materials,* 110
NFPA 496, *Standard for Purged and Pressurized Enclosures for Electrical Equipment,* 258
NFPA 704 hazard identification system, 251
   hazard levels in, 252
NFPA 704, *Standard System for the Identification of the Hazards of Materials for Emergency Response,* 251, 267
NFPA 1031, *Standard for Professional Qualifications for Fire Inspector and Plan Examiner,* 5, 6
NFPA 1143, *Standard for Wildland Fire Management,* 275
NFPA 2001, *Standard on Fire Clean Agent Extinguishing Systems,* 266
NFPA 5000, *Building Construction and Safety Code,* 42, 43, 45
NITMAM. *See* Notice of Intent to Make a Motion
Nitromethane, 246
Nonbearing wall, 26–27
Nonbulk storage vessels, 260–261
Noncoded alarm system, 160
Noncombustible construction, 19
Nonflammable gases, 246
Nonpressurized rail tank cars, 262, 264
Notice of Intent to Make a Motion (NITMAM), 317
Nozzle sprinklers, 189, 190
Nuclear energy, 59

O

Obscuration rate, 157
Occupancy, 42
   classification, 42
Occupant load, 119
   capacity for, 123–124
   factors, 120–122
   for multiple-use, 122–123
   for single-use, 122
One-or two-family dwelling, 43
Open-head drum, 260
Open sprinkler head, 189, 190
Ordinary construction, 19–20

Ordinary hazard contents, 89–90
Ordinary/moderate hazard area, 209
Organizational practices, 301–302
Ornamental sprinklers, 189
Outlets, 229
Outside stem and yoke (OS&Y) valve, 181
Outward-opening doors, 30, 32
Overcurrent protection, 223, 224
Overhead doors, 31, 33
Overheating, 224
Oxidizing gases, 247

P

Packing and shipping materials, 280
Panel doors, 30
Panic hardware, 135
Parallel chord trusses, 26
Parapet walls, 28
Partition wall, 28
Party wall, 27
Passive smoke management. *See* Compartmentation
Pendent sprinkler heads, 189
Performance-based code, 7
Performance based design, 113
Permit, 8
Pesticide bags, 260
Photoelectric smoke detectors, 153, 154
Pictograph labeling system, 208
Pipe arrangement, 181
Pipeline right-of-way, 264
Pipelines, 264–265
Pipe schedule system, 181
Pitched chord truss, 26
Pitched roof, 23
Pitot gauge, 177
Pitot tube, 82
PIV. *See* Post indicator valve
Placards, 249–251
Plan review process, 103
  application phase, 108
  commissioning, 113
  deficiencies and variances, 111–112
  review phase, 108–112
Plan sets, 103
  drawings in, types of, 105–107
Plan view, 107
Plastics, 17–18
Plate glass, 32
Platform-frame construction, 22
Plenum system, 233
Plug fuses, 225
Plume, 61
Polar solvents, 212
Polychlorobiphenol (PCB) fluid, 228
Polyethylene drums, 260
Portable fire extinguishers, 204
Post indicator valve (PIV), 181, 182
Post-inspection meeting, 93
Potential electrical hazards, 224

Pre-inspection process
  codes, 81–82
  construction and occupancy, classification of, 81
  scheduling, 86
  standard forms, 83–86
  tools and equipment, 82–83
Premise identification, 275
Prescriptive code, 7
Pressurization method, 234
Pressurized rail tank cars, 262, 264
Projected windows, 34, 35
Proprietary systems, 161
Public education, 308
Public information, 309
Public relations, 308
Pyrolysis, 17
Pyrophoric gases, 246–247

R

Raceways, 228–229
Radiation, 62
Rafters, 23
Railroad transportation for hazardous material, 262–264
Railway tank cars, 262–264
Range of authority, 6
Rate-of-rise heat detectors, 157
Recessed sprinklers, 189, 190
Record-keeping practices, 299
  FOIA, 300
  organizational practices, 301–302
  records retention requirements, 299–300
  reporting systems, 302
Regular glass, 32
Reinspections, 79
  report, 294
Relative humidity, 230
Remodeling of building, 92–93
Remote annunciator, 152
Remote station systems, 161
Renewable link cartridge fuse, 225
Reporting systems, 302
Report on Comments (ROC), 317
Report on Proposals (ROP), 317
Residential board and care occupancy, 45–46
Residential smoke alarms, 153–155
Residential sprinklers, 189
Residual pressure, 176
Review phase, 108–112
Revision cycles, 317–318
Revolving doors, 30, 32
ROC. *See* Report on Comments
Rollover, 68
Room contents, 66
ROP. *See* Report on Proposals

S

Safeguards
  container/gas hazard, 257
  for escaping gas, 257–258

Safety data sheet (SDS), 252–253, 267
Safety glass, 33
Saponification, 213
Scheduled inspections, 86
Secondary containment, 259
Sectional view, 107
Self-inspections, 80
Separated occupancy, 52
Severity, 310, 311
Shipping papers, 252
Shunt trip, 225
Sidewall sprinklers, 189
Signal words, 260
Simple beam, 29
Single-hung windows, 33
Single-pane glass, 32
Site plans, 105, 108–109
Slate tiles, 23
Sliding doors, 30, 32
Slope, 277
Smoke, 60
  control of, 235, 282–283
Smoke dampers, 235
Smoke detectors
  definition of, 156
  ionization, 153, 154
  obscuration rate, 157
  photoelectric, 153, 154
  types of, 154
Smoke management systems, 234–235
Smokeproof enclosure, 136
Smoke zone, 234
Solid-core doors, 30
Solid-fuel fire development, 64–65
Solids, 58, 59
Spalling, 15
Special-use railcars, 264
Specifications book, 106–107
Spontaneous ignition, 278
Spray booth, 280
Sprinkler piping, 181
Sprinkler systems
  combined dry-pipe and preaction systems, 185
  deluge, 185–186
  dry-pipe, 184–185
  preaction, 185
  risers, 181
  wet-pipe, 184
  zoning, 183
Stairwell smoke management systems, 235
Stakeholder, 311–312
Standard, 6–7
Standpipe systems, 192
  water flow in, 194
State and local law, 8
Static electricity, 230
Static pressure, 176
Steel, 15–16, 24
Steel utility drums, 260

Stop work order, 83
Storage occupancy, 49–50
Structural inspector, 236
Structural plans, 105, 109
Switch, 229

T

Tamper switches, 182
Tanks, 256
Technical assistance, 113
Technical Committee Report Session, 317
Technical committees, 316
Tempered glass, 16, 33
Temporary Interim Amendments (TIAs), 318
Thermal column, 60
Thermal conductivity, 14
Thermal layering, 68
Thermal radiation, 62
Thermoplastic materials, 18
Thermoset materials, 18
Three E's of fire prevention, 4
TIAs. *See* Temporary Interim Amendments
Total flooding carbon dioxide systems, 265, 266
Tote, 259
Toxic gases, 246
Traditional lettering system, 207–208
Transformers, 227–228
Trash/recycling, 279–280
Travel distance, measuring, 125, 133–134
Truss, 25–26
Tube trailers, 262, 263
Tunnel test, 72
Type I construction, 18–19
Type II construction, 19
Type III construction, 19–20
Type IV construction, 20
Type V construction, 20–22

U

UEL. *See* Upper explosive limit
Unannounced inspections, 86
Underground storage tanks (USTs), 259
Underwriters Laboratories, Inc. (UL), 207
Upper explosive limit (UEL), 69
Upright sprinkler head, 189
U.S. Department of Transportation (DOT), 249

V

Valves, 181
Vapor density, 69
Ventilation of spaces, 258
Vent pipes, 265
Vestibules, 235
Volatility, 69
Volts (V), 224

W

Wall post indicator valve (WPIV), 181, 182
Walls, 71
Water
  distribution pipes, 181
  spray systems, 265, 266
Water flow alarms, 183–184
Water-motor gong, 181
Water supply, 180
  control valves, 181–183
  dry-barrel hydrants, 174–175
  municipal water systems, 172–174
  wet-barrel hydrants, 174
Watt (W), 224
Weapons of mass destruction (WMD), 251–252
Wet chemical extinguishing agents, 213
Wet chemical extinguishing systems, 187
Wet chemical fire extinguishers, 213
Wildland interface, 277
Window frame designs, 33–35
Wired glass, 16, 33, 35
Wiring
  concerns, 162
  problems with, 228
WMD. *See* Weapons of mass destruction
Wood, 16–17
Wooden beams, 17
Wood-frame construction, 20–22
Wood panels, 17
Wood-supported floors, 22–23
Wood trusses, 17
WPIV. *See* Wall post indicator valve
Writing tools, 82
Written documentation, 290–291

Z

Zoned coded alarm, 160
Zoned noncoded alarm system, 160
Zoned system, 159
Zone of origin, 234

Photo Credits

Credits Page
You Are the Fire Inspector Courtesy of L. Charles Smeby, Jr./University of Florida

Chapter 1
01-05 © Toby Talbot/AP Photos.

Chapter 2
Opener © Softdreams/Dreamstime.com; **2-01** © AbleStock; **02-04** © AbleStock; **02-05** © John Foxx/Alamy Images; **02-06** © Michael Doolittle/Alamy Images; **02-08** © Ken Hammon/USDA; **02-12** © BelleMEdia/ShutterStock, Inc.; **02-14A** ©AbleStock; **02-14B** Courtesy of Pacific Northwest National Laboratory; **02-14C** Courtesy of Royal Dum/US Army Corps of Engineers; **02-14D** © AbleStock; **02-15A** © Paul Springett/Up the Resolution/Alamy Images; **02-15B** © Ron Chapple/Thinkstock/Alamy Images; **02-15C** © TH Photo/Alamy Images; **02-16** Courtesy of Captain David Jackson, Saginaw Township Fire Department; **02-20** Courtesy of Captain David Jackson, Saginaw Township Fire Department; **02-21** © M Stock/Alamy Images; **02-25** Courtesy of A. Maurice Jones, Jr.; **02-27** © AbleStock; **02-32** © Kathy deWitt/Alamy Images; **02-33A** © Emmanuel Lacoste/Alamy Images; **02-33B** © Photodisc/Creatas; **02-35** © Frank Naylor/Alamy Images; **02-38** © BONNIE WATTON/ShutterStock, Inc.; **02-39** © AfriPics.com/Alamy Images; **02-40** © mkimages/Alamy Images; **02-41** Courtesy of Securalldoors.com.

Chapter 3
3-01 © L. Barnwell/ShutterStock, Inc.; **3-02** © Karin Hildebrand Lau/Dreamstime.com; **3-03** © Daniel Raustadt/Dreamstime.com; **3-04** © Timothy R. Nichols/ShutterStock, Inc.; **3-05** © Konstantin Lobastov/Dreamstime.com; **3-06** © Jennifer Walz/Dreamstime.com; **3-07** © Studiosnoden/Dreamstime.com; **3-10** © littleny/ShutterStock, Inc.; **03-11** © pics721/ShutterStock, Inc.; **03-12** © AbleStock; **03-13** © Brandon Bourdages/ShutterStock, Inc.; **3-14** © Robert Elias/ShutterStock, Inc.; **3-15** © Winzworks/Dreamstime.com; **03-16** © enkrut/Dreamstime.com; **03-18** © Gail Johnson/Fotolia.com.

Chapter 4
Opener © Glen E. Ellman; **04-24** © Dennis Wetherhold, Jr; **04-27** © AP Photos

Chapter 5
5-04 © Kanwarjit Singh Boparai/ShutterStock, Inc.; **05-08** © Kirsz Marcin/ShutterStock, Inc.; **5-12** © Aprescindere/Dreamstime.com; **5-13** © Martin Shields/Alamy Images; **05-14** © Cyril Hou/ShutterStock, Inc.; **05-15** © Blaine Walulik/ShutterStock, Inc.; **05-17** © M. Niebuhr/ShutterStock, Inc.; **05-18** © Jones & Bartlett Learning. Photographed by Glen E. Ellman; **05-19** © Jones & Bartlett Learning. Photographed by Glen E. Ellman.

Chapter 6
06-02 Courtesy of the City of Fort Worth. Photographed by Glen E. Ellman; **06-03** Courtesy of the City of Fort Worth. Photographed by Glen E. Ellman; **06-05–06-10** Courtesy of the City of Fort Worth. Photographed by Glen E. Ellman.

Chapter 7
Opener © Paul McKinnon/Shutterstock, Inc.; **07-18** © haveseen/ShutterStock, Inc.; **07-22** © Ragne Kabanova/ShutterStock, Inc.; **07-23A** © Robert Ranson/ShutterStock, Inc.; **07-23B** © Jorg Hackemann/ShutterStock, Inc.; **07-24** © Cristi111/Dreamstime.com; **07-27** © Catalin D/ShutterStock, Inc.; **07-28** © vichie81/ShutterStock, Inc.

Chapter 8
08-06 © AbleStock; **08-08** © Brendan Byrne/age footstock; **08-09A** © rob casey/Alamy Images; **08-12** Courtesy of Fire Fighting Enterprises, Ltd.; **08-17** Courtesy of A. Maurice Jones, Jr.; **08-19** © Graham Taylor/ShutterStock, Inc.

Chapter 9
09-02 © Will Powers/AP Photos; **09-03** © Paul Glendell/Alamy Images; **09-07** Courtesy of American AVK Company; **09-08A** Courtesy of American AVK Company; **09-17** © Jones & Bartlett Learning; **09-19** Courtesy of Ralph G. Johnson (nyail.com/fsd); **09-27** Photo Courtesy of Liberty Mutual Commercial Markets; **09-33** © Damian P. Gadal/Alamy Images; **09-42** Courtesy of A. Maurice Jones, Jr.; **09-43** © AbleStock; **09-44** Courtesy of A. Maurice Jones, Jr.; **09-46** Courtesy of A. Maurice Jones, Jr.; **09-48** Courtesy of A. Maurice Jones, Jr.; **09-50** © Stephen Coburn/ShutterStock, Inc.; **09-51** Courtesy of Ralph G. Johnson (nyail.com/fsd); **09-52** Courtesy of A. Maurice Jones, Jr.

Chapter 10
10-04 Courtesy of Joe Sesniak; **10-05** © Andrew Lambert Photography/Photo Researchers, Inc.; **10-06** © Scott Leman/ShutterStock, Inc.; **10-08** NFPA 10 Standard for Portable Fire Extinguishers, 2013 Edition, page 36; **10-13** Courtesy of Amerex Corporation; **10-17** Courtesy of Amerex Corporation.

Chapter 11
11-02A Courtesy of A. Maurice Jones, Jr.; **11-02B** Courtesy of A. Maurice Jones, Jr.; **11-05** © jcjgphotography/ShutterStock, Inc.; **11-06** © Robert Asento/ShutterStock, Inc.; **11-08** © Henry Nowick/ShutterStock, Inc.; **11-09** © emel82/ShutterStock, Inc.; **11-10** © Chas/ShutterStock, Inc.; **11-12** © Shane Phelan/Alamy Images; **11-16** © ARENA Creative/ShutterStock, Inc.; **11-17** © Yury Kosourov/ShutterStock, Inc.; **11-19** © Muellek Josef/ShutterStock, Inc.; **11-20** © Gary Curtis/Alamy Images; **11-21** © Spencer Grant/Alamy Images;

Chapter 12
12-01 © Tom Oliveira/ShutterStock, Inc.; **12-03** © Knud Nielsen/ShutterStock, Inc.; **12-04** © Jones & Bartlett Learning. Courtesy of MIEMSS; **12-05** © Samuel Acosta/ShutterStock, Inc.; **12-06** © Calvin Chan/ShutterStock, Inc.; **12-07** © Rob Byron/ShutterStock, Inc.; **12-08** © Steve Allen/Brand X Pictures/Alamy Images; **12-09** © Pedro Nogueira/ShutterStock Inc.; -© Courtesy of DOT; **12-12** © Travis Klein/ShutterStock, Inc.; **12-13A** © Martin Muránsky/ShutterStock, Inc.; **12-13B** © iofoto/ShutterStock, Inc.; **12-13C** © Nikolaj Kondratenko/ShutterStock, Inc.; **12-14** © Luiz Rocha/ShutterStock, Inc.; **12-15** Courtesy of Justrite Manufacturing; **12-17A** © Jones & Bartlett Learning. Courtesy of MIEMSS; **12-16** Courtesy of Justrite Manufacturing; **12-18** © Peter Carroll/Alamy Images; **12-20** © Ulrich Mueller/ShutterStock, Inc.; **12-21** © USDA; **12-23** Courtesy of Cryofab, Inc.; **12-24** © Urban Zone/Alamy Images; **12-25D** Courtesy of Jack B. Kelley, Inc.; **12-25E** Courtesy of Jack B. Kelley, Inc.; **12-25F** Courtesy of Jack B. Kelley, Inc.; **12-25H** Courtesy of Eurotainer; **12-28** © Mark Gibson/Index Stock/Alamy Images; **12-29** © iStockphoto/Thinkstock; **12-30** © Andrei Kolyvanov/Alamy Images; **12-31** Courtesy of Veltre Engineering, Inc.

Chapter 13
13-01 © Chris Howes/Wild Places Photography/Alamy Images; **13-02** © Raymond Kasprzak/ShutterStock, Inc.; **13-09** © Mikael Karlsson/Alamy Images; **13-10** © SHOUT/Alamy Images; **13-11** © Gary Curtis/Alamy Images; **13-12** © vadim kozlovsky/ShutterStock, Inc.; **13-13** © francesco survara/Alamy Images.

Chapter 15
Opener Courtesy of Captain David Jackson, Saginaw Township Fire Department; **15-02** © Mark C. Ide; **15-07** Courtesy of Andrea Booher/FEMA; **15-08** © Mark C. Ide.

Unless otherwise indicated, all photographs and illustrations are under copyright of Jones & Bartlett Learning or have been provided by the authors.